CORROSION CONTROL BY COATINGS

Papers presented at a meeting on corrosion control by coatings held November 13-15, 1978 at Lehigh University, Bethlehem, Pennsylvania.

Edited by

Henry Leidheiser, Jr., Ph.D.

Professor of Chemistry and
Director, Center for Surface and Coatings Research
Sinclair Laboratory, Lehigh University
Bethlehem, Pennsylvania

Published by

SCIENCE PRESS

Princeton

Library of Congress Cataloging in Publication Data

Main entry under title:

Corrosion control by coatings.

1. Protective coatings--Congresses.
2. Corrosion and anti-corrosives--Congresses.
I. Leidheiser, Henry.
TA418.76.C67 620.1'1223 79-3990
ISBN 0-89500-018-0

Printed in the United States of America

DEDICATION

This book is dedicated to Howard Gerhart who gave unstintingly of his energies and time to advance organic coatings technology. He became ill shortly before the Discussion was held and passed away in January 1979.

TABLE OF CONTENTS

PREFACE

The date of conception of "Corrosion Control by Coatings" occurred in the summer of 1977 during a discussion with Philip Clarkin of the Office of Naval Research. These discussions culminated in an official expression of interest by Lehigh University in organizing the meeting and serving as a host institution. Invitations to speakers were extended beginning in January 1978 and the program was essentially complete in the early summer of 1978. Thirty-one of the 33 papers were available sufficiently in advance of the meeting so that the full text could be reviewed by attendees. Two of the papers were delayed because of administrative requirements of the institutions at which the authors were employed.

Authors were given until December 31, 1978 to make changes in their papers. Only one paper was excluded from the text because a revised version was not received by the deadline. After minor editorial revisions to obtain uniformity in style, the papers were presented to the publisher on January 10, 1979 and the book was ready for distribution on April 1, 1979. Minor inadequacies in the quality of the original figures were accepted in order to make the book available as soon after the meeting as possible. It was specially desired to have the book available sufficiently in advance of the August 1980 International Conference on Corrosion Control by Organic Coatings to be held at Lehigh University and sponsored by the National Association of Corrosion Engineers. It was anticipated that the participants at the 1980 NACE meeting could profitably use this book as background material.

I am particularly grateful to Raymond Myers of Kent State University, Ellis Verink of the University of Florida, Jerome Kruger of the National Bureau of Standards, A. C. Zettlemoyer of Lehigh University, B. Floyd Brown of American University, Werner Funke of the University of Stuttgart, Frank Graziano of Pre Finish Metals, Ind., Howard Lasser of the Naval Facilities Engineering Command, John W. Vanderhoff of Lehigh University, Robert Frankenthal of Bell Telephone Laboratories, and Gary W. Simmons of Lehigh University for their services as chairmen and reporters during the six sessions. My two colleagues, Katherine Chapman and Rosalie Freudenberger, will be remembered by attendees for their many kindnesses during the meeting and for making sure that the social periods occurred smoothly so that the social periods were proper extensions of the technical sessions. Mrs. Freudenberger handled the arrangements prior to the meeting and assisted in organizing the manuscripts prior to submission to the publisher.

Henry Leidheiser, Jr.
Lehigh University
January 28, 1979

I. GENERAL PAPERS

COATINGS—THE PROMISE AND THE CHALLENGE FOR ENHANCED NAVAL SHIP PERFORMANCE

H. S. PREISER

Herman S. Preiser and Stephen D. Rodgers

Nonmetallics Division, Ship Materials
Engineering Department
David W. Taylor Naval Ship Research
and Development Center
Annapolis, MD

ABSTRACT

This paper discusses the rationale and the Navy approach to obtain smooth hull coatings systems for reducing ship drag. It reviews some of the more critical coatings developments augmented in some cases by underwater hull maintenance techniques which when combined into a systems approach can offer substantial gains in ship performance enhancement. The role of micro-fouling (slimes) as it may affect drag is presented. The paper concludes with some predictions for trends in coatings development for the future.

BACKGROUND

In the area of ship performance enhancement, one is specifically concerned with ship bottom paints. Anticorrosion coating developments have now reached a point where not only do they prevent deterioration of the hull in seawater thereby preserving structural integrity, but such coatings can be applied in sufficient thicknesses over properly prepared surfaces to prevent peeling and roughening of the hull for periods in excess of ten years when augmented by cathodic protection techniques.[1] Chromate inhibited vinyl paint systems, the work horses of the Fleet in the 60's, are giving way to epoxy and epoxy-coal tar formulations, with various nonmetallic and metallic pigments, to provide tough primers for corrosion protection and suitable bases for overcoating with specialized topcoats.

Antifouling paint systems are undergoing extensive development to be effective against a wide spectrum of fouling organisms for extended exposure periods without serious ecological side effects. The prevention of macrofouling accumulation on a ship hull is a paramount requirement if severe fuel penalty and performance degradation are to be avoided. The concept of microscale hull smoothness based on paint texture is meaningless unless the surface of interest can be kept free of macrofouling. This point will be enlarged upon in greater detail elsewhere in the paper. Antifouling paint development has progressed beyond the conventional copper oxide pigmented systems (vinyl for Naval use) to the newer organometallic toxins used as additives or as chemically bonded elements onto various polymeric resins.[2] Work has also progressed to the point where low solubility, biodegradable, nonmetallic organic toxins are beginning to show promise as long life, safe, antifouling systems.[3]

The technology for selecting, testing, and modifying energy absorbing elastomeric coatings has progressed sufficiently to enable the coatings technologist to specify erosion resistant coatings for hull areas exposed to turbulent and separated water flows as well as debris impact. The use of polyurethanes, neoprenes, and flexible epoxies for these applications has shown promise. Erosion resistant coatings are sometimes designed to replace typical anticorrosion coatings and in some cases provide antifouling protection as well.[4]

Smooth hull coating systems are designed to perform all the above functions while simultaneously maintaining an acceptable microsmooth surface texture with optimum low drag characteristics for the given hull configurations. At times, periodic underwater maintenance techniques may be required to retain and renew the desired surface characteristics; these aspects will be reviewed subsequently as total systems approaches.

Rather than discuss the broad spectrum of coatings technology, which is a vast subject indeed, let us focus now on candidate hull coating systems and their contribution to performance enhancement. The goal is minimum fuel consumption to maintain a given speed, or the ability to achieve and maintain design speeds during extended operating cycles between overhauls.

SMOOTH HULL COATINGS

Before discussing the systems approach to achieving smooth hull coatings, it would be well to review briefly some pertinent ideas concerning effect of roughness on drag.

Hull Roughness

It is well documented that hull surface roughness increases frictional drag and results in both a degradation of ship performance and excessive energy requirements.[5,6,7,8] Most of this surface roughness, aside from structural discontinuities and disparities, can be attributed to biological fouling of the ship's hull by grasses and calcareous organisms. These growths are referred to as "macrofouling." Control of macrofouling can be achieved to a great degree by the use of antifouling coatings[9,10] in combination with underwater cleaning of the hull.[11] However, it has been found that a significant degree of frictional drag is still retained by a properly faired hull surface, even though all macrofouling organisms have been removed and the hull appears to be clean.[6,12] Indeed, the hull surfaces of newly built ships are not hydrodynamically smooth but have many minute surface irregularities which create what is known as "microroughness." Control of hull surface microroughness is especially important for today's high speed ships, since the degree of microroughness affects such operational parameters as speed, noise generation and fuel consumption.[1,13]

Hull surface irregularities arise from many causes. Aside from initial mill plate roughness and fabrication flaws, the various surface characteristics of paint systems employed by the Navy give rise to specific textural effects which may be classified as microroughness. These effects are further compounded by surface preparation procedures and painting techniques. Once a ship is in service, the paint surface is degraded in time by such factors as surface corrosion, erosion or pitting, as well as by chemical and biochemical changes in the coating surface due to the influence of slimes, weeds and barnacles. The influence of slime formation on the hydrodynamic drag of microrough surfaces will be discussed later in this paper. In addition, coatings are subject to mechanical damage from abrasion or by impact with debris.

Characterization and Measurement of Surface Roughness

A fundamental question in the study of skin friction is how best to define or characterize a rough surface. A review of the literature indicates that most of the hydrodynamic theory and a majority of the experiments have depended on a defined height of roughness measurement. However, it has become increasingly evident that no unique relationship exists between drag of different surface profiles since roughness is waveform dependent. Studies are going forward, both here and abroad, to correlate specific properties of wave profiles with the corresponding drag of a given surface.

The most widely accepted standard on surface texture is American National Standard B46.1-1962.[14] This standard establishes definite classifications for roughness, waviness and lay; details a set of symbols for drawings, specifications and reports; and provides specifications and requirements for precision reference specimens, roughness comparison specimens and tracer type instruments. Aside from surface contacting stylus type instruments[15] for laboratory measurement of surface microtopography, such as the Surfanalyzer and the Talysurf, there are various optical procedures and instruments[16,17] such as light scattering, multiple beam interference microscopes, optical sectioning, transmission electron microscopes (TEM) and scanning electron microscopes (SEM). Of all available techniques for measuring surface profiles, only stylus instruments are covered in ANS B46.1-1962. Development of stylus instrumentation has achieved a high degree of sophistication, with sensitivities approaching the fundamental limitations imposed by nature, i.e., single atom steps.[18] The equipment is rugged, simple to operate and rapid enough for routine work.[19] Computerization of stylus equipment has significantly increased its versatility. Although the stylus is probably the most popular instrument in use today, certain optical instruments, tactile and visual comparators are also widely used.

Surface roughness measuring devices presently available cannot be used underwater. To measure the roughness and drag of underwater surfaces of waterborne ships, it is necessary to make molds and replicas and bring them to the measuring instruments in the laboratory. The Center has recently completed the development of a submerged surface molding device affectionately known as a "hull sucker," which permits a diver to make an accurate mold of a ship surface underwater. The device utilizes an RTV molding compound which is introduced into the void created by a recessed piston after the device, positioned on the hull, has been evacuated of water. This device will permit the changes of hull roughness on an operating ship to be monitored as a function of time out of drydock and relate such changes to the ship drag behavior. A patent application has been filed describing the device.

Effects of Slime on Drag

The crucial issue in achieving significant drag reduction by control of surface microroughness of hulls is to define the role of microfouling on such surfaces. As mentioned earlier, the concept of microroughness control is predicated on the premise that macrofouling (grasses and shelled animals) can be prevented. Modern antifouling coatings in combination with periodic underwater mechanical cleaning[11,19] can eliminate to a large extent the problem of macrofouling attachment and growth, thereby allowing the underlying microroughness of the surface to determine its hydrodynamic drag characteristics.

Although toxic and/or mechanical control of macrofouling have been reasonably successful, the vast number of species of microorganisms present in the natural sea environment precludes the development of a universal antisliming agent. There is presently no toxicant available against all sliming organisms, nor is there likely to be any in the foreseeable future.

These microorganisms are prolific reproducers and settle rapidly on stationary and moving surfaces immersed in the sea. Usually, infestation is completed upon immersion in a matter of minutes or, at most, within a few hours. In many, but not all, cases the formation of these films may lead to the attraction and attachment of higher organic forms.

Although there is little likelihood that microfouling on submerged surfaces can ever be totally eliminated, investigators have found that natural repellent additives, such as tannin,[20] and the degree of wettability of organic materials[21,22] can inhibit the attachment of many microorganisms for reasonable periods of time. It is of primary importance to understand how such biological films or slimes affect hydrodynamic drag characteristics of various hull surfaces.

The effect of slime on drag is a critical element in the relationship between hull smoothing coatings and the improvement of ship speed. At present, this relationship and its effects are not precisely known or understood. However, there is some evidence that the change of drag brought about by slimes on surfaces is highly dependent on the initial roughness of the surface and the boundary layer characteristics of the flow regime. Speculation is that in flow regimes where drag is predominantly due to viscous effects as influenced by the Reynolds number, the deposition of slimes on hydraulically smooth surfaces would increase its drag. Confirmation of this concept has been demonstrated in our laboratory. However, initial drag may be significantly different than steady state drag since shear at the surface should cause marked change in the physical and geometric character of the slime layer. There is evidence that the viscoelastic strains on the slime layer introduce surface ripples and discontinuities which cause drag to rise in service.[23,24] However, for flow regimes where form drag predominates, deposition of slimes within the crevices of roughness elements should exert a smoothing effect with corresponding decreased drag effects. These aspects are being systematically explored at the Center and the initial work is nearing completion.

Candidate Smooth Hull Coating Systems and Techniques

Hull smoothness to a large degree has been dependent on the effective performance of antifouling and anticorrosive hull paint systems. The degree of inherent initial smoothness of the exterior coating film has been academic in the past because of the difficulty of maintaining this smoothness with rapid onset of slime and marine fouling. To date the use of coatings which are intrinsically smooth has been confined to icebreakers, and these coatings do not include antifouling topcoats. It now appears feasible to develop coatings which are inherently smoother because of the availability of longer lasting antifouling coatings and the availability of *in situ* maintenance techniques to extend service of those coatings. A search of the technical literature and of manufacturer's information has revealed a number of promising coating systems and methods that have potential merit in obtaining and maintaining a new level of hull smoothness. These candidate coating systems will be discussed after reviewing conventional paint practices and hull treatment systems in current use.

Conventional paint systems presently used by the Navy, such as Navy Vinyl 121 and 129 AF overcoated on Navy epoxy anticorrosive undercoats (F-150 series), have equivalent sand grain roughness of approximately 1,000 to 1,500 micro inches (μ in.) (25 to 38 microns) depending on paint application techniques.[2] These values of sand grain roughnesses are estimated to be equivalent to 100 to 125 μ in. average amplitude (AA). Under certain stringent conditions of application and light sanding, it is feasible to produce an initial smoothness of 60 μ in. or less AA. In a preliminary test conducted by Electric Boat Division of General Dynamics Corporation, a comparison was made of Navy vinyl antifouling coating (F-121) sanded to a smooth finish with an unsanded coating applied by airless spray. In time, after brackish river water immersion of about four months, the sanded paint surface originally at 61 μ in. AA approached the same value as the unsanded antifoul-surface, originally at 165 μ in. AA, reaching a roughness level of about 200 μ in. AA, which equates to 1600 μ in. equivalent sand grain roughness. This preliminary short-term experiment indicates, therefore, that it may be feasible to alter commonly observed roughness values on conventionally painted naval ships from the prevailing 200 to 1000 μ in. AA to surfaces which are smooth on the order of less than 30 μ in. AA. If such changes can be attained over a practical maintenance cycle, it would be possible to reduce the frictional drag of a high speed ship by 15 percent. It is cautioned that the above conversions of roughness equivalence is empirical and not fully verified experimentally.

There are currently three approaches to improved hull coating systems that warrant serious consideration:

- Hard, high gloss coatings
- Renewable surface coatings
- Controlled release coatings

In all cases the systems must perform on a ship hull exposed to a natural seawater environment, which is not only physically and chemically destructive to coatings but also biologically active in depositing slimes and other higher forms of visible plant and animal fouling on the coating surface.

Hard high gloss coatings - These coatings are smooth, self-levelling, hard and inert. They incorporate low leaching toxicants, such as organometallic polymers, chemical repellents, and natural enzymic additives which control microfouling and reduce slime accumulations. A variation of this coating type is smooth, self-levelling, hard and inert but has no antifouling additives. The latter system requires frequent mechanical intervention to remove growth if it is utilized without an antifouling topcoat. However, if the coatings prove to be compatible with Navy antifouling topcoats, they may form the basis of a smoother overall coating system since the smoothness of the base anticorrosive contributes to the overall smoothness of the system. Only 30 percent of final surface roughness is attributable to the application of standard Navy vinyl antifouling topcoats, on a freshly painted ship.[1]

The David W. Taylor Naval Ship Research and Development Center is developing an improved long life antifouling paint containing organometallic toxicants as mentioned earlier. Preliminary field tests indicate that these formulations resist microfouling to a greater degree than the conventional copper toxic antifouling coating.

Several firms are investigating the development of hard coatings containing enzyme substances and special chemical structures which resist fouling attachment. Another Center development in progress is the use of a fungicide,[25] Nopcocide, to impart antifouling properties to hard smooth coatings. Previously, this fungicide had been added to paints to achieve mildew resistant properties. Formulations containing Nopcocide are undergoing preliminary field tests and the indications are that a long life antifouling system can be developed. The U.S. Coast Guard is evaluating low friction coatings for icebreakers.[26] These coatings are designed to have optimum smoothness so that water and/or ice do not adhere to the surface. Thus far two solventless coatings, one of which is an epoxy and the other a polyurethane, have excelled in Coast Guard evaluations. Inerta 160,[27] a Finnish developed epoxy, and Zebron,[28] a U.S. developed polyurethane, are corrosion resistant but not fouling resistant coatings. Dr. J. Griffith of the Naval Research Laboratory has developed a fluorinated epoxy polyethylene coating which has low friction properties. These corrosion resistant, low friction coatings may be compatible with antifouling coatings and therefore could be utilized as base coats for antifouling. As mentioned previously, the smoothness of a hull coating system depends to some extent on the smoothness of the base coats. It also may be possible to incorporate organometallic toxics, fungicides, enzymes or other additives to make the low friction coatings antifouling without affecting smoothness.

Renewable surface coatings - Self-polishing copolymers (SPC) are antifouling coatings whose outer surfaces soften with time and are ablated and smoothed while the ship is underway.[29] A subcategory of this type is the high film thickness antifouling coating, which is mechanically rejuvenated and smoothed periodically by underwater scrubbing. The feasibility of polishing this type of coating with antisliming waxes and providing additional inherent smoothness should also be considered.

A European affiliate of the International Paint Company has developed and marketed in Europe a self-polishing copolymer (SPC) based on an acrylic resin.[30] The outer layers of this copolymer slough off by hydrodynamic shear forces, and a smooth surface is thus reportedly maintained. Turbulence is necessary for polishing to take place but release of the toxicant (tributyl tin) is constant even under static conditions. The International Paint Company intends to market the product in the United States after receiving EPA approval. The coating system consists of three coats (alternate colors) of 100 microns (4 mils) each and will have a service life of two years. Although the serviceable life of this coating is not substantially greater than present Navy antifouling coatings, it does not require any underwater scrubbing to retain smoothness. This, of course, is a great advantage as far as ship speed is concerned. The Ameron Company of Brea, California, has a similar experimental coating, a siloxane, which sloughs off in service. It expects to develop a marketable produce in the near future. Conversations with technical representatives of Hempel's Marine Paints reveal that they are also involved in the development of a self-polishing antifouling paint. Hence, there is considerable research by paint manufacturers in this area. They are designing coatings which are initially smoother or will become smoother with time in service. If a concept can be made to work for Navy ships, it may be feasible to develop longer life, self-smoothing systems.

Another type of renewable surface coating system utilizes periodic underwater mechanical brushing to rejuvenate the inactive antifouling paint surface developed while in service. The Seamaster System is a thick film version of conventional copper antifouling paints which can be rejuvenated by mechanical underwater scrubbing.[31] This paint system is about two and one half times as thick as the standard Navy vinyl formula 119/121 system and therefore can tolerate a greater number of *in situ* brush cleanings. It consists of the following:

a. One coat of zinc rich primer, 15 to 20 microns (0.6 to 0.8 mils)
b. Four coats of vinyl mastic anticorrosive 100 microns (4 mils) per coat
c. Four coats of vinyl antifouling (cuprous oxide toxic) on underwater sides but only two coats on the bottom, 75 microns (3 mils) per coat.

The Seamaster System is commercially available and is used in conjunction with periodic cleaning to reactivate the outer layers of antifouling paint. The supplier also provides the underwater cleaning services which utilizes the SCAMP® vehicle outfitted with special brushes. To achieve maximum effectiveness the cleaning process should be timed to reactivate the antifouling paint before gross macrofouling occurs. The firm claims service life of four to five years for this paint system. It is supplied in the United States by the Jotun-Baltimore Copper Paint Company. The use of antifouling and/or antisliming waxes to polish the surface to a smoother finish and rejuvenate the antifouling paint has a potential for development of longer lasting smooth antifouling finishes. Waxes are gaining increasing attention and efforts are underway at the Center to explore feasibility fully.

Controlled release antifouling coatings - Controlled release antifouling coatings are hydrophilic topcoats which provide smoothness and regulate toxic depletion.[6] An outgrowth of this development would be a modified conventional copper oxide Navy antifouling system which achieves lower critical toxic release rates without the formation of insoluble copper compounds on the functioning surface. Hydron, a product of Hempel's Marine Paints consists of a thick hydrophilic copolymer which is applied over a smooth one coat antifouling paint (cuprous oxide toxic).[32] The material swells by absorbing water up to 75 percent of its original weight. This system is claimed to form a virtual water-to-water interface, thereby counteracting turbulence and reducing skin frictional resistance and at the same time preventing excessive release of the toxic. An effective service life of two years is reported for this system. A disadvantage is that it cannot be reactivated by underwater cleaning. The system consists of one coat Dynamic Antifouling 100 microns (4 mils) over a coal tar anticorrosive. This is followed by one coat of Hydron of 10 microns (0.4 mils).

Modified cuprous oxide antifouling systems - The Center is planning the investigation of copper salts which dissolve more slowly than cuprous oxide in antifouling paints. The objective is to modify formulations which do not become completely insoluble on the surface after exposure to seawater. Also, the use of lower level loading of cuprous oxide may toughen and smooth present vinyl AF systems. Periodic rejuvenation of the surface to maintain optimum leach rates would be provided by mechanical cleaning similar to the Seamaster System described above. Table I summarizes the candidate coatings discussed above.

Screening and Selection Methods

A program is underway to expose a series of friction disc specimens at a natural fouling site in Miami. These friction discs painted with promising candidate coatings selected from the above described categories will be mounted in a specially designed dynamic testing facility. Roughness measurements will be made before exposure and then the test rig will operate intermittently under dynamic and static modes to simulate a typical submarine operating cycle. After each cycle of exposure, observations and measurements will be made of the fouling attachment, wear and roughness. These specimens will then be tested in the friction disc facility to determine drag. The specimens will be subjected to repetitive exposure cycles, cleaned and waxed as appropriate and retested for their drag characteristics.

FUTURE TRENDS

What then are the coatings trends for the next decade and beyond? The most easily identifiable trends are those arising from the continuing environmental and occupational health protection programs. In the next decade, shipboard painting is going to be revolutionized by the proposed EPA edict which will again redefine the permissible solvents that can be used

TABLE I.—CANDIDATE SMOOTH HULL COATINGS

TYPE	DESIGNATION	SPECIFIC PROPERTIES
CURRENT NAVY SYSTEMS		
• Vinyl	a.(1) 117/(4) 119 or (4) 120/ (2) 121/or (2) 129	150-500 μin. (AA) initial roughness; Can be mechanically reactivated; Maximum user experience; Baseline data
• Epoxy	(1) 150/(1) 151/(1)154 (2) 121 or (2) 129 or b.(2) 1020A	
• Hull Treatment	Rubber sheeting with joints grouted with polysulphide compounds and coated with (1) 133/ (1) 134	Discontinuities and waviness.
HARD, HIGH GLOSS COATINGS	———	———
Epoxies, Polyurethanes and Fluoropolymers (no additives)	INERTA® 160, ZEBRON® and Navy Fluoropolymer (experimental)	Low ice friction; Self-levelling.
Epoxies, Polyurethanes and Fluoropolymers (with additives)	Navy OMP Epoxy (experimental)	Low leach AF toxicant
	Navy Nopcocide Additive (experimental)	Low leach biocide
	Ameron Enzyme Additive (experimental)	Non-polluting
RENEWABLE SURFACE COATINGS	———	———
c. SPC Acrylic, OMP	International Paint Co. Hemple (experimental)	Antifouling; renews and smooths surfaces while ship is in motion
SPC Siloxane, OMP	Ameron (experimental)	
Mechanically Reactivated AF	Jotun SEAMASTER®	Extended AF service by periodic cleaning and smoothing
AF Waxes	Navy and Commercial (experimental)	Conceptual stage
CONTROLLED RELEASE AF COATINGS	———	———
Hydrophilic Copolymer	Hempel HYDRON®	Long life AF, slow degradation of surface with time.
Modification of F-121	Navy (experimental)	

a. Number of coats in parentheses followed by Navy formula number.
b. Currently used only for experimental purposes. Contains tributyl tin oxide in carboxylated vinyl resin.
c. Self-polishing copolymer or similar designation.

in paints. Already, the Navy and the paint industry are being advised by the California Air Resources Board (CARB) that, possibly by 1982, use of our present maintenance paints will no longer be allowed in that state. EPA is considering the CARB type rule (maximum of 235 gms of volatile solvent per liter of paint) for nationwide application in 1985. It can be anticipated that EPA will, within five years, ban the use of chromate-inhibitive pigments which will require reformulation of a great many of the Navy paints. It is likely that the use of many other heavy metals such as lead oxide and copper oxides will also be restricted. Substitute paint systems will be required. Environmental protection laws have restricted abrasive cleaning and other cleaning methods (acid cleaning) aboard ship. Coatings compatible with new, as yet undefined, cleaning procedures will be required. It can be anticipated that there will be a continuing requirement for new coatings as the medical-regulatory communities continue to identify carcinogenic material (epichlorohydrin and chloroprene are already suspect). In addition to these externally imposed restrictions which will require new Navy coating developments, there is the internal movement toward extended operating cycles in excess of five years. The Navy is seeking hull coatings, particularly antifouling coatings, capable of reliably guaranteeing five years unattended service, and anticorrosion coatings for ten-year periods and beyond.

The day will come when materials engineers, chemists and ship designers will crossfeed their multidisciplines into a systems approach hat will enable a material to be characterized so that the limitations and variations of its behavior and response to specific environmental phenomena are understood and properly measured. The engineer will be able to select optimum material systems based on proper design trade-offs of performance and service life. Further, materials engineers and coatings specialists will one day understand the internal, molecular parameters that govern polymer behavior and response characteristics to that molecular tailoring can proceed rationally to modify coatings to meet new performance requirements.

If we extrapolate into the future, coatings with ten-year life for seawater immersion service, capable of adhering to poorly prepared, wet surfaces, having indefinite antifouling properties, high degree of smoothness and stability and resistance against cavitation erosion could be a reality. Combining these materials with augmenting cathodic protection, waterborne grooming for hull smoothness, underwater touch-up and dockside fouling removal, we should see further progress in ship performance enhancement.

Pollution requirements and fire safety will accelerate the development of water-based coatings to replace solvent-based types in all areas of shipboard application.

The promise of coatings to enhance ship performance to usher in the future high speed, power-efficient Navy is a challenge worthy of our best scientific efforts. This symposimm should help smooth the way.

ACKNOWLEDGEMENT

The work in hull smoothness described in this paper is being performed under the sponsorship of the Naval Sea Systems Command. Acknowledgement is made for the assistance provided by several of our associates at the David W. Taylor Naval Ship Research and Development Center; to Messrs. A. Ticker and M. Acampora of the Coatings Applications Branch for their technical contributions.

The opinions expressed in this paper are solely those of the authors and do not necessarily reflect the official views of the Navy Department or the Naval Establishment at large.

REFERENCES

1. Ellengsen, P.A., et al, "Improving Fuel Efficiency on Existing Tankers," Proceedings Second STAR Symposium, pp. 475-479, SNAME 1977.
2. Yager, W.L. and Castelli, V.J., "Antifouling Applications of Various Tin Containing Organometallic Polymers," pp. 175-180, Organometallic Polymers, Ed. C.E. Carraher, Jr. et al, Academic Press, N.Y., 1978.
3. Dear, Hing, "The Design and Application of Antifouling Paint," Proceedings Third International Conference in Organic Coatings Science and Technolgy, State University of New York and The Greek Chemists Association, Athens, Greece, July, 1977.
4. Dick, R.J., et al, "New Marine Coatings Technology Applied to the Protection of Buoys," pp. 145-154, 4th International Congress on Corrosion and Fouling, Antibes, France, June 1976.
5. Lynn, W.M., "Trial Performance Results and Hull Surface Roughness Measurments for 18000-ton Deadweight Tankers, " B.S.R.A. Report No. 267, 1958.
6. Lackenby, H., "Resistance of Ships, with Special Reference to Skin Friction and Hull Surface Condition," *Marine Engineering and Shipbuilding Abstracts 25(5), 73 (May 1962).*
7. McCarthy, J.H., "Ship Resistance Due to Surface Irregularities: Waviness, Roughness and Fouling," 17th American Towing Tank Conference, California Institute of Technology, Pasadena, California (June 1974).
8. Canham, H.J.S. and Hyn, W.M., "The Propulsive Performance of a Group of Intermediate Tankers," *Ann. Trans. of the Roy. Inst. Nav. Arch. (British) 104(1) (Jan. 1962).*
9. "The Seamaster Method." Jotun Marine Coating, Sadefjord, Norway, Product Literature (Baltimore Coppers Paint Company).
10. "SPC-International Red Hand Marine Coatings: New Answer to the Fouling/Ship Performance Problem." *Tech Bull RSW/6-4-75/TB,* International Red Hand Marine Coatings, New York.
11. Preiser, H.S., et al, "Energy (Fuel) Conservation through Underwater Removal and Control of Fouling on Hulls of Navy Ships," *NSRDC Report 4543 (May 1975).*
12. "Effect of Bottom Maintenance on Frictional Resistance of Ships." *SNAME Tech. and Res. Report R-18 (Feb 1975).*
13. Brown, D.K., "Fouling and Economic Ship Performance," Proc. 5th Inter-Naval Corrosion Conference, New Zealand (April 1976).
14. "Surface Texture: Surface Roughness, Waviness and Lay." *Am. Std. ASA B46.1-1962,* Soc. Automotive Engrs. and Am. Soc. Mech. Engrs., 1962.
15. Young, R.D. and Teague, E.C., "Properties of Electrodeposits - Their Measurement and Significance," R. Sard, H. Leidheiser, Jr., and F. Ogburn, Eds., *The Electrochemical Society, 22 (1975).*
16. Young, R.D., "Eight Techniques for the Optical Measurement of Surface Roughness," *National Bureau of Standards Report NBSTR 73-219 (May 1973).*
17. Teague, E.C., "Surface Finish Measurements: An Overview," SME Technical Paper IQ 75-137 (1975).
18. Young, R.D., "Surface Microtopography," *Physics Today 24, no. 11 (Nov 1971).*
19. Preiser, H.S., et al., "Fouling Control Means Fuel Savings in the U.S. Navy," Proceedings Second Ship Technology and Research (STAR) Symposium, *Soc. of Naval Architects, 499-516 (1977).*
20. Chet, I., et al., "Repulsion of Bacteria from Marine Surfaces," *Applied Microbiology, pp. 1043-1045 (Dec 1975).*
21. Loeb, G., "The Settlement of Fouling Organisms on Hydrophobic Surfaces," *NRL Memo Report 3665 (Dec 1977).*
22. Dexter, S.C., Sullivan, J.D., Jr., Williams, J., III and Watson, S.W., "Influence of Substrate Wettability on the Attachment of Marine Bacteria to Various Surfaces," *Applied Microbiology 30, 298-308 (1975).*

23. Hanson, R.J. and Humston, D.L., "An Experimental Study of Turbulent Flow Over Compliant Surfaces," *J. of Sound and Vibration 34(3), 297-308 (1974).*
24. Minkus, A.J., *J. New England Water Works Assoc. 68, 1-10 (1954).*
25. Diamond Shamrock Company, Morristown, N.J. Trademark.
26. U.S. Coast Guard,"Low Friction Hull Coatings for Icebreakers." Report no. CG-D-32-76 (Mar 1976).
27. Tecknow-Malit Oy (Helsinki, Finland). "Inerta 160." Brochure.
28. Zebron Zerex Corp., Houston, Texas.
29. Olesen, G., "In-Water Maintenance of Ships World-Wide Today and Tomorrow," Paper presented at 16th Marine Coatings Conference of National Paint and Coatings Association, March 1976.
30. "Technical Bulletin on SPC." (Trademark of International Red Hand). Brochure. International Paint Company, June 1975.
31. Drydocking at Sea." Brochure. Jotun Marine Coatings, Division of A.S. Jotungrupper, Norway.
32. "Dynamic System." Brochure. Hempel's Marine Paints, December 1973.

PROPHETIC IMPERATIVES IN CORROSION CONTROL

H. L. Gerhart

The National Coatings Center
Carnegie-Mellon University
Pittsburgh, PA

CORROSION FOR THE CITIZEN

The most recent statement on loss due to the corrosion of metals places the cost at 70 billion dollars, annualized for 1975. (Supt. of Documents - Stock No. 003-003-01936-6). This loss represents 4.2% of the Gross National Product. Ten billion dollars worth could have been saved by better utilization of known procedures. The citizen is concerned at the cost of imported crude oil currently at 45 billion dollars. Except for the repair cost plus accelerated depreciation of the prematurely rusted family car, the citizen is relatively silent. In its many manifestations, corrosion is not visible in the cost of living. It is a securely hidden charge in the cost of goods and services. For its impact on the subject of this conference, we are justified in citing 40 billion dollars as the thought-irritating stimulant in planning for progress in the discovery and utilization of barrier coats for greater effectiveness.

WHO CARES?

The Metallurgist Cares — His desire is to know the rates of corrosion under specific conditions; to make precise laboratory measurements under the heading of "loss per year"; to recommend specific alloys; to monitor safe practices; to test organic inhibitors in storage tanks and fluid flow processes; add pipeline protection, preservation of ship hulls, chemical plants, refineries and oilwell equipment. For the most part, these activities may not require extensive use of barrier coats. Many problems are solved by the choice of alloys, the use of organic chemical inhibitors and extensive clinical technology in practice.

The Coatings Engineer Cares — He had commendably elevated the practice of paint making and coatings application from mixology to science. His struggle for perfection requires that he knows the contributions and craft in: (a) the selection of polymeric vehicles (also called "binders"), (b) the power of pigments to fight rust, reduce cost (inert ingredients) and decorate. The choices are legion. He is accountable to assure that the paint formulations comply with regulations; that application techniques are practical for each new change; that costs of product and application are equitable to the extent of improved protection. His new raw materials are frequently the result of research by chemical company suppliers but he is expected to innovate in terms of his own concepts.

The Consumers and Applicators Care — Their choices are circumscribed by balancing product cost against facility in application to structures of complicated design and frequently in hazardous locations. The local environment dictates the specification. They care about the endurance of the applied coating.

The Government Cares — Paints are regulated in terms of real or suspicious toxicity, contributions to the pollution of air, land, and water. Regulation is easy and frequently justified. But corrosion does not enjoy the support it deserves in terms of a systematic approach to finding solutions to its premature occurrence. There is a lot of attention to energy but no "Corrosion Bill" in any legislature. There is no master plan to coordinate the substantial skills and research results of the several scientific disciplines which undergird the problems. Most regrettably there is no locus with permanent staff to deal with the multidisciplinary aspects of the total problems responsible for the high cost of product failures in bridges, roads, mobile equipment, ships, aircraft, vehicles of defense, etc. In its diverse manifestations, corrosion is complex to organize for action, yet sufficiently damaging to deserve a cause celebre attack.

The Citizen Cares — He is inconvenienced by the closing of corroded bridges; by the cost of rustproofing the family car; by the detours necessary to replace rusted guard rails and highway repairs because reinforcing steel is deteriorated by salt; by the problems in maintaining a seaside home. Yet these are endured as inevitable, even normal and excused as is the common cold. Happily, there has been substantive progress by motor car manufacturers. But for many occasions corrosion simply insults the proprietor. He is hopelessly unconcerned that someone or some organization might be accountable to find a remedy.

The Scientist Cares — Scientific research of the highest order follows the course of *ad hoc* tradition. Professors report on theses of their students. Thus the causes of metal oxidation, fatigue, chemical attack, embrittlement, etc., are well documented. The precision of surface analysis is remarkable, though the relevance to the design of specific coatings for each event is essentially lacking. The fact is that the research metallurgist and the research coatings chemist are frequently isolated as country cousins.

Witness a summary (below) of "studies" supposed to correlate surface composition of metals with performance of coatings applied to delay oxidation. The Coatings Engineer is beset with expressions such as "good" and "bad" steel. He finds it difficult and expensive to match the objectives in polymeric surface coatings with advances in metal preparation pretreatment and composition at the interface. There has been progress and notably so in the more generous use of zinc to hide the iron.

Metals and Metal Pretreatment

The contribution of metals and metal pretreatment on the performance of paint systems is a very critical one which is difficult to deal with on a pragmatic day-to-day basis. It sometimes seems that one of the major functions of an industrial coating laboratory is laborious testing over a variety of metal and metal pretreatment substrates. There are great variations in the resistance to corrosion of cold rolled steel specimens from various sources when topcoated with paint systems. The variation in the quality of metal pretreatment also is very large. Pretreatment is a chemical process and unless it is well controlled, results can be variable. The coating formulator is asked to develop a system strong enough in corrosion resistance to overcome any shortcomings in the pretreatment process. The ability of various steels or other metals to accept pretreatment also appears to vary. Cases have been observed where corrosion resistance of a bare steel panel can be better than a pretreated panel. A demand for better coating systems should be enforced so that performance even over the worst of substrate is hopefully adequate for the expectation. Unfortunately this is often accomplished at some substantially higher cost. One wonders whether effort on more adequate pretreatment may not be more effective than massaging coatings for a deficient substrate. It is difficult to cure a metal or weak pretreatment by hiding the deficiencies with paints.

The reason for various metals and especially steel varying in their ability to resist corrosion, not only in conventional corrosion tests in dilute acids, but beneath films of coatings materials, is not very clear. The literature on the effect of minor impurities in metals and the effect on corrosion is quite limited. Some work reports that copper in small amounts has a beneficial effect, but the behavior of other trace elements does not appear to be substantiated. Grossman has attributed the wide variations in corrosion resistance on both untreated and phosphated steel to carbon contamination resulting from decomposition of oils on the surface during annealing. The nature of such oils used in annealing could also be a factor in the corrosion resistance of the ultimate product. A series of review articles by J. L. Prosser covers the subject of metals and metal pretreatments in a thorough fashion. It appears that there have not been substantial improvements in metal pretreatment or in improving the basic quality of ordinary steel in terms of corrosion resistance in recent years. The trend not to use chromic acid rinses, while it may be desirable from ecological standpoint, is a step backward to the ultimate consumer in securing maximum service for his coated article.

Metallurgical Effects on the Corrosion Resistance of Coated Steel

LaQue and Boylan studied the effect of an alkyd urea coating system on iron, 0.2% copper steel, and a high-strength low-alloy Cr-Ni-Si-Cu-P steel. Exposed to marine and industrial atmospheres as bare, coated, and phosphated and coated specimens, there was consistently better durability for the low-alloy steel over the copper steel, and for the copper steel over the iron. In a similar study, Copson and Larrabee used oil based paints over carbon steel, copper steel, and a similar low-alloy steel, with increasingly good corrosion resistance in both industrial and marine atmospheres.

Grossman studied the effect of steel surface conditions on paint performance, and found wide variations in corrosion resistance of both untreated and phosphated steel from various sources. He attributed this to carbon contamination resulting from decomposition of oils on the surface during annealing.

Data on the effect of trace elements on the atmospheric corrosion of steel is mostly restricted to exposure studies of uncoated steel. However, there is good evidence, including the above, that the corrosion resistance of coated steel correlates with that of bare steel. These data can be summarized in terms of different types of elements. Copper has a well-documented, beneficial effect on the corrosion resistance of steel. Up to 0.2%, its contribution is dramatic, and is related to the amount of residual sulfur in the steel and the resultant formation of copper sulfide rather than ferrous sulfide. Other metallic alloying elements - chromium, manganese, nickel, and molybdenum, also increase corrosion resistance, while the mechanism of such is not well understood. Many synergistic effects are present, and the explanation involves the nature of the rust film. There is evidence that tin and lead cause poor surface conditions for coating steel.

Regarding non-metallic elements, sulfur is known to be harmful. The effects of boron, carbon, silicon, phosphorus, and nitrogen are less clear.

They are known to be extremely harmful to corrosion by acids, but their effect on atmospheric corrosion is either neutral, or beneficial synergistically in the presence of copper, chromium, and nickel, as is the case with silicon and phosphorus. On the question of steel processing differences, and why older steels might be more resistant to corrosion, there is little relevant data. Problems of different atmospheric conditions and getting representative samples have created conflicting results. The problem is defined well by LaQue, in a paper discussing the whole spectrum of corrosion testing.

Once the facts about surface variations are firm and meaningful there may be an opportunity to place the desirable surface modifiers by ion implantation. It is an area to be watched by the Coatings Engineer.

THE ADVENT

If the above appears to be accusative, it is not evidence of criminal wrongdoing. Rather, it is a statement of the fact that reporting research results in *ad hoc* parcels is normal and can be justified for its value as archival literature. The force of its utility resides in our ability to coordinate the separate events into a full service concept. The horrendous costs demand face-to-face contact of the interdisciplinarians. Progress in metallurgy is significant; so is progress in polymers and paint formulation. The thrust must be to innovate at the interface, for that is where corrosion occurs. With that as a professional accountability, we can utilize the "month-to-month resuscitation" in the relevant journals. Co-joining the disciplines can derive solutions which are greater than the sum of the parts.

Is it absurd or wrong to create an accountability to develop protection for massive structures (such as bridges) which provide for repaint schedules at intervals of 20 years? Guaranteed protection is given by manufacturers for siding applied to residential and industrial buildings. Is it worth the cost of research? Certainly so for Allegheny County, PA, where one major bridge is condemned every year and where there are no funds to repaint the others on an adequate schedule.

There is some motion toward the compulsion to do some good. The crisis has inspired some professors to orient thesis studies to attack corrosion problems at the microdimensional level of interfacial phenomena. The inducement for survival is not sufficient. Acceleration can come from some external force which can impact the professors' abstractions upon the shirt-sleeves practicality of the inventor's imagination. The genius of our national history is the evidence of successfully focusing the targets upon both components of the equation to find solutions. At the moment the dream matters as much as the achievement provided the accountability is accepted.

To this end, I tabulate six prophetic imperatives (with credit to many colleagues) which are intended to inspire confidence in diversified further research on polymet coatings.

DETECTING NEW SIGNALS - PROPHETIC IMPERATIVES

(a) Adhesive Cement - Nature's Protective Genius

In principle, a perfectly "glued" surface coating can be reasoned as the ultimate protectant, provided and if the adhesion is perfect and enduring even allowing for absorbence and transport of nature's corrodants. One clue to such perfection is the observation of the invertebrate *Byssys*. Specifically, the common barnacle *Arca Byssus* secretes a fluid with extraordinary properties.

The liquid monomer is manufactured within the cells of the invertebrates and is secreted as a fluid which spreads over the surface contacted and crosslinked to a high polymer within five minutes at ambient temperatures. The molecular weight is over 100,000. The monomer has not been isolated because of the fast crosslinking mechanisms (five sequential mechanisms are postulated). No inhibitors have been found. The polymeric material is homogenous and bonds extremely well to most and moist surfaces. Bond strength has been measured from 600 to 1,200 psi depending on the condition of the substrate.

The color of the polymer is dark brown, and it is insoluble to all known solvents, resists most acids, but can be hydrolyzed in basic solutions. It is stable to 200°C., at which point it begins to decompose. Only very small traces of iron and chlorine have been found within the polymer. Copper is absent. The surface tension of the polymer was measured from contact angle experiments:

λ^d due to London dispersion forces ≈ + 24 dynes/cm.
λ^h due to hydrogen bonding forces ≈ + 30 dynes/cm.
λ *Byssus* = 54 dynes/cm.

In the crosslinked state, it is difficult to analyze. Spectrophotometric techniques used in chemical structure and functional group characterization are of little use. The precursor to the *Byssus* is thought to have free thio groups on the polypeptides which upon reacting with water, oxygen and/or enzyme, crosslink to form a ladder-type polymer.

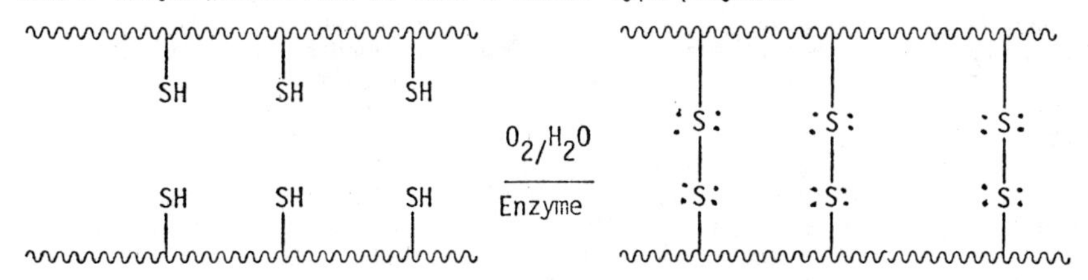

This in turn, forms a spiral thread-like system. Upon digesting the *Byssus* in Aq HCl, the following amino acids have been identified:

	Percent
Glycine	25.60
Alanine	6.10
Proline	7.30
Valine	4.91
Leucine	1.87
Isoleucine	2.41
Phenylalanine	3.12
Aspartic Acid	10.70
Glutamic Acid	2.64
Cysteic Acid	10.71
Histidine	2.97
Lysine	4.85
Methionine	0.72
Serine	3.35
Threonine	1.91
Throsine	9.60
Arginine	3.52

The object lesson from the above rather insecure signal is the fact that the *Arca Byssus* polymer is a corrosion shield. Not incidental to the explanation is the fact that the polymer is *renewed* during the lifetime of the barnacle. Further, the excitement to the polymer synthesist is the fact of the rapid polymerization initiated by enzymatic action plus the ability to occur in sea water in short time even at moderately cold temperatures.

To be confirmed are the facts about the actual degree of protection on bare steel. The advantageous concept of polymer renewal is restated in a specific proposal later in this paper.

(b) Efficient Placement

The substantial contribution made by electrodeposition of formulated organic primers is attributed (but not limited) to the fact that corrosion inhibitive membranes can be applied to "hidden" areas. Examples are lap-welded joints, box sections, channels and complex shapes.

The initiation of intensive technology can be attributed to a patent issued to Gilcrist. Applied studies in the laboratory of Ford Motor Company monitored by Dr. George Brewer pushed the horizon and demonstrated the virtues of the concept. During a 10-year period of extensive fundamental and applied research, the optimization of charged synthesized polyions concentrated upon anodic species. A multitude of complicated metal shapes are now coated by automated processes and in a manner not previously possible.

In a keynote talk before attendees of a conference on coatings research, I had the good fortune to state: "Anodic deposition may be second best when compared to potential advantages of placement on a cathode surface. This was based in part upon the fact that where the steel is anodic, the iron atom is oxidized, by leaving its lattice. The soluble ion then migrates toward the oncoming charged polymer. The total reaction was then understood in terms of first principles and is more involved than needed to be explained here. The premise for the prognostication was the fact that not only the iron atoms but also some portion of iron pretreatment "came loose" from the surface; that consequently, a portion of the phosphated and chromate-washed surface suffered a loss in effectiveness. This was supplemented by the knowledge of proprietary results, then existing, which indicated the superiority of utilizing positive charged macro-ion species applied to the cathode. So today, the advantage of specific cathodic deposition is verified especially in its employment in motor cars on four continents. The anti-corrosion effectiveness is stated in quasi-quantitative term as the greatest progress in 10 years."

Another constructive concept and useful approach is the practice of factory-applying a protectant to steel coil. The registered designation Zinchrometal® supplies a heat-cured coated surface which provides the combined protective power of zinc and chromate ion on a formable metal surface.

But the competition for value/cost effectiveness extends beyond the present horizon. The capital costs of electro-deposit systems are justified on utilization at high volume and continuous operation. This is one inducement for the autophoretic® application systems. It is in theory and to an advancing degree in practice, designed to apply protection without the application of impressed EMF. Research at Carnegie-Mellon University deals with understanding of fundamentals related to the mechanism of chemically oxidizing ferrous surfaces to supply ferrous and ferric ions. These involve themselves in the deposition of latexes as protecting priming surfaces. We have researched chemiphoresis and have derived models which elaborate the mechanisms and the distinctions *vis a vis* the electro processes.

Chemiphoresis utilizes latexes instead of macro-ions and latexes are avoided in electrodeposition for the sufficient reason that they have not been investigated in plenary programs. Yet certain reasons can be postulated as advantages to justify research studies on electro- and chem-deposition of latexes.

- Wide choice of copolymers to provide a scan of properties and service in rust-fighting.
- Coalescence temperatures are moderate and may conserve energy.
- Molecular weight can be higher than in macro-ions.
- Co-resin crosslinking may not be required.
- Simplified design of automated equipment is predicted for chemiphoresis.
- Chemiphoresis is conceived to replace the spray gun by reasons of efficiency and automation.

(c) Inhibitors in Polymeric Membranes

The traditional scheme of formulating metal primers with inhibitive pigments which supply molybdate, chromate, metaborate, phosphate and other ions of the metal/polymer interface is eminently effective. The mechanism depends upon the slow but interminable solubility of the salts in water in its transport from the atmosphere to the metal. After the salts have dissolved, the effectiveness "wears out" and renewal of the inhibitor requires blasting the coating followed by replacement.

Organic corrosion inhibitors are seldom used in a corresponding way. Yet a large body of experimental facts suggest this as a field to be investigated. Benzatriazoles protect copper and iron against attack by acids. Anti-corrosion engineers have a bag of chemical potions as additives to ward off attack by corrosive reagents over a range of adverse circumstances. For instance, a pinch of arsenic tames muriatic acid in iron containers, aqueous tension renders chlorine benigh to titanium. Deactivators remove oxygen by chemical reaction with reducing agents such as sulfite or hydrazine. A simple assault is mounted on the corrosion reaction by denying access of one corrodant.

Adsorptive inhibitors function as polar organic chemicals which layer upon the metal to exclude the adsorption of hydrogen ions. Also, there are agents such as amides, preferential wetting agents and polar organics which in an oil phase form an oil-wet surface to exclude the electrolyte.

The design of organic corrosion inhibitors is splendidly elucidated by Foroulis in terms of *adsorption* of organic molecules on metal-solution interfaces and their effect on the kinetics of corrosion reactions. There is a staggering body of technology with scientific underlayment, in the arsenal of anti-corrosion engineers who experience the effectiveness of adding homopathic doses of organics to water and other chemicals to stationary storage or fluid flow streams. The key to effectiveness is to select or design organics for their role in terms of electrostatic attraction, chemisorption and adsorption resulting from π-bond orbital interaction with the metal.

The inhibitive mechanism in effect operates through a very thin protective coating, since the adhered molecules interfere with mass transport of corrosive reactants toward and away from the metal-solution interface.

A primary necessity is that the barrier film have a thickness at least equal to a monolayer of the chemical. Polymeric carriers are not necessary.

It is appropriate to consider the rationale of incorporating the inhibitory organic moiety as a component of a polymeric barrier film. This rationale is defended on theoretical considerations and amplified by the suspicion that certain inorganic "poisons" may soon be regulated. Building the organic functional group into the polymeric membrane as pendant moiety can assure longevity by retarding leaching, exudation or loss by mechanical effects. The literature provides numerous examples of *ad hoc* experiments in which membrane carriers for organic inhibitors are studies. It is difficult to develop guidelines of relative effectiveness or to assess clearly the benefits of the carrier. Plans for experimentation have been documented. There is a fundamental question as to whether the "diluting" presence of the polymeric carrier will prevent close packing of the organic moiety upon the metal. It must be allowed that incorporation of surface-adhering chemicals in polymeric metal treatment or primer coatings will require a great deal of experimentation for leads. There is precedence in testing variations on this theme. The literature (50 references) presents a chaos of opportunity from which selections can be made for systematic study.

Conceivably, the adhesive attachment of the anti-corrosives (as a metal pretreatment) to a membrane might confine, protect and replenish the identical or alternative chemical species. At all events the high probability of regulating chromates justifies the discovery that the extensive skill of corrosion (sic) engineers can be incorporated in the design of new membranes for comparable effectiveness. It is an overdue idea whose time may be forced upon us - and sooner than we may appreciate.

(d) Zero Transport of Nature's Corrodants - Controlled Permitivity

This goal can be stated as an impossible dream. Impossible? Because an organic membrane can, in terms of Paul Flory's concept, behave like agitated spaghetti - having holes due to thermal motion. For instance, the transport of oxygen and water on many systematic studies is never zero. Professor Funke has a meritorious clarification of the problems which undergird the concept of designing a totally impervious membrane. (See paper in this volume).

Controlled Permitivity

In terms of signals for investigation, there are hints of organic polymers which have low absorbence, i.e., solubility. Advances can be made through use of poly(sulfones) and more advantageous inorganic surface pretreatment. Recall that steel in a sealed glass bottle will not corrode. A challenging prospect is the observation that *fluorinated* polymers possess advantageous characteristics relative to the common types because of their very low water absorption. An absorption of 0.35% (wt.) after six months' immersion is observed -- a tenfold decrease over that for a typical, structurally-related, unfluorinated species. At atmospheric conditions, there is a bonus advantage attributed to their exceptionally low surface tension and anti-wetting abilities for surface-condensed water. The two types reported are fluorinated epoxy and poly(urethane) classes.

In concept, the heavily fluorinated epoxy and poly(urethane) resins are pigmented with powdered poly(tetrafluoroethylene) at 40% (volume). The latter absorbs less than 0.1%(wt.). Absolute exclusion of water at the corrodible interface is not assured but the effect has the result of reducing the lingering of water.

Certain formulation tricks are potentially exciting. Glass flake in overlapping parallel orientation have been dispersed in liquid polymers. This mixture is applied as a coating, hoping for benefits derivable from the principle of *tortuosity*. Such have been applied to the walls of tunnels to seal against moisture seepage. With no attempt to presage a counsel of doom, we nevertheless warn that in such labyrinth arrangements the primary novelty (even allowing that the flakes have no pinholes) is primarily to increase the apparent distance of the polymeric compound in the membrane. By the known laws of molecular transport, the time required for diffusion of water may be increased. The adsorbence and ion exchange co-efficients are unaltered and the condensed water has every opportunity to reach the interface.

Thus, the problem is to design a construction in which the dispersed flakes have perfect lap joints while still being glued by the polymeric binder, a quite uncertain probability.

More plausible is the idea of using metallized flakes in which the metal is deposited in a fusible polymer of low Tg followed by cautious fusion. Extrapolating from this thought is the idea of placing the pinhole-free metal foil as the component of a laminated tape. If the latter is chosen from among the carbon pigmented poly(olefins) with extraordinary exterior desirability, it can effectively constitute a top coat with improved zero-transport attributes.

Of course, if the metal is sacrificial in electrochemical action, the advantages are multiplied but one would caution to contemplate first the advantages over zinc-rich concepts. I have, however, for some time felt that the use of tapes designed for specific protection is a much neglected area.

(e) Self-healing Films

Scratches, stone bruises and parking lot malfeasance produce abrasions which are the source of early corrosion. It is common widsom that metal primers should be hard and tough. These are properties which invite permanent breaks in the film which prompt development of anodic domains. Self-healing membranes have been invented. When these are cut with a knife edge, the two edges of the slit seal together at temperature above about 15°C. Such composition can be considered as a midcoat layer even though a Coatings Engineer will have concerns about the behavior of a rigid top coat over a surface having a high coefficient of expansion. This attitude is not proof that the idea will not work.

(f) Deactivating Nature's Corrodants

A separate report has been prepared on the potentials in formulating barrier films containing chemicals which are expected to prevent water, oxygen, chloride ion, sulfate ion and sulfur dioxide from reaching the interface. This subject will thus not be discussed here.

ACTION PLAN - CATHODIZING

Having thus documented the above thoughts as possible candidates for approaches to *finding new solutions* we launched a specific program in 1977. This program is based on two desires which were impressed by the study of the above prophetic case histories. From the barnacle we chose the target of renewing the inhibitory mechanism. There is no present process to do this, save removing the worn-out coating and repainting. We then experimented with a second idea, namely, to provide an interface which is *entirely cathodic*. This is on the premise that a cathode is a reducing electrode and cannot "rust". For this objective we looked to well known cathodic protective mechanisms such as the use of sacrificial anodes to protect underground pipes in trenches on land and ships' hulls on the sea. A second target was also provided, namely - "make a small ratio of renewable inhibitor go a long way". It appeared to us that these two targets can provide an (albeit radical) mechanism which would circumvent the more conservative approaches in the list of imperatives.

The initial experimentation fortified our confidence:

A small spot of zinc was pressed against a relatively large steel plate. The bimetallic couple was immersed in sodium chloride solutions in a range of concentrations. At above 0.1% the steel remained uncorroded for several days. At lower concentrations the steel rusted in one hour. Then followed a systematic study of immersion of "spotted" steel in other electrolytes. These were enlightening for their variety of specific effects and will be reported elsewhere.

We coined the term *cathodizing* for which the objective was defined in terms of three goals:

(1) A relatively large surface area of steel can be stabilized as a cathodic surface, (under specified conditions) by attaching a spot of zinc (or other less noble metal) to the iron surface as sacrificial anode. The possibility is thus suggested to conserve zinc in contrast to galvanizing or zinc-rich techniques. In practical terms, "a little zinc can go a long way".

(2) The possibility exists to *renew* the sacrificial anode in practice.

(3) The entire ferrous surface can be made a homogeneous cathode in the galvanic circuit and is therefore more easily protected with anti-corrosion coatings than is the normal condition where iron contains adjacent anodic/cathodic regions or domains.

The interpretation of the protective mechanism as demonstrated in the immersion tests is presented in Figure 1.

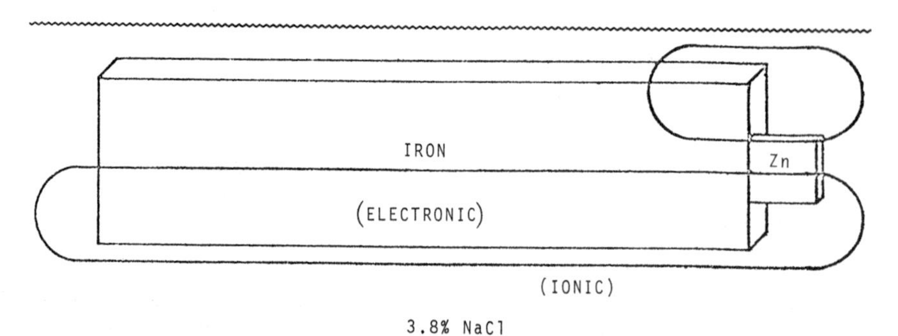

FIELD LINES THROUGH EVERY POINT IN CATHODIC PROTECTION

Fig. 1.—The immersed panel presents a circuit having two components. There is the obvious metal-to-metal electronic contact between the zinc spot and the steel. Second is the electrolytic (or ionic) circuit via ionized salt which connects the zinc to every point on the steel surface. The latter is graphically presented by field force lines.

Based on the above demonstrations and by extrapolation, cathodizing is an electrochemical system similar in many respects to a battery. The circuit engages an anode and cathode to separate the electrochemical half-reactions. Hence oxidation occurs only at the zinc or other sacrificial anode. The entire iron surface is a reducing electrode, i.e., the cathode. There are no anodic domains. The result is that the iron is not oxidized or deteriorated. Experimentation in progress is designed to optimize the use of a low ratio of zinc for such protection to provide a method to renew the anode. Emphasis is on metal conservation and renewability, having in mind especially the protection of bridges, storage tanks and other massive structures *in atmospheric environments*.

Having demonstrated the operability of the primary objectives there remained the desire to optimize the electrochemical properties of the electrolytic metal pretreatment for specific purposes. Beforehand we fortified our courage by demonstrating that previously rusted steel panels can be protected against further oxidation by cathodizing. We confirmed that a 1/4 inch diameter zinc spot at one end of a 30-inch bar offers protection. Such results were expected from experience with ships' hull cathodic protection which extends over many square feet. We also experienced some pleasant surprises such as the favorable effects of treating the steel with organic inhibitors prior to applying the electrolytic membrane.

Electrochemical measurements began with a determination of the potential of the metal surface beneath the membrane as a function of distance from the zinc spot. The potential ranged from -970 millivolts adjacent to the zinc to -670 millivolts at the most distant point. The variations corresponded to an estimation of observed corrosion products visible on the surface as cathodic protection declined. A study of pH ranges confirmed the composite Pourbaux diagram for iron and zinc metals which predicts that the iron/zinc system is more effective at higher pH.

Thus we have developed mature signals which presage the ability to lift cathodic protection "out of the sea and the earth" and adapt the corresponding protection to the ambient environment. To consummate the goals in practice we are prepared to count the above observations as opportunities to make precise measurements accordingly as preamble to demonstrations of operability.

A. Important Factors and Effects in Cathodizing

1. *Determination of protection limits, conditions and interactions*

a. The maximum effective and optimal cathodic to anodic surface areas ratios must be determined. At high zinc-iron surface area ratio, the iron electrode becomes more negative than necessary for adequate protection and results in extravagant consumption of zinc. At low zinc-iron surface area ratio, the zinc may not provide sufficient protection for the iron, zinc may dissolve too quickly, and reaction products may slow the reaction rate of zinc below a useful rate. Polarization curves are useful in the determination of the optimum zinc to iron surface area ratio.

b. Related to zinc-iron surface area ratio is specimen shape and geometry of both the zinc and the iron. For a given ratio, the length of time a zinc electrode provides protection will be directly proportional to the thickness of the zinc (for galvanic protection only). Also, the relative geometry of the cathodized system is important in determining whether oddly-shaped objects can be protected by cathodizing. The most important factor may be ohmic losses which may limit the distance of protection.

c. The maximum distance at which zinc can protect iron is an important consideration. If galvanic protection is to be provided by spots of zinc metal, the effects of placement distance must be determined. Again, ohmic losses may be a minimum distance necessary to prevent premature deterioration at the zinc-iron interface.

d. The effects of ionic conductivity and mobility are important. The product of current density and resistivity must be fairly low (tens of millivolts). Usually, high conductivity is required, but it is not necessary at very low current densities. These ohmic effects are quite important. In addition, ionic mobility and diffusion are important since these processes prevent concentration polarization which can cause a potential loss similar to ohmic loss.

e. Temperature effects of the cathodizing system must be determined. Reaction rates and transport processes are effected by temperature. In addition, below freezing temperatures may have a detrimental effect on the proposed hydrophylic membrane.

f. Corrosion of zinc metal is a possible problem since the process consumes the zinc without providing any protection for the iron. The environment of the zinc is important in regard to zinc corrosion. That is, zinc is consumed in the cathodizing process, however, the consumption of zinc must provide electrochemical protection for the iron surface; cathodic reactions on the zinc surface must be avoided.

g. It may be possible that ferrous iron may be reduced on the zinc surface and "plate out" on the surface of the zinc metal. This has the undesirable effect of promoting zinc corrosion by providing a cathodic surface on the zinc metal. This effect must be prevented.

h. An important question which needs to be considered is whether a cathodizing system promotes other types of corrosion, i.e., pitting, occluded cell, etc. Since the cathodizing conditions considered for this project are not too severe, these are not expected to be a problem.

2. *Chemical effects*

a. One of the most important parameters if the effect of pH on the system. Pourbaux has provided detailed studies showing how pH affects the equilibrium of the electrochemical system. The dynamics of the system are also affected by pH, as is the stability of other materials, e.g., paints. Optimum values of pH, as well as pH control, are important to cathodizing.

b. The effects of specific amounts of oxygen at the iron metal interface must be considered. Iron is the cathode in the cathodizing system and it must have a cathodic reactant, usually oxygen. Insufficient oxygen may cause problems, while excess oxygen will cause the cathodic reaction to proceed more rapidly and zinc consumption will increase.

c. Alloying materials can have a profound effect on electrode behavior. Alloys of zinc should be evaluated for improved performance. Most importantly, alloys upon which the cathodic reaction is impeded would increase the efficient use of zinc as a galvanic protector since corrosion of the anode material would be reduced. In addition, the cathodizing system must be tested on various ferrous alloys to determine the suitability and compatibility of the electrode material. Specifically, reaction products from alloying elements may enhance or degrade the system performance.

d. The effects of the chemical environment need evaluation, particularly with respect to reactions which can destroy the galvanic system. For example, carbonate ions can react with zinc ions to form an insoluble compound which could degrade the anode performance. Alternatively, cations such as calcium or magnesium may alter the membrane properties and degrade the electrolytic properties of the coating. Also, chemical interactions should be evaluated by actual field testing of the cathodizing systems, including effects of atmospheric substances such as carbon dioxide, sulphur dioxide, and particulate matter.

3. Electrochemical effects

a. Tests should be performed to evaluate the potential of the metal versus distance from anode, orientation of electrodes, and geometric design. The accepted maximum potential for galvanic protection of iron is about -0.60 volts versus a standard hydrogen electrode. Conditions in which the potential rises above -0.60 volts constitute significant corrosion rates under certain conditions, i.e., absence of inhibition or passivation. The behaviors of various electrode shapes must be determined to provide criteria for design characteristics of cathodizing systems.

b. Polarization measurements are very important to development of cathodizing systems. From potential-current data acquired by polarization techniques, diagrams (Evans-type), can predict the effective life of the system as well as other characteristics. These measurements provide a useful method of rapidly evaluating many parameters mentioned above.

c. When appropriate conditions have been determined, a polymeric material may be chosen as a likely candidate for the membrane and a synthetic electrolyte may be formulated for polarization tests. This electrolyte would contain monomeric material as well as electrolytes and other additives. This method will provide a convenient method for testing various formulations.

4. Some desired characteristics of the anode

In addition to the electrochemical requirements specified in the previous section, there are other criteria which would be useful in the design of an anode for a cathodizing system. Some of these are: (1) The anode should have a somewhat constant surface area over the entire life of the cathodized system. This characteristic should provide the most effective use of a given quantity of zinc metal. (2) The cathodizing system should be designed so that renewal for regeneration of the anode material is possible without the necessity of completely removing consumed material, e.g., zinc salts, or degraded paint materials. Possibilities for fulfilling this objective include "bolt-like" anode material that can be attached to the iron surface mechanically at calculated time intervals, or an outer coating of zinc which can be added to without stripping off the sublayers. (3) It may be possible to convert this system to an impressed current cathodizing system utilizing an inert anode material. Such systems have been effectively used in marine and underground protection systems.

5. *Design configurations for a cathodizing system*

In addition to the system in which the entire surface is coated with a hydrophylic membrane, another possible system would consist of coating the zinc anode and its immediate surrounding area with a hydrophylic membrane capable of turning the electrolytic current ON and OFF in response to humidity changes. The coating over the remainder of the surface would consist of a highly conductive material which need not be responsive to humidity changes. Another possible system could utilize some of the features of the zinc-rich concept where the zinc metal is dispersed in a continuous surface coating as a powder component of a paint. The advantage would be that iR (ohmic) losses would be reduced and that potential-distance limitations due to ohmic losses would be nearly eliminated. One type would consist of a primer or undercoat having specific electrolytic properties, and electrolyte, and possibly cathodic inhibitors. The coating applied over the primer would contain zinc metal in an appropriate binder. Finally, a top coat may be applied to give good appearance and greater durability.

6. *Inhibitors*

One of the advantages of a cathodizing system utilizing zinc is that zinc ions produced by anode dissolution have an inhibiting effect. Specific zinc salts might be added directly to the electrolyte coating to take advantage of this inhibiting effect. Inhibition is the decrease in the reaction rate caused by the addition of specific chemical components to the system. If the reaction rates are very low, a small amount of zinc can cathodically protect a large surface area of iron for a long period of time. One way to achieve this is to reduce the reaction rates by use of cathodic inhibitors on the iron surface. Inhibitors such as 3,1-tolylthiourea and bicyclohexyl ammonium nitrite can be used to decrease the rate of oxygen reduction on the iron surface. Also, high hydrogen overpotentials can be achieved on iron by use of antimony, tin, or arsenic salt which will "plate-out" on the iron surface and decrease the rate of hydrogen ion reduction. By use of such inhibitors, current densities for cathodic protection can be decreased substantially.

7. *Estimate of appropriate current density of cathodizing*

To determine the approximate electrochemical rates for a cathodizing system, a calculation was made assuming the equivalent amount of zinc necessary to provide a 0.003-inch solid zinc layer, a 100% reaction efficiency and galvanic protection by this amount of zinc for one year. The result of this calculation shows that at current density of approximately 5 microamperes per square centimeter, the zinc will be consumed in one year. This value indicates that a small amount of zinc cannot provide galvanic protection for iron for longer periods of time unless the iron surface can be polarized below --.6 volts (SHE) by a current density of 5 microamperes per square centimeter or less. Hence, the reaction rate on the iron surface must be the electrochemical equivalent of less than microamperes per square centimeter. This rate would be possible only if the surface of the iron were significantly inhibited in some manner, e.g., chemical inhibitor or primer coating impervious to moisture, oxygen, and other cathodic reactants. Note: Inhibitors can decrease reaction rates 2, 3, or even 4 orders of magnitude; thus they may permit proportionally less zinc to be used. Another consequence of working toward a system functional at low current densities is that the conductivity requirements are dramatically lowered. Hence, iR limitations, as well as the associated potential-distance limitations become more tenable, i.e., they are more easily worked out.

We have confidence that present signals and Kitty Hawk experiments are secure and relevant. The design of model membranes is underway having in mind utilization on highway bridges and steel storage tanks. The action plan is to proceed in two directions simultaneously (a) membrane design in electrochemical specifications, (b) membrane synthesis and in-use testing. We are open ended in studying whether the electrolytic interface should be a form of metal pretreatment, surface primer, or another and unconventional technique.

There is another option. It is conceivable that scheme such as the diagram may be a penultimate advance. The predicted advantage can be the fact that the problem of zinc oxidation product can present a problem in its removal. So we replace the zinc couple with an external battery. Conceivably the power source can be a photovoltaric system. The germane fact is that in such systems an electrolytic conductor as membrane or interface pretreatment is mandatory. The research can serve both approaches.

It is the hope that the illustration (Cathodizing) in this manuscript can serve as a model for more aggressive innovative research to move mountains in corrosion protection.

PRINCIPLES OF FORMULATING CORROSION PROTECTIVE COATINGS

Zeno W. Wicks, Jr.

Polymers and Coatings Department
North Dakota State University

ZENO W. WICKS, JR.

ABSTRACT

The most critical consideration in formulating a coating for corrosion protection is to assure adhesion in the presence of water and hydroxyl ions. Factors affecting adhesion are discussed. Other factors which should be considered are pigment volume effects, mechanical properties, water permeability, oxygen permeability and perhaps electrical conductivity.

There are two broad classifications of possible film failures associated with corrosion resistant coatings which must be minimized by formulation. If the film is ruptured-scratched, gouged, etc. - the remaining film must maintain adhesion. Second, the intact film must resist blistering.

First, let us consider the situation with a ruptured film. The iron is exposed to water and oxygen and corrosion begins. Other speakers will be considering means of passivation which will minimize this corrosion but even under the best of conditions some corrosion will occur.

If we consider the overall corrosion equation:

$$2\ Fe + O_2 + 2\ H_2O \longrightarrow 2\ Fe^{++} + 4\ OH^-$$

we see we are generating hydroxide ions. We must have a coating that will maintain its adhesion in the presence of water and hydroxide ions. Among other things, the polymer in the film must, therefore, resist saponification. It is surprising how often corrosion studies of coatings are done forgetting this basic consideration. Experience shows that polymer systems whose backbones are subject to saponification will not give as good corrosion resistance as those which cannot saponify. Epoxy-amine coatings have only C-C, ether and amine linkages - there is no possibility of saponification and they are generally good binders for corrosion resistant coatings. Phthalic alkyds are readily saponified and generally provide a lower level of corrosion protection. Some times for other reasons, for example cost, one wishes to use an alkyd or a polyester. What can one do? First, one should avoid esters of dibasic acids such as orthophthalic acid whose hydrolytic resistance is very poor due to the anchimeric effect of the two carboxyl functions on ortho positions. Isophthalic esters, terephthalic esters, adipic esters are much more resistant to hydrolysis. Second, one should use polyols which provide maximum substitution in the 6th and 7th position counting the carbonyl oxygen of the ester as number one.

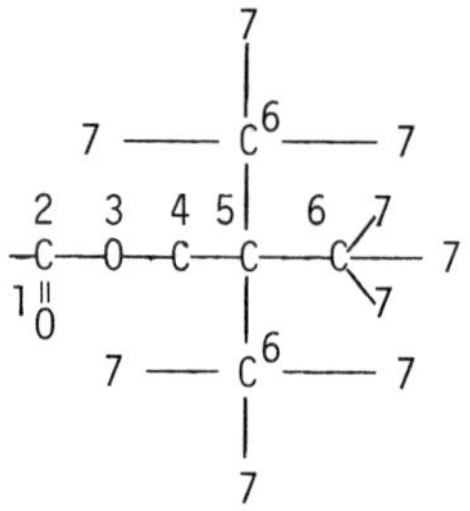

```
                     7
                     |
              7 ---- C6 ---- 7
                     |
      2   3   4   5  |   6   7
     -C---O---C---C------C<--- 7
    1 ||             |       7
      O              |
              7 ---- C6 ---- 7
                     |
                     7
```

Thus neopentyl glycol polyesters, for example, are more resistant to hydrolysis than are ethylene glycol polyesters.

Hydrolytic stability is a necessary but not sufficient characteristic of a vehicle. Polyethylene is completely resistant to hydrolysis but would not provide adequate adhesion even in the absence of the hydroxyl ions. This leads us into the very complex and poorly understood question of factors which control adhesion. Based on dispersion forces one can calculate that the adhesive forces between polyethylene and an iron oxide surface on steel are greater than the observed adhesion forces. These calculations assume complete wetting of the surface, which is a reasonable assumption between a very low surface tension material like polyethylene and a high surface energy substrate. While the cumulative effect of dispersion forces is great, each individual attractive force is small. Therefore, it is possible for a material such as water, which can associate with the iron oxide surface with much greater attractive force, to displace the $-CH_2$ - groups from the surface. It is common experience that adhesion of polymers with polar groups is better. Groups like -OH; -COOH, -CONH-, -NH-, are more strongly associated with the metal oxide surface and hence tend to be less readily displaced by water.

While this generalization may have some validity, degree of polar substitution of polymers does not always correlate with adhesion in the presence of water. Recent work at NDSU by Hill and Lindstrom provides an example which illustrates the complications which can be encountered. They compared two acrylic copolymers one a copolymer of methyl methacrylate, butyl acrylate, hydroxyethyl methacrylate and acrylic acid (I) and the other a similar copolymer not containing the carboxylic acid (II). The acid containing copolymer might be expected to provide better adhesion due to increased hydrogen bonding to the substrate and possible salt formation. Adsorption of the two copolymers labeled with C^{14} from dilute solutions was determined. The amount of copolymer I adsorbed on a steel substrate from dilute solution was much lower than of copolymer II. Apparently the presence of the carboxylic acid groups resulted in a flat extended conformation of adsorbed polymer resulting in a relatively low amount adsorbed. In contrast copolymer II without the acid groups showed higher adsorption resulting from few adsorption sites per molecule so that loops or ends of the polymer chain projected up from the surface. Then coatings formulated from these polymers were evaluated for adhesion on steel. It was surprising to find that greater force was needed to pull coatings made with copolymer II off a panel than with copolymer I. One must be cautious in interpreting the data since the adsorption studies were done from dilute solutions while the concentration is high in the coatings. However, a possible explanation of the results is that in the case of copolymer I adsorption to the steel interface was strong but the flat conformation minimized the opportunity for crosslinking and intermolecular attractive forces with the main layer of the coating leading to cohesive failure between the adsorbed layer and the bulk of the film. In contrast copolymer II has loops and ends extending into the bulk of the film permitting crosslinking and intermolecular attraction between the bulk of the film and the adsorbed layer resulting in better adhesion with the less tenaciously adsorbed polymer.

Another factor which may be important in corrosion is the electrical conductivity of the paint film. There have been many studies reporting possible relationships between film conductivity and corrosion resistance based on the idea that low conductivity may reduce current flow through the film connecting anodic and cathodic areas on the substrate. In interpreting such data, one should bear in mind that variations in polymer structure which could effect conductivity could also affect polymer adsorption and ease of displacement of polymer from the metal-polymer interface by water. Some of the results which have been attributed to film conductivity may really be caused by differences in adhesion in the presence of water and hydroxide ions. The need for further research is obvious but I feel the importance of achieving and maintaining adhesion for corrosion control cannot be over emphasized.

Still greater adhesive forces could result from formation of covalent linkages between the substrate and the polymer molecules. Especially if such a bond were hydrolytically stable, adhesion in the presence of water would be improved. While formation of such bonds with glass and cellulosic surfaces has been experimentally demonstrated with reactive silanes and substantial adhesion improvements are possible, there have not been equally clear demonstrations with metal surfaces. Certainly in some cases silane or titanate additives can provide improvements in adhesion.

Before proceeding further, the other mode of failure - blistering and that specialized case of blistering, filiform corrosion should be considered. It is well known that all organic polymer films permit the permeation of at least some moisture vapor and oxygen through the film. When water and oxygen reach the substrate, corrosion starts. The corrosion reactions result in the formation of soluble ionic products dissolved in the water at the interface. Now we have an osmotic cell driving the permeation of additional water. If the adhesion of the coating to the substrate in the presence of water and hydroxyl ions is insufficient, a blister will form and expand. If the adhesion is uneven, the pocket of water can disrupt the adhesion at one spot on the circumference of an initial blister leading to filiform corrosion. Again, we see that adhesion plays a critical role in corrosion protection. It would be of interest to investigate the possibility of an accelerated corrosion test based on this phenomenon. It is well known that when one applies coatings over steel which has soluble salt residues on its surface and then runs water immersion or high humidity exposure tests, blisters will result. A more stringent test might be to prepare panels with traces of alkali purposely on the surface, coating over them and then testing blister formation.

Thus far I have mentioned only a few of the factors which can affect adhesion. Cleaning of the surface before applying a coating is of critical importance. If there is surface contamination with oil, for example, in most cases adhesion will be poor. If there is loose rust on the surface, adhesion can be expected to be poor. Cleaning methods such as sand blasting can be very beneficial - not only is oil and loose rust removed but also effective surface area is increased by roughening the surface. However, there are times either due to economics or to inaccessibility when it is necessary to paint over improperly cleaned surfaces - loose rust for example. It is known that some coatings are more effective over loose rust than others. The ability of a coating to wet and spread on a substrate is an important criteria in achieving adhesion. The coating must have a lower surface tension than the substrate, but it must be remembered that this is a thermodynamic expression and does not deal with the kinetics of wetting. Drying oil coatings, especially fish oil coatings are often recommended for use over rusty surfaces. Why might they be superior? The vehicle must penetrate through the rust layer to the substrate and completely surround the rust particles rather than only wetting the surface of the rust layer. What controls penetration through such a porous layer? The variables can be seen by looking at the equation for the rate of penetration into a capillary tube.

$$u = \frac{R\sigma \cos \theta}{4 \eta L}$$

The variables available to the coatings formulator for any given surface are $\cos \theta$, σ and η. Contact angle should be 0° so $\cos \theta$ will reach its maximum value of 1; this requires that the surface tension be less than the critical surface tension of the substrate. On the other hand σ should be as near this critical level as possible since high surface tension gives high penetration rate. This combination leaves little latitude, so the variable over which the formulator has greatest control is viscosity. In this case we must consider viscosity of the continuous phase of the vehicle not the bulk viscosity of the coating since the pigment will be filtered out and only the vehicle can penetrate. The viscosity should be as low as possible and furthermore stay low until penetration is achieved. Some solvents will evaporate relatively rapidly leading to viscosity increase. Drying oil paints contain less solvent and usually use a relatively high boiling solvent (the high boiling solvent can be tolerated because the films dry so slowly that evaporation of even low volatility solvents before the top surface of the film is dry is possible). Therefore, they can penetrate well. Most so-called synthetic paints use higher viscosity polymers and more volatile solvents, hence the opportunity for penetration is less. The drying oils are esters hence are sensitive to hydrolysis but over rusty steel this disadvantage may be offset by better penetration. Fish oil is highly unsaturated hence when it dries, a lower fraction of the crosslinks is ester and it has better saponification resistance than say linseed oil. It would be interesting to design a synthetic vehicle which cannot hydrolyze and which crosslinks slowly enough so that low volatility solvents could be used with it, to see if it would be more effective yet over rusty steel. As far as I know, the trend has been to try to achieve fast dry which is perhaps the wrong direction to go from this specific standpoint. The problem with latex coatings can be even more severe since the latex particles can also be filtered out and in some cases only the polymer-free water can penetrate, so that little binder gets through a rust layer to the substrate.

Another generalization is that baked coatings show better adhesion and better corrosion resistance than air dry coatings. Why should this be? Penetration into fissures, and irregularities in the surface can improve adhesion and these are affected by viscosity. As a baking coating is being heated before crosslinking takes place, the increased temperature reduces the viscosity and permits more rapid penetration. Another factor which should be considered is that with the same system more complete crosslinking can be achieved at a higher temperature. Initial stages of curing of a film are controlled by the rate of reaction of functional groups. As crosslinking proceeds, mobility of the functional groups decreases and the extent of reaction is controlled by the possibilities of functional groups reaching other functional groups. At higher temperatures more crosslinking can occur before the reduced mobility stops or reduces the rate of reaction. This phenomenon has been established experimentally but why should it lead to better adhesion and corrosion resistance? Frequently, greater crosslinking gives greater toughness and greater toughness may increase the scratch and gouge resistance of the film hence reduce the chances for breaking through to the substrate. In general, permeability decreases with crosslink density. Another factor, which has not to the best of my knowledge been studied, could be the effect of crosslink density on blister growth. Let us assume there has been permeation of some water through a film and that a very small oxmotic cell has been established. Earlier it was pointed out that poor adhesion would permit growth of a blister. But it is also logical to assume that the mechanical properties of the film could affect blister growth. Let's illustrate the idea with an extreme example. Let us take two osmotic cells one with a concentrated salt solution in an elastomeric balloon. The shell of the cell expands easily as water permeates into the cell to dilute the salt solution. At the other extreme, let us take a cell fabricated from a rigid high strength plastic. The water permeates in but the cell cannot expand significantly. In a short while the pressure inside the cell matches the osmotic pressure and net diffusion stops. In a coating extremely rigid films cannot usually be used due to other requirements such as flexibility, low temperature impact resistance, etc., so there may well be an optimum crosslink density. What mechanical properties should a film have to resist blister formation? Again the need for research is evident.

Another factor that is very important to adhesion and corrosion is pigmentation. Other papers will be dealing with one aspect of the problem - passivation. Obviously in many cases color is a requirement. Such factors are affected by the chemical composition of the pigments. But there is also the physical effect of having a heterogeneous system on the physical properties of the coating. Most primers for metals are formulated to be very near critical pigment volume content (CPVC). At CPVC, the pigment volume is such that there is just enough vehicle in the dry paint film to provide a layer of polymer adsorbed on the pigment surface and to fill the interstices between the polymer-covered pigment particles. Above CPVC, voids are left in the film. Having a very few voids can help intercoat adhesion between topcoat and primer by permitting interlocking between the layers. However if there are too many voids, a porous film will result.

Tying back to the previous discussion of the mechanical properties of the film and adhesion, pigment volume content can have profound effects on such properties. In many cases, the tensile strength of a film reaches a maximum at CPVC. As certain pigments are added, tensile strength increases - presumably by adsorption of polymer molecules on more than one particle resulting in a type of crosslink. However, beyond CPVC the presence of voids leads to a sharp fall off of tensile strength. In general, it appears sound to formulate primers at about CPVC. There are only two classes of primers that I know of where this is not the case. Electrodeposition primers are formulated below CPVC, since it has not been possible to get uniform electrophoretic behaviour with such highly pigmented systems and zinc-rich primers are formulated substantially above CPVC because their activity requires electrical and electrolytic contact between the zinc particles and with the steel substrate. As a sidelight - one of the limitations of attempts to reduce solvent content of primers to meet emission standards is that by definition the viscosity at CPVC is infinite. Many evaluations of pigments in primers are flawed by the failure of the evaluators to take into consideration the pigment volume factors. Critical pigment volume content is affected by particle size distribution, CPVC must be calculated or determined for each different combination of pigments.

Another consideration of potential importance in pigmentation is its effect on permeability of films to water and oxygen. There has not been enough research to permit generalizations. In some cases, it has been demonstrated that increasing pigment concentration leads to decreased permeability until CPVC is reached. However, in at least some cases it has been reported that increased concentration of polar pigments leads to increases at least in water vapor transmission. The whole field of the relation between permeability and corrosion has been inadequately investigated in the past. Relatively early work led to the conclusion that the permeabilities of paint films to both water and oxygen was so high that they could not be controlling factors in corrosion control by paints. Fortunately, Werner Funke and his colleagues in Stuttgart have undertaken a major research program in the area. I am sure Professor Funke will include some reference to the work in the paper he is presenting at this meeting. They have already published a major result of the research - the importance of film-substrate adhesion in the presence of water that is detectable in water permeability studies. These results are a major reason I have placed so much emphasis on adhesion in my presentation today. Anyone interested in the relationship between corrosion control and coatings should impatiently await further results from Stuttgart. Meanwhile, coatings formulations with reduced water and oxygen permeability should be sought.

Obviously I have mentioned only a very few of the factors which must be considered in formulating a coating. For example, if a coating is to be exposed to an acid environment, acid resistance will obviously become important. I have tried to bring out those which bear most directly on the question of corrosion resistance. I have particularly emphasized the importance of adhesion in the presence of water and hydroxide ions and the need to consider pigment volume relationships. Practical paint formulators have made real progress in preparing coatings with good adhesion and saponification resistance. However, compared to other aspects of the corrosion problem relatively little basic research has been devoted to the relationships between adhesion, the mechanical properties and corrosion. In these days when drastic reformulations are being required on a time table that does not permit slow evolution of coatings formulas, the need for fundamental understanding of the mechanism of action of protective coatings is great.

CORROSION TESTS FOR ORGANIC COATINGS—USEFULNESS AND LIMITATIONS

W. Funke

Institut für Technische Chemie der Universität Stuttgart,
Forschungsinstitut für Pigmente und Lacke,
Stuttgart, Germany

WERNER FUNKE

ABSTRACT

Various exposure tests and methods for classifying the protective performance of organic coatings are reviewed. Permeability to water and oxygen together with the adhesion on exposure to high humidity, are proposed as to evaluate the corrosion protection and to explain differences in coatings subjected to conventional corrosion tests and to outdoor exposure. Finally, some special and still unsolved problems are considered.

INTRODUCTION

As organic coatings are mostly applied for decorative and protective purposes, quality requirements are predominantly concentrated on gloss and color changes rather than corrosion tests. In the past two decades testing methods to classify coatings according to these optical properties have been brought to a remarkably high accuracy and reliability, in contrast to classification according to their protective quality. Despite the numerous efforts to evaluate corrosion protection quantitatively with electrical and electrochemical measurements or to correlate it with properties involved when a paint film functions as a barrier, the final decision on the application of protective coatings is still based on the classical corrosion tests and results of outdoor exposure. These corrosion tests are the only ones which have already been standardized in many industrial countries.

Exposure Tests and Classification of Corrosion Protection by Coatings

Among all exposure tests for organic coatings (Table 1) the salt spray test probably is the most common one for testing corrosion resistance of coatings. On exposure to salt spray at moderately elevated temperatures, coated samples may be classified qualitatively by visual inspection or semi-quantitatively by measurement of the degree of undercutting started from a scratch deliberately applied before exposure. There is little doubt that conditions in salt spray tests approximate well with practical exposure of painted steel, salt water on the roads during the winter or to sea water. The salt spray test is commonly considered as a general method to determine the tendency of a coating system to resist corrosion. The salt and elevated temperature accelerate corrosion and undercutting. As salt spray tests commonly are carried out at 35 or 40°C, the results may not be representative

TABLE I.—CONVENTIONAL CORROSION TESTS AND CLASSIFICATION OF THE PROTECTIVE QUALITY OF COATINGS

Exposure Tests	Classification	
Salt spray test Humidity cabinet Dew test Outdoor exposure Sulfur dioxide	Qualitative	(by visual inspection of underfilm corrosion and blistering)
	Semiquant.	(scratch test - width of undercutting)
Immersion tests (salt water, seawater, deionized water)	Quantitative	(Electrical measurements: Resistance, Capacitance, Polarization curves, Potential)

for temperatures near the freezing point. Within this temperature range, many paint films have a glass transition and then permeability to water and gases significantly change. Moreover some doubts exist whether the salt spray test results remain valid for exposure to rain or dew. For osmotic reasons water uptake of coatings is significantly higher in pure, deionized water than in sea water or salt water.

Exposure to humid air containing SO_2 has been controversial for a long time and discussion still goes on. The permeation rate of SO_2 through paint films may be easily determined by titration of the water at the downstream part of a permeation cell. It is observed that the pH remains constant there until SO_2 passes through the membrane and enters the water, a fact which is indicated by a rapid drop of the pH (Figure 1). The time at which the pH changes as well as the slope of the pH-time curve are related to the SO_2 permeability of the film. When pure SO_2 at atmospheric pressure is present in the upstream part of the permeation cell, rather high rates are observed (Table II). It may be supposed, therefore, that sulfate ions may easily be formed at the coating/metal interface and stimulate corrosion without requiring permeation of sulfate anions. Some evidence exists that too high a concentration of SO_2 not only accelerates corrosion but in some cases may also destroy the binder chemically. Therefore data are needed for reasonably lower upstream concentrations of SO_2. Originally SO_2 exposure has been considered to be typical for an industrial atmosphere, however, with the strong increase of SO_2 emission from oil burners used for heating purposes, relatively high concentration of SO_2, at least locally, may also occur in residential or even rural areas. Similar arguments are also pertinent in considering immersion tests.

Along with the qualitative classification based on visual inspection, semiquantitative means of evaluating the corrosion protective properties of a coating, as the scratch, are very common. A scratch is applied to the

TABLE II.—THE RATE OF PERMEATION OF SO_2 THROUGH PAINT FILMS AT 23°C.

Rate of Permeation SO_2 $(g \cdot cm^{-2} \cdot d^{-1})$	
EC No. 6	$1{,}01 \cdot 10^{-2}$ (36 μm)
CN	$4{,}05 \cdot 10^{-2}$ (100 μm)
AR	$2{,}0 \cdot 10^{-2}$ (88 μm)
PVCA	$1{,}1 \cdot 10^{-3}$ (77 μm)
Rate of Permeation* SO_2 $(g \cdot cm^{-2} \cdot d^{-1})$	$5 \cdot 10^{-2}$ - $0{,}3 \cdot 10^{-6}$

*s. Lit., Polymer Handbook, 2nd. Edition

test samples before exposure and information is obtained as to how easily a film is undercut and detached by corrosion, which starts from the exposed steel at the scratch. There is considerable agreement among paint technologists that corrosion resulting from mechanical paint defects is more important than any other types of corrosion encountered in organic coating systems. On the other hand, despite standardization, the scratch test is prone to ex-

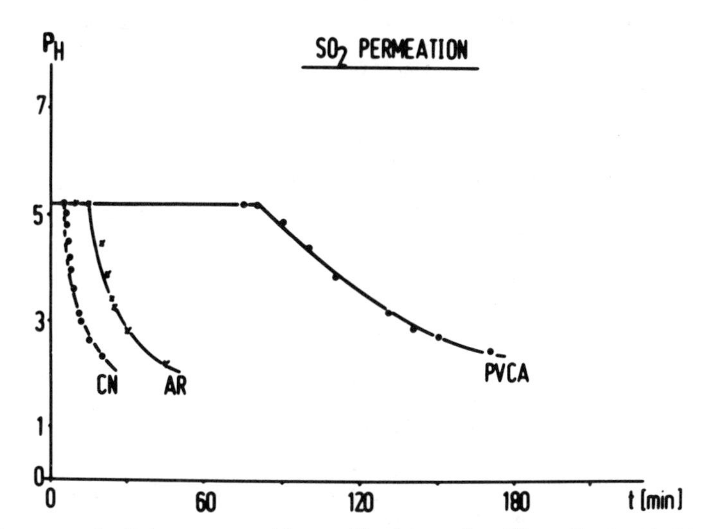

Fig. 1.—Drop of pH in a permeation cell when using SO_2. Temperature = 23°C.

perimental errors, e.g., different pressures and velocity of scratching the coated panels, abrasion of the scratching tool, lifting the coating far beyond the area of the scratch. The last factor depends on the viscoelastic properties and thickness of the coating[1]. The scratch test is useful to classify roughly coating systems differing considerably in their corrosion protection. However, it usually fails to discriminate small differences reliably.

Electrical and electrochemical methods are often used for quantitative classification of coatings. A great body of literature exists on the measurement of resistance, capacitance, polarization, and potential to evaluate the corrosion protective property of organic coatings. An excellent survey by H. Leidheiser critically describes the present state of these methods[2]. Despite many efforts in this field of corrosion testing, as yet none of these methods has found unequivocal approval and none have been standardized.

Whatever may be said about conventional and newer methods to classify the corrosion protective quality of coatings, these tests principally compare only coatings. All further conclusions and explanations of the differences found are analogy arguments or, bluntly expressed, rather more or less speculative. The optical appearance of a coated sample after exposure to corrosive environments is always a complex result of the influence and action of a number of primary film properties, which are discussed below.

Measurement and Significance of Primary Properties Involved in the Protective Function of Organic Coatings

The most important requirement for corrosion protection by organic coatings is adhesion to the metallic substrate (Table III). Normally adhesion is

TABLE III.—THE PRIMARY PROPERTIES OF COATINGS INVOLVED IN CORROSION PROTECTION BY COATINGS

Primary Properties Involved in Corrosion Protection	Action	Measuring Methods
Adhesion on exposure to water	Prerequirement for protection	Comparative absorption, Mechan. adhesion measurement, Sound absorption?
Water permeability	Directly involved in the cathodic reaction and solvent for electrolyte	Gravimetric, volumetric, capacitance
Oxygen permeability	Directly involved in the cathodic reaction: $H_2O + \frac{1}{2} O_2 + 2e \rightarrow 2OH^-$	Volumetric, barometric, oxygen electrode
Ion permeability	Stimulating agent and electrolyte for local corrosion element	Radioactive tracing, conductivity, ion sensitive electrodes (Cl^-)
SO_2 permeability	Stimulating agent and electrolyte for local corrosion cells	Titration P_H-measurements

roughly classified by the cross-cut test or measured quantitatively by the vertical tear-off method. However, both methods have drawbacks, especially if the coated samples are to be measured when exposed to water or high humidity. Since it is critical that the coating maintain adhesion on exposure to water, some other method of determining the loss of adhesion on exposure to high humidity has been investigated by measuring the difference between water absorption by free and supported films[3]. Loss of adhesion on exposure to high humidity is indicated by a crossover of the water absorption curve for the supported film over that of the corresponding free film (Figures 2 and 3). The "cross-over" time thus indicates the time after which the loss of

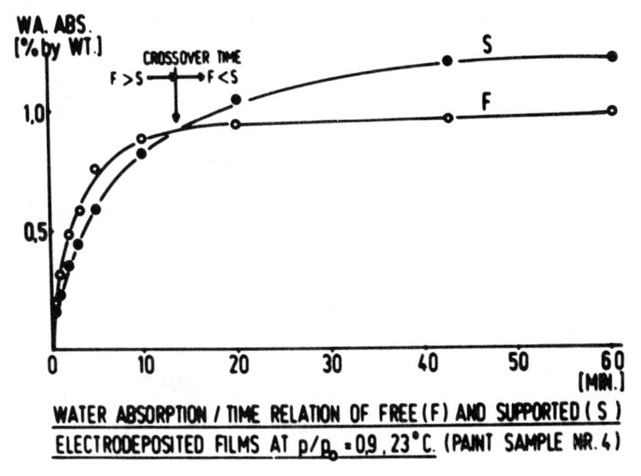

Fig. 2.—Water absorption/time relation of free (F) and supported (S) electrodeposited films at p/p_o = 0.9 and 23°C. Paint Sample No. 4.

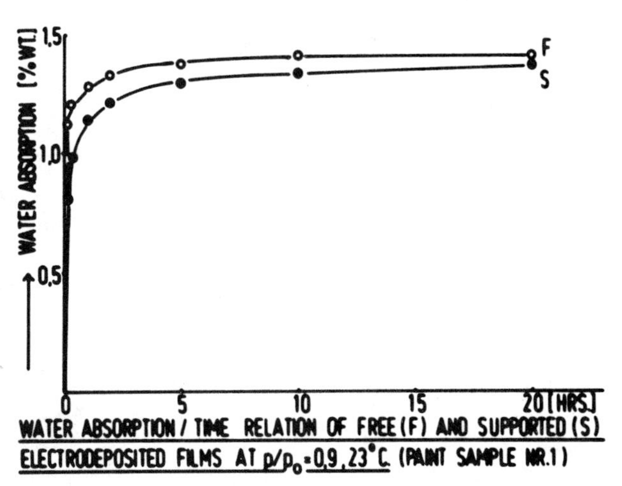

Fig. 3.—Water absorption/time relation of free (F) and supported (S) electrodeposited films at p/p_o = 0.9 and 23°C. Paint Sample No. 1.

adhesion becomes significant by the accumulation of water in the coating-substrate interface. As this method requires some skill, there is a need to simplify it or to find other methods for measuring adhesion of coatings on exposure to high humidity or water.

As both water and oxygen are required for the anodic reaction in the corrosion of steel, the permeabilities for these agents directly express the barrier properties of organic coatings in suppressing the corrosion. Moreover water serves as solvent for the electrolytes present in the local corrosion cells. A series of methods is available for measuring both water and oxygen permeabilities of organic coatings (Table III). It is generally agreed that permeability of coatings for water is relatively high and sufficient to provide an adequate supply for the corrosion reaction[4]. On the other hand, oxygen permeability of organic coatings may be in about the same range or even below what is needed for the corrosion reaction to proceed at a significant rate.

Electrolytes and especially chloride and sulfate ions, which are already present at the coating/metal interface or enter this area by diffusion, are also important as stimulating agents of the corrosion of steel (Figure 4).

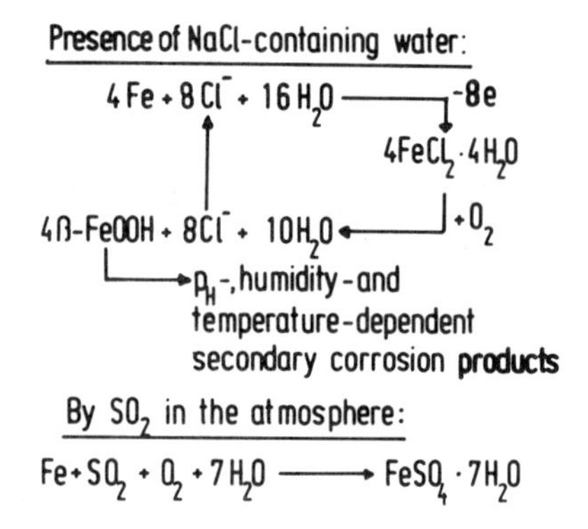

Fig. 4.—Mechanism of corrosion stimulation by chloride and sulfate anions.

It is, therefore, important to know the permeability of coatings for these ions. Methods to measure ion permeability of free films are also indicated in Table III. The more reliable data published in the literature, however, show that protective coatings of customarily used thickness have extremely low anion permeabilities if no external electrical voltage is applied (Table IV). One may assume, therefore, that initiation of the corrosion on exposure to salt solutions mainly takes place at injured or imperfect parts or spots of the film. Stimulation below a perfect film, which completely covers the substrate with a reasonable thickness, must be due to impurities, possibly of the same kind but already present on the metal surface before the coating is applied. Perfect coatings are slightly permeable to ions and the consequent osmotic migration of water causes blistering on exposure of coatings to high humidity or liquid water. In order for this mechanism to operate, soluble material such as chlorides or sulfates must be present at the metal surface, and the coating film must be impermeable or slightly permeable to these impurities. Finally it should be considered that sulfate ions may be formed *in situ* after permeation of SO_2 through the film to the metal substrate. According to my own results, SO_2 permeability is very high if coatings are exposed to the pure gas at atmospheric pressure. As stated earlier, it cannot be excluded that some chemical reaction with the binder takes place under these conditions.

TABLE IV.—CHLORIDE ION PERMEABILITY OF SEVERAL COATINGS

Binder	Permeation rate $[mg.cm^{-2}d^{-1}]\cdot 10^{-3}$	Concentration gradient (NaCl-n)	Film-thickness [μm]
Cellulose nitrate-alkyd resin (9)	0.35	0.053	50
Neoprene (9)	0.85	0.053	50
Epoxy resin (9)	0.71	0.053	50
Cellulose acetate (9)	22.0	0.053	50
Pentaerythritol alkyd resin, modif. with linseed oil (8)	0.13 to 0.16	0.53	75

Primary Film Properties and Classification

Many efforts have been made to correlate water permeability data with the practical corrosion protective efficiency of organic coatings without being able to offer unequivocal and convincing statements. Progress has been made, however, since the permeability oxygen is now considered[5] which formerly, due to lacking data, was almost neglected in discussions about the corrosion protection by coatings. A comparison of permeabilities with the protective behaviour of coatings both in corrosion tests and in practical performance make it soon apparent that adhesion of a paint film when exposed to high humidity plays a decisive role in the evaluation of corrosion protection. Primary film properties are very important. It is well known, that adhesion of many paint films significantly decreases and even is lost on prolonged exposure to high humidity or liquid water[6]. As long as adhesion persists, there is no possibility for corrosion. If water not only enters the coating but displaces the film from its substrate, the rate of corrosion depends on the rate of oxygen diffusion to the interface (Figure 5). Therefore the rate determining step for the loss of adhesion is related to water permeabilities, provided that water disturbs the film/substrate interaction. If adhesion is lost, the rate of corrosion should depend on the oxygen permeability. If corrosion stimulators are present at the interface, the need for oxygen supply increases and oxygen permeability even more significantly governs the rate of underfilm corrosion.

If this approach to the problem of evaluating the practical protective performance of organic coatings is correct, it should be possible to explain differing corrosion protection of various coating systems by these primary parameters: "wet" adhesion, oxygen permeability and water permeability. Results of some first experiments with a series of electrodeposited coatings are encouraging. As may be seen from Table V EC-systems No. 7 and 3 have the lowest values for water and oxygen permeability but are in the medium range regarding adhesion under the influence of high humidity (cross-over

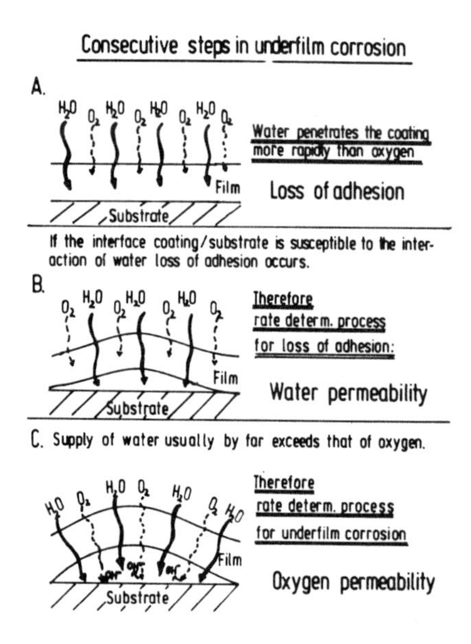

Fig. 5.—Consecutive steps in underfilm corrosion.

time). Because of the low permeability for water, it takes relatively long for these systems until adhesion is challenged. Then corrosion may occur, however slowly, due to the low oxygen permeability. On the other hand, in the EC-system No. 6 it is obvious that adhesion strongly resists the attack of water. Therefore, despite the relatively poor rating for water permeability and its moderate oxygen permeability the good protective properties of this system are not impaired, as confirmed by its rating in the salt spray

TABLE V.—CLASSIFICATION OF ELECTROCOATING SYSTEMS ACCORDING TO WATER AND OXYGEN PERMEABILITIES AND CROSS-OVER TIME COMPARED WITH THE CLASSIFICATION ACCORDING TO THE SALT SPRAY TEST

Method	EC - system number
Water permeation rate	7 < 3 < 4 < 5 ≃ 6 < 1 < 2
Oxygen permeation rate	3 < 7 < 6 < 5 < 4 < 1 < 2
Cross-over time	6 ≃ 1 > 2 ≃ 7 > 3 > 4 ≃ 5
Salt spray test (DIN53167)	3 ≃ 6 ≃ 7 > 1 > 4 > 2 ≃ 5

test. In this connection, it should be remembered that according to the permeability equation $P = D \cdot S$, a high water uptake may be at least partly compensated by a low diffusion coefficient because both coefficients do not always correspond to each other (Figure 6). The experimental technique to measure the primary properties involved in the protective mechanism must be improved and the results of these measurements must be corroborated by more extended studies with other coating systems. Nonetheless, the large advantage of this approach is that it explains why a coating protects well or poorly, whereas conventional testing methods, at best, only tell how *to classify* coatings according to their corrosion protective quality.

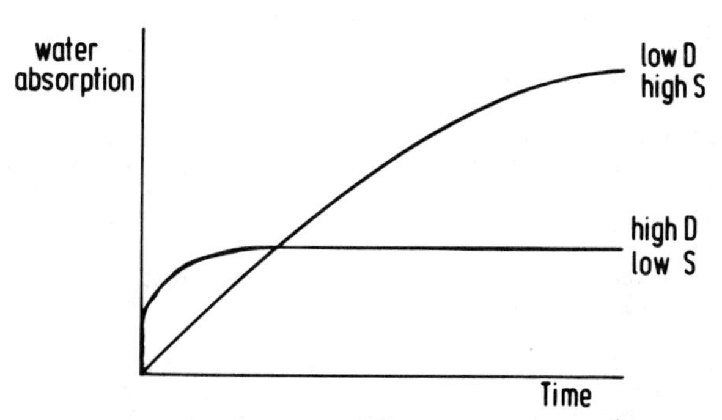

Water absorption rates of coatings depending on diffusion coefficient (D) and solubility coefficient (S).

Fig. 6.

Special and Unsolved Problems

Considering the three-parameter approach to estimate corrosion protection of coatings, it should be mentioned, that in the case of coatings systems containing anticorrosive pigments, water and oxygen permeability lose much of their relevance, but adhesion on exposure to high humidity still remains important. For anticorrosive pigments to act, a certain amount of water is required and the vehicle should be able to absorb some water. On the other hand, by their oxidative character most of the anticorrosive pigments probably dominate over the effect of molecular oxygen in inhibiting as well as in reactions related to the corrosion process.

Some controversy still exists as to whether permeability data obtained with free films are also representative of supported ones. Film formation and drying depends on access of oxygen and when the release of residual solvent from lower regions is retarded, properties such as the permeability of supported films may differ significantly from those of free ones.

Attention also should be paid to the temperature dependence of permeabilities which often are quite significant. It is common practice to measure permeabilities at room temperature so that the use of these values to explain corrosion protective performance of coatings in the salt spray test at 35 - 40°C or in practical performance in winter time at temperatures little above the freezing point may be unjustified and misleading, the more so as the glass transition temperature of the coating frequently lies within the range of practical performance.

Oxygen permeability is only important when the coating is exposed to high humidity and one must ask how it is influenced by the water uptake of the coating. Some data presently available[7] indicate that in practical coating systems oxygen permeability is not much influenced by the degree of absorbed water[2]. Another point, to which attention should be paid, is the change of film composition on exposure to liquid water or even to water vapor. The leaching of soluble coating components and residual solvents may result in considerable changes in permeabilities.

Despite much interesting work about ion exchange properties, electrical charge and influence of electrical resistance of coatings on their protective behaviour, viz. by Mayne and his co-workers[7], these relations have still obtained only little significance and attention in the practical world of protective coatings. It is expected that further research in this field may help to clarify some still unsolved questions concerning the effect of electrical charges on the protective action of paint films. Finally, a complaint must be made that all experiments and methods used in testing the corrosion protective quality of coatings are greatly dependent on the perfection of the coating. The problem of obtaining or preparing free as well as supported films without defects, especially in the range of lower film thickness is always present. Results reported without adequate information on the number of parallel runs or, better, on the reliability of the tests, based on a statistical treatment, should always be considered with reservation.

As far as the three-parameter approach to evaluate corrosion resistance by coatings is concerned, we certainly may expect that it will provide a more sound and scientific basis for this very old testing problem. However, it seems realistic to say that we probably never shall be able to renounce practical corrosion tests for getting a quick and direct differentiation of coatings with regard to their corrosion protective quality. Being able to tell why they differ, means progress even when using these conventional testing methods.

ACKNOWLEDGEMENT

Appreciation is expressed to Howard Lasser for his assistance in editorial revisions.

REFERENCES

1. Funke, W., *Farbe u. Lack 84, 380 (1978).*
2. Leidheiser, H., Jr., Progr. Org. Coatings, December 1978, in press: these proceedings.
3. Funke, W. and Zatloukal, H., *Farbe u. Lack 84, 584 (1978).*
4. Funke, W., Machunsky, E., and Handloser, G., *Farbe u. Lack 84, 493 (1978).*
5. Baumann, K., *Plaste u. Kautschuk 19, 455 (1972).*
6. Walker, P., *J. Paint Technol. 31, 15, 22 (1967).*
7. Mayne, J.E.O. and Mills, D.J., *J. Oil Colour Chem. Assoc. 58, 155 (1975).*
8. Svoboda, M., Kuchynka, D., and Knapek, B., *Farbe u. Lack 77, 11 (1971).*
9. Glass, A.L. and Smith, J., *J. Paint Technol. 39, 490 (1967).*

II. SACRIFICIAL COATINGS FOR CORROSION CONTROL

GALVANIZED STEEL AND ZINC PIGMENTED PAINTS

Albert R. Cook

International Lead Zinc Research
Organization, Inc.
New York, NY

ABSTRACT

The use of galvanized reinforcement for concrete exposed to marine conditions is recommended. Conditions are described for obtaining good paint adhesion on forming and freedom from "spangle cracking" of zinc coatings. The use of a "duplex" system where paint is applied directly to galvanized steel without primer is described and suitable types of paint are mentioned. Anodizing of zinc is recommended for extreme longevity for galvanized steel and die cast zinc in contact with sea water. The merits of organic and inorganic zinc-rich paints are discussed. Best performers in half tide immersion tests of paint coatings over sprayed zinc are cited. Mention is made of the outstanding performance of triphenyl lead acetate as an antifouling agent in elastomeric and water-based paints.

FOREWORD

We at ILZRO are very pleased to acknowledge the close cooperation we enjoyed with the U. S. Navy. To mention just a few examples, the Navy allowed us to examime galvanized reinforcement installed years ago in the Longbird Bridge, Bermuda. We worked closely with the Navy on behalf of the zinc industry in establishing the zinc anode specification MIL-A-18001, and our scientists enjoy your hospitality at the Naval Research Laboratories in a cooperative effort to improve the lead-acid battery for propulsion purposes. We want to take this opportunity to express our thanks to the U. S. Navy for their unfailing cooperation.

Corrosion protection has always been important to the Navy. It is, therefore, at the risk of redundancy that we emphasize the national importance of corrosion protection.

The recent report to Congress by the National Bureau of Standards, "Economic Effects of Metallic Corrosion in the United States", makes important reading. This study shows that 3½% of the total U. S. energy demand was generated because of metallic corrosion. Total annual losses are assessed as high as $70,000,000,000. The federal government's share of this loss is placed at $8,000,000,000, or approximately 2% of the federal budget.

Energy considerations will play an important role in the choice of materials in the future. Reduction in material thickness can reduce energy needs. Additional corrosion protection must then be provided to ensure satisfactory life. Increased attention to improved design can make a worthwhile contribution to corrosion resistance and therefore to energy conservation. To this end the corrosion engineer must be consulted at the design stage to ensure correct design and correct choice of materials.

As a nonprofit organization, we at ILZRO stand ready to make our contribution to ensure that zinc, lead and cadmium can be used effectively and economically in all applications.

The objective of this paper is to discuss zinc coatings, within the context of the broad needs of the Navy, and to highlight those areas where further development of zinc's potential is possible. Perhaps it is reasonable to see the Navy's needs as relating to ships, vehicles and shore and offshore structures and piping, and that seems to cover just about everything. Basic materials for shore or offshore structures are reinforced concrete and steel.

GALVANIZED REINFORCING STEEL FOR CONCRETE

The basic problem with reinforced concrete is that concrete has negligible tensile strength. Steel reinforcement provides the tensile strength, but in-service cracking of the concrete will occur as the steel takes up the load. It is the extent and progression of the cracking which is usually the limiting factor in the life of the concrete structure when exposed to marine conditions. The mechanism of failure with black steel reinforcement is as follows:

1. When placed under stress, the concrete cracks because it lacks appreciable tensile strength; its structural integrity is then maintained by the reinforcement.

2. These cracks may allow:

   a. ingress of pollutants and carbon dioxide. These can neutralize the alkalinity which inhibits the corrosion of the steel. This incursion leads to active corrosion of the reinforcement.

   b. chlorides to gain access. These ions are very effective depolarizers for steel, and rapid corrosion of the reinforcement can take place.

3. Once the steel corrodes, its voluminous corrosion products set up extreme pressure within the concrete. This results in further cracking and progressive structural deterioration.

ILZRO-sponsored research has demonstrated that galvanizing extends the life of such structures. Laboratory and accelerated test data [1-7] are important, but field studies of existing structures are the most convincing.

On the island of Bermuda, the Public Works Department does not allow the use of black reinforcing steel. They have specified hot dip galvanized steel for the past 30 or 40 years. Table I summarizes the results of extensive ILZRO tests which have involved taking concrete cores from bridges and docks, including the oldest available structures in Bermuda, to determine the condition of the galvanized reinforcement and estimate its life in relation to the chloride content of the concrete. Detailed reports [8,9] and further advice are available on request.

Photo 1.—The Anodic Platform, said to be the largest concrete platform in the world, carries 2,000 tons of galvanized steel rebars in the ceiling of its crude oil reservoirs. Picture was taken while the platform was being towed towards Norway for further construction.
**Courtesy of Ing. J.F.H. van Eijnsbergen of Stichting Doelmatig Verzinken.*

Photo 2.—Closeup of two ducts used for cooling water for a power plant in northern Holland on the edge of the Wadden Sea. These square ducts include galvanized steel rebars for mechanical strength.
**Courtesy of Ing. J.F.H. van Eijnsbergen of Stichting Doelmatig Verzinken.*

TABLE I.—EXAMINATION OF REINFORCED CONCRETE STRUCTURES ON THE ISLAND OF BERMUDA

Sample	Age, Yrs.	Lbs.Cl^-/cu.yd. Concrete at Level of Steel	Average Corrosion Layer Thickness, mils	% of Coating Remaining
St. George (SG 17)	7	5.0	0.1	98
Bermuda Yacht Club (BYC 3)	8	6.1	none	100
Hamilton (H 22)	10	3.2	0.2	95
Hamilton (H 26)	10	6.0	0.3	96
St. George (SG 10)	10	7.7	0.2	99
St. George (SG 9)	12	10.7	0.5	92
Longbird (LB 20)	23	7.3	0.2	98

ZINC-RICH PAINT

Zinc-rich paints are superior protective coatings for many marine applications. They provide excellent barrier protection. Pores in the coating are rapidly filled with zinc corrosion products produced by the galvanic protection of the zinc. An impermeable barrier to corrosion is thus provided. When a break in the coating develops, further action by the zinc pigment tends to repair the damage. Thus the galvanic action of the zinc is not a wasting asset, but is held in reserve until needed.

At Cape Kennedy important use has been made of zinc-rich paints. The inorganic coatings provide heat resistance, outstanding corrosion resistance and excellent abrasion resistance. The "inorganics" are more demanding than the organic based zinc-rich paints as regards surface preparation. The organic zinc-rich paints are tolerant to deformation of the basis steel. Current ILZRO research in this area is to ensure the compatibility of topcoats applied to zinc-rich paint for camouflage, or to provide additional protection.

At the present time, a total of 549 steel test specimens, coated with 11 types of zinc-rich paint and various experimental topcoats, are being tested under several environmental exposure conditions. The zinc-rich coatings represent five types:

1. water-reducible, inorganic, zinc-rich primers, self-curing;
2. solvent-reducible, inorganic, zinc-rich primers, self-curing;
3. inorganic zinc-rich primers, self-curing;
4. organic zinc-rich paints; and
5. galvanized and metallized steel.

The exposure environments include accelerated salt spray testing, mild industrial atmospheric corrosion, severe industrial chemical exposure and severe marine atmospheric exposure. Five types of topcoats were used, including vinyl, epoxy, chlorinated rubber, coal tar epoxy and latex topcoats. Approximately 200 panels with zinc-rich paint systems were exposed up to 6,000 hours in salt fog before many started to fail, while others were still in perfect condition. To date, outdoor exposures have indicated only a few systems with gross incompatibilities, either from blistering or peeling along a scribe mark. Outdoor exposure tests are continuing to identify systems showing incompatibility or early signs of failure.

Scanning electron microscopy is being used to relate surface conditions and nonconductive areas to corrosion failures. The potential of the surface compared with a silver/silver chloride electrode is being related to its observed corrosion resistance. Protective zinc-rich primers have shown stable electrode potentials whereas poor coatings become increasingly anodic during 30-day immersion tests. Lithium/sodium and potassium-based zinc silicate primers have given excellent protection in North Sea and tropical waters for over 10 years in dry film thicknesses of 100-200 μm (4-8 mils). These products were 2 component, self-curing and contained approximately 87-90% of zinc in the dry film.[10]

The use of chromate inhibitors in conjunction with zinc dust has been a dramatic development in the automobile industry over the past few years. Much used by the automobile industry as a protection against road salt corrosion of automobile bodies, the technique is very effective and low in cost. Galvanized coatings can be supplied in greater thicknesses and can therefore provide increased life, but Zincrometal should be carefully considered as a cost effective treatment, especially for manufacturing applications.

GALVANIZED STEEL

Galvanized steel for structural applications needs little description here. Its service life is predictable from ASTM tests. A rule of thumb which has been used for many years is that protection is proportional to the thickness of the zinc coating.

PAINTED GALVANIZED

Where demands are made for extended life under marine conditions, painted galvanized should be specified. In this area ILZRO has been concerned to improve the paintability of zinc. It can rightly be maintained that galvanized steel is one of the best substrates for painting. In fact, there is a synergism involved in the use of painted galvanized. The composite coating will last 1.5-2.3 times as long as might be calculated, when the life of the zinc coating and the paint coating are considered separately. It therefore pays to paint sooner rather than later.

When problems have arisen with paint adhesion, they have been due to improper procedures or the inability, perhaps for economic reasons, to follow strictly the best recommended pretreatment practice. A few years ago there occurred occasional cases of loss of paint adhesion in the coil coating process. These were not due to outdoor exposure but occurred on bending or deformation soon after the painted material had left the coil coating line. Outstanding work by Dr. Leidheiser, at Lehigh University, has pinpointed one variable, not hitherto identified, as being important in providing greater production latitude for the coil coater. The steel industry is now actively working to apply this research to their production process.

What Dr. Leidheiser and Mr. Dong Kim found was that when the basal plane of the crystal in the zinc coating is parallel to the surface, this short-term paint adhesion is at its best, as shown in Figure 1.[11]

It is also becoming clear that chemical activity and corrosion resistance are related to crystallographic orientation.

Microscopic cracks may appear on the surface of zinc and galvanized steel on bending, especially at low temperatures. Such spangle cracking, as it is called, can cause paint coatings to appear unsightly and probably has an adverse effect on corrosion resistance. The Lehigh investigators have unfortunately demonstrated that when the coating provides optimum paintability, it also provides minimum susceptibility to spangle cracking.[12]

We are now concerned to expand this work to embrace the long term exposure performance of all painted galvanized steel and to investigate how the inherent corrosion resistance of galvanized coating can be improved.

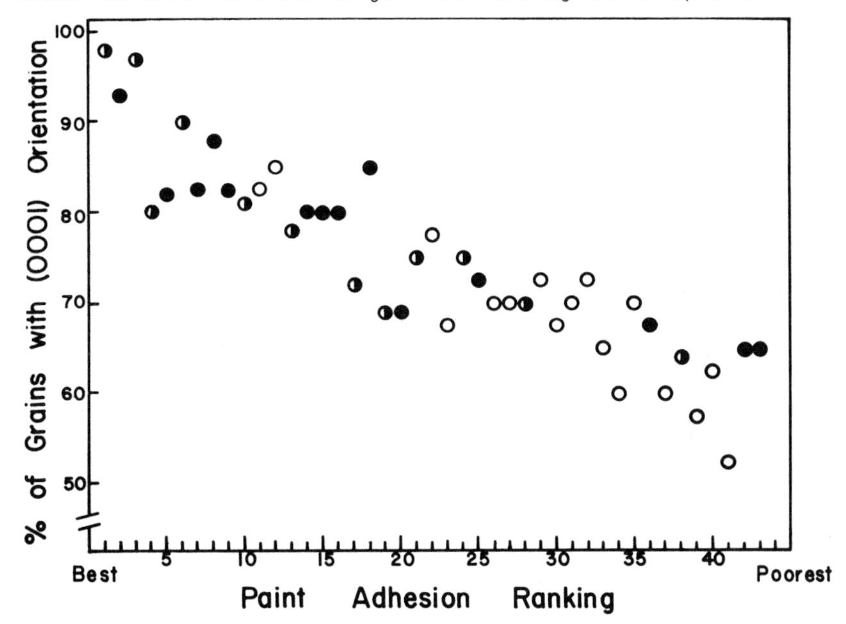

Fig. 1.—The relationship between paint adhesion ranking and the degree of (0001) preferred orientation of commercial galvanized steel as determined by the chemical test. Open circles--bright spangle; half-solid circles--minimized spangle; solid circles--temper rolled (skin passed).

COLORED GALVANIZED

Camouflage considerations and the need for increased corrosion protection sometimes require that galvanized steel be coated. The merits of painting galvanized steel prior to installation are not sufficiently appreciated. It is extremely difficult to ensure good quality control in the field, and in many cases weather considerations make it virtually impossible to follow specification. Offshore structures in the North Sea, for example, provide only a few days per year when conditions are acceptable for good quality painting.

DIRECT PAINTING ON GALVANIZED STEEL

Paints have now been developed which can be directly applied to zinc surfaces. In Holland, for example, practically all galvanized light poles are airless sprayed within 2 hours of galvanizing to take advantage of a clean active surface. A coating of 100 μm of polyurethane/micaceous iron oxide paint is applied in air with relative humidity less than 50% to ensure optimum drying, reaction of the two paint components, and surface hardness after 24 hours in the drying room. Poles are stacked for a further 1 or 2 days after shipment.[13]

Two-component epoxy-based paints applied directly to galvanized by airless spray in thickness of 75 to 125 μm dry film thickness provide excellent weathering and sea water resistance. They must have a pigment to volume ratio of 24 to 34% to ensure good adhesion. Drying must take place above 10° C to ensure sufficient crosslinking, and of course recommendations are available [14,15,16] to ensure application of good, low-cost systems for general application to galvanized steel.

ANODIZING

Where outstanding corrosion resistance and abrasion resistance is required, zinc and galvanized steel can be anodized.[17] This technique is the subject of U. S. Navy Specification MIL-A-81801, January 14, 1971. Testing has been carried out by Naval Ship Research and Development Laboratory (Report 8-812, June 4, 1971). A measure of its effectiveness is shown by its resistance to salt spray. Coatings withstood 1,500 hours of 5% NaCl salt spray, ASTM B117. The anodizing process was made available by ILZRO and is now an established commercial procedure.

The colors available range from a matte black to a camouflage green. There is scope for further research to produce a wider range of colors and a transparent coating. Perhaps A.C. rather than D.C. anodizing would be a fruitful line of investigation. It may be possible to produce a color anodized finish by incorporating metallic elements into the anodizing bath. We are not funding any such work at the present time and would of course welcome any encouragement you might want to give to get us started in this area.

WET STORAGE STAIN

All engineering metals suffer, to one degree or another, from wet storage stain. Destructive corrosion of sheet material can take place in a matter of days when the sheets are piled in contact and without air circulation under humid conditions. Chromatic conversion coatings provide excellent low-cost protection, but it seems possible that environmental consideration will restrict the use of chromates. A satisfactory low-cost alternative is not available today. Research is needed in this area, and it might be that anodizing in an appropriate electrolyte might be a satisfactory solution. The anodizing may be necessary to ensure satisfactory treatment in the few seconds available on a continuous galvanizing line.

SPRAYED ZINC COATINGS

The thickness of galvanized coatings is limited by the production process. Sprayed zinc can be applied in any required coating thickness. Buildup can be localized where desirable, and there is no possibility of heat distortion from the coating process. Such coatings provide excellent performance when correctly painted.

ILZRO studies have defined the optimum organic coatings for sprayed zinc. The contractor, Paint Research Station of England, concluded that flame sprayed zinc can be successfully painted to give excellent protection to structural steelwork, even under the most severe marine conditions. Specimens were exposed to half-tide immersion at Tilbury in the Thames River Estuary (10% salinity), and best results were achieved with zinc, rather than aluminum, spray using coal tar epoxy and two pack polyurethane systems. These were applied directly to the sprayed zinc in three coats in a typical case. The quality of the sprayed zinc affects the performance of the system, and recommendations were made to ensure a good quality sprayed zinc coating.

Zinc coatings retain their position as the most cost effective material for structural purposes. Correct use of painting and anodizing can further increase this cost effectiveness, and increased use of painted galvanized steel in the future is expected.

Antifouling paints are of course of considerable interest. Reports are available from ILZRO describing elastomeric and water-based paints containing triphenyl lead acetate which prevent fouling for periods up to 5 years at Miami, Florida, and Pearl Harbor, Hawaii.

REFERENCES

1. Tonini, D.E., American Hot Dip Galvanizers Association, Inc., and Dean, S.W., Jr., Air Products and Chemicals, Inc., "Chloride Corrosion of Steel in Concrete," published as ASTM Special Technical Publication 629, June 1977.
2. Radtke, S.F. and Cook, A.R., ILZRO, "Recent Research on Galvanized Steel for Reinforcement of Concrete," published in ASTM Special Technical Publication 629, June 1977.
3. Cook, A.R., ILZRO, "Recent Research on Galvanized Steel for Reinforcement of Concrete," presented to Transportation Research Board of the U.S. National Academy of Sciences, Committee on Corrosion (A2G05) and Performance of Concrete-Chemical Aspects (A2E02), January 19, 1976.
4. Okamura, H. and Hisamatsu, Y., "Effect of Use of Galvanized Steel on the Durability of Reinforced Concrete," *Materials Performance (U.S.), July 1976*.
5. Okamura, H. and Hisamatsu, Y., U. of Tokyo, "Effect of Use of Galvanized Steel on the Durability of Reinforced Concrete," *Materials Performance*, official publication of the National Assn. of Corrosion Engineers, *July 1976*.
6. Zinc Institute, Inc. and ILZRO, Inc., "Galvanized Reinforcement for Concrete," November 1970.
7. Bresler, B. and Cornet, I., "Galvanized Reinforcement in Concrete," presented at the VII Congress IABSE Session V, Theme V-c, August 13, 1964.
8. Stark, D. and Perenchio, W., "The Performance of Galvanized Reinforcement in Concrete Bridge Decks," Project No. ZE-206, Final Report, July 1974-October 1975.
9. Stark, D., "Galvanized Reinforcement in Concrete Containing Chlorides," Project Ze-247, Final Report, October 1976-April 1978.
10. van Eijnsbergen, J.F.H., Stichting Doelmatig Verzinken, private communication dated May 1, 1977.
11. Leidheiser, H. and Kim, D.K., Tech. Paper No. 780185, *Soc. Automotive Engrs., 6 (March 1978)*.
12. Leidheiser, H. and Suzuki, I., to be published.
13. van Eijnsbergen, J.F.H., Stichting Doelmatig Verzinken, private communication dated February 28, 1978.
14. Zinc Institute, "Painting Galvanized Steel."
15. Livingston, J.J. and Hughes, H.J., "Paint Systems for Galvanized Steel," *Paint and Varnish Production, Nov. 1971*.
16. Cook, A.R., "Pigmentation of Coatings for Zinc and Lead Substrates," Pigment Handbook, Vol. I - Properties and Economics, published by John Wiley & Sons, New York.
17. Mansfield, F.M., "Zinc Anodize-New Finish for Industry," The Tool & Manufacturing Engineer, July 1969.

ALUMINUM-ZINC ALLOYS AS SACRIFICIAL COATINGS

J. B. Horton

Homer Research Laboratories,
Bethlehem Steel Corporation
Bethlehem, PA

JAMES B. HORTON

ABSTRACT

Of the 1 to 70% Al-Zn alloys, the 55% Al-Zn coating has the best combination of atmospheric corrosion resistance with galvanic protection of sheared edges. The general corrosion resistance of the 55% Al-Zn alloy coating is superior to that of galvanized steel. Specifically, the commercial 0.8 mil 55% Al-Zn alloy-coated product should last two to four times longer than conventional galvanized with about the same coating thickness (0.9 mil). In all but the severe marine atmosphere the best galvanic protection for cut edges is provided by the galvanized, with the 55% Al-Zn coating being next best and with the aluminum coating providing little or no protection. Since the 55% Al-Zn coating combines some of the best properties of both galvanized and aluminum coatings, it is inherently of great commercial importance as a new type of protective metallic coating for a wide range of sheet and other steel products.

INTRODUCTION

Among the various metallic coatings used for protecting steel products from corrosion two have been outstanding in terms of performance and widespread usage. These are hot-dip zinc and aluminum coatings.

Zinc coatings have been used for 135 years to protect steel products. About 5 million tons of galvanized sheet are produced each year in the U. S. and several million tons of other galvanized steel products also. Galvanized coatings provide good protection to steel in the atmosphere, fresh water and soil. Zinc is outstanding in its ability to protect steel by galvanic action where the coating is mechanically damaged, such as at sheared edges of a sheet. On the other hand, galvanized coatings don't last as long as we would like in some environments, especially sulphate-bearing industrial atmospheres. Zinc coatings show a linear corrosion-time relationship, indicating that an increasingly protective oxide film does not form with increasing time of exposure. Low-alloy steels and aluminum do form more-protective films with increasing exposure time in the atmosphere.

Aluminum coatings are more corrosion-resistant than zinc because of the protective aluminum oxide film formed in the atmosphere and show outstanding high-temperature oxidation resistance. The major drawback of aluminum coatings is their lack of galvanic protection to the steel base. As a result, aluminum-coated steel shows unsightly rust staining and rust growths, or pox, at points of mechanical damage such as sheared edges or scratches through the coating.

In 1961 Bethlehem Steel Corporation undertook a research program to find a better coating material than either zinc or aluminum. One of the many directions the research took was to explore the Al-Zn alloy system for improved hot-dip coating alloys. We were successful with the Al-Zn alloy system in combining the best properties of galvanizing with those of aluminizing. Indeed, Bethlehem here in the U. S. and Lysaght in Australia are producing a 55% Al-Zn-alloy-coated sheet in large quantities today. Bethlehem's product is called Galvalume sheet.

In this paper the physical metallurgy of the Al-Zn alloy system will be reviewed first, and the corrosion resistance and galvanic properties of 1 to 70% Al alloys will be discussed. The properties and corrosion resistance of the 55% Al-Zn alloy which has the optimum properties as a coating for steel products will then be detailed. The atmospheric corrosion mechanism will also be briefly discussed. The results that will be reported were provided by my co-workers at the Homer Research Laboratories of Bethlehem Steel Corporation. This Al-Zn alloy coating development is covered by several U. S. and foreign patents assigned to the Bethlehem Steel Corporation.[1]

PHYSICAL METALLURGY OF AL-ZN ALLOY SYSTEM

A review of the Al-Zn equilibrium phase diagram will be helpful in understanding the corrosion behavior of Al-Zn alloys. Currently, the Presnyakov, *et al.*,[2] diagram shown in Figure 1 is most widely accepted, although some areas are not yet reliably established. Here are some of the features of the equilibrium diagram:

1. There is a eutectic at 5% Al.
2. There is very limited solid state solubility of Al in the beta-zinc terminal phase.
3. The terminal alpha-aluminum phase ranging from 100% to 30% Al is very extensive at intermediate temperatures. The solubility of beta-zinc in alpha-aluminum decreases from 30% to 5% on cooling from 525°F to room temperature.
4. There is a peritectic transformation in which the gamma phase is formed, which in turn undergoes a eutectoid transformation at 22% Al on further cooling.
5. At room temperatures the equilibrium structure comprises alpha-aluminum and beta-zinc. There are no intermetallic compounds or ordering reactions in the Al-Zn alloy system. An intermetallic layer is actively formed between the Al-Zn melt and a steel surface. Small additions of silicon are added to the Al-Zn melt to control this reaction.[1]

Upon cooling a liquid 55% Al-Zn alloy under near-equilibrium conditions, one would expect to get an alpha-aluminum matrix with a spherodized beta-zinc phase, as shown in Figure 2 for furnace-cooled 55% Al-Zn alloy. Under rapid air-cooling conditions typical of commercial hot-dip coating of Galvalume sheet (Figure 3) the structures are nonequilibrium and more complex. The first solid formed in alpha-aluminum at about 80% Al leading to a cored dendritic structure in which the final liquid to freeze is substantially lower in aluminum content (zinc-rich) and whose structure is quite fine and not well defined. As the aluminum content of the Al-Zn alloy increases to 70% Al, the dark-etching zinc-rich phase decreases in volume and disappears at 60-70% A.

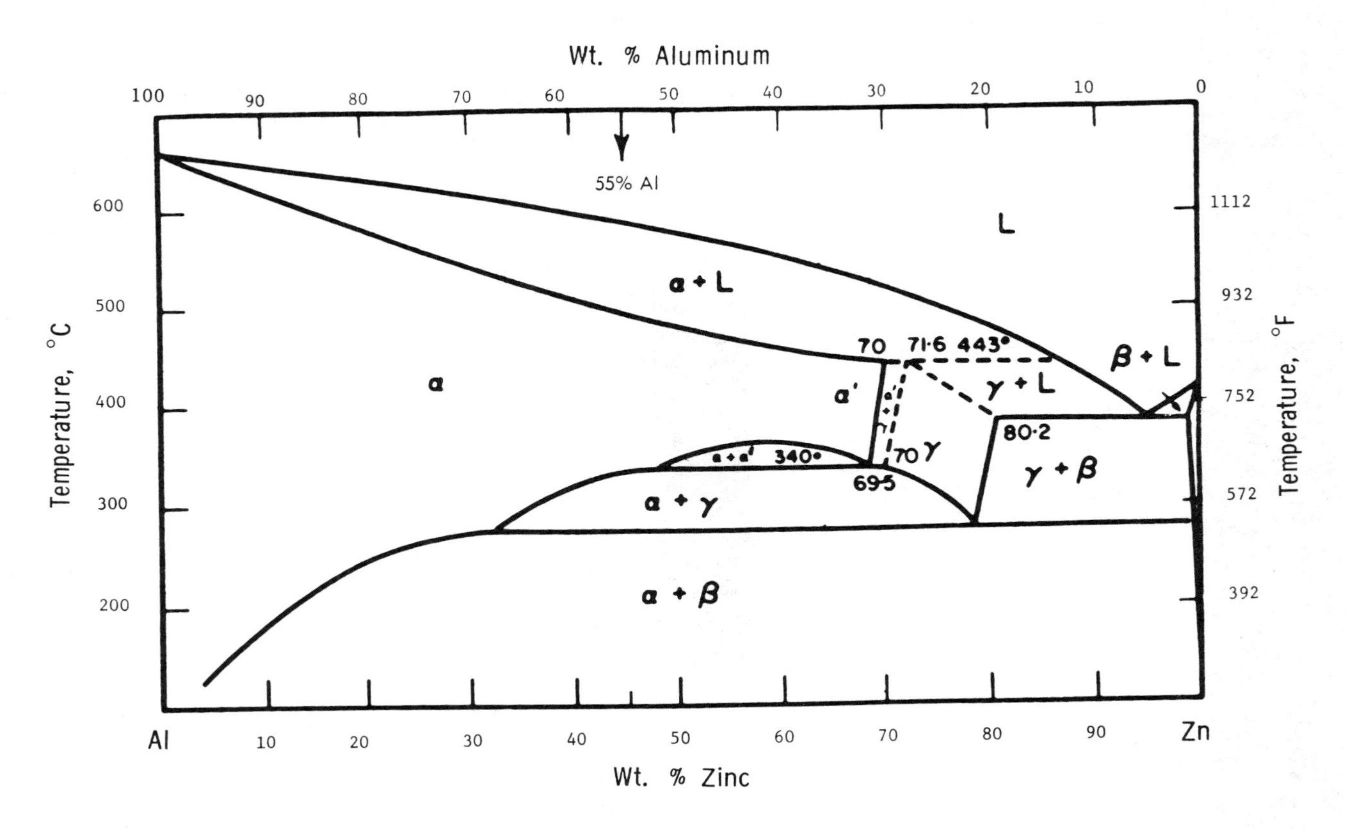

Fig. 1.—Phase diagram of Al-Zn system.

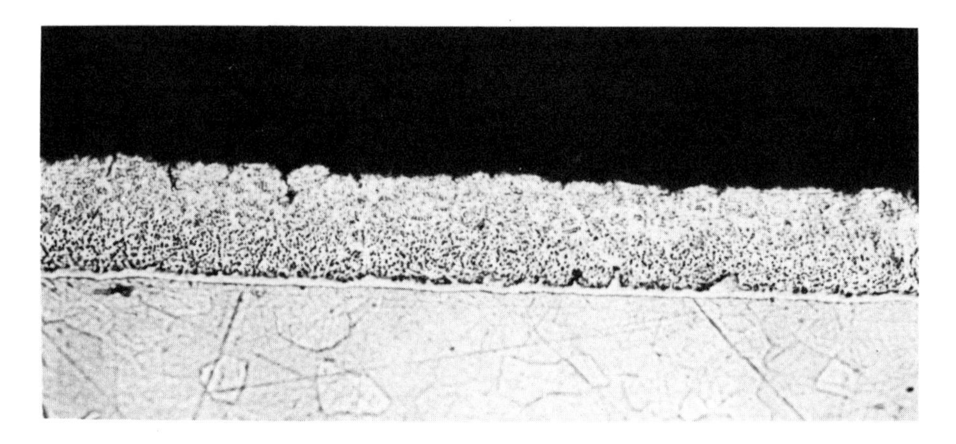

Fig. 2.—Near-equilibrium structure of 55% Al-Zn coating (X500, Amyl-Nital Etch). Furnace cooled from 750°F.

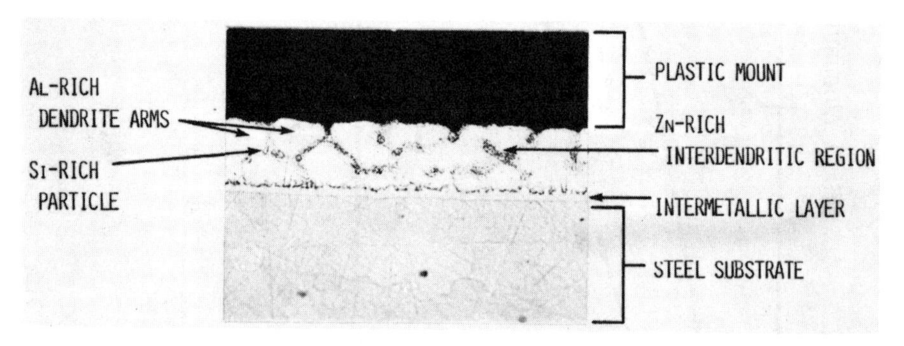

Fig. 3.—Random cross section of a 55% Al-Zn coating (X500, Amyl-Nital Etch).

From this brief discussion of the physical metallurgy of the Al-Zn alloy system, it is evident that the structure of these alloy coatings and consequently their corrosion behavior are a function of the aluminum content and to some extent the freezing and subsequent cooling rate of the coating. As explained in the following discussion, the optimum corrosion resistance with adequate galvanic protection is obtained with the 55% Al-Zn alloy, which gives a coating of mostly alpha-aluminum phase with an interdendritic zinc-rich phase.

ATMOSPHERIC CORROSION RESISTANCE OF 1 TO 70% AL-ZN ALLOY COATINGS

Al-Zn alloy compositions from 1-70% Al were prepared. These alloys were applied as hot-dip coatings at about 0.8 to 1.0 mil average thickness to 0.018" steel sheet. These were then corrosion tested in the atmosphere to determine the optimum aluminum content based on corrosion resistance and galvanic protection of sheared edges.

Figure 4 shows the atmospheric corrosion losses after five years' exposure at one rural, two marine, and two industrial test sites.[3] Because the density of the various alloys varies with aluminum content, the corrosion weight losses have been converted by calculation to average loss of thickness using the density of the Al-Zn alloy for up to 21% Al coatings and the density of 22% Al-Zn alloy for 25 to 70% Al-Zn alloy coatings.

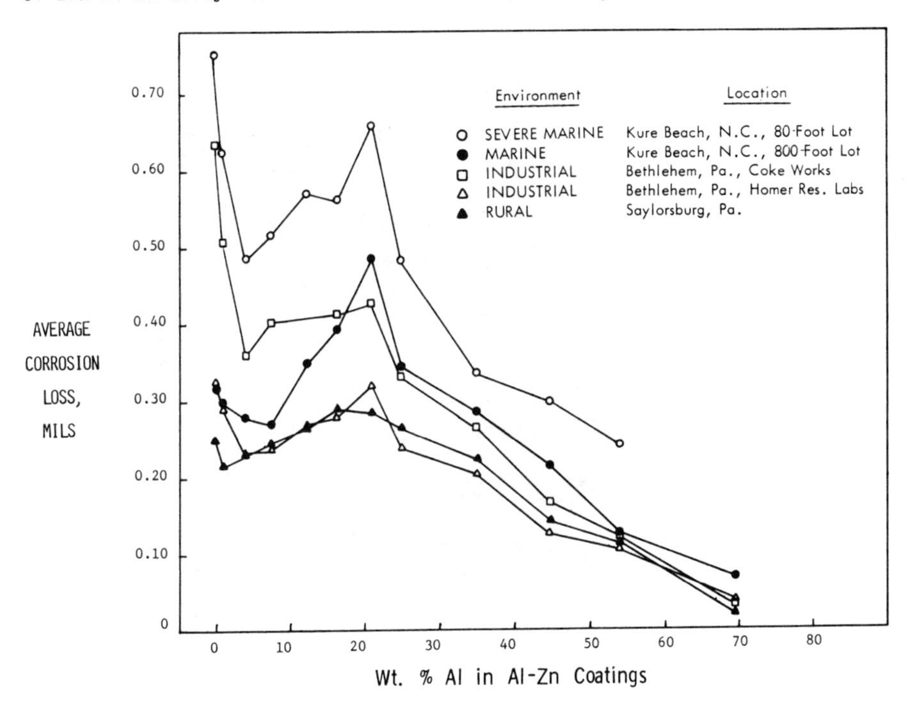

Fig. 4.—Effect of aluminum content on the corrosion performance of Al-Zn alloy coatings after 5 years' exposure in various atmospheres.

The corrosion rate compared to that of galvanized steel decreases with increasing aluminum content to about 4-7% Al, then rises to a maximum at 21% Al, near the eutectoid composition, then decreases progressively to 70% Al. Thus, additions of aluminum above 22% increase the corrosion resistance of zinc markedly.

The ability to provide galvanic protection to the steel base is also an important quality of a coating alloy. Inevitably, coated steel products are subject to mechanical damage of the coating during processing or in service. Scratches through the coating or sheared edges of sheet can be subject to unsightly rust staining if the coating does not provide at least minimal galvanic protection.

During the test programs over a period of years we examined the sheared edges of sheet samples with 1 to 70% Al coatings for evidence of galvanic protection of the sheared edges of 0.019" coated sheet during atmospheric exposure. Visual appraisal of rust staining on the faces of the panels and microscopic examination of the sheared edges for rust and rust growths show that the 1 to 70% Al alloys are about as good as galvanized coatings in their ability to protect sheared edges and are clearly superior to aluminum coatings in this respect. This protection is especially evident in the case of the 55% Al coating, which was found to be the optimum composition in terms of providing both excellent corrosion resistance and good galvanic protection of sheared edges. These results may be clearly seen in Figure 5, which shows the 7.5X magnified appearance of the sheared edges of 55% Al-Zn-coated sheet, galvanized, and aluminum-coated sheet after six years' exposure at Bethlehem, Pa., a moderate industrial atmosphere. The aluminum-coated sheet edges show

AL COATED
(2 MIL)

GALVANIZED
(1 MIL)

55% AL-ZN
(1 MIL)

Fig. 5.—Condition of sheared edges of thin sheets after 6 years' exposure at Bethlehem (industrial). 7.5X.

heavy rusting and rust growths, whereas galvanized and 55% Al-Zn-coated sheet show a small amount of local rusting but no edge roughness due to rust growths. The same behavior was also observed at Saylorsburg, a rural site (Figure 6). This behavior can be explained by the fact that in industrial and rural atmospheres galvanized and 55% Al-Zn coatings do provide galvanic protection to sheared edges, whereas aluminum coatings do not because the Al_2O_3 film formed on aluminum prevents galvanic protection to steel in these atmospheres. On the other hand, in marine atmospheres where the chloride ion breaks this film down, galvanic activity by aluminum toward steel is free to occur and prevent rust growths from developing on the sheared edges of aluminum-coated sheet (Figure 7). Thus, in marine atmospheres all three coatings galvanically protect sheared edges.

The above observations on galvanic behavior are based on the long-term performance of thin sheets such as those used for roofing on pre-engineered buildings.

We need to acknowledge that as the steel thickness increases and therefore as the area and distance over which galvanic effects must be exerted

Fig. 6.—Condition of sheared edges of thin gage sheets after 6 years' exposure at Saylorsburg (rural). 7.5X.

Fig. 7.—Condition of sheared edges of thin gage sheets after 5.5 years' exposure at 80 foot lot, Kure Beach (marine). 7.5X.

also increase, 55% Al-Zn coatings become less effective than galvanized coatings in edge protection. However, in this respect they remain quite superior to aluminum coatings.

TABLE I.—AL-ZN ALLOY COATING COMPOSITIONS

Nominal Aluminum Content	Actual Chemical Analysis, wt % Al	Zn	Si
1%	1.0	99.0	.04
4	4.0	95.8	.1
7	7.4	92.4	.2
12	12.2	87.4	.3
17	16.6	83.0	.4
21	21.0	78.4	.6
25	24.9	74.3	.8
35	35.1	63.7	1.0
45	44.6	53.8	1.3
55	53.9	44.0	1.6
70	69.6	27.7	2.1

LONG-TERM CORROSION RESISTANCE OF 55% AL-ZN ALLOY COATINGS

On the basis of early atmospheric and laboratory accelerated corrosion test results such as salt fog and based on the above observations of galvanic behavior, the 55% Al-Zn coating alloy was selected for commercialization as having the best combination of properties. Coatings of this alloy have now been on atmospheric exposure for 14 years, with excellent results being observed. Figure 8 shows the appearance of coated sheet in comparison with G90 galvanized (1 mil) and type 2 aluminum (2-mil)-coated sheet after 13 years' exposure at four test locations.[4]

Before describing the condition of the surfaces in detail, we should note that the small, localized discolorations at the vertical edges of some specimens were points which contacted ceramic insulators.

- In the severe marine atmosphere the galvanized coating, which had started to rust after four years' exposure, is now heavily rusted. In contrast, panels with 55% Al-Zn and 2-mil aluminum coatings are still in good condition, although some corrosion products are starting to creep inward on the faces of panels from cut edges.

- In the marine atmosphere of the 800-foot lot all three types of coatings are still in good condition.

- In the industrial atmosphere most of the galvanized coating has been corroded away, and more than three-quarters of the steel surface is rusted. The Al-Zn- and aluminum-coated panels exhibit superficial light-brown oxide stain due to particulate fallout from nearby steelmaking operations but are otherwise in good condition.

- In the rural atmosphere all three materials are in good condition, but there is some rust staining apparent along the edges of the 2-mil aluminum-coated panels.

Figures 9 through 12 are the corrosion-time curves for these same three materials after 13 years' exposure.[4] In order to facilitate a comparison of the corrosion resistance of a 55% Al-Zn coating with that of a conventional galvanized coating, we calculated the ratio of the 13-year corrosion losses for the two coatings, as shown on page 70.

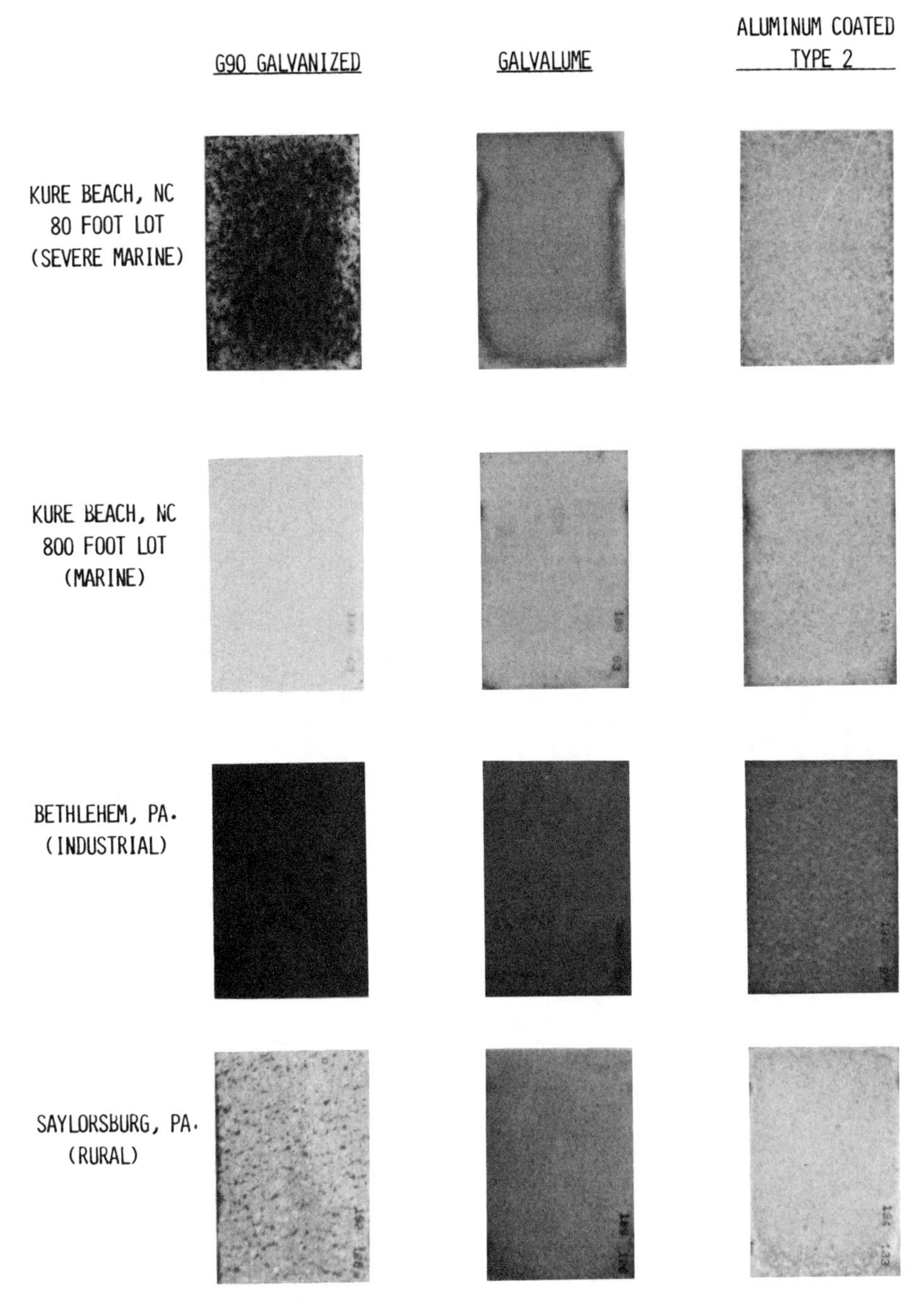

Fig. 8.—Appearance of coated steels after 13 years in the atmosphere (skyward surface).

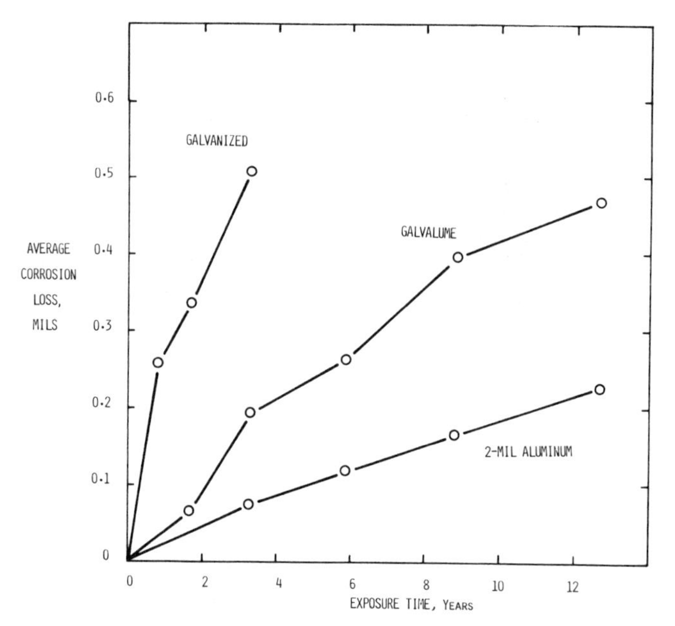

Fig. 9.—Corrosion performance of Galvalume, galvanized and 2-mil aluminum-coated steels in severe marine atmosphere (Kure Beach, 80 foot lot).

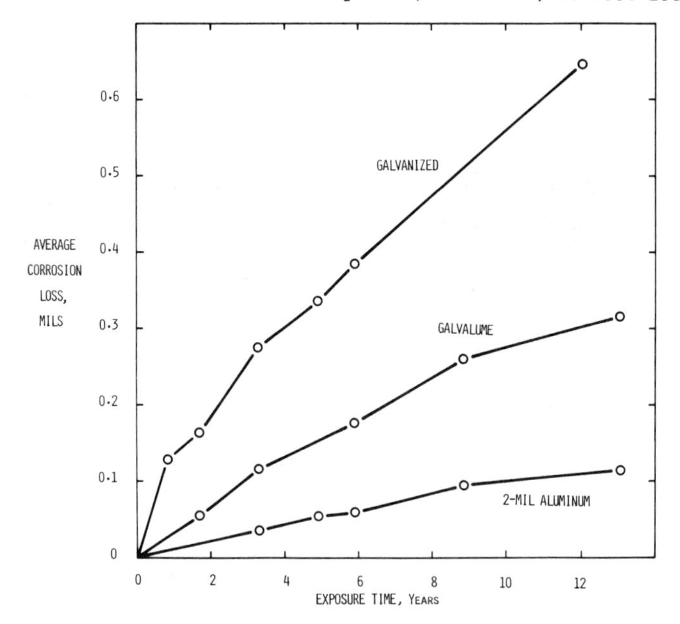

Fig. 10.—Corrosion performance of Galvalume, galvanized and 2-mil aluminum-coated steels in marine atmosphere (Kure Beach, 800 foot lot).

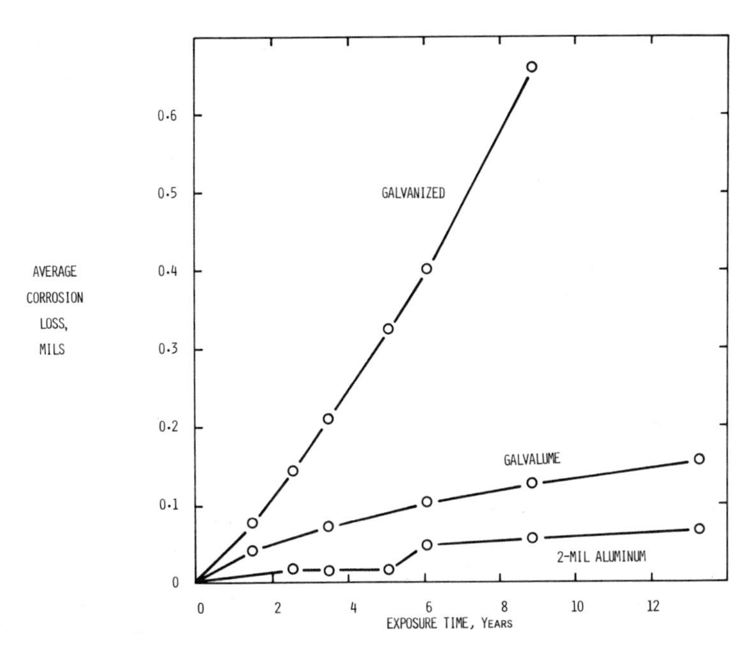

Fig. 11.—Corrosion performance of Galvalume, galvanized and 2-mil aluminum-coated steels in industrial atmosphere (Bethlehem, Pa.).

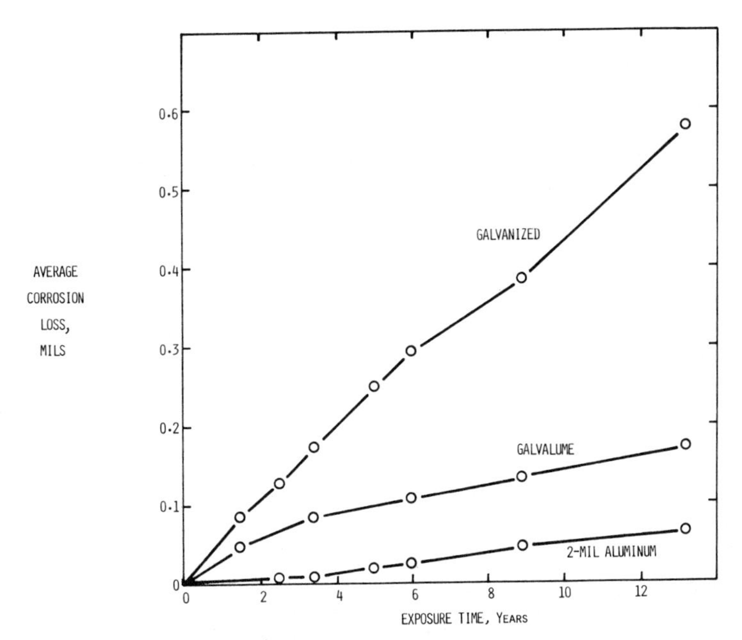

Fig. 12.—Corrosion performance of Galvalume, galvanized and 2-mil aluminum-coated steels in rural atmosphere (Saylorsburg, Pa.).

Site	Ratio of 13-Year Corrosion Losses Galvanized/55% Al-Zn
Kure Beach, N.C., 80-ft. lot	4.2
Kure Beach, N.C., 800-ft. lot	2.0
Saylorsburg, Pa.	3.4
Bethlehem, Pa.	6.2

Thicknesses of commercially available coatings are typically 0.9 mil for G90 galvanized and 0.8 mil for 55% Al-Zn. Accordingly, it can be predicted that the commercially available 55% Al-Zn coating will outlast G90 galvanized by at least two to four times in a wide range of atmospheric environments.

The concentration of zinc in that part of the coating which has corroded (Figure 13) is initially about 90% and decreases with time. In the rural and

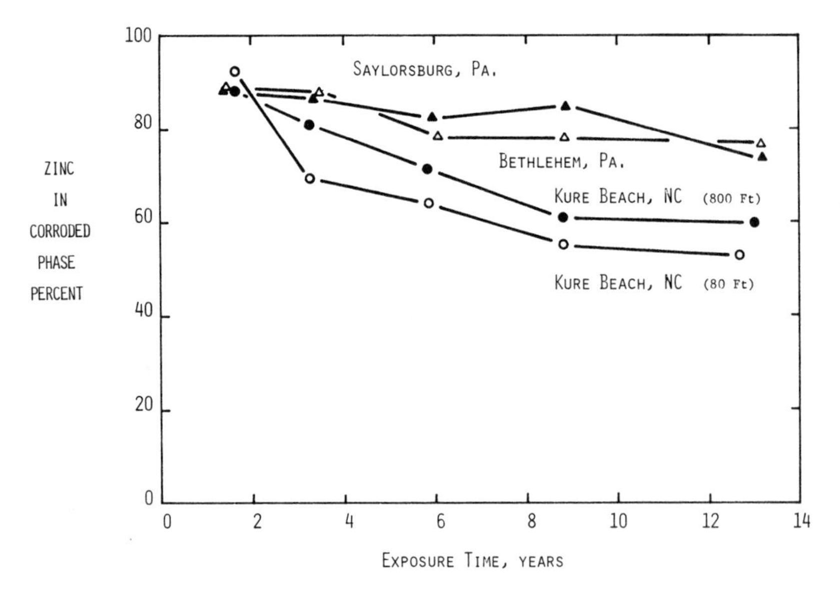

Fig. 13.—Zinc content of the corroded portion of 55% Al-Zn coatings.

industrial environments the composition of the corroded phase seems to level off after nine years at roughly 80% zinc, whereas in the marine environments it continues to decrease.[4]

ATMOSPHERIC CORROSION MECHANISM

The results of the atmospheric corrosion testing now going into the fourteenth year have demonstrated that the 55% Al-Zn alloy coating has excellent corrosion resistance. On the basis of these results, the commercial 0.8 mil 55% Al-Zn alloy coating is expected to outlast the G90 galvanized coating (about 0.9 mil) by at least two to four times in most atmospheres and be more effective than aluminum-coated in resisting rust stain at cut edges. To account for the exceptionally good performance of the 55% Al-Zn coating, it is useful to consider the mechanism and morphology of the corrosion process.

The time dependence of the corrosion potential for 55% Al-Zn coatings exposed to laboratory chloride or sulfate solutions is shown schematically in Figure 14. Subsequent to first immersion (State 1) the coating exhibits a

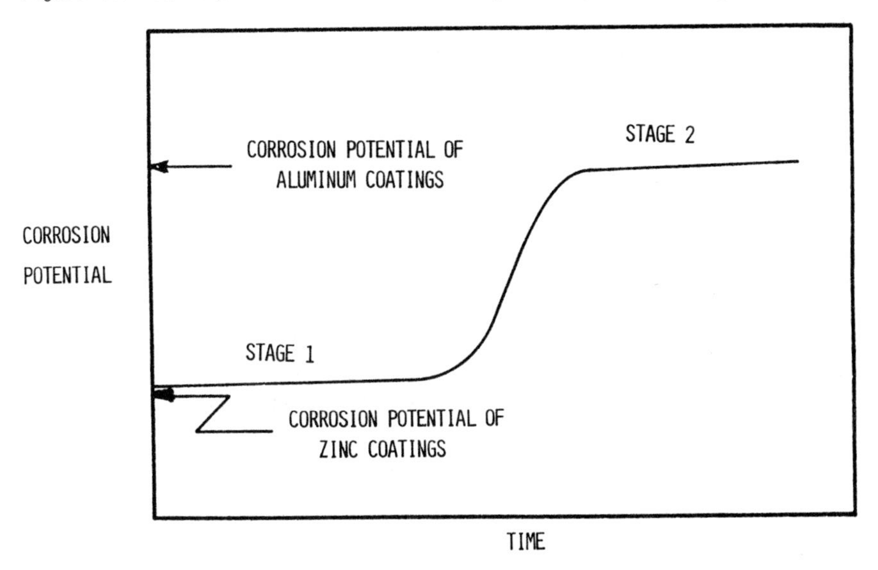

Fig. 14.—Schematic time dependence of corrosion potential of 55% Al-Zn coating in aqueous solutions.

corrosion potential close to that of a zinc coating exposed under identical conditions, generally about -1.0 to -1.1 V (SCE). During State 1 the zinc-rich portion of the coating dissolves preferentially, and the coating, like zinc, is anodic to steel. Stage 1 persists until the zinc-rich interdendritic portion of the coating is consumed, the exact time depending on the thickness of the coating (mass of available zinc) and the severity of the environment (rate of zinc corrosion). Following depletion of the zinc-rich fraction, the corrosion potential rises and approaches that of an aluminum coating, generally about -0.7 V (SCE). During this period (Stage 2) the coating behaves like an aluminum coating, passive in sulfate environments, but anodic to steel in chloride environments.

The behavior of the 55% Al-Zn coating during atmospheric exposure appears to proceed in a manner analogous to that observed in laboratory solutions, although the time scale is greatly extended. The zinc-rich interdendritic portion of the coating corrodes perferentially, as evidenced by the composition of the corroded phase (Figure 13). During this period the coating is sacrificial to steel, and the cut edges of thin steel sheet are galvanically protected. The initial over-all rate of corrosion of the Al-Zn coating is less than that of a galvanized coating because of the relatively small area of exposed zinc.

As the zinc-rich portion of the coating becomes gradually corroded, the interdendritic interstices are filled with zinc and aluminum corrosion products. The coating is thus transformed into a composite comprised of an aluminum-rich matrix with zinc and aluminum corrosion products mechanically keyed into the interdendritic labyrinth. The zinc and aluminum corrosion products should offer continued protection as a physical barrier to the transport of corrodents. In addition, as others have also reported,[5] these products may act as a cathodic inhibitor and thus, as they are gradually leached from the coating, serve to provide continued protection at cut edges. The decreasing corrosion rates with time in the rural and industrial environments (Figures 11 and 12) appear to reflect a gradual change from active, zinc-like behavior to passive, aluminum-like behavior.

The intermetallic layer is generally cathodic to the steel substrate as well as to the other components of the coating. Accordingly, this layer appears to function as a barrier that prevents corrosion of the steel substrate subsequent to interdendritic corrosion of the overlay.

ACKNOWLEDGEMENT

The author thanks: the Bethlehem Steel Corporation for permission to publish these results; his co-workers, A. R. Borzillo, H. E. Townsend, J. C. Zoccola, L. Allegra and H. J. Cleary, who have so freely given their time and effort to this development and provided the results summarized here; and B. S. Mikofsky, Technical Editor, for his valuable assistance in the preparation and editing of this manuscript.

REFERENCES

1. Borzillo, A.R. and Horton, J.B., U.S. Patents 3,343,930 and 3,393,089.
2. Presnyakov, A.A., Gorban, Yu.A. and Chervyakova, V.V., *Zhur. Fiz. Khim. 35(6), 1289 (1961).*
3. Zoccola, J.C., Townsend, H.E., Borzillo, A.R. and Horton, J.B., *ASTM STP 646, 165 (1978).*
4. From a forthcoming publication by H.E. Townsend and J.C. Zoccola.
5. Evans, U.R., "Metallic Corrosion, Passivity and Protection," 2nd ed. Edward Arnold and Son, London, 1946, p. 536.

TINPLATE AS A SACRIFICIAL COATING

Ronald E. Beese

American Can Company
433 N. Northwest Highway,
Barrington, IL

RONALD BEESE

ABSTRACT

Tinplate as it is known today and factors which affect its performance in container applications are examined. Discussion includes the following: (1) the effect of steel chemistry, (2) the protection afforded the tin surface by surface passivation treatments, (3) the continuity of the tin and the tin-iron alloy layers, (4) the mechanisms by which tin provides cathodic protection to steel in plain and enameled cans and the factors which affect this mechanism. These discussions include a description of the tin and steel electrode reactions and the influence of complexing ions and stannous ion on performance.

INTRODUCTION

Tinplate is a mature steel base material primarily used for the packaging of both food and non-food products in steel based containers. It has been used since the early 1800's and evaluated by steel producers, can manufacturers and the academic community for over a hundred years. A wealth of information has been generated. Factors which affect performance of electrolytic tinplate produced in the 1970's will be discussed; however, many aspects merit a more complete review than can be given in this limited space. References should provide deeper insight into aspects of tinplate which may not be discussed fully. *The Technology of Tinplate* by Hoare, Hedges and Barry[1] is an excellent tinplate reference.

Quantity of Tinplate Used

The world usage of tinplate has grown from one million tons in 1900 to 14 million tons in 1974[2] for the canning of foodstuffs, beverages, and non-food items. An annual growth rate of 2 to 3% is anticipated until 1985[3] which represents a mature market growing only in relation to the needs of our growing world population. In 1900, all tinplate was made by the hot dip process where sheets of steel were dipped into molten tin baths. This plate had a heavy tin coating, usually more than was required for performance. Today most all tinplate is made by the electrolytic plating process and attention is given to provide only the level of tin required for designed performance. Most current programs involving tinplate studies deal with cost reduction and are concerned with the following:

- reduction of base weight
- reduction in tin coating
- improved inside and outside lacquering of cans
- improved container stability and integrity with container design
- increased production speeds to gain efficiency

Tinplate Manufacture

Electrolytic tinplate is made by electroplating a thin layer of tin on a special, low carbon steel. The steel strip is cold rolled to the desired thickness, cleaned, annealed and temper rolled to produce single reduced electrolytic tinplate. The steel for double reduced electrolytic tinplate is given a 30 to 40% cold reduction in place of the temper rolling pass.

The surface is then recleaned in an alkaline bath to remove rolling oils, pickled in an inhibited dilute mineral acid, usually with an applied electric current, plated with tin, passed through a melting tower to melt or reflow the tin coating, chemically treated to stabilize the tin surface to prevent tin oxide growth and lubricated with a thin layer of synthetic oil. The tin coating can be applied by any one of four processes: plating from the alkaline tin stannate bath; the acid Ferostan process which deposits tin from a stannous sulfate, sulfuric acid bath; the duPont Halogen tin process which plates tin from a stannous chloride-hydrochloric acid bath; and the fluborate process which plates tin from a stannous fluorborate electrolyte. These processes are described in detail by Lowenheim,[4] and essentially provide the same product - a pure tin coating electrodeposited on a clean low carbon steel strip. Minor differences between the electrolytic tinplate produced by these processes will be discussed in the section concerning the performance of heavily coated electrolytic tinplate. A schematic cross section of 55 lbs. double cold reduced electrolytic tinplate is shown in Figure 1.

Tinplate Specifications

Tinplate can be purchased in a wide range of tempers, basis weights, chemistry, tin coating weights and surface passivation treatments. Temper and basis weight* determine the strength of the container and the chemistry, tin coating weight and surface treatment determine corrosion performance. Specifications are given in ASTM publications.[5] Tinplate can be purchased in 18 different basis weights which are summarized in Table I. Intermediate basis weights also can be purchased.

The temper of the single reduced base steel is determined by chemistry, processing and the 1 to 2% reduction during temper rolling after the anneal but before tinplating. It is a measure of the stiffness and hardness of the steel. Single reduced electrolytic tinplate is available in five temper ranges. Double reduced electrolytic tinplate was introduced to lower cost[6] of the lighter basis weights. A second 30 to 40% cold reduction is given after the anneal which imparts generally greater hardness and tensile strength, a loss of ductility and an increase in directionality. These factors make it more difficult to form containers but provide a stronger container for a given plate thickness. This plate is available in three tempers. Tempers available for electrolytic tinplate are summarized in Table II. Intermediate tempers are produced.

---

* Basis Weight - The base weight is the nominal weight in pounds for one base box of plate. A base box is a unit of area defined by 112 sheets, 14 in. by 20 in., which is equivalent to 31,360 square inches of plate or 62,720 square inches of surface.

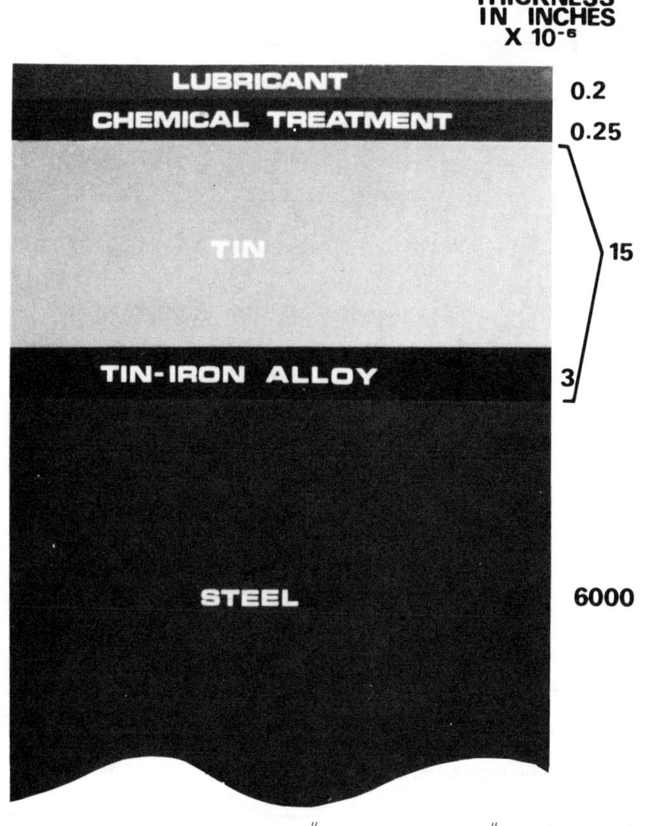

Fig. 1.—Schematic cross section of 55# 2CR tinplate (#25 tin coating).

The chemistry requirements of the base steel for electrolytic tinplate are as important today as when they were introduced in the 1940's. The requirements are determined by a large number of test packs.[7,8] Specifications are given for Type MR, Type L and Type D steel. The MR chemistry is used for most applications; however, when improved corrosion resistance is required, Type L chemistry is specified. The Type D steel is designed for improved ductility for deep drawing applications.

The relatively new continuous casting process provides a base steel with exceptional cleanliness and formability; however, the deoxidizing process requires the addition of aluminum or silicon to kill the steel. Experience with Type D steels suggest that aluminum additions will not cause a corrosion problem. Laubscher and Weyandt[9] have shown that the silicon found in silicon-killed continuous cast electrolytic tinplate will not adversely affect the corrosion performance of plain cans packed with mildly acid food products. Data with regard to enamel cans is not definitive. Additional data is needed to determine whether or not silicon can reduce corrosion performance in enameled, heavily coated tinplate cans. The steel chemistry requirements are summarized in Table III.

TABLE I.—NOMINAL BASE WEIGHT THEORETICAL THICKNESS

Nominal Weight lb./Base Box	Theoretical Thickness Inches	Theoretical Thickness Millimeters
45	0.0050	0.127
50	0.0055	0.140
55	0.0061	0.155
60	0.0066	0.168
65	0.0072	0.183
70	0.0077	0.196
75	0.0083	0.211
80	0.0088	0.224
85	0.0094	0.239
90	0.0099	0.251
95	0.0105	0.267
100	0.0110	0.279
103	0.0113	0.287
107	0.0118	0.300
112	0.0123	0.312
118	0.0130	0.330
128	0.0141	0.358
135	0.0149	0.378

TABLE II.—TEMPER DESIGNATIONS FOR ELECTROLYTIC TINPLATE

Plate Type	Temper Designation	Rockwell Hardness Range HR30T*	Characteristic
Single Reduced	T-1	49 ± 3	Soft for Drawing
Single Reduced	T-2	53 ± 3	Moderate Drawing
Single Reduced	T-3	57 ± 3	General Purpose
Single Reduced	T-4	61 ± 3	Increased Stiffness
Single Reduced	T-5	65 ± 3	Stiffness - Ends
SR - Continuous Annealed	T-4-CA	61 ± 3	Stiffness - Ends
SR - Continuous Annealed	T-5-CA	65 ± 3	Increased Stiffness
Double Reduced	DR-8	73	Round Can Bodies and Ends
Double Reduced	DR-9	76	Round Can Bodies and Ends
Double Reduced	DR-9M	77	Beer & Beverage Ends

* Values are based on a diamond anvil.

TABLE III.—CHEMICAL REQUIREMENTS FOR TINPLATE STEEL

Element	Cast Composition, Maximum Percent Type MR	Type L	Type D
Carbon	0.13	0.13	0.12
Manganese	0.60*	0.60*	0.60
Phosphorus	0.020	0.015	0.020
Sulfur	0.05	0.05	0.05
Silicon*,†	0.020	0.020	0.020
Copper	0.20	0.06	0.20
Nickel	-	0.04	-
Chromium	-	0.06	-
Molybdenum	-	0.05	-
Other Residual Elements, each	-	0.02*	-‡

* Unless otherwise agreed upon between the manufacturer and the purchaser.
† Where strand case steel is furnished, the silicon maximum may be increased to 0.080 unless expressly prohibited by the purchaser.
‡ Aluminum-killed.

Tin is electroplated on the surface of cleaned and pickled steel strip. Tinplate is available with ten different tin coating thicknesses. The number designation is equivalent to the pounds of tin on a base box of plate. The ASTM[5] designation, shown in Table IV, indicates that half the tin is on one surface and the remaining half is on the other surface.

TABLE IV.—ELECTROLYTIC TINPLATE COATING WEIGHT

Designation Number	lb. Tin/Base Box One Each Side	Nominal Tin Thickness x 10^6 Inches
10	0.05/0.05	6/6
25	0.125/0.125	15/15
50	0.25/0.25	30/30
50/25	0.25/0.125	30/15
75	0.375/0.375	45/45
75/25	0.375/0.125	45/15
100	0.50/0.50	60/60
100/25	0.50/0.125	60/15
100/50	0.50/0.25	60/30
135/25	0.675/0.125	81/15

The differentially coated tinplate, for example 50/25, has a different weight of tin coating on each surface. The first number is the tin coating weight on the top surface or the outside surface of a coil. The weight of the alloyed tin in the tin-iron alloy layer is included when tin coating weights are identified. Methods to determine tin coating weights are given in ASTM Standard A630-68.[5]

Immediately after plating the matte tin, the tin deposit is flow brightened by heating the strip above the melting point of tin followed by quenching in water. The resulting bright surface is more attractive, more resistant to abrasion and more easily soldered than the original matte tin deposit. For some special applications, however, tinplate is used in the matte condition. During the flow brightening, a layer of tin-iron alloy ($FeSn_2$) is formed between the tin coating and the base steel.

After flow brightening, tinplate is given a chemical treatment to control the growth of tin oxide during storage and subsequent baking operations. It also improves the organic enamel coating adhesion. The chemical treatment is applied by passing the tinplate strip through a chemical solution with or without electric current. Five treatments are available:

1. Cathodic Sodium Carbonate Treated Tinplate

   Tinplate is made the cathode in a dilute sodium carbonate bath. The cathodic current reduces the tin oxide formed during melting but provides no additional protection. It provides the least passivation to control tin oxide growth. It is mainly used for evaporated milk can stock. No chromium is present.

2. Chromic Acid Treated Tinplate

   Tinplate is immersed in a dilute chromic acid bath, rinsed and dried. Tin oxide formed during melting is dissolved and chromium oxide is deposited with an aim not to exceed 250 μg chromium/ft^2. The chromium on the surface provides a moderate passivation against tin oxide formation. The treatment provides limited storage stability and is used where a highly passivated tin surface is detrimental to the end use.

3. Sodium Dichromatic Dip

   Tinplate is immersed in a dilute sodium dichromate solution. The desired chromium oxide level is 150 μg of chromium/ft^2 and is similar to the chromic acid treatment.

4. Cathodic Sodium Dichromate Treated Tinplate

   Tinplate is made a cathode in the sodium dichromate bath. Tin oxides are reduced and chromium and chromium oxides are deposited. The desired chromium level is 500 μg of total chromium/ft^2. The resulting highly passivated surface has the following properties:

   a. Minimum tin oxide growth during prolonged storage.

   b. Minimum discoloration during baking required for organic coating and lithography.

   c. Minimum discoloration during soldering.

   d. Provides some resistance to tin sulfide discoloration from some sulfur-bearing food products.

   A report distributed to the AISI and the CMI by D. T. Smith[10] in 1971 suggests that both metallic chromium and chromium oxide are deposited from the sodium dichromate solution.

Rauch and Steinbicker[11] report that a large portion of the cathodic dichromate film consists of chromium in the metallic state. They show that chromium deposition occurs because the main reaction, hydrogen evolution, has an unusually high over-voltage on tin. Rauch and Steinbicker[12] also demonstrate that the oxidation resistance of cathodic dichromate tinplate is directly related to the amount of metallic chromium in the passivation film. Subsequent studies by Aubrun and Rocquet[13] confirm that the high hydrogen overvoltage on tin makes possible deposition of metallic chromium during the cathodic dichromate treatment. S. C. Britton[14] reports two main chromium containing constituents on cathodic sodium dichromate treated tin surfaces. The metallic chromium is dissolved by sulfuric acid or anodic oxidation. The oxides of chromium are soluble in hot concentrated alkali.

The quantity of metallic chromium deposited is small and there is concern over the validity of the observations. Recent studies of the cathodic dichromate treatment by Soepenberg, Vrijberg and Spruit[15] describe the kinetics of film formation. Three stages were found. The first involves an incubation step during which tin oxide reduction takes place. The second is a logarthmic growth to an electrochemical surface chromium content of 6-10 mg/m^2. The third step involves a square root growth upon further increase of electrochemical chromium. These kinetics are consistent with the diffusion of a reacting species through a porous layer. Leroy, Servais, and Habroken[16] evaluate the cathodic dichromate film on tinplate with secondary ion mass analysis, Auger and photoelectron spectroscopy. The results of this preliminary study indicate that metallic chromium contents are much less than those determined by wet chemistry. Aubrun and Pennera[17] report a technique to measure the metallic chromium content in the cathodic dichromate film on tinplate. Their technique involves anodic dissolution of the metallic chromium with constant current in a pH 5 phosphate solution after a 1 minute dip in 90°C 9N NaOH solution.

It is concluded from the evidence that metallic chromium is deposited on the tin surface during the cathodic dichromate treatment. Whether this film affects cathodic protection afforded iron by tin is not known.

The cathodic dichromate treatment is the most important chemical treatment. It is specified for most of the tinplate used in food and non-food containers.

5. Cathodic- Anodic Chromate Phosphate Treated Tinplate

The tinplate surface is treated with cathodic and anodic currents, in a chromate phosphate solution to deposit a desired level of 150 μg chromium/ft^2 and 250 μg of phosphate/ is ft^2. This treatment provides poor resistance to tin oxide growth and is generally used only for milk can stock and jar cap closures.

Oiling - Electrolytic tinplate is lubricated on both sides with 0.10g/BB to 0.40g/BB of lubricant. Dioctyl sebacate (di (-2-ethyl-hexyl) sebacate) is commonly used to lubricate tinplate; however, other lubricants may be used.

Corrosion Performance of Tinplate

The corrosion performance of tinplate depends on the tinplate selected and the environment in which it is expected to perform. Tinplate exposed to a moist atmosphere or to natural waters exhibits corrosion suggested by the standard oxidation - reduction potentials for the tin and iron half cells:

$$Fe = Fe^{++} + 2e \qquad -440\ mV$$

$$Sn = Sn^{++} + 2e \qquad -136\ mV$$

The tin coating is more noble than the bare steel. Iron is the anode and corrodes at pores, fractures or distontinuities in the tin surface. The best that can be said is that tin provides a reasonably corrosion resistant barrier limiting iron corrosion to sites where iron is exposed through the tin coating. If this were the case inside a tinplate can, tin could not be used to preserve foodstuffs, beverages and non-food items. Successful packaging in tinplate containers is dependent on the unique ability of tin and stannous ion to protect steel from corrosion in air-free, mildly acid systems.

OVERVIEW

The corrosion performance of tinplate essentially is a function of the cathodic activity of the base steel, the area of steel exposed through the tin-iron alloy and tin layers, and the stannous ion concentration. When tin coating weights are No. 25 or less, the relative area of steel exposed through the tin coating is large and insufficient tin is present to protect the steel for a reasonable container life, such as 30 months. Thus, the mechanisms proposed are related to the corrosion performance of electrolytic tinplate with No. 50, or heavier, tin coatings. The No. 25 or lighter tin coatings are employed where corrosion is not a problem or when the surface is protected with organic enamel coatings.

The industry recognizes two special types of heavily coated electrolytic tinplate. They are defined in ASTM A623[5] and include J-plate and K-plate. J-plate is electrolytic tinplate with No. 50, or heavier, tin coating, with improved corrosion performance measured by the following special property tests:

1. Pickle Lag with an aim property of 10 seconds maximum.

2. Iron Solution Value with an aim property of less than 20 μg iron.

3. Tin Crystal Size with an aim property of crystal size ASTM No. 9, or larger.

The alloy layer of J-plate is normally light in color, which is characteristic of the acid tinning processes. K-plate provides additional corrosion protection and has all the special properties of J-plate, plus the condition measured by the Alloy Tin Couple test which has an aim maximum of 0.12 $\mu A/cm^2$. Good mill practice has demonstrated the ability to obtain average ATC values of 0.05 $\mu A/cm^2$ or less over an extended period of production.

The corrosion performance of J and K electrolytic tinplate for moderately acid foods such as citrus products, tomato products, peaches, pears, pineapple, etc., in plain, unenameled cans and for products typified by red sour pitted cherries, green beans, tomato products, and pickled beets in enameled cans can be described using the schematic cross section of tinplate shown in Figure 2. The basic corrosion reactions in air-free, mildly acid systems are listed. The schematic cross section of the tinplate surface

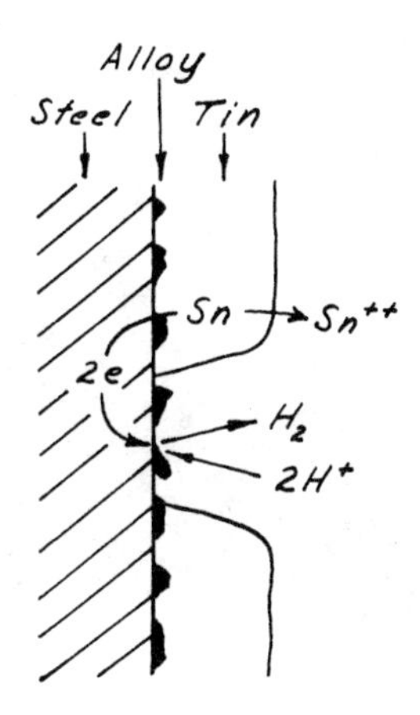

Fig. 2.—Schematic cross section of tinplate surface indicating corrosion reactions in an acid food product.

shows the steel partially covered by a discontinuous layer of tin-iron alloy which in turn is partially covered by a tin coating containing pores. Figure 3 is a SEM photograph of a tinplate pore taken from a packed can. The corrosion mechanism to be discussed in detail can by outlined as follows:

1. Tin is relatively inert to direct chemical attack but it is anodic to both steel and tin-iron alloy.

2. Dissolution of tin is primarily caused by the galvanic couple action with steel and/or alloy exposed through pores in the tin coating.

3. Steel and alloy receive essentially complete cathodic protection from the electrochemical reactions, i.e., steel and alloy are polarized to the potential of tin.

4. Stannous ion, the result of tin dissolution, makes steel and alloy easier to protect cathodically and slows the galvanic corrosion of tin.

5. The $FeSn_2$ alloy is an inefficient cathode as compared to steel. The galvanic corrosion of tin coupled to steel is approximately fifty (50) times as great as for tin coupled to tin-iron alloy.

6. The rate of tin dissolution is determined by the area and cathodic activity of the steel exposed through the combined layers of tin and alloy.

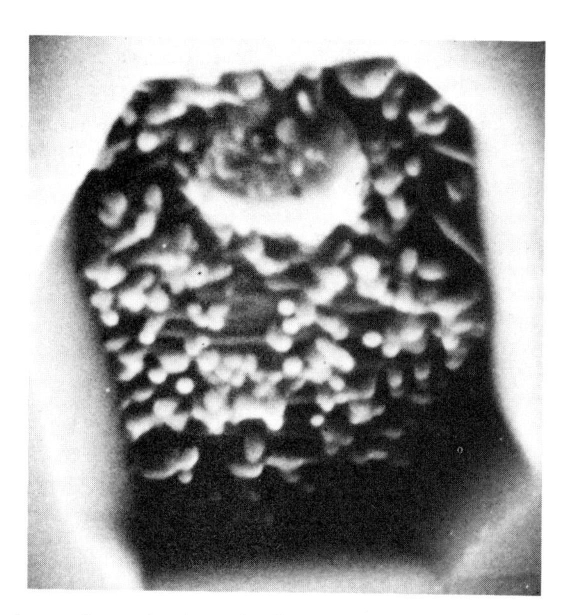

Fig. 3.—Tinplate surface showing tin-iron alloy exposed in a pore 45,000X.

7. Closely packed, continuous alloy covering the base steel results in low tin corrosion rates and superior corrosion resistance.

Kohman and Sanborn[18] and Lueck and Blair[19] first demonstrated that tin affords electrochemical protection to steel in food containers. Reports by Culpepper and Moon[20], Hoar[21] and Morris and Bryan[22,23] contribute to a better understanding of the mechanism. They show that steel is protected by the preferential dissolution of tin and that stannous ion resulting from this dissolution further inhibits steel corrosion. Vaurio[24], Koehler[25], Willey, Krickl and Hartwell[26], and Koehler and Canonico[27] discuss the mechanism of tinplate corrosion in plain tinplate cans. Studies of Kamm et al[28,29,30,31] identify the importance of the tin-iron alloy layer to provide corrosion resistant electrolytic tinplate.

The Tin Surface

The tin electrode corrodes at a very slow rate in most air-free, mildly acid systems. The Sn/Sn^{++} electrode is more negative than the reversible hydrogen potential. Thus, both the tin dissolution and hydrogen evolution reactions are possible on the tin surface. The exchange current for the hydrogen evolution reaction on tin is small. Quintin and Hagymas[32] measured a 10^{-11} A/cm^2 exchange current for the hydrogen evolution reaction on tin in 0.5M H_2SO_4. Rates of tin dissolution in citric acid solutions are of the order of 10^{-8} A/cm^2 [33].

Corrosion of the tin surface on tinplate is a function of the size of the tin crystals. Tin crystal size is evaluated by etching the tin surface and estimating the tin crystal size by comparison with ASTM micro-grain size number standards at 1X magnification (ASTM Methods E 112). A No. 9, or larger crystal size is acceptable. Smaller tin crystals usually indicate electrolytic tinplate with poor corrosion resistance. Procedures are described in the ASTM Standard A623-77[5]. This empirical test may relate to observations by several researchers which indicate a substantial difference in reactivity among different tin crystal faces. Seiler[34] measured potential differences of only about 5 mV among different crystal faces in grapefruit juice; however, Asano and Oyagi[35] measured galvanic currents of up to 0.67 $\mu a/cm^2$ in couples between single crystals of different orientation in citric acid.

When oxygen or other oxidants are present, the rate of tin dissolution is markedly increased. The cathodic reaction is now oxygen reduction:

$$1/2\ O_2 + H_2O + 2e^- \longrightarrow 2OH^-$$

Canning procedures are designed to limit the quantity of air in canned foods. In plain unenameled cans, most of the oxygen trapped during the packing process reacts rapidly with tin because of the higher temperatures required for the thermal process to control bacterial spoilage. After oxygen is consumed, the remaining tin dissolves slowly in air-free conditions. Other oxidants such as nitrate can cause rapid detinning in plain cans. Oxygen and other oxidants are found in packed enameled cans. It is believed that the organic enamel coatings control the reaction between oxidants and tin.

Sherlock and Britton[33] show that the tin corrosion potential is a function of the stannous ion concentration as defined by the Nernst equation:

$$E_{(M/M^{z+})}/M^{z+} = E^o + \frac{RT}{zF}\ ln\ M^{z+}$$

Willey[36] indicates that the potential of tin is constant as soon as a few parts per million of stannous ion are added to air-free acid solutions. The tin electrode potential drifts without stannous ion. The free stannous ion concentration is small in most solutions because stannous ion is complexed by hydroxyl ion through the hydrolysis of water. The following equation describes the hydrolysis reaction:

$$Sn^{2+} + H_2O \rightleftharpoons SnOH^+ + H^+$$

Thus, pH significantly influences the tin electrode potentials. It is quite evident that other anions which complex tin ion, such as the citrate ion, will lower the stannous ion concentration and thus affect the tin potential. Results reported by Willey[36] on this subject are shown in Figure 4. The data illustrate the effect of stannous ion concentration, pH, and citrate ion concentration on the potential of tin. Willey also demonstrates that the tin potential is cathodic to steel in weak stannous ion complexing systems such as perchloric, phosphoric and acetic acid systems. Tin is anodic to steel in dilute acids which complex tin such as >0.2 M hydrochloric acid, oxalic acid, and citric acid. The tin and steel potentials are very close in tartaric acid and malic acid systems; however, in most cases tin is the anode. The large potential difference between tin and steel found when a strong complexer is present, such as oxalic acid, results in greater current flow between tin and steel which causes increased rate of tin dissolution. Morris and Bryan[23] report that rapid detinning can be expected in plain rhubarb and spinach cans in which oxalic acid occurs.

The passivation treatments on tinplate can affect the tin electrode is asparagus[37]. In this case the passivity of the cathodic dichromate film limits tin dissolution and thus permits the formation of a black tannin compound. The black reaction product between iron and tannin does not form when tin is active and stannous ion is present. A chromic acid dip provides

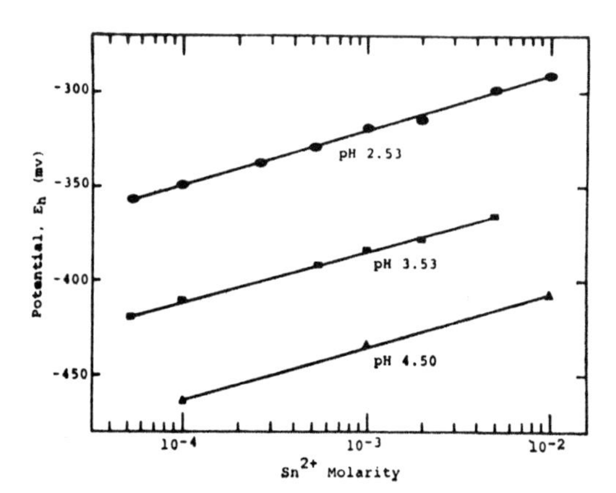

Effect of Sn^{2+} on potential of tin in 0·1 M citrate at different pH values

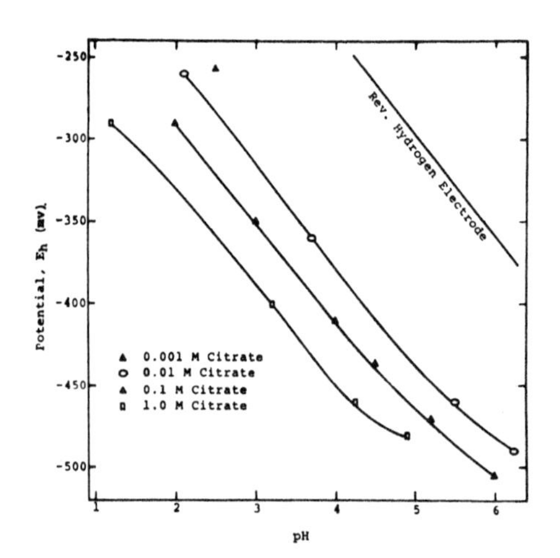

Effect of pH and citrate concentration on potential of tin in presence of 10^{-3} M Sn^{2+}

Figure 4

a mildly passivated tin surface which can dissolve, protect steel, and inhibit the formation of the black compound. The cathodic dichromate surface treatment does not appear to affect tinplate performance in cans packed with more corrosive products in which 10 to 40 parts per million tin dissolves during the thermal process.

One must conclude that the tin electrode potential and tin corrosion rate is a function of the environment within the packed can.

The Iron Surface

The electrochemical reactions on the iron surface are not as straightforward as on the tin surface. The iron electrode potential is affected by pH, stannous ion, and possibly the concentration of dissolved monatomic hydrogen. Willey[36] reports that the ferrous ion does not usually affect the iron electrode potential, which fact suggests that the potential at the anodic sites is not the controlling factor determining corrosion potential. He suggests that the iron potential is probably the result of a hydrogen electrode potential at the cathodic sites displaced to more negative potentials by cathodic polarization occurring on the corroding iron surface. Uhlig[38] has examined this hypothesis in detail.

Kamm[31] and Willey[36] show that the measured iron potential is a function of both stannous ion and pH, respectively, as shown in Figure 5. The role of stannous ion parallels that of pH and may be related to the fact that stannous ion catalyzes the entry of hydrogen in the steel lattice. Kamm[28] shows that when tin is coupled to steel in 3N HCl in the absence of stannous ion, high cathodic currents provide only partial cathodic protection to the steel and very little hydrogen diffuses through the steel. The addition of stannous ion to the solution decreases the current required for cathodic protection and increases the rate of hydrogen diffusion through the steel four-fold. The second observation which indicates that hydrogen atoms in steel affect the corrosion potential is the uncoupling potential shift reported by Koehler[25] and also demonstrated by Kamm[28]. Koehler reports that upon uncoupling a steel sample, which had been coupled to tin, the potential of the steel shifts rapidly in the positive direction for less than a minute and then shifts slowly back to the original uncoupled potential. These potential changes suggest that the hydrogen charged into the steel surface by the cathodic reaction affects the corrosion potential. The return of the uncoupled steel to the original corrosion potential is believed to be the result of hydrogen diffusing from the iron surface. This interpretation is consistent with the potential change observed as the solution pH is decreased.

Hoar and Havenhand[39] attribute the shift of the steel potential observed because of the addition of stannous ion to anodic inhibition which presumably affects the polarization of local steel anodes. Kamm[28] suggests that stannous ion inhibition is related to the entry of hydrogen atoms into the steel lattice. Sherlock[40] concludes that the inhibition by stannous ion is due to the deposition of metallic tin on iron by a chemical replacement reaction. He cites the work of Buck, Heyn and Leidheiser[41] which shows that iron immersed in boiling 0.2M citric acid containing stannous ions becomes coated with several monolayers of metallic tin and tin-iron alloy and that corrosion resistance is due to the formation of $FeSn_2$ which limits iron corrosion. Kamm[28] reports that tin plated on steel shifts the potential of steel in a direction opposite to that observed with either the uncoupling potential shift or the addition of stannous ions. It is interesting to note that less than 10 ppm stannous ion will inhibit steel corrosion in citric acid solutions but not in phosphoric acid solutions. Thus, either the

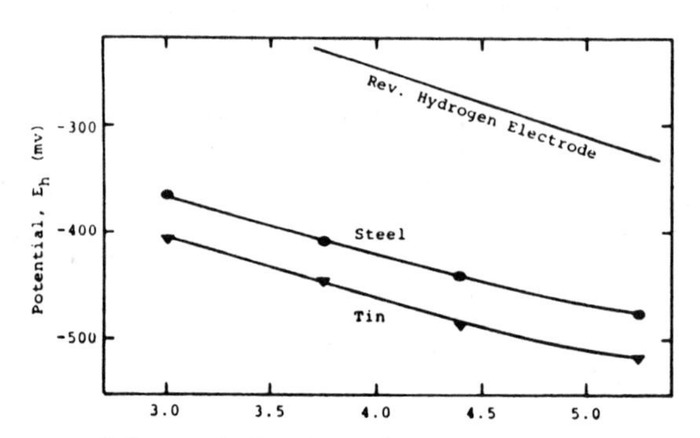

Influence of pH in 0·2 M citrate on potentials of tin and steel at 38°C after 10 h exposure

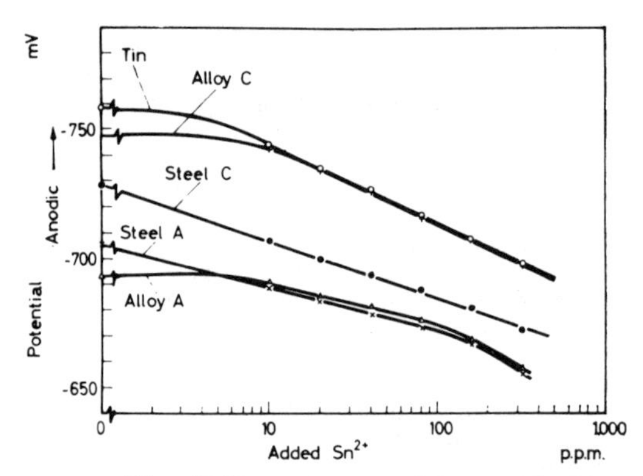

Effect of Sn^{2+} concentration on relative potentials of tin, alloy and steel in de-aerated 0·1M citrate buffer at pH 3·2

Figure 5

anion or complexing by the anion affects the inhibition of steel corrosion caused by stannous ion. Whatever the mechanism, stannous ion is an important corrosion inhibitor which affects the performance of tinplate containers.

The corrosion resistance of the base steel for tinplate is also influenced by minor constituents in steel. Hartwell[7,8] reports that increasing silicon, phosphorus and sulfur contents adversely affects corrosion resistance. In steels containing low amounts of phosphorus and sulfur, an increase in copper content greatly reduces steel corrosion resistance in red cherries or prune cans, improves corrosion resistance in loganberries cans and had no effect in peach cans. Copper has no consistent effect for all products; however, lowering copper generally improves corrosion resistance. Shelf life of cherry cans indicate that the total carbon, manganese, sulfur, and phosphorus are important. Additional studies indicate that control of nickel, chromium and molybdenum makes more corrosion resistant plate for strongly corrosive products. These observations resulted in the chemistry specifications established for tin mill products.

The surface of the steel can affect corrosion performance. Willey[36] demonstrates that annealing gas affects the corrosion resistance of the steel surface in 90°C 6N HCl. Koehler[42] reports that grain boundry oxides formed as the result of annealing in DX atmospheres containing nitrogen, hydrogen, oxides of carbon and water cause pickle lag. Dry hydrogen gas or HNX annealing gas consisting of dried nitrogen with 4 to 8 percent hydrogen gas an an annealing atmosphere can reduce the grain boundry oxides and thus reduce pickle lag. Koehler[25] describes tin-steel couple studies with high pickle lag steel for which tin could only provide partial protection to steel. Willey[25] reports that tinplate made from high pickle lag steel has poorer corrosion performance than tinplate made from low pickle lag steel in cans packed with prunes and stored at 100°F. Details to perform the pickle lag test are given by ASTM Standard A623 [5].

Area of Iron

The porosity of the tin coating and the tin-iron alloy coating determines the area of iron to be protected by tin in plain tinplate cans. Iron is polarized to the potential of tin, and the current required for a given steel is a function of the iron area exposed to the environment. Thus, the tin dissolution rate increases above the rate for pure tin in a direct relation to the iron area exposed. The area of steel exposed at pores in a No. 100 electrolytic tin coating is estimated by Sherlock[43] to represent one thousandth of the total tinplate surface area. The number of pores counted is in the range of 300 to 2,500 per cm^2.

The polarization technique utilized by Sherlock[43] assumes that the cathodic activity of the base steel exposed through tin pores has the same cathodic characteristics as the general steel surface after removing the tin and tin iron alloy coatings. In general, Sherlock reports that, "Estimates of exposed steel area were parallel with the estimates of cathodic activity from which they were derived by application of measured cathodic polarization characteristics of the steel base plates." Tin steel couple studies and cathodic steel polarization measurements show that the cathodic activity of steel for tinplate varies from lot to lot. However, the simple measure of the current measurements or polarization techniques do not correlate with test pack performance. Thus, both the area and the cathodic activity of the base steel are required to estimate test pack performance.

The industry standard test to evaluate the area and the cathodic activity of the base steel exposed through the tin coating is the ISV test. The test is described in the ASTM Standard A623 for tin mill products[5] and was originally suggested by Willey[26]. The iron solution value (ISV) is the micrograms iron dissolved from a 3.14 in.^{2} tinplate surface exposed two hours at 80°F to 50 ml of a mixture of dilute sulfuric acid, hydrogen peroxide and ammonium thiocyanate. The total iron in solution is determined directly using a spectrophotometer set at 485 millimicrons.

Test packs show that low ISV values under 20 do not necessarily provide highly corrosion resistant tinplate. The tin-iron alloy is an important factor in controlling the increasing iron area exposed as tin dissolves. The schematic cross section of detinning tinplate as suggested by Kamm[31], shown in Figure 6, indicates that as tin dissolves, the area of iron and

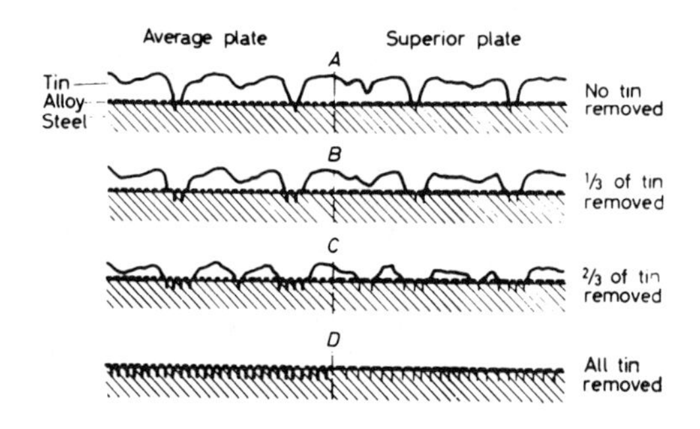

Fig. 6.—Schematic drawings of tinplate surface cross sections representing the tin and alloy layers and their effect on the amount of steel exposed as tin is corroded from the plate surface.

tin-iron alloy exposed through the tin pores increases. Further, the area of iron exposed through the alloy layer ultimately determines the area requiring protection. The tin-iron alloy layer as viewed by a direct chromium shadow replica technique described by Ebben[44] with an electron microscope provides excellent detail of the alloy crystals and any steel surface exposed through discontinuities in the alloy layer. Figure 7 illustrates a typical alloy layer on a Type J electrolytic tinplate. Figure 8 identifies a typical alloy on Type K electrolytic tinplate. The concept of alloy coverage is readily apparent.

Tinplate produced by the alkaline tinning process usually has a close-packed layer of relatively large grains of tin-iron alloy. Tinplate made on acid tinning lines usually has a discontinuous alloy which results in Type J electrolytic tinplate. The producers of electrolytic tinplate from acid tinning baths have developed techniques to produce the close-packed tin-iron alloy grain structure necessary for Type K electrolytic tinplate properties.

The area of iron and the cathodic activity of iron can be measured with the alloy-tin couple (ATC) test described in the ASTM Standard[5]. The

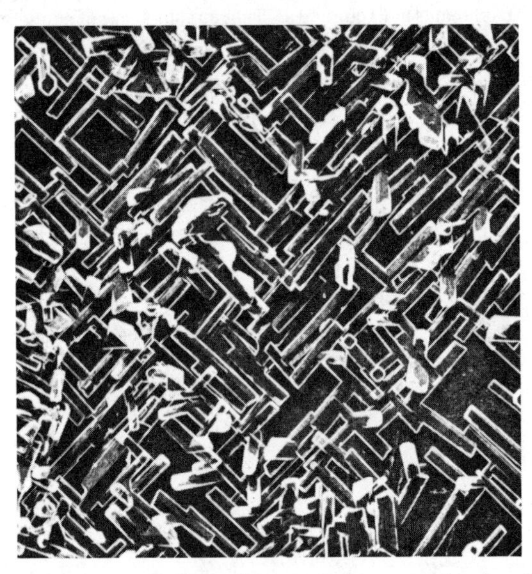

Fig. 7.—Alloy layer on tinplate having average corrosion resistance.

Fig. 8.—Alloy layer on tinplate having superior corrosion resistance.

test as first described by Kamm et al[29,30] measures the steady state current required to polarize an alloy area to the potential of tin in air-free 80°F grapefruit juice containing 100 ppm stannous ion. The ATC value is the current density at the tin-iron alloy surface after 20 hours which flows between a relatively large pure tin electrode and the tin-iron alloy surface.

Correlation between ATC and average performance in an accelerated grapefruit juice test pack is shown in Figure 9. The authors also suggest a progressive

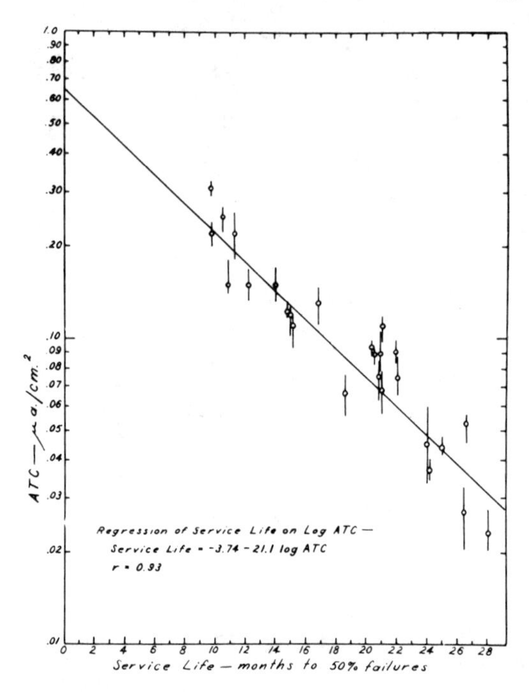

Fig. 9.—Relationship between log ATC and average performance in grapefruit juice pack stored at 100°F.

ATC test which measures detinning rate of a tinplate surface at various stages of detinning in grapefruit juice containing 5 ppm stannous ion. The calculated rate of tin dissolution closely parallels the rate of tin pickup in actual test pack cans. The progressive ATC test considers both the porosity of the tin and the tin-iron alloy layers. Unfortunately it is a time-consuming procedure and thus cannot be considered as a control test.

Enamel Coated Tinplate

The corrosion performance of organic enamel coated No. 50 or heavier electrolytic tinplate cans is similar to that found in plain uncoated cans with two exceptions. The area of tin available to provide protection is obviously reduced, and the stannous ion concentration is also decreased. Tin still provides cathodic protection to the steel exposed through the combined enamel, tin and tin-iron alloy coatings. Detinning usually is localized at fractures in the enamel coating. Organic coatings are usually used to reduce tin dissolution and the effect of oxidants in the product on the tin coating. Products packed in enameled No. 25 or less electrolytic tinplate cans are either noncorrosive when single enamel coatings are specified or very corrosive when two or more enamel coatings are employed. In both cases, tin does not afford significant cathodic protection to steel. The enamel coatings are designed to control iron corrosion.

SUMMARY

The sacrificial protection provided by tin to steel is a function of the tin coating weight, the quality of the tinplate provided by the manufacturer as related to the special property tests, the container construction, the composition of the product, and the packing procedure. Deviation from accepted practice in any one of these areas will cause container failure and the loss of product.

REFERENCES

1. Hoare, W.E., Hedges, E.S. and Barry, B.T.K., "The Technology of Tinplate," St. Martin's Press, New York, 1965.
2. LaSpada, A., "World Tinplate Statistics: Production, Trade and Apparent Consumption," International Tinplate Conference, First, London, Oct. 5-8, 1976 Proceedings.
3. Lang, J.B., "The U.S. Tinplate Industry," International Tinplate Conference, First, London, Oct. 5-8, 1976 Proceedings.
4. Lowenheim, F.A., "Modern Electroplating," Second Edition, John Wiley & Sons, Inc., New York, 1963.
5. "Annual Book of ASTM Standards," 1978, Part 3, A623-77, A624-76, A626-68 and A630-68, Philadelphia American Society for Testing Materials.
6. Brighton, K.W., Reister, D.W. and Braun, O.G., *National Canners Association Information Letter No. 1909, 61-64 (January 31, 1963)*.
7. Hartwell, R.R., "Corrosion Resistance of Tinplate Influence of Steel Base Composition on Service Life of Tinplate Containers," Symposium on Surface Treatment of Metals, *Am. Soc. Metals, 69 (1941)*.
8. Hartwell, R.R., "Certain Aspects of Internal Corrosion in Tinplate Containers," Advances in Food Research, 3, Academic Press, New York, 1951, pp. 327-383.
9. Laubscher, A.N. and Weyandt, G.N., *J. Food Sci. 35, 823 (1970)*.
10. Smith, D.T., "Improved and Better Control of Surface Treatments for Tinplate," American Iron and Steel Institute, July 15, 1971 Report.
11. Rauch,S.E.,Jr. and Steinbicker, R.N., "A Study of Surface Chromium on Tinplate," *J. Electrochem. Soc. 120, 735 (1973)*.
12. Rauch, S.E., Jr. and Steinbicker, R.N., "Mechanism of Cathodic Dichromate Passivation of Tinplate," *Plating 62, 246 (1975)*.
13. Aubrun, Ph. and Rocquet, P., "The Mechanism of Metallic Chromium Electrodeposition in the Cathodic Dichromate Treatment of Tinplate," *J. Electrochem. Soc. 122, 861 (1975)*.
14. Britton, S.C., "Examination of the Layer Produced by Chromate Passivation Treatments of Tinplate," *Br. Corros. J. 10, 85 (1975)*.
15. Soepenberg, E. Nagel, Vrijburg, H.G. and Spruit, A.C., "Investigation on the Kinetics and Mechanism of the Formation of Cathodic Dichromate Passivation Films on Electrolytic Tinplate," International Tinplate Conference, First, London, Oct. 5-8, 1976, Proceedings.
16. Leroy, V., Servais, J.P., and Habraken, L., "Secondary Ion Mass Analysis, Auger and Photoelectron Spectrometry of Passivation Layers on Tinplate," International Tinplate Conference, First, London, Oct 5-8, 1976, Proceedings.
17. Aubrun, Ph.S. and Pennera, G.A., "Coulometric Determination of the Metallic Chromium Content in Tinplate Passivation Films," International Tinplate Conference, First, London, Oct. 5-8, 1976, Proceedings.
18. Kohman, E.F. and Sanborn, N.H., *Ind. Eng. Chem. 20, 76 (1928)*.
19. Lueck, R.H. and Blair, H.T., *Trans. Am. Electrochem. Soc. 54, 257-92 (1928)*.
20. Culpepper, C.W. and Moon, H.H., *Canning Age 9, 461 (1928)*.
21. Hoar, T.P., *Trans. of the Faraday Soc. 30, No. 157, Part 6 (1934)*.
22. Morris, T.N. and Bryan, J.M., Food Investigation Special Report, No. 40, 1931 (London, D.S.I.R.).
23. Morris, T.N. and Bryan, J.M., Food Investigation Special Report, No. 44, 1936 (London, D.S.I.R.).
24. Vaurio, V.W., *Corrosion 6, 260 (1950)*.
25. Koehler, E.L., *J. Electrochem. Soc. 103, 486 (1956)*.
26. Willey, A.R., Krickl, J.L. and Hartwell, R.R., *Corrosion 12, 433t (1956)*.
27. Koehler, E.L. and Canonico, C.M., *Corrosion 13, 227t (1957)*.

28. Kamm, G.G. and Willey, A.R., *Corrosion 17, 77t (1961).*
29. Kamm, G.G., Willey, A.R., Beese, R.E. and Krickl, J.L., *Corrosion 17, 84t (1961).*
30. Kamm, G.G., Willey, A.R. and Beese, R.E., *Mater. Prot. 3, 70 (1964).*
31. Kamm, G.G. and Willey, A.R., "First International Congress on Metallic Corrosion," London, April, 1961. London Butterworths, 1962, p. 493.
32. Quintin, M. and Hagymos, G., *J. Chim. Phys. 61, 541 (1964).*
33. Sherlock, J.C. and Britton, S.C., *Br. Corros. J. 7, 180 (1972).*
34. Seiler, B.C., *Food Technol. 22, 1425 (1968).*
35. Asano, H. and Oyagi, Y., *Tetsu-to-Hagane 55, 184(1969).*
36. Willey, A.R., *Br. Corros. J. 7, 29 (1972).*
37. Vosti, D.C. and Hernandez, H.H., *Food Technology 17, 100 (1963).*
38. Uhlig, H.H., *Proc. Natl. Acad. Sci. 40(5), 276 (1954).*
39. Hoar, T.P. and Havenhand, D., *J. Iron and Steel Inst. 1, 239P (1936).*
40. Sherlock, J.C., *Br. Corros. J. 10, 144 (1975).*
41. Buck, W.R., Heyn, A.N.J. and Leidheiser, H., *J. Electrochem. Soc. 111, 386 (1964).*
42. Koehler, E.L., *Trans. Amer. Soc. Met. 44, 1076 (1952).*
43. Sherlock, J.C., Hancox, J.H. and Britton, S.C., *Br. Corros. J. 7, 227 (1972).*
44. Ebben, G.J. and Lawson, G.J., *J. Appl. Phys. 34, 1825 (1963).*

III. BARRIER COATINGS FOR CORROSION CONTROL

SOME FURTHER STUDIES ON POROSITY IN GOLD ELECTRODEPOSITS

R. J. Morrissey and A.M. Weisberg

Technic, Incorporated,
Providence, RI

RONALD MORRISSEY

ABSTRACT

X-ray diffraction data and plots of porosity versus deposit thickness are presented for bright, semibright, and unbrightened gold deposits plated from a citrate-phosphate electrolyte. The data indicate that at very low deposit thicknesses, both the crystallography and the porosity of the deposits are strongly dependent on the surface characteristics of the underlying substrate. At greater thicknesses, the slope of the porosity-thickness curve is controlled by parameters relevant to the deposit itself. Between these two regimes there is a sharp, well-marked transition region in which the porosity of the deposit falls extremely rapidly. It is shown that the thickness at which this sharp transition occurs varies with the deposit grain size. The form and position of the porosity-thickness plots are shown to be affected by the deposit grain size, the crystallographic orientation, and the ratio of nucleation rate to rate of grain growth, which in turn controls the average grain size of the deposit at any given thickness.

INTRODUCTION

In a previous paper[1] we showed that for bright gold electrodeposits on copper, both the covering power and the rate of pore closure with increasing thickness are related to the crystallographic orientation of the deposit. In the present work we examine the porosity versus thickness curves for bright, semibright, and unbrightened gold deposits plated from the same basic electrolyte system, and attempt to evaluate these in terms of the grain size and crystal orientation of the various deposits.

EXPERIMENTAL

The electrolyte employed for gold deposition in this work is a mildly acid citrate-phosphate system made up, per liter of solution, as follows:

123.6 g tripotassium citrate
11.2 g citric acid
44.9 g monopotassium phosphate
12.3 g potassium gold cyanide
(8.2 g gold metal)

The resulting solution is buffered at pH 5.7. Solutions were made up using reagent grade chemicals and deionized water, and were treated for 30 minutes with activated carbon at 60°C and filtered prior to addition of the gold salt. After makeup, the solutions were allowed to stand for at least 48 hours prior to plating.

Details of the procedure for preparing and plating Hull cell specimens have been given previously[1]. As in the previous work, our general approach has been to prepare Hull cell specimens from the bath of interest under various plating conditions, and to examine these by x-ray diffraction at selected values of the indicated current density. Specimens for the determination of porosity versus thickness curves were prepared by barrel plating various lots of small copper electrical terminals (AMP, Inc., No. 30697). At various times during each run, the plating was interrupted, the barrel withdrawn from the solution, and a number of plated specimens removed at random from the top of the barrel load. These were rinsed, marked, and set aside, and the plating operation was then resumed. A series of groups of specimens of various plating thicknesses was thus accumulated for each run. These were assembled into subgroups of five terminals each, and porosity values for each subgroup were determined by corrosion potential measurement in 0.1 NH_4Cl electrolyte[2]. Deposit thicknesses of the individual specimens were determined by beta-backscatter, and the average thickness was calculated for each subgroup.

Prior to plating, the surfaces of the copper terminals were prepared by degreasing in trichloroethylene, cathodic cleaning, rinsing, and bright-dipping at room temperature in a solution consisting, by volume, of

4 parts concentrated H_2SO_4
2 parts concentrated HNO_3
3 parts deionized water

with approximately 1 ml. concentrated HCl added per liter to control the etching rate. In order to assure as consistent a surface finish as possible, all of the specimens for a series of comparison plating runs were thus prepared in one batch, dried, and carefully mixed at random before being segregated into individual barrel loads. Except as otherwise noted, all plating operations were carried out at 60 ± 2°C, and at a current density of 1 mA/cm^2.

RESULTS AND DISCUSSION

Pulsed-Plating Experiments

The porosity versus thickness curve for the bath of interest under d.c. conditions at 62°C has been given previously[1] and is shown in Figure 1. The plot consists of two segments, with a sharp increase in the rate of pore closure occurring at thicknesses beyond approximately 0.5 micron. This transition in rate of pore closure was at first tentatively ascribed to a transition from (111) to (311) preferred orientation of the deposit with increasing thickness, which has been shown[3] to occur in this system. Such, however, has proven not to be the case. Extension of the data to greater plating thicknesses yields the porosity-thickness curve shown in Figure 2, in which *three* segments now appear. X-ray diffraction patterns were then run on specimens from each group of this series, and the results obtained are shown in Table II. Owing to the small size and complex geometry of these specimens, diffraction patterns obtained from them are less precise than those obtainable from Hull cell panels, but it is clear from Table II that the gold deposits obtained in this experiment show *no* clearly preferred

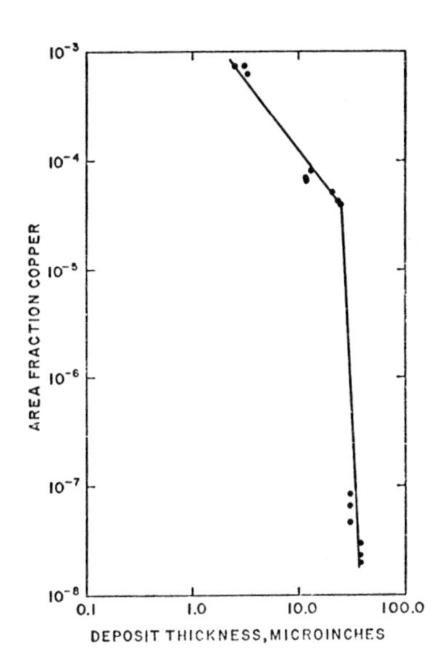

Fig. 1.—Porosity vs. thickness. Unbrightened gold on copper.

orientation, and there are no sharp crystallographic changes corresponding to the transitions in slope of the porosity versus thickness curve shown in Figure 2.

The multisegmented nature of the plots shown in Figures 1 and 2, and their apparent lack of correlation to crystallographic orientation, led to a reconsideration of factors which might affect the morphology and grain size of the deposits. It had been shown by Jernstedt[4] that periodic reversal of the plating current greatly reduced the grain size in copper electrodeposits from cyanide electrolytes. Hickling and Rothbaum[5] studied the effects of pulsed d.c. and periodically reversed current on the electrodeposition of silver. Following the introduction of practical circuitry for pulsed d.c. plating by Avila and Brown[6] in 1970, interest in this technique grew rapidly, particularly in the deposition of precious metals. Rehrig[7] reported that the porosity of unbrightened gold deposits plated with pulsed d.c. was markedly lower than that of d.c. plated deposits from the same bath. More recently, Rehrig, Leidheiser and Notis[8] studied the effects of varying pulse parameters on the crystallography and morphology of unbrightened gold deposits plated at constant average current density, and found an almost linear decrease in the maximum grain size of the deposit with increasing peak pulse current density, i.e., with decreasing duty cycle. The application of pulsed techniques to electrodeposition in general has been reviewed by Wan, Cheh, and Linford[9]; and to gold electrodeposition in particular by Raub and Knodler[10].

For the present study, Hull cell panels prepared as previously outlined[1]

TABLE I.—X-RAY DIFFRACTION DATA: HULL CELL PANELS. INTENSITY RATIOS NORMALIZED TO (111).

Conditions	Peak	Current Density, mA/cm^2				
		2.0	4.0	6.0	10.0	15.0
D.C., no additives	111	100	100	100	100	100
	200	4	2	3	8	15
	220	1	1	3	12	8
	311	7	7	12	20	39
Pulsed, 1 msec "on" 1 msec "off"	111	100	100	100	100	100
	200	3	2	4	6	6
	220	1	0	2	9	7
	311	6	8	10	28	24
Pulsed, 1 msec "on" 2 msec "off"	111	100	100	100	100	100
	200	3	1	1	2	5
	220	1	0	0	1	7
	311	6	4	8	14	18
Pulsed, 1 msec "on" 4 msec "off"	111	100	100	100	100	100
	200	7	2	1	0	0
	220	1	0	0	0	0
	311	6	9	5	5	5
Pulsed, 1 msec "on" 9 msec "off"	111	100	100	100	100	100
	200	5	2	1	1	0
	220	0	1	0	0	0
	311	6	8	6	7	4
Pulsed, 1 msec "on" 19 msec "off"	111	100	100	100	100	100
	200	12	4	2	1	1
	220	2	1	0	0	0
	311	9	14	13	14	11

TABLE I.—(continued)

Conditions	Peak	Current Density, mA/cm^2				
		2.0	4.0	6.0	10.0	15.0
D.C., 0.5 ppm Tl	111	100	100	100	100	100
	200	57	49	46	14	6
	220	10	18	13	238	24
	311	30	23	18	21	21
D.C., 2.0 ppm Pb	111	100	100	100	100	100
	200	108	102	81	43	26
	220	10	8	15	34	15
	311	20	20	20	14	15
D.C., 5.0 ppm Bi	111	100	100	100	100	100
	200	1	0	0	10	24
	220	0	0	0	2	9
	311	1	0	0	7	19

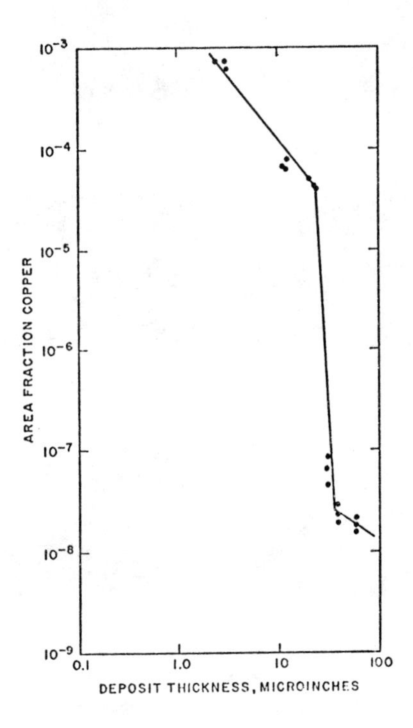

Fig. 2.—Porosity vs. thickness. Unbrightened gold on copper (extended data).

were plated using a Nova Tran Model DP 20-30-100 pulse plating power supply. An "on" time of 1 millisecond was employed for all experiments. For the several panels, the "off" time was set to zero (d.c.), 1, 2, 4, 9, and 19 milliseconds, the peak pulse current being adjusted in each case so as to obtain an average current of 0.5 ampere. All panels were plated for 10 minutes at 60 ± 2°C. A photograph of the panels thus obtained is shown in Figure 3. These deposits were then analyzed by x-ray diffraction, with results as shown in Table I.

The diffraction data shown in Table I are in general agreement with those obtained by Rehrig, Leidheiser and Notis[8]. An increase in the peak pulse current density produces an increase in the degree of (111) preferred orientation of the deposit at pulse duty cycles (ratio of pulse "on" time to pulse interval) down to 10%. At 5% duty cycle, a transition is observed at which the occurrence of higher-order peaks again begins to increase. Rehrig et al[8] had noted such a transition leading to an increase in (100) preference. In the present case the effect is less marked, and the higher-order peak most strongly observed is the (311).

Porosity vs. thickness curves were then generated by barrel-plating various lots of copper terminals, in each case at an average current density of 1 mA/cm^2, with pulsed current, the pulse parameters for the various lots

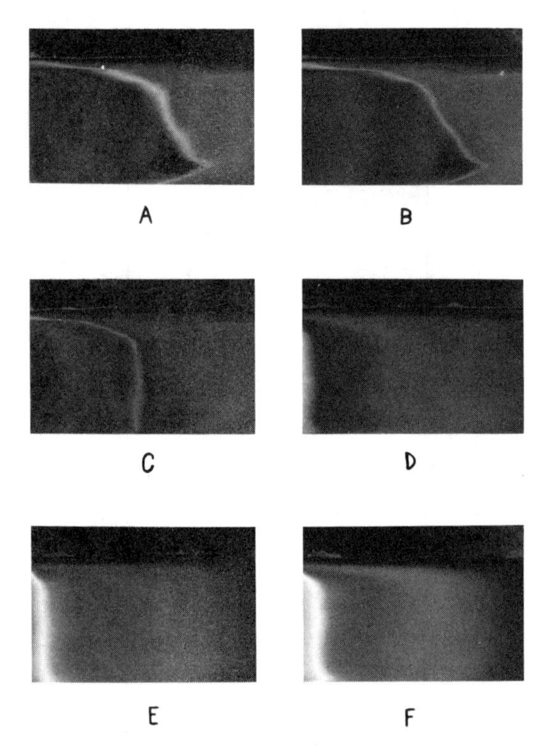

Fig. 3.—Hull cell panels. Pulse-plated gold. Pulse "on" time 1 millisecond. Pulse "off" time as follows: A - zero (d.c.); B - 1 msec.; C - 2 msec.; D - 4 msec.; E - 9 msec.; F - 19 msec.

being set to values corresponding to those used for preparing the Hull cell panels. The porosity-thickness curves thus obtained are shown in Figure 4. The curves in Figure 4 are all of the same form, but are displaced downwards and to the left as the duty cycle is decreased.

X-ray diffraction patterns were then run of specimens from each group of the barrel lot plated at 5% duty cycle (1 msec "on", 19 msec "off"), and these are shown in Table II. Contrary to what would be expected on the basis of the Hull cell results, the barrel-plated deposits are observed to be highly random in orientation and essentially comparable to those obtained at d.c. This immediately suggests a very strong effect of the substrate surface morphology on the characteristics of the gold deposits. Rolled copper exhibits a strong (110) texture, which was confirmed by x-ray measurements on unplated copper terminals, but does not appear as an epitaxial effect in even the thinnest gold deposits examined. It thus becomes apparent that the bright dip employed in preparing the copper terminals produces an essentially random *surface* texture on the copper, which texture is carried over into the gold deposits, whether produced under pulsed or d.c. conditions in this system. On the basis of this evidence, we postulate that the displacement observed in the porosity-thickness curves of the pulse-plated deposits is due solely to reduction of the deposit grain size.

TABLE II.—X-RAY DIFFRACTION DATA: BARREL-PLATED SPECIMENS. INTENSITY RATIOS NORMALIZED TO (111).

Conditions	Peak	Plating time, minutes						
		1.0	2.0	4.0	8.0	16.0	24.0	32.0
D.C., no additives	111		100	100	100	100	100	100
	200		50	114	121	128	152	113
	220		40	55	67	47	38	11
	311		10	105	52	147	59	45
Pulsed, 1 msec "on" 19 msec "off"	111		100	100	100	100	100	100
	200		60	67	121	74	88	77
	220		80	58	121	60	59	47
	311		40	58	79	67	69	58
D.C., 0.5 ppm Tl	111		100	100	100	100	100	100
	200		89	125	120	150	118	
	220		67	125	97	117	103	
	311		56	108	87	133	40	
D.C., cobalt-brightened	111	n/a	n/a	100	100	100	100	100
	200	n/a	n/a	46	29	20	18	20
	220	n/a	n/a	...	13	12	10	11
	311	n/a	n/a	...	16	19	14	14

n/a - *not* analyzable due to thinness of deposit

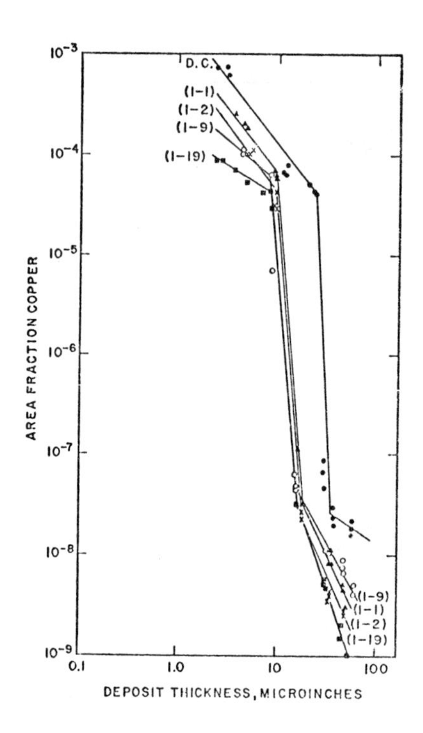

Fig. 4.—Porosity vs. thickness. Pulse-plated gold on copper.

Effects of Heavy-Metal Additives

The patent literature of the past decade contains numerous references [11,12,13,14] to the use of heavy metal ions which, when added in trace quantities (parts per million) to gold electroplating baths, induce a marked cathodic depolarization which extends the range of current densities over which smooth, fine-grained deposits can be obtained. Such systems have been studied at length by McIntyre and Peck[15] who showed that in slightly alkaline phosphate electrolytes the most effective additives comprise the family Hg, Tl, Pb, Bi, which lie immediately adjacent to gold in the periodic table, and which exhibit a strong tendency to form an adsorbed monolayer on gold and platinum electrodes at potentials positive to those at which their cathodic deposition as bulk metals would begin, i.e., at underpotentials. In a subsequent paper, McIntyre, Nakahara and Peck[16] examined the morphological characteristics of gold deposits obtained by the use of such additions, and observed that the deposits were of very fine and highly uniform grain size. It was suggested that adsorbed heavy metal species on the cathode surface could act as nucleation centers for gold, thus accounting for the reduction in grain size.

Photographs of Hull cell panels obtained from the citrate-phosphate electrolyte at pH 5.7 with additions of 0.5 ppm Tl (as Tl_2SO_4), 1.5 ppm Hg (as $Hg(NO_3)_2$), 2.0 ppm Pb (as $Pb(OOCCH_3)_2$), and 5.0 ppm Bi (as $Bi(NO_3)_3$) are shown in Figure 5. The brightening effectiveness of the various additives in this system is observed to be in the order Tl > Pb > Bi >> Hg, which is the inverse of the order of their electron work functions, as shown in Table III. This correlation had been noted previously by McIntyre et al[16].

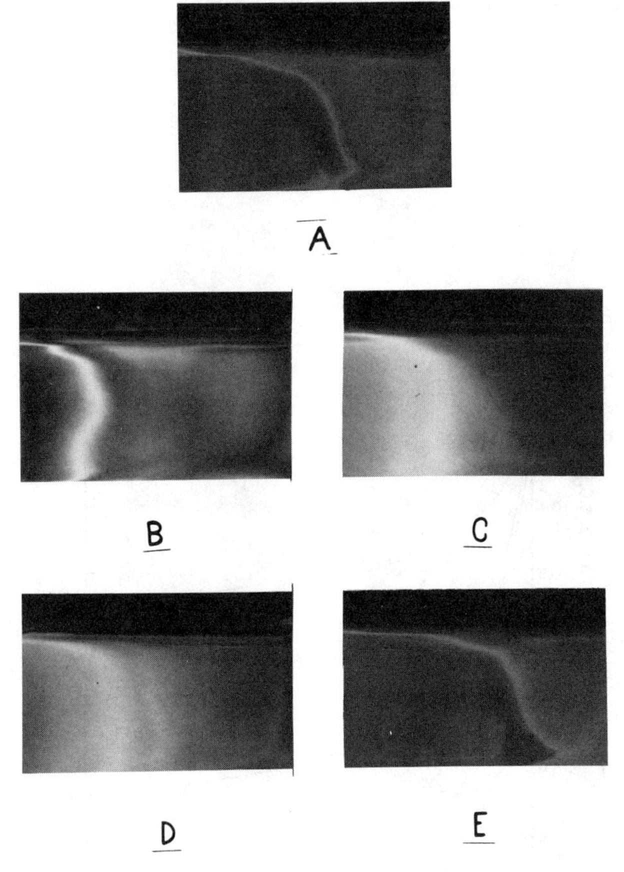

Fig. 5.—Hull cell panels. Heavy metal-brightened golds. A - no additives; B - 0.5 ppm Tl; C - 2.0 ppm Pb; D - 5.0 ppm Bi; E - 2.0 ppm Hg.

One may note from Table III that the work function of mercury is very close to that of gold itself. The addition of 1-2 ppm of mercurous ion to the citrate-phosphate system at pH 5.7 is observed to cause no significant effect either as a brightener or as a depolarizer, although McIntyre and Peck were able to observe both effects in a phosphate electrolyte at pH 8.0.

Figure 6 shows a series of cathodic polarization curves of 99.99% gold in the citrate-phosphate electrolyte at pH 5.7, both unmodified (curve "O") and with additions of 0.5 ppm Tl (curve "T"), 2.0 ppm Pb (curve "L"), and 5.0 ppm Bi (curve "B"). The addition of 2 ppm Hg yields a curve substantially identical to curve "O". These curves were obtained under unstirred conditions at 60 ± 2°C in an apparatus which has been described previously[18]. Sweep rate in each case was 50 millivolts per minute. Again,

TABLE III.—WORK FUNCTIONS OF VARIOUS ELEMENTS[17]

Element	Work Function, eV
Au	4.82
Hg	4.53
Bi	4.25
Pb	3.97
Tl	3.68

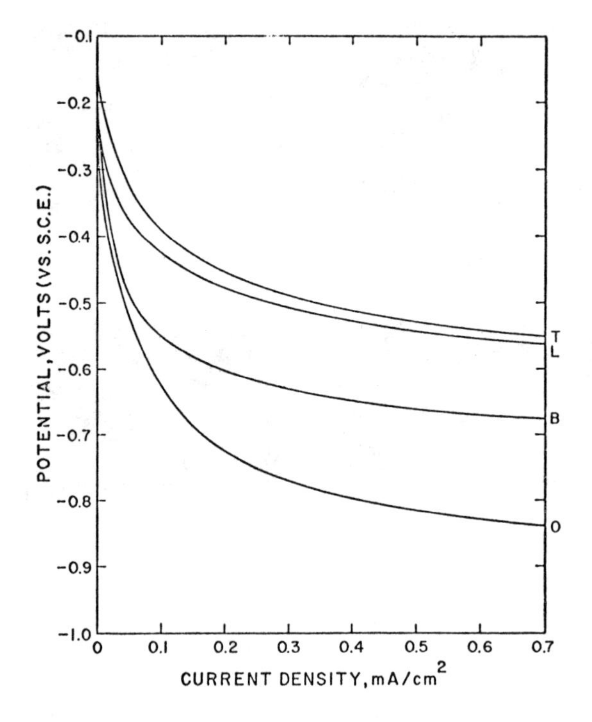

Fig. 6.—Cathodic polarization curves of 99.99% gold in citrate-phosphate plating bath. O - no additives; T - 0.5 ppm Tl; L - 2.0 ppm Pb; B - 5.0 ppm Bi.

the order of depolarization observed is Tl > Pb > Bi >> Hg.

X-ray diffraction data for Hull cell specimens prepared with additions of 0.5 ppm Tl, 2.0 ppm Pb, and 5.0 ppm Bi are shown in Table I. The addition of Tl is shown to result in a highly randomly oriented deposit, whereas the addition of Pb yields a (200) orientation, and, at low current densities, the deposit from the bath with added Bi is almost a pure (111).

These data would appear to indicate that the mode of adsorption at the cathode interface varies considerably among the elements of this group.

Porosity-thickness curves were then generated from specimens barrel-plated in the citrate-phosphate electrolyte with additions of 0.5 ppm Tl, 2.0 ppm Pb and 5.0 ppm Bi, and these, together with the d.c. curve obtained from the unmodified bath, are shown in Figure 7. Here we observe, in the case of additions of Tl and Pb, an apparent deviation from the z-shaped form of the porosity-thickness curves seen previously. It is highly probable that in barrel plating experiments of the type described here the earliest stages of the porosity-thickness curve may be passed through before the barrel can be stopped for withdrawal of the first specimens. It is difficult to stop the barrel very early in the plating run because at very low plating thicknesses (0.025 micron or less) the thickness distribution among the various specimens tends to be quite poor.

In any event, it is evident that for the case of the Tl and Pb additions the rapidly-falling portions of the porosity-thickness curves lie well to the left of that for specimens from the unmodified bath; and this we take to indicate that, at least at the early stages of deposition, the grain sizes of deposits from baths containing such additives are indeed very fine. It might also be noted that at thicknesses above about 0.25 micron, deposits pulse-plated from the unmodified citrate-phosphate bath achieve somewhat lower porosities than those plated at d.c. from the baths containing added heavy metals. On the basis of the porosity-thickness curves,

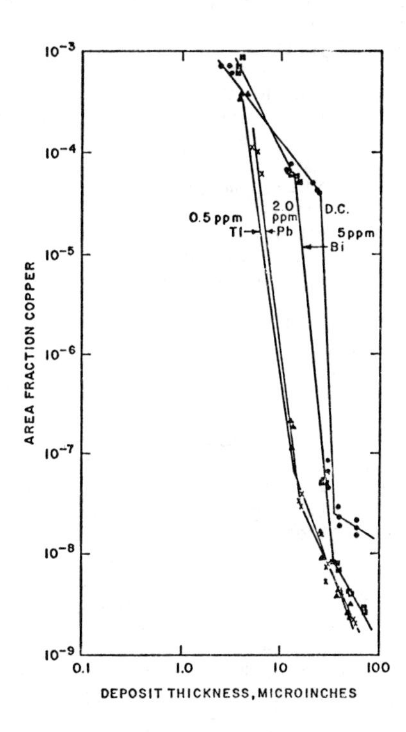

Fig. 7.—Porosity vs. thickness. Heavy metal-brightened gold on copper.

it would appear that pulsed deposition is more effective than the heavy metal additives at maintaining fine grain size in the deposits at thicknesses ordinarily encountered, although at very low thicknesses, the grain size appears to be finer in deposits placed from baths containing Tl and Pb. Of the heavy metal additives investigated, the order of effectiveness in porosity reduction is Tl > Pb > Bi, which is the same as their order of effectiveness as brighteners and as depolarizers.

Effects of Striking and Cobalt-Brightening

The porosity and x-ray data presented thus far lead to the formulation of a working hypothesis which may be stated as follows: In the early stages of deposition, both the porosity and crystal orientation of a gold electrodeposit are very strongly dependent on the surface condition of the substrate. This condition persists up to a limiting deposit thickness, after which there ensues a very rapid fall in porosity and a transition to a second regime in which the slope of the porosity-thickness curve is determined by parameters relevant to the deposit itself. In the case of bright gold deposits on copper, the dominant such parameter has been shown to be the deposit crystal orientation. As the grain size of the deposit is reduced, the transition from substrate-dominated to deposit-dominated porosity-thickness behavior occurs at progressively lower deposit thicknesses.

One stratagem commonly employed to obtain an extremely fine-grained deposit at the substrate interface is the use of a strike bath. On the basis of the working hypothesis expressed above, one would expect that the porosity-thickness curve of a deposit struck prior to plating should be of a form similar to that obtained by the use of heavy-metal additives, but displaced farther downwards and to the left from the curve of deposits obtained from the unmodified bath. In order to test this hypothesis, a barrel-plating experiment was conducted in which the specimens were struck for 90 seconds at 3 mA/cm^2 in a phosphate-citrate strike bath consisting per liter, of

89.8 g monopotassium phosphate
17.4 g tripotassium citrate
12.5 g citric acid
3.0 g potassium gold cyanide
(2.0 g gold metal)

The strike bath was operated at pH 4.3 and a temperature of 50°C. Following the strike step, the specimens were rinsed and plating was then continued at 1 mA/cm^2 in the unmodified citrate-phosphate bath as described previously. The porosity-thickness curve obtained from specimens produced in this experiment is shown in Figure 8. The form and the position of the porosity-thickness curve are as predicted on the basis of the working hypothesis.

Finally, it may be recalled that the porosity-thickness curves originally reported by us[1] for cobalt-brightened gold deposits on copper were of a simple linear form when plotted on logarithmic coordinates. In view of the data presented above, it would be expected that such plots should actually prove to be segmented if examined at sufficiently low deposit thicknesses. Accordingly, the citrate-phosphate electrolyte was modified by reducing the pH to 4.5 with citric acid and adding 1.06 g cobalt per liter in the form of a porprietary chelate. Specimens were then barrel-plated in this modified electrolyte at 2 mA/cm^2 at 33°C, the first interruption of the operation taking place after only 60 seconds of plating.

The porosity-thickness curve of specimens obtained in this experiment is shown in Figure 9. As expected, the plot is segmented, and owing to the

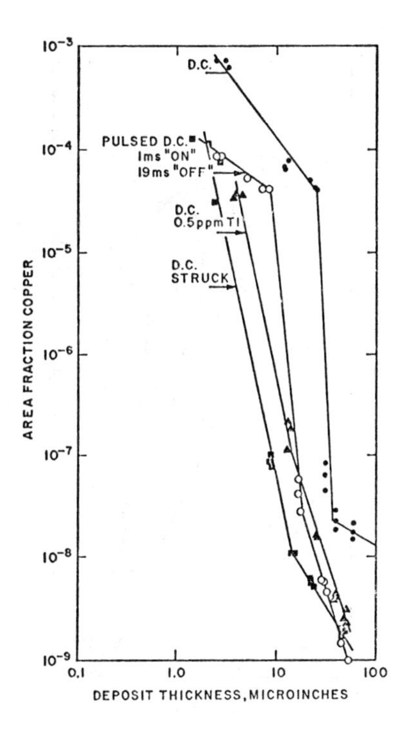

Fig. 8.—Porosity vs. thickness. Comparison of various conditions.

extremely fine grain size of the cobalt-brightened deposits[19], it is displaced far downwards and to the left from that of specimens from the unmodified bath. X-ray diffraction data for the cobalt-brightened deposits are shown in Table II. As noted previously[1], cobalt-brightened gold deposits rapidly acquire a strong (111) preferred orientation and hold this strong (111) preference even when plated to considerable thicknesses.

REMARKS

Grain size determinations were not performed on specimens plated in this work. The small size, complex geometry, and micro-roughened surface finish of the specimens used for barrel plating in this work render virtually impossible the removal of thin foils for examination by transmission electron microscopy. Grain size data have been cited[8,16,19] by other workers employing similar electrolyte and brightener systems. From the standpoint of the present porosity-thickness results, such data are likely to be approximate and useful primarily for establishing a rough order or decreasing grain size for the various deposits. The marked discrepancy observed in the x-ray diffraction data for Hull cell and barrel specimens pulse-plated in this work indicates clearly that even under closely controlled conditions

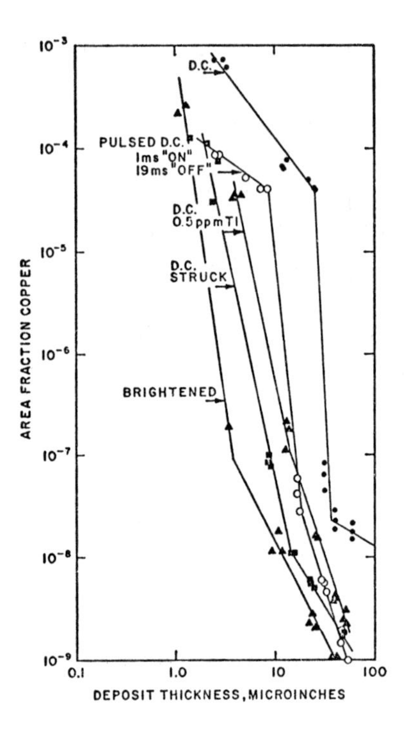

Fig. 9.—Porosity vs. thickness. Comparison of various conditions.

the characteristics of deposits plated onto smooth, well-prepared surfaces are not necessarily commensurate with those obtained on an industrial finish.

The porosity-thickness curves presented herein uniformly support the model or working hypothesis stated previously:

1. that at low deposit thicknesses both the porosity and crystallographic characteristics of gold deposits on copper are very strongly affected by the characteristics of the substrate surface,

2. that at greater thicknesses the slope of the porosity-thickness curve becomes determined by the characteristics of the deposit itself,

3. that between these two regimes there exists a sharp, well-marked transition region in which the deposit porosity falls with extreme rapidity, and

4. that the thickness at which this rapid transition takes place varies with the grain size of the deposit.

The nature of the processes leading to the region of very rapid porosity transition are of considerable interest both fundamentally and practically. At present, these are yet to be specified. Porosity-thickness curves showing essentially identical features had been shown previously by Clarke and Subramanian[20]. At one point it was feared that the evidence for a rapid transition might actually represent an artifact of the electrolytic porosity test, i.e., a rapid dewetting of test electrolyte from the pore channels when these reached a certain minimum size. It can, however, be seen from Figure 8 that in the case of specimens struck prior to plating we were able to obtain experimental data at several points along the transition, as have Clarke and Subramanian. This evidence renders the dewetting hypothesis highly improbable, and the existence of the rapid transition is evidently real. Elucidation of the processes occurring in this region will probably require considerable additional experimentation.

Finally, it might be asked how the present data agree with the well known model for pore closure advanced by Clarke[21]. It is evident from the porosity-thickness plots shown here that at low deposit thicknesses, i.e., up to and including the region of rapid transition, the porosities obtained are not easily expressible in terms of a simple analytic function of deposit thickness. At thicknesses beyond the transition region, agreement to the model is reasonable. From the slopes of the plots shown in Figure 9, we obtain a range of functions from $P \alpha t^{-0.6}$ to $P \alpha t^{-3.3}$, all of which correspond to a generalized single pore the walls of which increase in steepness with increasing deposit thickness, i.e., a volcano pore.

Empirically, we have thus far identified two significant factors, grain size and crystallographic orientation, which are contributory to determining the position and slope of the porosity-thickness curves. A third factor may be inferred from a comparison of the results obtained from the pulse-plated and heavy metal brightened specimens. On the basis of the porosity-thickness curves, it would appear that at low deposit thicknesses the grain sizes of deposits from baths containing added Tl and Pb are smaller than those of deposits pulse-plated from the identical electrolyte without additives, yet at thicknesses above about 0.25 micron the latter achieve porosity values which are equal or slightly superior. It is well known[22] that during outward growth of an electrodeposit the processes of agglomeration and growth of existing crystallites and nucleation of fresh crystallites occur simultaneously. From these facts, it can be inferred that the ratio of nucleation rate to growth rate is highly significant in determining the average grain size of the deposit, and hence the form and position of the porosity-thickness curves.

ACKNOWLEDGEMENT

The authors are indebted to Professor Henry Leidheiser, Jr., of the Center for Surface and Coatings Research at Lehigh University, Bethlehem, Pa., for performing the x-ray diffraction measurements in this and previous work.

REFERENCES

1. Morrissey, R.J. and Weisberg, A.M., *Trans. Inst. Met. Finishing 53, 9 (1975)*.
2. Morrissey, R.J., *J. Electrochem. Soc. 117, 742 (1970)*.
3. Craig, S.E.Jr., Harr, R.E., Henry, J. and Turner, P., *J. Electrochem. Soc. 117, 1450 (1970)*.
4. Jernstedt, G.W., *Proc. Am. Electroplaters' Soc. 36, 63 (1949); 37, 151 (1950)*.
5. Hickling, A. and Rothbaum, H.P., *Trans. Inst. Met. Finishing 34, 53 (1957)*.
6. Avila, A.J. and Brown, M.J., *Plating 57, 1105 (1970)*.
7. Rehrig, D.L., *Plating 61, 43 (1974)*.
8. Rehrig, D.L., Leidheiser, H.Jr., and Notis, M., *Plating and Surface Finishing 64, 40 (1977)*.
9. Wan, C.C., Cheh, H.Y. and Linford, H.B., *Plating 61, 559 (1974)*.
10. Raub, Ch.J. and Knodler, A., *Gold Bulletin 10, 38 (1977)*.
11. Duva, R. and Simonian, A., U.S. Pat. 3,562,120 (1971).
12. Smith, P.T. and Fletcher, A., U.S. Pat. 3,644,184 (1972).
13. Reinheimer, H.A., U.S. Pat. 3,833,487 (1974).
14. Winters, E.D., U.S. Pat. 3,873,428 (1975).
15. McIntyre, J.D.E. and Peck, W.F.Jr., *J. Electrochem. Soc. 123, 1800 (1976)*.
16. McIntyre, J.D.E., Nakahara, S. and Peck, W.F.Jr., presented at the Atlanta, Ga., Meeting of the Electrochemical Society, October 1977. *Cf. J. Electrochem. Soc. 124, 302C (1977)*.
17. Weast, R.C., Ed., "Handbook of Chemistry and Physics," 56th edition, CRC Press, Cleveland, O., 1975, pp. E81-E82. Values are for polycrystalline materials.
18. Morrissey, R.J., *J. Electrochem. Soc. 119, 446 (1972)*.
19. Cleghorn, W.H., Crossley, J.A., Lodge, K.J. and Gnanasekaran, K.S.A., *Trans. Inst. Met. Finishing 50, 73 (1972)*.
20. Clarke, M. and Subramanian, R., *Trans. Inst. Met. Finishing 52, 73 (1974)*.
21. Clarke, M., *Trans. Inst. Met. Finishing 51, 150 (1973)*.
22. Fischer, H., *Plating 56, 1229 (1969)*.

CORROSION CONTROL WITH TIN NICKEL AND OTHER INTERMETALLICS

M. Antler

Bell Laboratories
Columbus, OH

MORTON ANTLER

ABSTRACT

Intermetallics are metal-metal compounds. They include solid materials with binary or multicomponent compositions, and both ordered and disordered structures. The first discovery of intermetallics was from corrosion studies when it was recognized that there is a discontinuity in the action of acids on alloys of copper and zinc at the equiatomic composition now called beta brass, CuZn. Although many intermetallics are more corrosion resistant than their components, this is not universally true. Considerable confusion also existed in the early days of metallurgical study, before x-ray crystallography, where phase diagrams developed from thermal analysis were sometimes incorrectly interpreted to signify the existence of certain intermetallic materials. Likewise, equiatomic SnNi, which is the subject of this paper, does not appear in the phase diagram of the metals (Hansen, 1958). Today, however, thousands of intermetallics are recognized. Other general properties of intermetallics are their extreme hardness and brittleness, consequences of directionality in their atomic bonding. The treatise edited by Westbrook (1967) should be referred to for a general review of the properties and structures of these materials.

Only a few intermetallics have been used industrially as coatings for corrosion control, with electrodeposited tin nickel being the most important. This material has also been the subject of much scientific study. Accordingly, this paper will primarily review the tin nickel system and then will consider a few related compounds that have special academic or industrial importance.

SYNTHESIS OF TIN NICKEL INTERMETALLICS

Ni_3Sn, Ni_3Sn_2, Ni_3Sn_4 According to the phase diagram, intermetallics formed between tin and nickel are Ni_3Sn, Ni_3Sn_2, and Ni_3Sn_4. Preparation of these compounds is, therefore, accomplished by vacuum melting stoichiometric amounts of tin and nickel, followed by cooling.

Coatings of Ni_3Sn and Ni_3Sn_2 have been made by a conventional pack diffusion process (Dean and Ennis, 1972). Nickel or nickel plated steel specimens were surrounded with a mixture of powdered Sn (40%), alumina (50%), and

NH_4Cl (10%) and heated in a reducing atmosphere at 700-900°C. Coatings 0.03-0.05 mm thick were produced at 700-800°C, and 0.20 mm thick at 900°C in about 7 hours. The coatings consisted of an outer layer of Ni_3Sn_2, and an inner one of Ni_3Sn. Further heat treatment resulted in thickening of the coating because of additional diffusion of tin into the substrate, with transformation of the surface material entirely to Ni_3Sn and a nickel-tin solid solution underneath. Corrosion studies at room temperature showed Ni_3Sn_2 coatings to be little degraded on prolonged contact with 5% HNO_3 or aerated 5% H_2SO_4, behavior comparable to that of SnNi electroplate. However, Ni_3Sn and nickel-tin solid solution were more readily attacked by HNO_3 than pure nickel.

SnNi Although there are early reports of the preparation of electrodeposits containing nickel and tin, the basis for the commercially successful process was work by Parkinson (1951) at the Tin Research Institute. Brenner (1965) reviewed the literature of baths for electroplating alloys of nickel and tin containing about 65 weight percent Sn (i.e., equiatomic SnNi), and Dutta and Clarke (1958) described compositions, structures (by X-ray diffraction) and other properties of deposits from some of these baths. The literature is replete with different bath formulations, and new systems or variations of the old are reported regularly (e.g., Enomoto and Nakagawa, 1978). SnNi is readily plated from the bath in Table I. Bath replenishment is accomplished by additions of anhydrous $SnCl_2$; the nickel comes from dissolution of the anode.

TABLE I.—BATH FOR PLATING SnNi

	Grams per Liter	
	Makeup	Control Limits
$SnCl_2$, anhydrous	49	
(Sn^{+2})	30	26-37.5
$NiCl_2 \cdot 6H_2O$	300	
Ni metal	73.5	60-83
NH_4HF_2	56	
Total F^-	37.5	34-45
(maintain as high or higher than Sn^{+2} + Sn^{+4})		
NH_4OH	As needed to pH = 2.0 - 2.5	
Temperature	68 ± 3°C	
Cathode current density	1 - 3A/dm^2	
Cathode efficiency	~100%	
Anodes	Nickel, Nylon bagged	
Anode current density	To 5A/dm^2	

(Lowenheim, 1974; with permission of Wiley-Interscience)

SnNi can be deposited in a wide range of stress, from 2500 kg/cm^2 tensile to 2800 kg/cm^2 compressive, according to bath temperature and chloride content of the solution (Clarke, 1961). Spontaneous cracking may occur when the stress exceeds 500 kg/cm^2 tensile. However, addition of 50 g/l of NH_4Cl to a fresh bath insures that the initial deposit will be in compressive stress. The chloride content is maintained at a desirable high level by the $SnCl_2$ replenisher. Undercoatings for SnNi are ordinarily not used, except that copper or bronze attenuate the corrosion of steel substrates if the SnNi has some porosity.

It is a characteristic of Parkinson's and of most alternative baths that they deposit single phase SnNi having a composition that closely approximates the stoichiometric value. However, at extreme plating conditions deposits with as much as 69 weight percent or as little as 57% tin can be obtained, also as single phase material. Recently, however, Antler, et al. (1976) obtained a bright, coherent deposit containing 65.1% tin which was multiphase SnNi, Ni_3Sn_4, and possibly Ni_3Sn_2 from the Parkinson bath using a fluoride to total tin ratio of 0.8, which is less than the recommended value of 1.0 or more. The deposit was indistinguishable from the single phase material in its excellent corrosion resistance (see below). This was gratifying since there was virtually no prior knowledge of the corrosion resistance of intermetallics of tin and nickel other than the equiatomic compound, and this finding suggested that these other phases in the deposit were not undesirable for corrosion control.

Another way of obtaining equiatomic SnNi is by sputter deposition (Augis and Bennett, 1977). It had heretofore been believed that it could be obtained only by electroplating. However, by using a target of electroplated SnNi and molybdenum or copper substrates maintained below 84°C, the single phase material readily formed having a structure and thermal properties virtually indistinguishable from that of the electrodeposit.

CORROSION RESISTANCE OF SnNi ELECTRODEPOSITS

An exhaustive study of the chemical behavior of electrodeposited SnNi was reported by Parkinson, et al. (1951), who found it to be resistant to corrosion at room temperature by alkalis, neutral solutions, HNO_3, and by other acids at pH values greater than 1.2. It resisted tarnishing in severely polluted atmospheres, being unaffected by SO_2 or H_2S, although steam at 98°C caused visible tarnish to appear after 6 hours. The material was unaffected by salt spray tests or by exposure in marine environments. Additional corrosion studies were conducted by Lowenheim and Rowan (1952), Britton and Angles (1953) and Lowenheim, et al. (1958), particularly by comparison with nickel-chrome coating -- which at one time was thought could be replaced by SnNi. These authors concluded that SnNi behaved similarly to nickel-chrome at comparable thicknesses.

Another important finding, as expected, was that for corrodible substrates to be resistant to chemical attack, the SnNi coating must be free of porosity. The required thickness of SnNi depends on the roughness and on the other characteristics of the substrate surface, but ranges from 25-50 μm on steel at the most severe conditions of exposure, to only a few μm for service in relatively unpolluted environments.

More recently, Antler, et al. (1976,1977), studied the corrosion behavior to a number of reagents, primarily gaseous, and from exposures in the field to determine whether a galvanic couple persists at pores in thin (0.43 μm) electrodeposited gold on SnNi on a copper substrate. The objective was to test the prediction (Britton and Clarke, 1963) that the gold/SnNi couple would be nonreactive. It was found that the corrosion resistance of SnNi was not degraded by the gold plating.

Of considerable interest was an investigation of the kinetics of the development of passivity of SnNi (Antler, et al., 1977). It had been realized that its corrosion resistance was attributable to the presence of a protective surface oxide, but previous investigators generally did not consider that passivity may develop slowly - depending on the environment and the conditions of exposure of SnNi prior to it being subjected to chemical attack. This study was made with: (a) freshly plated SnNi, (b) SnNi that had been freed of its oxide by etching in 18% Hcl, (c) surfaces freshly exposed by fracturing SnNi electroforms, and (d) deposits that were abraded to remove surface oxide. Typical results are given in Figure 1 and

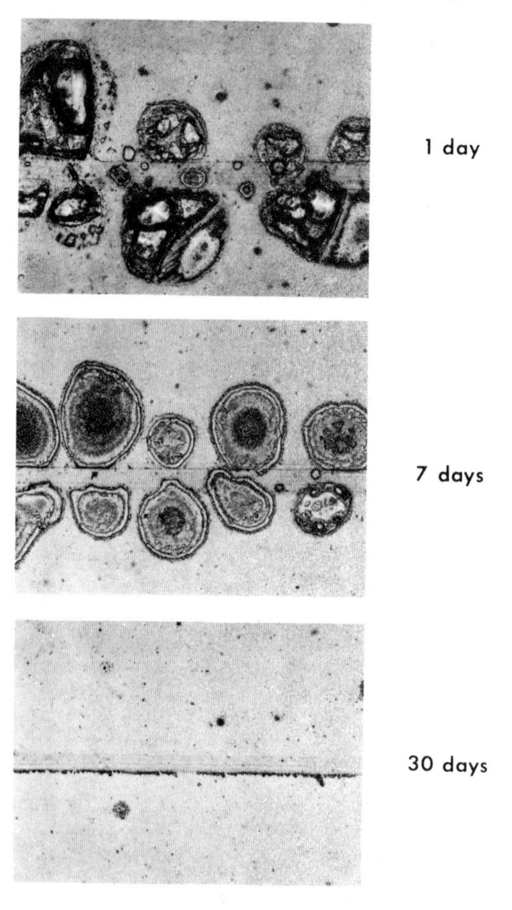

Fig. 1.—Surface of thick SnNi electrodeposit abraded with a sharp diamond stylus. One pass at 20g, then aged in air for the lengths of time indicated, followed by exposure to HNO_3 vapor for 1 hr. Severity of corrosion diminishes with aging time. Original magnification, 460X. (With permission of The Electrochemical Society, Inc.).

Table 2 which show that the passivity of SnNi increases with time. It is significant that there was no observable difference in corrosion behavior among the various SnNi surfaces, although, as will be shown later, it is now believed that the passive film on SnNi is not identical in all cases. A model of a plating of thin gold on SnNi on copper is proposed in Figure 2.

TABLE II.—CORROSION OF WORN SnNi PLATING[a,b]

Environment	Temp, °C	% RH	Time of Exposure	Aging Time in Air Before Exposure (days)		
				1	7	30
1. HNO_3 Vapor (Conc. sol.)	25		1 hr.	D	D	A
2. 10% SO_2	25	85	24 hrs.	D	D	A
3. Air	25	88	27 days	B	B	A
4. Sulfur Vapor	50	85	7 days	A	A	A
5. Our-of-doors, Sheltered, Columbus, OH	-10 to 15 (est.)	45 to 100 (est.)	30 days	A	A	A
				27	33	57
6. Electroplating plant; proxi mate to copper and gold plating tanks	25	50-70 (est.)	55 days	B	B	A
7. Factory; light assembly area	25	10-30 (est.)	55 days	A	A	A

[a]Worn at 20 and 30g by single pass abrasion with sharp diamond stylus. Results identical at both loads.

[b]The severity of corrosion is expressed on a scale of A-D, from low to high, based on visual and microscopic examination of the panels after exposure.

A. no corrosion at 460X;
B. corrosion observed at 460X;
C. corrosion observed at 70X;
D. corrosion evident with unaided eye.

(With permission of The Electrochemical Society, Inc.)

STRUCTURE AND THERMAL STABILITY OF SnNi

Determination of the structure of electrodeposited SnNi is due to Rooksby (1951) who, by Debye-Scherrer X-ray analysis, found it to be the hexagonal close packed nickel arsenide type ($B8_2$). Ni_3Sn_2 has the same

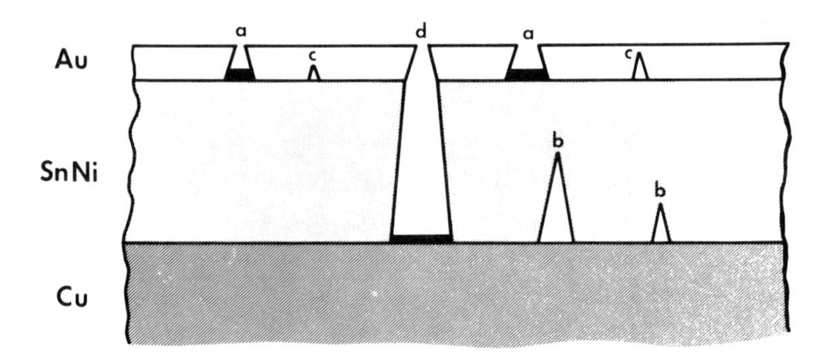

Fig. 2.—Model of a plating of thin gold on SnNi alloy on a copper substrate: (a) designates random asplated porosity in the gold to the underplate; (b) and (c) are internal pores in the SnNi and gold platings; (d) is the superposition of a pore in gold on one in SnNi to the substrate (adapted from Clarke and Chakrabarty). A film at the bottom of the (a) pores inhibits corrosion of the plating. This film may be similar to that on bare SnNi electrodeposits. The film at the bottom of pore (d) is nonprotective in aggressive atmospheres. There is no (or little) oxide at the original interface between the gold and SnNi. (a)-type porosity is presumed to exist when the gold plating is thin. (With permission of The Electrochemical Society, Inc.).

structure, and Rooksby interpreted this to mean that additional tin is incorporated into lattice sites of Ni_3Sn_2 which are unoccupied to give the equiatomic composition with changes in structural dimensions. Average crystallite size from the breadth of the x-ray lines was estimated to be very small, between 10-50 nm, ignoring effects which could be due also to a strained lattice. Augis and Bennett (1978) later estimated average crystallite size to be about 15 nm. Nickel arsenide structures are adaptable to changes in axial ratio with changes in composition (Robinson and Bever, 1967). Sputter deposited SnNi had essentially the same structure as the electrodeposited material and Ni_3Sn_2, and also was very fine grained (Augis and Bennett, 1977). Lo (1978b) proposed that SnNi is Ni_3Sn_2 with excess tin at the grain boundaries; his view is consistent with analytical studies of the passive film on the equiatomic material (see later).

Rooksby (1951) and subsequent investigators, e.g., Smart and Robbins (1960), Dutta and Clarke (1968), Wynne, et al (1972), showed that the equiatomic compound transformed irreversibly to Ni_3Sn_2 and Ni_3Sn_4 when it was heated. This finding, coupled with its nonappearance in the phase diagram, indicated that SnNi was metastable, but a contrary view (Dutta and Clarke, 1968) based on measurements of heats of formation was advanced that SnNi is in fact, a stable phase at the low temperatures, from 65-100°C, at which it is formed by electrodeposition. Diffusion below 300°C was reasoned to be so extremely slow in this system that a phase appearing by transformation near room temperature would have escaped detection in the studies which established the accepted phase diagram. However, a differential thermal analysis (DTA) and X-ray diffraction study (Bennett and Tompkins, 1977) found the transformation to be exothermic at 350-380°C, and the evolution of heat is generally interpreted to mean that a material is energy-rich and metastable at lower temperatures. Earlier, partial transformation was detected on heating for 150 hours at 250°C (Dutta and Clarke, 1968) and on longer heating at still lower temperatures (Hornig and Bohland, 1977). A sample of electroplated SnNi stored at room temperature for over 10 years showed no detectable transformation. Some of the uncertainty can be resolved from the work by Wynne, et al (1972), who found that the temperature for rapid transformation of the electrodeposited single phase material depended on its composition, coming at lower temperatures as the Sn/Ni ratio decreased from the equiatomic level.

Augis and Bennet (1978), using DTA and differential scanning calorimetry on electrodeposited and sputtered SnNi, made a kinetic study of the transformation and analyzed their data by the Avrami (1939,1940) technique. This method permits a prediction to be made of the fraction of material transformed from data available at different times and temperatures. Figure 3 summarizes their results. It is evident that the rate of transformation

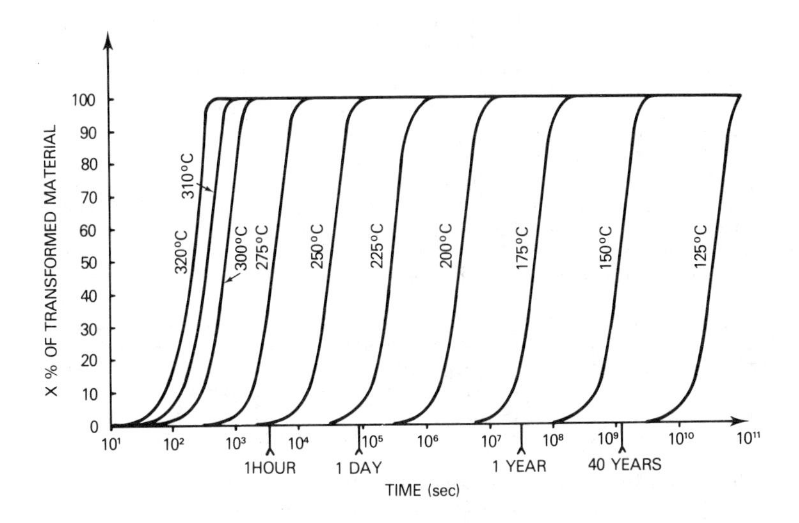

Fig. 3.—Transformation of electrodeposited SnNi to Ni_3Sn_4 and Ni_3Sn_2 as a function of aging time at various temperatures. (With permission of The Electrochemical Society, Inc.).

is sharply affected by temperature. The practical result is that the metastable alloy will not revert to the stable state at 125°C or less in times of interest. In terms of the corrosion resistance of SnNi coatings, this means that the small volume change which accompanies the transformation, and which can cause cracking of the deposit with possible exposure of the substrate, is unlikely to occur in any practical application to 125°C, or at higher temperatures for shorter times. For example, Rao, et al (1974), observed no change in a 6 μm deposit (overplated with 2.5 μm of gold) on copper in thermal shock at 260°C for 6 seconds in a wave soldering machine.

DIFFUSION OF SnNi

Limited atom mobility in a solid can be advantageous for corrosion resistance. Clarke and Britton (1963) have argued, for example, that low mobility will cause any oxide which forms to remain thin, with, consequently, less tendency to become highly stressed and cracked. Also, a structure with immobile atoms tends to oppose the passage of metal from the lattice into a potentially corrosive solution. Thus, the reaction product of any local action would be the final stable form, presumably a thin passive oxide. This reasoning applies, obviously, to film growth originating in cation, not anion, mobility.

There is experimental evidence that the atoms in SnNi do, indeed, have low mobility:

a. prolonged storage of SnNi plated on copper, brass, nickel and steel at 50°C gave no new layer, while pure tin on the metals formed a substantial compound layer at the interface (Thwaites, 1959). A study with an X-ray microporbe by Rao, et al (1974), of gold plated SnNi on copper heated at 165°C for 15 hours, led to the same conclusion. Golyzhnikov and Smirnova (1974) heated gold plated copper, with and without a SnNi underplate, at 220°C for one-half hour. By X-ray diffraction analysis Cu_3Au was found to be present at the interface between gold and copper, while no observable changes occurred when the underplate was used. Likewise, Leidheiser, et al (1978) found that electrodeposits on copper of tin nickel containing more than 50 atom percent tin resulted in the formation of a copper-tin intermetallic at the interface, whereas 1:1 SnNi did not give this interfacial layer. Analysis was by Mössbauer spectroscopy of deposits that were obtained from the Parkinson bath (similar to that in Table I) having a nickel content that was varied through a wide range.

b. the rates of increase of contact resistance of SnNi at 25°C and at 100°C were identical for at least 1000 hours; oxide film thicknesses differing by more than a few tenths of a nanometer have been readily observed by the contact resistance measuring method that was used (Antler, 1978). On prolonged aging at 100°C, however, contact resistance did eventually rise.

c. the diffusion of basis metal or underplate to the surface of a thin noble metal coating such as gold can lead to the formation of an insulating oxide with increase in contact resistance. A study of contact resistance showed that 3337 hours of heating in air at 150°C of a 0.5 μm pure gold deposit on 5 μm of SnNi underplate on copper did not result in a measurable change of contact resistance, i.e., <0.1 mΩ, (Antler, unpublished), while only 1320 hours of heating the same thickness of gold plate on copper at 125°C caused contact resistance to increase significantly (Antler, 1970).

It should not be surprising that atom mobility in SnNi is less than that in pure tin or nickel; this is a consequence of the directional bonding of intermetallics. Robinson and Bever (1967) stated that binding in nickel arsenide structures is mixed ionic, covalent, and metallic. The bulk electrical resistivity of SnNi, 140×10^{-6} ohm/cm (Tin Research Institute, unpublished), is 10-20 times greater than that of tin and nickel because its bonding is less metallic than that of the pure elements.

MECHANICAL PROPERTIES OF TIN NICKEL INTERMETALLICS

SnNi is hard, its level depending to some extent on the conditions of electroplating (Ramanathan, 1973). Bright deposits from fresh Parkinson baths have a value of 710 kg/mm^2, obtained with a Vickers diamond pyramid indenter. Ductility of the SnNi electrodeposit, on the other hand, is quite low, with a fracture strain of 0.25% for 16 μm and 0.38% for 1 μm thick coatings by a four point bend test (Lo, 1978a). Ramanathan (1957) showed that cracks developed around Vickers microhardness indentations at loads above 50g. A consequence of low ductility is that parts coated with SnNi cannot be bent without risk of cracking the deposit, and thereby loss of corrosion resistance. Wilson (1972) doped the Parkinson bath with a variety of additives claimed to improve the ductility of SnNi coatings obtained from them without success.

Qualitative observation of Ni_3Sn_2 and Ni_3Sn_4 show them to be brittle solids.

The wear resistance of platings is largely controlled by their hardness and ductility (Briscoe, et al, 1974), except in sliding at unlubricated conditons where adhesive wear may predominate (Antler and Drozdowicz, 1979). To determine the wear resistance of SnNi electrodeposit on copper at abrasive conditions, a sharp diamond was drawn across its surface at a series of loads. Failure was considered to occur when corrosion products formed during exposure of the abraded sample to an atmosphere which was aggressive to the substrate. Provided time is allowed for the repassivation of SnNi (since its protective oxide is also removed by scratching), such exposures are quite effective in revealing minute cracks in the deposit (Antler, et al, 1977), as shown in Figure 4. Alternately, reagents can be used which do not attack SnNi coatings devoid of their natural film, but which corrode the substrate. Testing by exposure to flowers of sulfur at 75% relative humidity (RH) is satisfactory for SnNi plated copper (Antler and Drozdowicz, 1978). Figure 5 illustrates wear cracks in SnNi oriented at right angles to the direction of sliding, a characteristic of fracture wear in brittle solids.

The anomalously high wear rate of electrodeposited SnNi reported by Raub and Müller (1967) in testing with emery-oil suspensions, is, therefore, explained by the low ductility of the coating. On the other hand, if pressure is low, for example, by sliding at light loads with conformal surfaces or with SnNi mated to soft opposing members, its rate of wear is small. SnNi was also reported to retain an oil film better than steel, which may make it a desirable coating for lightly loaded bearing applications where it is impractical to use other than small amounts of lubricants and to relubricate. The selection of SnNi on watch parts is based, in part, on this property (Robins, 1967).

It is evident that coatings of SnNi and other brittle materials resist corrosion only when brittle fracture is avoided. Ductility, like atomic mobility discussed earlier, can be related in part to atomic bonding of a material, covalent and ionic solids being in general less ductile than metals.

NATURE OF THE PASSIVE FILM ON TIN NICKEL INTERMETALLICS

The first significant discussion of the nature of the passive film on electroplated SnNi is due to Clarke and Britton (1963) who suggested that in aqueous solutions it might contain tin and nickel in equal amounts, perhaps as mixed oxides and hydroxides, and that its limiting thickness is due (as discussed earlier) to low atomic mobility in the crystal lattice. Further electrochemical studies of the passivity of SnNi were published by Clarke (1966) and Clarke and Elbourne (1968).

1 cm

Fig. 4.—Worn thick SnNi electrodeposit on copper after exposure to 10% SO_2 at 85% RH for 24 hrs. Panel hung vertically. Upper row, unlubricated wear. Lower row, worn after coating with mineral oil. Lubricant removed after wear before each aging interval and before SO_2 exposure. Loads for tracks 1, 5, 9: 100g; 2, 6, 10: 50g; 3, 7, 11: 30g; 4, 8, 12: 20g. SnNi becomes cracked at 50 and 100g, but only its natural oxide is removed at 20 and 30g. Groups of wear tracks were aged in air before SO_2 exposure as follows: 1-4, 1 day; 5-8, 7 days; 9-12, 30 days. All tracks display corrosion films except 11 and 12. Note that corrosion in tracks 3, 4, 7, and 8 is of SnNi; in tracks 1, 2, 5, and 6 is of both substrate copper; and SnNi; and in 9 and 10 is of copper only through fracture cracks in the SnNi. Original magnification, 2X. (With permission of The Electrochemical Society, Inc.).

Hoar (1970) proposed that an oxide film formed on alloys at ambient temperature, whether by reaction with atmospheric oxygen or by anodic passivation, must contain the different cations in the same proportion as they exist in the metal. He further suggested, based on potential measurements made earlier by Clarke and coworkers, that there could be several films on SnNi according to the solutions in which they were formed, and that they were, perhaps, nickelous stannite ($NiSnO_2$), nickelous stannate ($NiSnO_3$), and pernickelic stannate ($NiSnO_4$). Clarke and Elbourne (1971) agreed with this view and further suggested that the root of the stability of the passive film was tin, since a number of other tin intermetallics (see later) were found also to be corrosion resistant.

The first investigation of films on electrodeposited SnNi (Hoar, et al, 1973) by a physical technique, Auger spectroscopy, showed the air formed film to have a Sn/Ni atom ratio of 3/1 with O/ (Sn + Ni) = 0.7. In aqueous solution at pH = 1.5, a film formed anodically, without preliminary removal

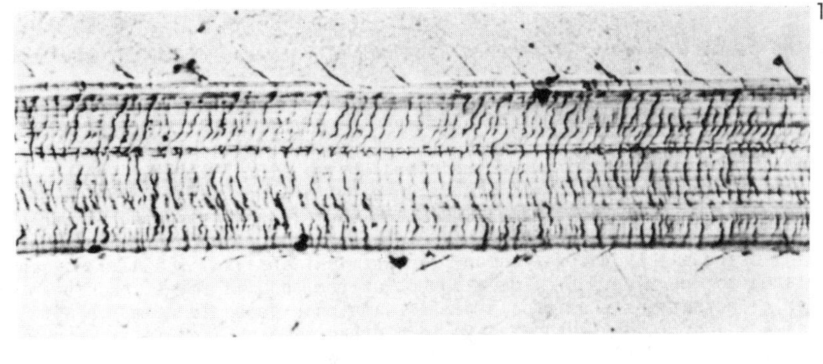

0.1 mm

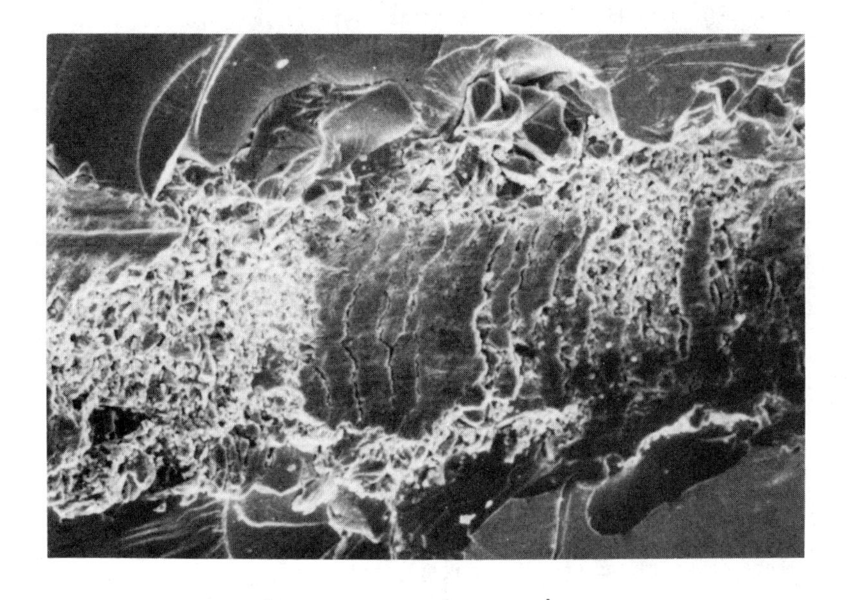

.01 mm

Fig. 5.—Wear track in thick SnNi electrodeposit on copper made with a diamond stylus at 100g shown at different magnifications.

of the air formed film, had a SnNi ratio of 3.1-4.5, with O/(Sn + Ni) = 1.3-1.6. Anodic films produced in alkaline solutions were tin deficient. This confirms the notion that there can be different types of passive films on SnNi depending on how they were formed; but the earlier view that the Sn/Ni ratio must be unity appears to be wrong. The authors also suggested that the passive oxide films were glassy polystannates containing water and -OH.

A series of papers (Tomkins and Bennett, 1976; Thomas III, 1977; Sharma and Thomas III, 1977; Thomas III and Sharma, 1977; Nelson, 1978; Schubert, 1978; and Tompkins, Wertheim and Sharma, 1978) followed which reported studies by a battery of surface analytical methods of air formed passive films on electrodeposited SnNi. Of these, the last paper is the most interesting since it was a study of the time dependence of film composition using electron spectroscopy for chemical analysis (ESCA), and can be related to the kinetics of development of the passive films described earlier. The authors examined as-plated samples exposed in room air with ages of 6 hours, 14 days and 62 days, and other samples aged in air for 2 minutes, 5, 8 and 52 days after prior etching in HCl to remove the "natural" oxide. They found as-plated films of any age to be an amorphous tin hydroxide with the same O/Sn ratio, about 3, and having the same thickness about 2 nm. The Ni/Sn ratio, however, increased from 0 to 0.11 on aging, with the nickel postulated to be an hydroxide. Another effect of aging seemed also be be the development of crystallinity in the tin compound. Films that grew on the HCl etched samples increased in thickness from 1.2 to 2 nm; the O/Sn ratio increased from 5.5 to 6.5/1; and the Ni/Sn ratio varied from 0-0.13. The nickel was considered to be an hydroxide, with the bound tin containing considerable water of hydration, at least during the early stages of oxidation. In summary, the films which existed on the as-plated samples that were reasonably young (14 days) were significantly different from the films which grew after the native oxide had been removed. All films contained primarily tin components. A small amount of nickel hydroxide formed as the samples aged. The change of the film on the etched sample could be described as growth; for the as-plated sample the major change on aging was in its chemical composition.

The large tin to nickel ratio in films formed on SnNi remains unexplained. Lo (1978b) suggested that this could be attributed to the presence of tin at grain boundaries in the SnNi. Since it is characteristic of both electrodeposited and sputtered SnNi to be very fine grained, this material may actually be Ni_3Sn_2 grains with enough tin at the grain boundaries to give the equiatomic composition by bulk analysis. This view assumes that the surface of SnNi as it is synthesized is tin rich. SnNi freed of its film by etching in HCl could reacquire a tin rich surface by diffusion of tin from its grain boundaries across the surface.

We can now reexamine some passivity data (Antler, et al, 1977) in the light of these findings. Although qualitative, there appeared to be no significant difference in the corrosion susceptibility of as-plated, fractured, HCl etched, and SnNi which had been abraded to remove its air formed film. These observations were for strongly acidic gaseous environments (HNO_3 vapor and 10% HCl at 85% RH). Passivity was time dependent, and with HCl etched samples increased for at least 60 days, although the severity of chemical attack and the time required for the development of passivity are obviously dependent on the reagent. The passive films which form on SnNi are not unique, but change according to the chemical environment in which they are formed, and this is in accord with traditional thinking (e.g., Fontana and Greene, 1978).

The contact resistance of SnNi electrodeposits can also be reviewed. Figure 6 (Antler, 1978) illustrates the time dependence of contact resistance on aging HCl etched SnNi in air. It appeared to increase for at least 10^3 - 10^4 hours, although at a different rate in dry air compared to air at 15-88% RH. Contact resistance depends, of course, on both film

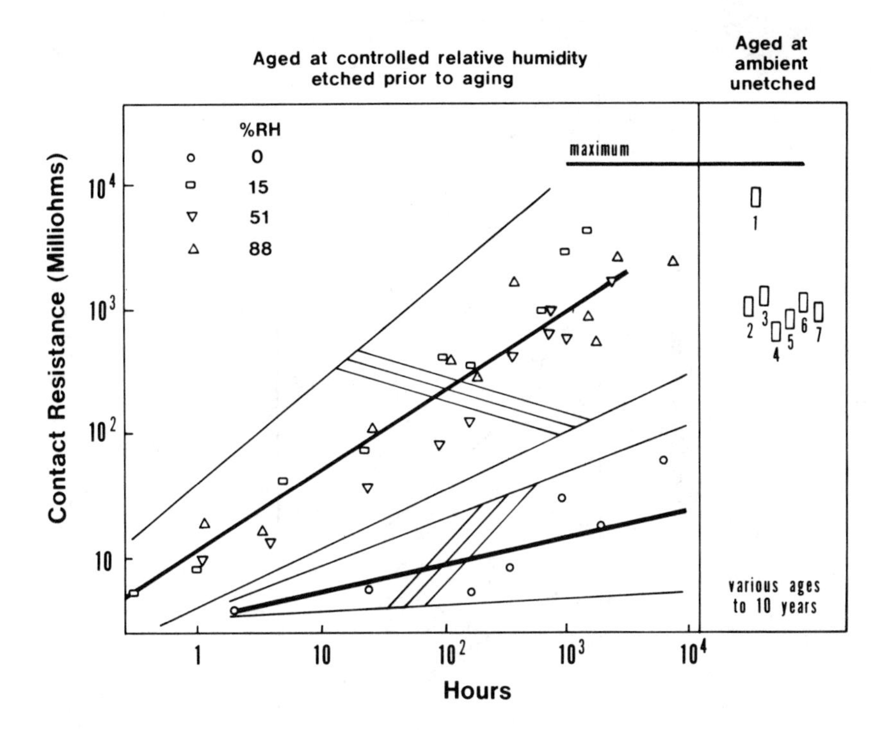

Fig. 6.—Contact resistance at 100g of smooth SnNi electrodeposits. Left: Aged in air at room temperature and controlled RH. Samples were 12 μm thick on copper. Etched in 18% HCl prior to aging. Right: Aged indoors at ambient uncontrolled RH. Samples 1, 2, 3, and 7 were coupons of 12-15 μm SnNi on copper or brass; 4, 5, and 6 were printed circuit boards with 5-12 μm SnNi on copper. Not etched. Sample 1, 3 months old; sample 2, 6 months old; sample 3, 8 months old; sample 4, 18 months old; sample 5, 18 months old; sample 6, 72 months old; and sample 7, 120 months old. Probed with a hemispherically ended gold rod. Each point is the median of 9 or more measurements. 95% of all measurements are within the bands, with a maximum value of 15Ω from 500 determinations. (With permission of The Electrochemical Society, Inc.).

thickness and its electronic properties, and does not provide unequivocal information on film structure. It is noteworthy, however, that the contact resistance increase occurs in the same time frame as that required for the development of passivity. The contact resistance rise on aging HCl etched SnNi is consistent with the finding of film growth (Tompkins, Wertheim, and Sharma, 1978). Growth can also explain the change of slope with time of the contact resistance-force characteristic curves, Figure 7, which have negative slopes that change with duration of air exposure, from -0.6 to -1.3 in 0.2 to 1512 hours. Contact resistance theory (Holm, 1967) teaches that for clean metallic contacts

$$CR \propto \left(\frac{1}{F}\right)^{-\frac{1}{2}}$$

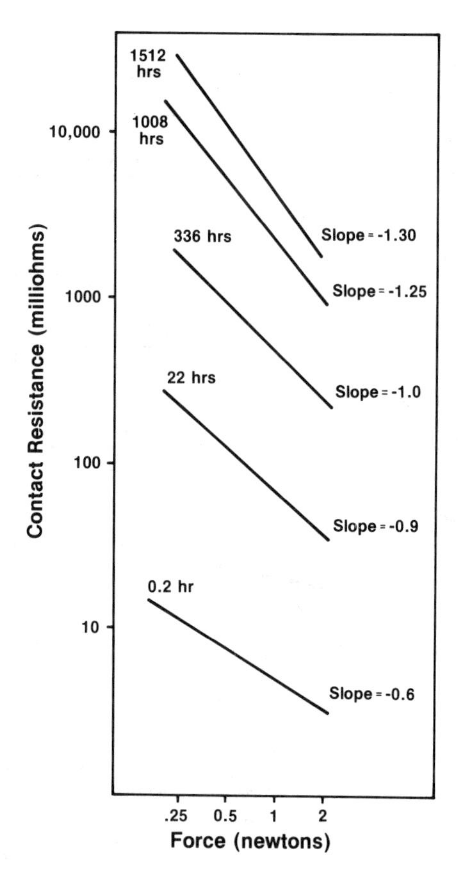

Fig. 7.—Contact resistance-force characteristic curves for smooth SnNi electrodeposit. Aged in air at room temperature and 15% RH. Deposit thickness 12 µm; plated on copper. Etched in 18% HCl prior to aging for times shown on the curves. Probed with a hemispherically ended gold rod, 1.6 mm radius. All determinations on same specimen.

where CR = contact resistance and F = force. When the surface is covered with a thin film, conduction occurs by electron tunnelling, and

$$CR \; \alpha \; F^{-1}$$

At larger thicknesses, film fracture may occur as force is increased, with the slope of the contact resistance curve becoming more negative than F^{-1}.

One additional surface study should be mentioned. Tompkins and Bennett (1977) determined the profiles of films on SnNi, Ni_3Sn_2, Ni_3Sn_2, and Ni_3Sn by Auger spectroscopy with depth profiling. The samples were sputter cleaned, then aged at atmospheric pressure in air. Samples were also heated at 156°C for 5 days. Although a detailed analysis was not made, the profiles were interpreted as being similar, and not to change on heating. An inference could then be made that the four intermetallics would be similarly corrosion resistant.

INDUSTRIAL USES OF SnNi COATINGS FOR CORROSION CONTROL

Equiatomic SnNi electrodeposit is used or has been proposed for a number of specialized applications. A few of them will be mentioned where corrosion resistance was the determining reason for its choice.

In the electronics industry, it is employed as a resist in printed circuit board manufacture (Wilson, 1972; Robinson, 1976) because of its resistance to etchants employed for copper removal. The deposit is ordinarily coated on the edge contacts of the board and may serve as-is as a contact material for other than low voltage applications, where the passive oxide would cause an undesirable increase in contact resistance. Water soluble active fluxes are necessary for good solderability. Where low voltage applications are involved, or where inactive solder fluxes are required, a flash of gold on the SnNi has been found desirable. Wear resistance of the contact finish may be adequate provided precautions mentioned above are followed.

There has been some interest in a finish of thin gold on SnNi underplate on connector contacts for service in aggressive atmospheres to replace more costly thick gold because of the corrosion resistance of the gold-SnNi composite finish at porosity which is inevitable in thin noble metal deposits (Antler, 1975; Antler, et al, 1976). For example, Schiff and Schnable (1978) found 1 μm of gold on 2 μm of SnNi to be equivalent in corrosion and wear resistance to 3 μm of gold on 2 μm of nickel underplate in a German DIN standard connector, where their testing included exposures for 14 days to an aggressive mixed gas environment involving 0.2 ppm each of H_2S, SO_2, and NO_2 at 75% RH. One μm of gold on 2 μm of nickel was inferior because of pore corrosion.

In other applications, SnNi is used for surface protection on a large variety of parts including: cases, gears and other parts of watches (Csuthy, 1978), thermostatic mixing valves, automotive ignition components, kitchenware and household accessories, electrical drying apparatus, and scientific equipment.

TIN COBALT INTERMETALLICS

Bright tin cobalt electrodeposits, which are SnCo (Clarke, et al, 1972) or mixtures of SnCo and Sn_2Co (Miyashita and Kurihara, 1970), have been obtained from a variety of baths. Although intermetallics with these compositions are in the phase diagram (Hansen, 1958), the crystallographic structures of the electrodeposits may be somewhat different from those of the metallurgical alloys. Like SnNi electrodeposit, tin cobalt platings are hard, brittle, and far more corrosion resistant than the parent metals (Tsuji and Ichikawa, 1971). Tin cobalt plating is commercially exploited (Yamada, et al, 1975) as a corrosion resistant decorative finish on nickel for appliance knobs where it is reported to have processing advantages and superior durability to nickel chrome. In these applications, the substrate is usually a molded plastic.

COPPER TIN INTERMETALLICS

Electrodeposited copper tin alloy containing 40-45% Sn, or speculum, resembles silver, but being considerably more tarnish resistant is preferred for many decorative indoor applications. It consists largely of the intermetallic, Cu_6Sn_5, with the nickel arsenide structure, but there is some disagreement as to what other phases are codeposited (Smart, et al, 1961). The plated material is different from the phase diagram at the indicated composition, but is kinetically stable to about 200°C. Clarke and Britton (1963) showed it to be corrosion resistant in many electrolytic solutions, and suggested that its passivity was due to a protective surface oxide. This intermetallic is brittle and hard (575 kg/mm^2 Vickers at the 40% Sn composition). Electroplated copper tin bronze containing 15% Sn is more corrosion resistant than speculum out-of-doors and is also more ductile. Brenner (1963b) reviewed the history of the development of copper tin plating solutions.

Cu_6Sn_5 occurs spontaneously with tin and tin lead coated copper (Clarke and Britton, 1963). On prolonged storage at 50°C Thwaites (1959) found a layer of Cu_6Sn_5 over a thinner underlayer of the intermetallic, Cu_3Sn. Kay and Mackay (1976) reported the kinetics of growth of C_6Sn_5 and Cu_3Sn from 70-170°C at the interfaces between tin and various tin-lead solders. Clarke and Elbourne (1971) found only Cu_3Sn on aging tin plated copper at 220°C. Both Cu_3Sn and Cu_6Sn_5 were reported by them to be highly corrosion resistant in aqueous electrolytes.

$FeSn_2$ INTERMETALLIC

$FeSn_2$ is a factor in the protection of steel by tin plate, as in the tin can. Leidheiser (1971) reviewed this system and pointed out that the effectiveness of a tin coating in reducing corrosion is dependent upon four factors: (1) by mechanically shielding the steel with a slowly corroding metal, (2) by sacrificially preventing attack of exposed steel, (3) because tin ions in solution sometimes inhibit the corrosion of steel, (4) from formation of $FeSn_2$, which is a barrier between the overcoat and steel substrate.

$FeSn_2$ forms spontaneously and rapidly when the two metals are in contact underneath electrolytic tin plate (Frankenthal and Loginow, 1960). Even when the plate is not continuous because of retraction of tin during flow brightening, the layer of $FeSn_2$ bridges the gaps at pores in the tin plate.

$FeSn_2$ is more inert than either iron or tin in all but the strongest oxidizing environments. It is believed to serve as a barrier which reduces greatly the surface area of the iron that needs to be sacrificially protected. The intermetallic is stable to 496°C, at which it decomposes into FeSn and Sn. Hoar (1970) suggested that its passive film is ferrous distannite ($FeSnO_2$) or distannate ($FeSn_2O_4$).

SUGGESTIONS FOR FURTHER RESEARCH

1. The limited ductility of equiatomic SnNi is a handicap to its use in mechanical applications. The ductility of electrodeposited non-equiatomic material, e.g., containing excess tin, has not been determined, and might be superior to that of usual electrodeposit.

2. There is only limited information on the time dependence of the development of passivity of intermetallics and of their relationship to the environments in which they are aged. Studies such as those cited in this paper for SnNi could be extended to other intermetallics. This would, in particular, include visual and gravimetric determinations after various exposures, and the use of analytic methods for surface examination. Determinations of contact resistance are a usful adjunct to such studies, and can be made relatively inexpensively.

3. Additional information of the kinetics of film growth and of composition of SnNi is needed, involving newly plated material,

surfaces obtained by fracture, samples freed of films by abrasion, by chemical etching, and by argon ion sputtering. ESCA appears to be an especially powerful analytical method. Ellipsometry has not yet been tried.

It would be especially interesting to follow film growth by analytical methods with SnNi in dry air, at about 50% RH, and above 90% RH. Freshly etched SnNi has been found to be corroded at 88% RH (Table II; see also Figure 6).

SnNi aged in air acquires a passive film which becomes progressively more crystalline (Tompkins, Wertheim, and Sharma, 1978). This is surprising, since traditional views (Hoar, 1970) hold that glassy films are desirable for corrosion resistance. Further studies of the structure of the passive films are warranted.

4. Potential new methods of synthesis of intermetallic coatings can be examined. Ion implantation (Hirvonen, 1978) and laser glazing (Weinman, et al., 1978) may be useful in this regard.

CONCLUSIONS

Intermetallics are readily formed by solid state diffusion between metal couples. Many are corrosion resistant and a few are important for corrosion control, notably in the tinned can. Methods of applying intermetallics on the surfaces of metals are limited, but alloy electrodeposition can be used in some cases. The most studied of these is equiatomic SnNi, which was not known prior to the development of a plating process by Parkinson because it is metastable and not in the phase diagram of the metals. It is, however, kinetically stable to at least 125°C for all practical purposes. Its corrosion resistance is due to a tin rich oxide or hydroxide which in air grows to 2-3 nm, and its passivity to some corrodents improves over 30-60 days. Intermetallics tend to be brittle, which may limit their wear resistance. There have been few systematic studies of the corrosion properties of intermetallics, of their electrochemistry in metal couples, and of coating techniques for them. Were more information available along these lines, it is likely that new and useful protective coatings for corrosion control would be developed.

REFERENCES

1. Antler, M., *Plating 57, 615 (1970).*
2. Antler, M., *I.E.E.E. Trans. on Parts, Hybrids, and Packaging, PHP-11, 216 (1975).*
3. Antler, M., *J. Electrochem. Soc. 125, 420 (1978).*
4. Antler, M. and Drozdowicz, M.H., *Plating and Surf. Fin. 65, 39 (1978).*
5. Antler, M. and Drozdowicz, M.H., *Bell Syst. Tech. J., (Feb. 1979)*; see also "Proc. Ninth International Conference on Electrical Contact Phenomena," Ill. Inst. of Tech., Chicago, Sept. 11-15, 1978, pp. 125-141.
6. Antler, M. and Drozdowicz, M.H., and Hornig, C.F., *J. Electrochem. Soc. 124, 1069 (1977).*
7. Antler, M., Feder, M., Hornig, C.F. and Bohland, J., *Plating and Surf. Fin. 63, 30 (1976).*
8. Augis, J.A. and Bennett, J.E., *J. Electrochem. Soc. 124, 1455 (1977).*
9. Augis, J.A. and Bennett, J.E., *J. Electrochem. Soc. 125, 330, 335 (1978).*
10. Avrami, M., *J. Chem. Phys. 7, 103 (1939).*
11. Avrami, M., *J. Chem. Phys. 8, 212 (1940).*
12. Bennett, J.E. and Tompkins, H.G., *J. Electrochem. Soc. 123, 999 (1976).*
13. Brenner, A., "Electrodeposition of Alloys," Academic Press, New York, 1963a, Vol. 1, p. 497.
14. Brenner, A., "Electrodeposition of Alloys," Academic Press, New York, 1963b, Vol. II, p. 315.
15. Briscoe, B.J., Eyre, T.S. and Gologan, V.F., *Wear 28, 271 (1974).*
16. Britton, S.C. and Angles, R.M., *Trans. Inst. Met. Fin. 29, 26 (1953).*
17. Clarke, M., *Trans. Inst. Met. Fin. 38, 5, 186 (1961).*
18. Clarke, M., *Corr. Sci. 6, 1 (1966).*
19. Clarke, M. and Britton, S.C., *Corr. Sci. 3, 207 (1963).*
20. Clarke, M. and Chakrabarty, A.M., *Trans. Inst. Met. Fin. 50, 11 (1972).*
21. Clarke, M. and Dutta, P.K., *J. Phys. D: Appl. Phys. 4, 1652 (1971).*
22. Clarke, M. and Elbourne, R.G.P., *Corr. Sci. 8, 29 (1968).*
23. Clarke, M. and Elbourne, R.G.P., *Electrochimica Acta. 16, 1949 (1971).*
24. Clarke, M. and Elbourne, R.G.P., and Mackay, C.A., *Trans. Inst. Met. Fin. 50, 160 (1972).*
25. Csuthy, B., *Prod. Fin. 42(9), 56 (1978).*
26. Cuthbertson, J.W., Parkinson, N., and Rooksby, H.P., *J. Electrochem. Soc. 100, 107 (1953).*
27. Dean, A.V. and Ennis, P.J., *J. Inst. Met. 100, 322 (1972).*
28. Dutta, P.K. and Clarke, M., *Trans. Inst. Met. Fin. 46, 20 (1968).*
29. Enomoto, H. and Nakagawa, A., *Met. Fin. 76, 34 (1978).*
30. Fontana, M.G. and Greene, N.D., "Corrosion Engineering," 2nd ed., McGraw-Hill, New York, 1978, p. 319.
31. Frankenthal, R.P. and Loginow, *J. Electrochem. Soc. 107, 920 (1960).*
32. Golyzhnikov, A.I. and Smirnova, E.M., *Protection of Metals 10(15), 561 (1974).*
33. Hansen, M., "Constitution of Binary Alloys," McGraw-Hill, New York, 1958.
34. Hirvonen, J.K., *J. Vac. Sci. Technol. 15(5), 1662 (1978).*
35. Hoar, T.P., *J. Electrochem. Soc. 117, 17C (1970).*
36. Hoar, T.P., Talerman, M. and Trad, E., *Nature Phys. Sci. 244,* 241 *(1973).*
37. Holm, R., "Electric Contacts," 4th ed., Springer-Verlag, New York, 1967, p. 40.
38. Hornig, C.F. and Bohland, J.F., *Scripta Met. 11, 301 (1977).*
39. Kay, P.J. and MacKay, C.A., *Trans. Inst. Met. Fin. 54, 68 (1976).*
40. Leidheiser, H. Jr., "The Corrosion of Copper, Tin, and Their Alloys," Wiley, New York, 1971, p. 306.

41. Leidheiser, H. Jr., Czáko-Nazy, I., Varsanyi, M.L. and Vértes, A., accepted for publication in *J. Electrochem. Soc.,(1978).*
42. Lo, C.C., *J. Electrochem. Soc. 125, 400 (1978a).*
43. Lo, C.C., private communication. (1978b).
44. Lowenheim, F.A.,"Modern Electroplating," 3rd ed., Wiley-Interscience, New York, 1974, p. 533.
45. Lowenheim, F.A. and Rowan, W.H., "46th Annual Tech. Proc. of Am. Electroplaters' Soc.," 1959, p. 205.
46. Lowenheim, F.A. and Sellers, W.W. and Carlin, F.Y., *J. Electrochem. Soc. 105, 338 (1958).*
47. Miyashita, H. and Kurihara, S., *J. Met. Fin. Soc. Japan 21, 79 (1970).*
48. Nelson, G.C., *J. Electrochem. Soc. 125, 403 (1978).*
49. Parkinson, N., *J. Electrodep. Tech. Soc. 27, 1 (1951).*
50. Parkinson, N. and Britton, S.C. and Angles, R.M., *Sheet Metal Industries, 957 (1951).*
51. Ramanathan, V.R., *Trans. Inst. Met. Fin. 34, 1 (1957).*
52. Ramanathan, V.R., *Met. Fin., Oct. 1973, p. 66.*
53. Rao, M.V., Mitchell, M. and Anderson, R.N., *Solid State Technol. 17, 47 (1974).*
54. Raub, E. and Müller, K.,"Fundamentals of Metal Deposition," Elsevier, New York, 1967, p. 168.
55. Robins, D.A., *Industrial Lubrication, June 1967.*
56. Robinson, G.T., *Products Fin. 38 (Nov. 1976).*
57. Robinson, P.M. and Bever, M.B.,chapt. 3 in "Intermetallic Compounds," J.H. Westbrook, Ed., Wiley, New York, 1967.
58. Rooksby, H.P., *J. Electrodep. Tech. Soc. 27, 153 (1951).*
59. Schiff, K.L. and Schnabl, R., "Proc. Ninth International Conf. on Electric Contact Phenomena," Ill. Inst. of Technology, Chicago, 1978, p. 309.
60. Schubert, R., *J. Electrochem. Soc. 125, 1215 (1978).*
61. Sharma, S.P. and Thomas, J.H. III, *Anal. Chem. 49, 987 (1977).*
62. Smart, R.F., Angles, R.M. and Robins, D.A., *J. Inst. Met., 349 (1961).*
63. Thomas, J.H. III, *J. Electrochem. Soc. 124, 677 (1977).*
64. Thomas, J.H. III and Sharma, S.P., *J. Vac. Sci. Technol. 14, 1168 (1977).*
65. Thwaites, C.H., *Trans. Inst. Met. Fin. 36, 203 (1959).*
66. Tompkins, H.G. and Bennett, J.E., *J. Electrochem. Soc. 123, 1003 (1976).*
67. Tompkins, H.G. and Bennett, J.E., *J. Electrochem. Soc. 124, 621 (1977).*
68. Tompkins, H.G., Wertheim, G., and Sharma, S.P., *J. Vac. Sci. Technol. 15, 20 (1978).*
69. Tsuji, Y. and Ichikawa, M., *Corrosion 27, 168 (1971).*
70. Weinman, L.S., Tucker, T.R. and Kim, C., and Metzbower, E.A., *Applied Optics 17, 906 (1978).*
71. Westbrook, J.H., Ed., "Intermetallic Compounds," Wiley, New York, 1967.
72. Wilson, G.C., *Trans. Inst. Met. Fin. 50, 109 (1972).*
73. Wynne, B.E., Edington, and Rohwell, G.P., *Met. Trans. 3, 301 (1972).*
74. Yamada, T., Fueki, S. and Ohsawa, K., *Jitsumu Hyomen Gijitsu (Practical Surface Technology) 252 (1975).*

CORROSION RESISTANCE OF PAINT FILMS FROM ANODIC AND CATHODIC RESINS

George E. F. Brewer

Coating Consultants
11065 East Grand River Road
Brighton, MI

GEORGE BREWER

ABSTRACT

Assuming that an actively rusting area is an anodic site, then the surrounding area is relatively cathodic and should attract film formers which contain residual cathodic groups to adhere more firmly. Thus, it is postulated that cathodic polymers will provide inherently higher corrosion protection compared with anodic resins. Various test and field results are presented, lending support to the above theory.

INTRODUCTION

A peculiar type of corrosion was described in 1944 in an article entitled "Filiform Underfilm Corrosion of Lacquered Steel Surfaces"[1]. Thread-like tunnels, about 1mm wide and 3mm to about 5 cm long lift the paint film about 1mm high, as shown in Figure 1.

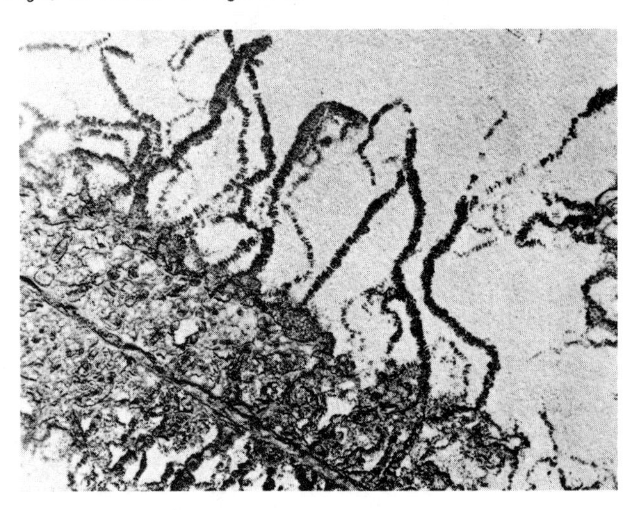

Fig. 1.—Filiform corrosion.

Filiform corrosion is rarely observed as a major corrosion problem. However, the trained eye finds approximately 5mm long filiform marks frequently on all types of merchandise. The study of filiform corrosion resulted in the formulation of a theory[2] for improved use of paints which contain ionizable groups. These thoughts and their subsequent verification through the study of performance data of various paints are reviewed in this paper.

FILMFORMING MACRO-IONS

Linseed oil and other drying oils have been used as components of film formers since ancient times. Various synthetic unsaturated acids containing alkyl and aryl groups have been developed during the last 50 years and are used for the synthesis of improved resins. The demand for these synthetic resins has grown rapidly since their carboxylic acid groups, when reacted with alkali, make these resins water dispersable to form highly desirable waterborne paints. Therefore most films formed from waterborne paints contain carboxyl groups which at least to a minor degree tend to form organic ions ($RCOO^-$) in the presence of water or humidity, a reaction which has no inherent value for corrosion protection. On the other hand the presence of macro-ions and their behavior in a galvanic cell had been made the basis of a widely used painting method.

THE ELECTROCOATING PROCESS

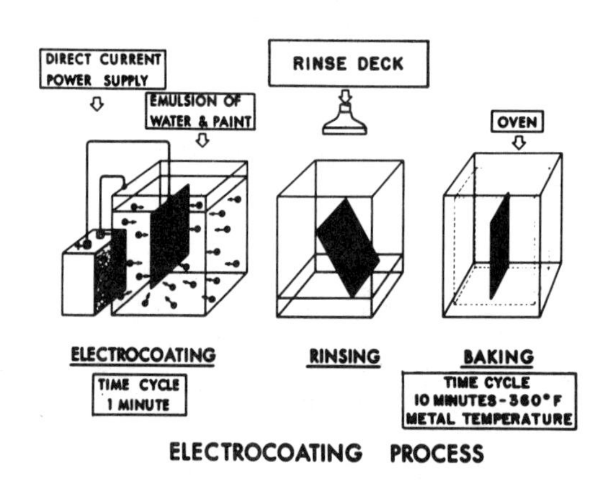

Fig. 2.—Electrocoating process.

During the 1960's a novel coating process was introduced, and is variously called electrocoating, electrodeposition, electro-painting, elpo, hydrocoating, etc.[3] The process resembles metal plating superficially, inasmuch as an aqueous dispersion of polymeric ions is placed between two electrodes, where the macroions, through the action of an impressed polarity (see Fig. 2). The freshly deposited, semisolid coat is then cured or baked into a durable paint film.

The macro-ions used for electrodeposition show an electrical equivalent weight of approximately 1600, or approximately 60 coulombs of electricity result in a deposited coat which weighs 1.0 gram. These macro-anions can be symbolized as $RCOO^-$, or more accurately as $R(COOH)_n(COO^-)_m$. The radical "R" may contain a preponderance of epoxy, acrylic, or other desired linkages which predetermine the properties of the finished coat. During the manufacture of these anodic paints, their acidic resins are partially neutralized to form aqueous dispersions:

$$R(COOH)_{n+m} + mOH^- = R(COOH)_n(COO^-)_m + mH_2O$$

During electrodeposition on the anode, the COO^- groups are reconverted to COOH, while little or no Kolbe-type decarboxylation takes place. The carboxyl groups in the paint film, however, seem to be oriented toward the metallic substrate[4].

The finished, anodically deposited paints show a superior wear resistance when compared with spray paints and dipcoats even after many years in the field[5].

FILIFORM CORROSION

Subsequent to the original description of filiform corrosion[1], studies by several scientists, notably W. H. Slabaugh and co-workers[6,7,8] clarified the mechanism by which filiform grows. A concise analysis of the chemistry involved in the growth of filiform corrosion was presented by H. H. Uhlig[9]: The forward, tunneling end (or head) of the filament contains ferrous salts, due to water and oxygen diffusion through the paint film. The diffusion process results in a differential aeration cell with cathodic areas and OH^- ions at the paint/metal interface. Thus, the growth of the filament depends upon high air humidity, and explains why filiform corrosion is not generally observed in ambient weather. The optimal conditions for filiform growth are 25°C at 85% relative humidity, and are used in laboratory tests[10].

The mechanism of filiform corrosion seemed related to corrosion failure of painted merchandise, and the question was asked whether the insight gained through the studies of filiform corrosion could be used to improve corrosion protection in general[2].

RUST SPOTS ON PAINTED SURFACES

It is generally agreed that the formation of rust on steel can be symbolized by the overall equation

$$4\,Fe^{++} + 3\,O_2 + 2\,H_2O = 2\,Fe_2O_3 \cdot H_2O$$

which can be interpreted as the stoichiometric sum of two half cell reactions plus subsequent oxidation of Fe^{++} to Fe^{+++} [11]:

$$\text{I} \qquad Fe^0 = 2\,Fe^{++} + 2\,e^-$$

$$\text{II} \qquad 1/2\,O_2 + H_2O + 2\,e^- = 2\,OH^-$$

$$\text{III} \qquad Fe^{++} + 3/4\,O_2 + \tfrac{1}{2}\,H_2O = \tfrac{1}{2}\,Fe_2O_3 \cdot H_2O$$

If we assume that the anodic process of metal dissolution takes place on an anodic site surrounded by a relatively cathodic area, then the location of the half cell reactions (I) and (II) can be represented as shown in Figure 3.

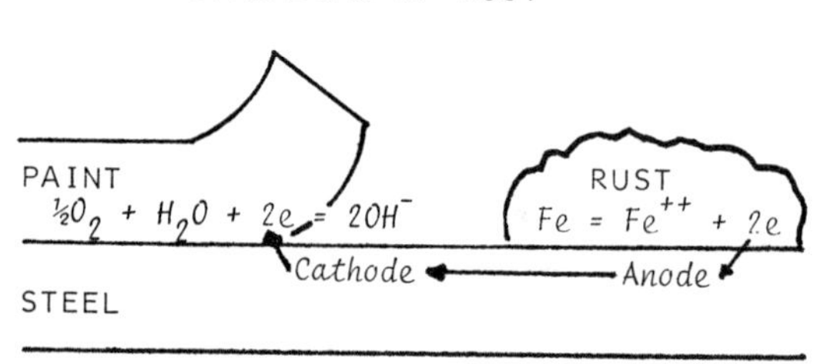

Fig. 3.—Spreading of rust.

The theory that an actively rusting area is surrounded by a cathodically protected area, is supported through inspection of test panels after salt spray tests. When an adhesive tape is applied to the rusted area, and then lifted (Figure 4), some of the paint adheres to the tape, revealing steel without a trace of corrosion.

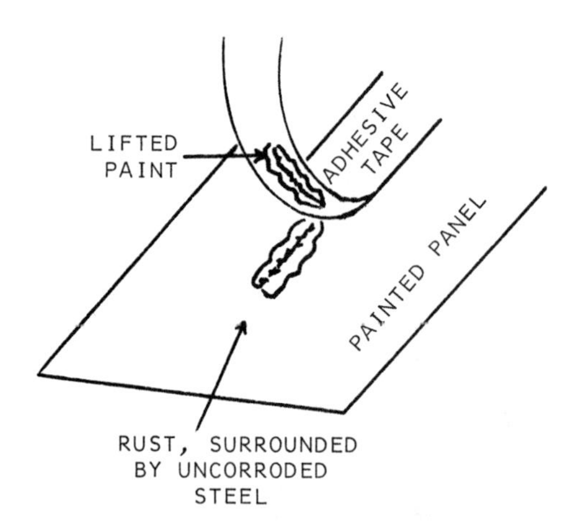

Fig. 4.—Rust, surrounded by uncorroded steel.

Returning to theoretical considerations, the cathodic vicinity of a rusting area should resist corrosion, and the presence of OH^- ions at the paint/metal interface should convert COOH groups into COO^- groups, which are repelled by cathodic areas. Conversely, paint resins which contain residual cationic groups should be attracted to the cathodic area in the vicinity of rust, and provide higher corrosion protection.

CATIONIC FILMFORMERS

The development of filmformers which deposit on the cathode (merchandise) in a voltaic cell was announced by scientists of Dow Chemical Co.[12], and of PPG Industries[13]. Paints for cathodic deposition manufactured by PPG Industries are now widely used[15], these resins may be symbolized as R_3N, and are solubilized through the action of acids:

$$R_3N + HX = R_3NH^+ + X^-$$

TEST RESULTS

The electrodeposition of *anodic* paints results in higher corrosion protection than obtained by use of *spray* paints or dipcoats[15]. In the case of anodically deposited paints this is largely due to the uniform film thickness even in deeply recessed areas. These cannot be uniformly coated by use of the spray technique. Dipcoats, while reaching into recessed areas, are subject to the formation of "wedges," tears," and "runs" which are not only unsightly, but through difficulties in curing, and differences in the thermal expansion coefficients of paint versus metal, are prone to cracking.

Tests carried out by use of *cathodic* electrocoats show superior corrosion protection as predicted from theory. Figure 5 shows high resistance against filiform corrosion when compared with anodic electrocoats.

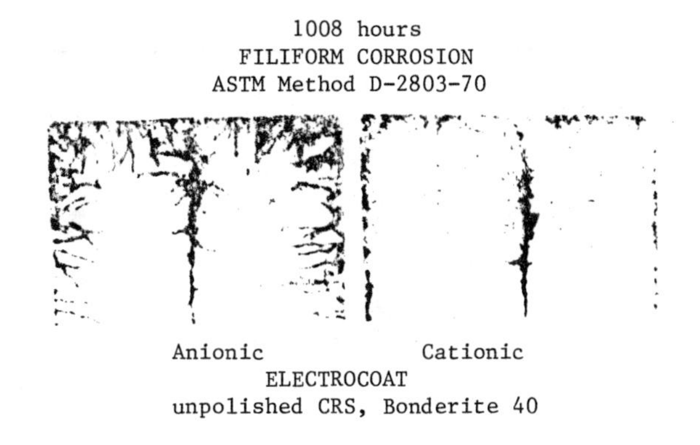

Fig. 5.—1008 hours. Filiform corrosion, ASTM Method D-2803-70.

Figure 6 shows the improved saltspray resistance on phosphated steel, and much reduced rust bleeding.

One of the measures for the corrosion resistance of paints is the adhesion in the vicinity of a scribe mark when exposed to a saltspray test[16]. The distance between the scribe mark and the still well adhering paint is called "creep," and lower corrosion resistance expresses itself as wider "creep." Let us consider an automobile body made from steel, with the lowest part of the door frame (rocker panel) made from galvanized steel, and then spray painted with an anionic paint. The zinc-coated rocker panel is now the anode with respect to the welded-on steel fender (cathode). Corrosion of the fender is very frequently observed under these circumstances (see Figure 7).

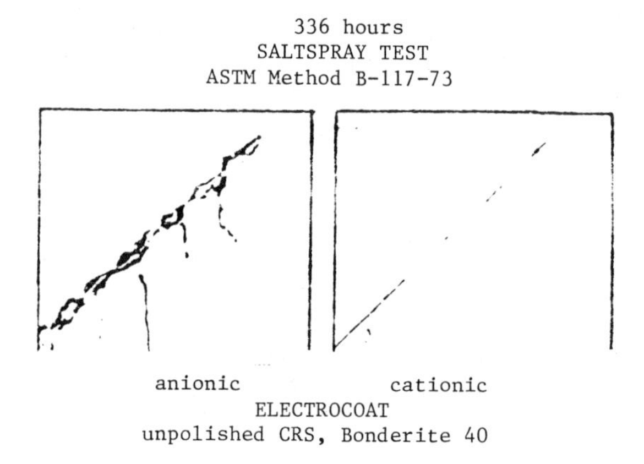

Fig. 6.—336 hours. Saltspray test, ASTM Method B-117-73.

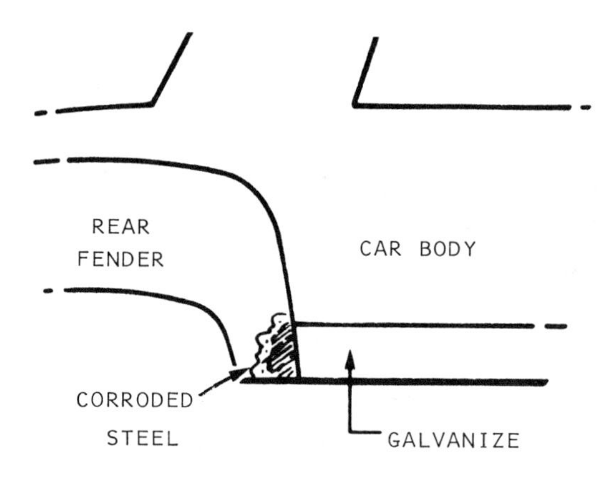

Fig. 7.

Galvanized (zinc-coated) steel poses an interesting problem: suppose steel is exposed by scribing or scoring through the zinc coat which is then the anode in this corrosion cell (see Figure 8). Thus, the theory predicts than an anodic paint will exhibit higher corrosion protection on galvanized steel then on plain steel. Conversely, a cathodic paint will exhibit higher corrosion protection on steel, rather than on galvanized, and that is what has been found in a recent study[17]. Some of the data presented in that paper are compiled in Table I.

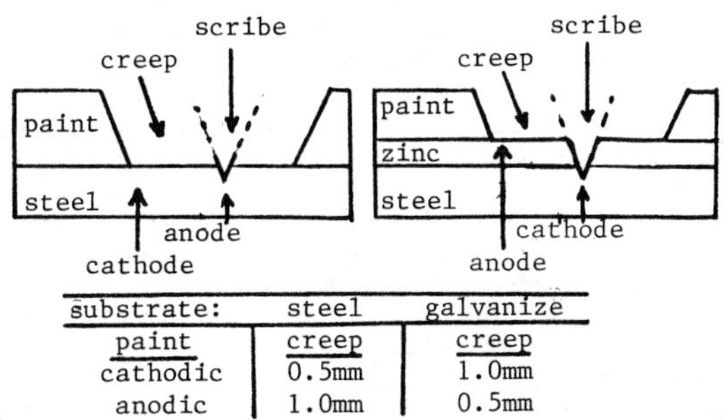

Fig. 8.—Creep damage at scribe.

TABLE I.—CREEP DAMAGE AT SCRIBE

336 h Saltspray exposure
Zincphosphated panels

substrate:	steel	galvanize
paint	creep	creep
cathodic	0.5mm	1.0mm
anodic	1.0mm	0.5mm

DISCUSSION

Residual ionization in polymeric films affects corrosion resistance and can be utilized to increase corrosion protection.

The use of cationic and anionic polymers is not confined to electrodeposition, they may be formulated to lend themselves to spray, dip, roll, and other applications, thus offering a variety of opportunities to reduce corrosion.

Bimetallic joints, or dissimilar metals in vicinity of each other are frequently encountered. In these situations as increased corrosion protection should be achieved through coating the relatively base metal with a coat of anionic polymers, while protecting the relatively noble metal with a cationic material.

ACKNOWLEDGEMENT

Thanks are due to Mr. Frederick M. Loop, PPG Industries for photographs of certain test panels.

REFERENCES

1. Sharman, C.F., *Nature 153, 621 (1944).*
2. Brewer, G.E.F., *Metal Finishing 72 (8), 49 (1974).*
3. Brewer, G.E.F., *J. Paint Technol. 45 (587) 35 (1973).*
4. Koral, J.N., Blank, W.J. and Falzone, J.P., *J. Paint Technol. 40 (519) 156 (1968).*
5. Burnside, G.L. and Brewer, G.E.F., *Metal Finishing 71 (11), 62 (1973).*
6. Slabaugh, E.H. and Grotheer, M., *Industrial and Engineering Chemistry 46, 1014 (1954).*
7. Slabaugh, W.H. and Kennedy, G., *Official Digest 34 (454), 1139 (1962).*
8. Slabaugh, W.H., Dejaper, W., Hoover, S.E. and Hutchinson, L.L., *J. Paint Technol. 44 (556), 76 (1972).*
9. Uhlig, H.H., "Corrosion and Corrosion Control," end Edition, J. Wiley & Sons, N.Y., 1971, pp. 252-56.
10. ASTM Test Method D-2803-70
11. Evans, V.R., "The Corrosion and Oxydation of Metals," Chapter 21, Arnold, London, 1960.
12. Wessling, R.A., Gibbs, D.S., Settineri, W.J. and Wagener, E.H., ACS Div. *Org. Coatings & Plastics Chem. 31 (1), 299-302 (1971).*
13. Wismer, M., and Moffat, T.W., "Electrocoat 72," Paper No. 14, Nat'l. Paint & Coatings Assoc., Washington DC, Oct. 72.
14. Loop, F.M., SME Technical Paper FC 77-641.
15. Burnside, G.L. and Brewer, G.E.F., *Paint & Varnish Production 60 (4), 46 (1960).*
16. ASTM Method B-117-73
17. Loop, F.M., *Plating & Surface Finishing 65 (7), 32 (1978).*

ELECTRICAL AND ELECTROCHEMICAL MEASUREMENTS AS PREDICTORS OF CORROSION AT THE METAL-ORGANIC COATING INTERFACE

Henry Leidheiser, Jr.

Center for Surface and Coatings Research,
Lehigh University, Bethlehem, PA

HENRY LEIDHEISER, JR.

ABSTRACT

Five types of electrical and electrochemical measurements are reviewed: D. C. electrical properties, A. C. resistance, capacitance, corrosion potential, and polarization curves. When the coating resistance decreases with time and falls below 10^6 *to* 10^7 *ohms/cm*^{2}*, corrosion beneath the coating is occurring. A low electrical resistance requires (a) ion and water penetration into the coating, (b) ready motion of ions through the coating, and (c) ongoing ionic/electron transfer reactions (corrosion) at the polymer/metal interface. Care must be taken in making electrical measurements that the measurement itself does not affect the corrosion rate.*

INTRODUCTION

The coatings industry has been greatly affected by the recognition that some of the important and common constituents in paint formulations no longer can be indiscriminately used because of the hazardous nature of these constituents. Particularly affected are components used as corrosion inhibitors such as chromates and lead compounds. Greater recognition is also being accorded to constituents that might be carcinogenic during some step in manufacture or might be irritants during manufacture or during use. The formulator remains continuously apprehensive that constituents used for many years will suddenly become unacceptable for unanticipated reasons. Since many formulations are based on long service experience, it is frightening to conceive that one of the essential components in the formulation may have to be removed and a substitute material must be found. The need for rapid laboratory tests that aid in predicting service experience becomes acute, because a large number of formulations may have to be studied in a short time.

This review has as its purpose to summarize those electrical measurements that have been used, and are being used, to make laboratory evaluations of the ability of a formulation, or a formulation plus a pretreatment, to provide effective corrosion protection to the base metal. The review is not meant to be exhaustive and only a limited number of papers have been chosen to illustrate specific points.

The review will discuss the following five types of electrical and electrochemical measurements: (1) D. C. electrical properties, (2) A. C.

resistance; (3) capacitance, (4) corrosion potential, and (5) polarization curves.

D. C. Electrical Properties

One of the leading workers concerned with corrosion of painted metals, J. E. O. Mayne, makes the following statement: "In order for a polymer film to protect a metallic substrate against corrosion, the film should possess low permeability to ions, that is to say it should have a high electrolytic resistance"[1]. Several papers that provide background for this conclusion and others that support this conclusion will be summarized in this section.

The first application of D. C. resistance measurements of organic coatings to an understanding of corrosion of painted metals appears to have been made with Wirth[2-6]. He measured the currents which flowed between steel-zinc and steel-silver couples immersed in solution when the steel was coated with a protective film. Similar measurements were made when the film separated the two electrodes but was detached from the steel surface. Measurements of the cell potentials and the current flowing permitted an estimate of the film resistance.

Bacon, Smith and Rugg[7] carried out an extensive study in which they measured the resistance of over 300 organic coatings on metals immersed in electrolytes as a function of time. Concurrently the extent of corrosion was determined. In every case good protection (no corrosion) was observed when the resistance of the coating exceeded 10^8 ohms/cm^2 and poor protection (visible corrosion) was obtained when the resistance of the coating was less than 10^6 ohms/cm^2. In their words: "... the predictions of coating protective merit from the resistance behaviors during immersion have always conformed with the visual protective ratings assigned to the corresponding test specimens after prolonged laboratory exposure." A schematic description of the relative behaviors of coatings with good, fair, and poor corrosion protection properties on the basis of the resistances is given in Figure 1. Good coatings were characterized by a high resistance that fluctuated about a mean value. The resistance values also provided information useful in determining the binder to use, compatible pigment-binder systems, and optimum coating thickness.

A summary of selected studies relating corrosion behavior and the D. C. resistance of organic coatings on steel substrates is given in Table I. It is apparent that the boundary between corrosion and the absence of corrosion is of the order of 10^7 ohms/cm^2. When the resistance is greater than 10^7 ohms/cm^2, corrosion is not observed. When the resistance is less than 10^7 ohms/cm^2, corrosion is observed.

Recently, good correlations have been reported between laboratory measurements of electrical resistance of painted steel after 15 days exposure to 3% NaCl and corrosion behavior during extended outdoor testing[12]. Paint systems which exhibited high values for the electrical resistance exhibited good performance in outdoor tests.

The salutary effect of a phosphate pretreatment coating between the metal substrate and the organic coating is well known. The beneficial effect of phosphate in promoting corrosion protection also shows up in D. C. conduction studies. Thin primer films exhibit D. C. conduction immediately or within several hours after immersion in aerated 5% NaCl, whereas phosphate pretreatment, primer, plus topcoat systems showed "no ionic conduction for the duration of the 3000 hour immersion test."[8]

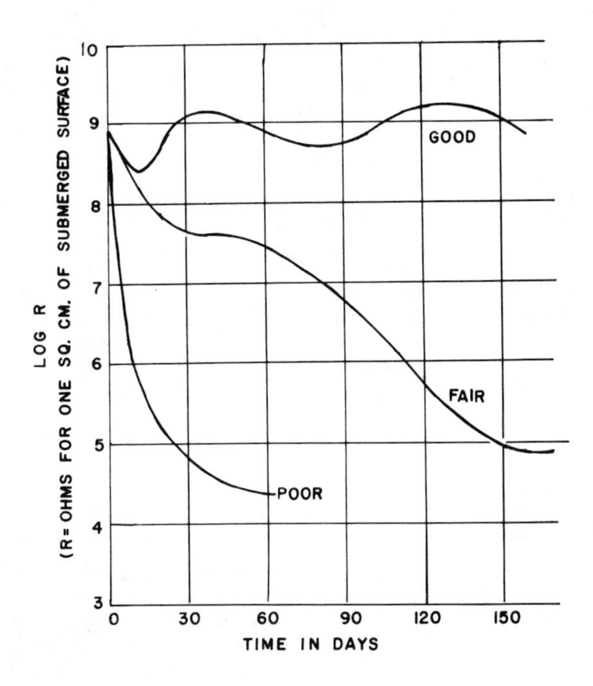

Fig. 1.—Schematic representation of behavior of resistance for "good," "fair," and "poor" coatings after exposure to electrolyte for many days. Figure taken from Bacon, Smith, and Rugg[7].

TABLE I.—SUMMARY OF SEVERAL INVESTIGATIONS IN WHICH RESISTANCE OF COATING WAS CORRELATED WITH CORROSION BEHAVIOR OF STEEL SUBSTRATE

Corrosion Medium	Corrosion Performance In Terms of Resistance		Reference
	Corrosion Noted Below	Corrosion Not Noted Above	
5% NaCl, aerated	10^7 ohms/cm^2	10^7 ohms/cm^2	(8)
Sea Water	10^6	10^8	(7)
	10^6	10^8	(9)
Dilute HNO_3	Dropped to 10^1	10^7	(10)
Dilute HCl	Dropped to 10^1	10^7	
3.5M KCl	10^7	10^{11}	(1)
15% HNO_3	10^6	10^7	(11)

As might be intuitively suspected, the electrical properties of different sections of the same film may differ greatly. An excellent example of this is the work of Kinsella and Mayne[13]. These workers cast films, 20x10 cm, and measured the resistances of different 1 cm^2 sections. Two types of behavior were observed. In one type, the resistance of the film decreased when the concentration of KCl was greatly increased, whereas in the other type, the resistance of the film increased when the concentration of KCl was greatly increased.

A very direct correlation between coating resistance and corrosion has been obtained by Mayne and Mills[1]. These workers coated mild steel with various lacquers to thicknesses of the order of 75-100 μm and exposed the coated metal to 3.5M KCl for 1 week at 22°. The corrosion behavior was recorded. The film was then removed from the substrate and cut into 25 squares of 1 cm^2 size and the electrical resistance determined. An example is shown in Figure 2 for any epoxy coating. It will be noted that the resistance was of the order of 10^{11} to 10^{12} ohms/cm^2 in those areas where no corrosion was observed and was 10^6 to 10^7 ohms/cm^2 in those areas where corrosion was noted.

• — Dark blue blisters
— Area of adhesion loss

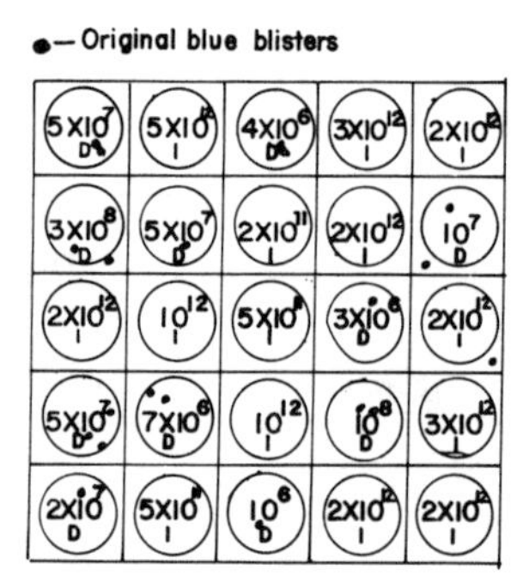

Fig. 2.—Correlation between corrosion and electrical resistance in different areas of a thick epoxy coating on a steel substrate after exposure to 3.5M KCl for 1 week at 22°. Drawing on right shows the resistance of each sq. cm and the drawing on left shows where the corrosion was observed. Note that the resistance was appreciably less in those squares in which corrosion was observed below the coating. The resistance was measured after separation from the steel substrate. Figure was taken from Mayne and Mills[1].

A rapid and continuous method for studying the protection properties of organic coatings on metallic substrates is based on the use of a very thin foil mounted on glass with an adhesive[14]. The foil is generally less than $2x10^{-3}$ cm in thickness. Riedel and Voight[15] have modified this method by combining it with D. C. resistance measurements. Thin films of iron, either evaporated on glass or in the form of foil, were coated with polymer and exposed to a corrosive environment. The change in resistance parallel to the surface was indicative of the fraction of the film that was converted to a high-resistance material, i.e., rust. An example of the use of this method in determining the efficacy of different inhibitors is shown in Figure 3 for an acrylic latex pigmented with TiO_2. Haruyama and Tsuru[16] have used a similar method to study the rate of formation and thickness of the passive film formed during anodization of a vapor-deposited iron film.

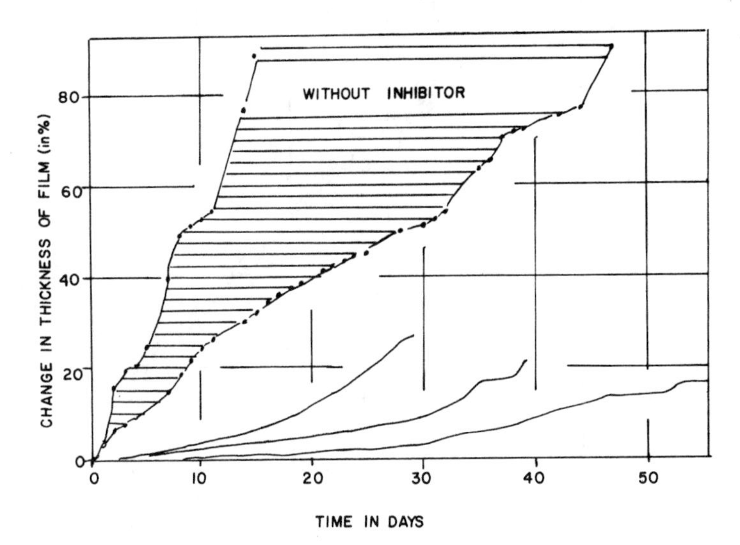

Fig. 3.—The change in thickness, as determined by resistance measurements, of a thin iron substrate coated with an acrylic latex pigmented with TiO_2 and exposed to condensing water vapor at 40°. The three lower curves show the lower rate of corrosion obtained when the latex was formulated with three different inhibitors. Figure taken from Riedel and Voigt[15].

A. C. Resistance Measurements

A. C. resistance measurements on polymeric coatings on metals have the advantage over D. C. resistance measurements in two respects: (1) the resistive component resulting from polarization at the solution/polymer and polymer/metal interfaces is avoided in A. C. measurements; and (2) the ability to make measurements as a function of frequency provides information useful in understanding the corrosion process.

The common method of making A. C. measurements of the resistance is to determine the response of the coating to an applied oscillating field $V_0e^{J\omega t}$ by bridge balancing of the response to that of an equivalent parallel resistor. Most authors present such information in the form of the real part of the permittivity (ε'') or in terms of the experimentally determined resistance. These two values are related by:

$$\varepsilon'' = \frac{d}{A\varepsilon_o \omega R_{x_p}} \tag{1}$$

where d is the coating thickness, A is the apparent area over which the measurement is performed, ε_0 is the absolute permittivity of free space, ω is the frequency in radians, and R_{x_p} is the equivalent parallel resistance.

The contrasting behavior of different coatings is illustrated by the work of Menges and Schneider[11,17,18]. Figures 4-7 show the change in resistance as a function of time at various frequencies for coatings exposed

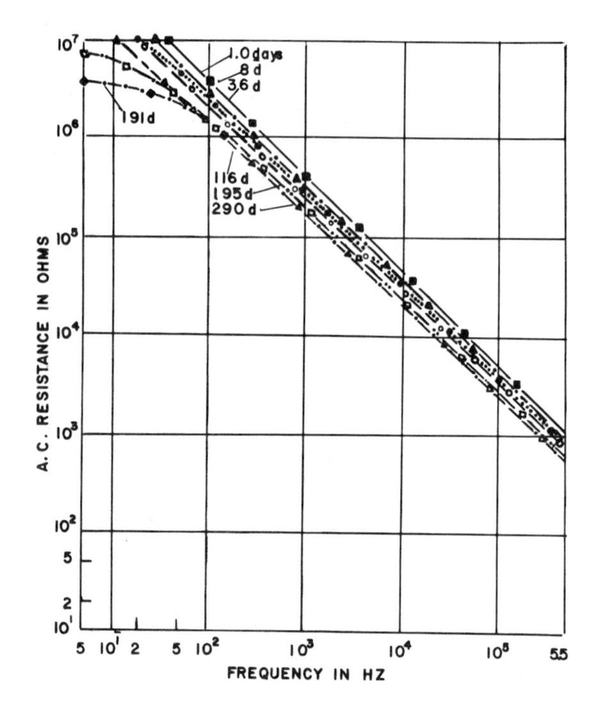

Fig. 4.—The change in resistance as a function of frequency for a phenol/formaldehyde coating, 260 μm thickness, on a steel substrate after exposure to 15% HNO_3 at 23°. Figure taken from Menges and Schneider[18].

to 15% HNO_3 at 23° and 60°. The phenolformaldehyde coating (Figure 4) exhibited only a small change in electrical properties after exposure to 15% HNO_3 for 290 days at 23°. The resistance at the lowest frequency was in excess of 10^6 ohms per 15 cm^2 of measured area. On the other hand, the same coating at 60° (Figure 5) exhibited a steep drop in resistance at frequencies below 10^2Hz after exposure to 15% HNO_3 for only 1 day. The constant resistance at frequencies of 10^3Hz and below is indicative of one or more highly conductive paths through the coating.

The studies of Menges and Schneider were carried out on coatings 100-900 μm in thickness and the test reagent was the very aggressive 15% HNO_3 at temperatures as high as 60°. The work of Kendig and Leidheiser[19] was carried out on thin coatings 4-29 μm in thickness and the medium was 0.52M NaCl. The transparent nature of the coatings used by Kendig and Leidheiser permitted correlations between the electrical properties, corrosion and delamination. Representative data for the change in the real part of the permittivity, ε", with time is shown for a range of frequencies for several types of coatings in Figures 8-11. It will be noted in Figures 8, 10 and 11 that the log ε"-log frequency curve approaches a -1 slope at low frequencies. Such a behavior is indicative of conductive paths through the coating as shown in Figure 12 where the -1 slope is developed by intentionally pricking a polybutadiene coating with a fine-pointed needle. The -1 slope indicates D.C. conductivity since the value of $\frac{d}{A\varepsilon_0 R_{x_p}}$ in Equation (1) may be con-

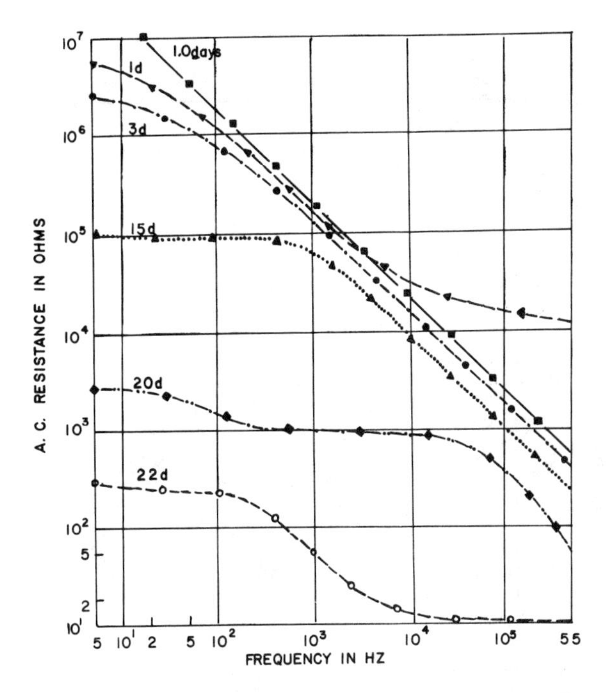

Fig. 5.—The change in resistance as a function of frequency for a phenol/ formaldehyde coating, 250 μm in thickness, on a steel substrate after exposure to 15% HNO_3 at 60°. Compare with Figure 4. Figure taken from Menges and Schneider[18].

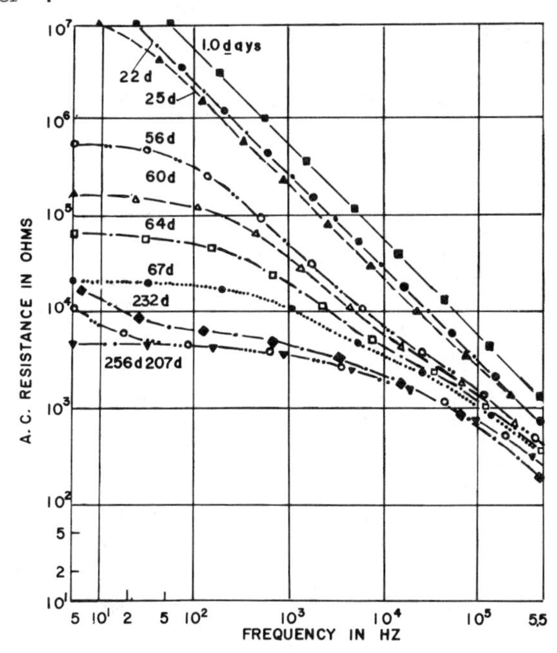

Fig. 6.—The change in resistance as a function of frequency for polyurethane coating, 260 μm in thickness, on a steel substrate after exposure to 15% HNO_3 at 23°. Figure taken from Menges and Schneider[18].

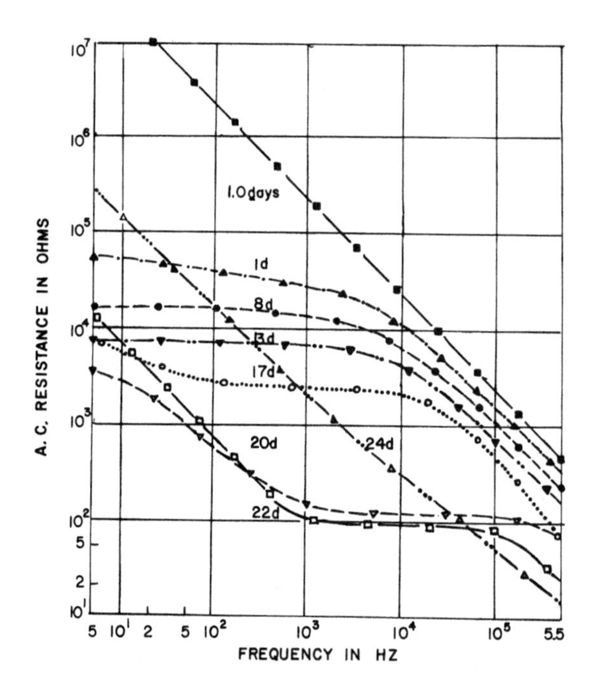

Fig. 7.—The change in resistance as a function of frequency for a polyurethane coating, 130 μm in thickness, on a steel substrate after exposure to 15% HNO_3 at 60°. Figure taken from Menges and Schneider[18].

sidered a constant and thus log ε" = -log ω + constant. The resistance at low frequencies is no longer a function of frequency.

Capacitance[20]

In most types of organic protective coatings on metals the system is designed to be subjected to an aqueous environment whether it be underground, submerged, or exposed to the atmosphere. The corrosion process requires water and thus the penetration of the water through the coating to the coating/substrate interface is of vital concern. Electrical measurements have provided a means to follow water uptake as a function of time on coatings ranging in thickness from 5 μm to 3,000 μm.

When a polymer-coated metal is exposed to aqueous solutions, the measured capacitance of the coating increases with time. Representative data are given in Figure 13 for thin polybutadiene coatings on steel. A similar but less dramatic effect is observed for a thick bituminous coating on galvanized steel. This increase in capacitance has been associated with the penetration of water into the coating with consequent change in the dielectric properties. There is some doubt as to the 1:1 relationship between the change in capacitance and the amount of water taken up by the coating, but it does appear that the major portion of the capacitance change is caused by water uptake. The two major difficulties in making a comparison between

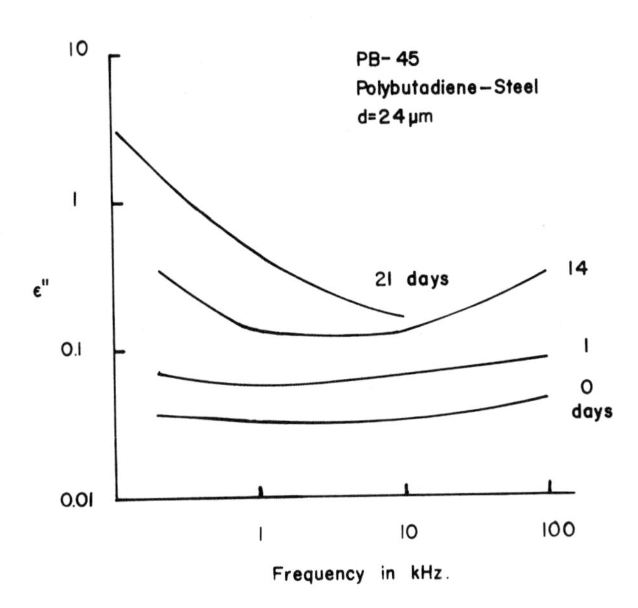

Fig. 8.—A plot of log permittivity loss vs. log frequency for a polybutadiene-coated steel exposed for 0, 1, 14 and 21 days to 0.52M NaCl at room temperature. Figure taken from Kendig and Leidheiser[19].

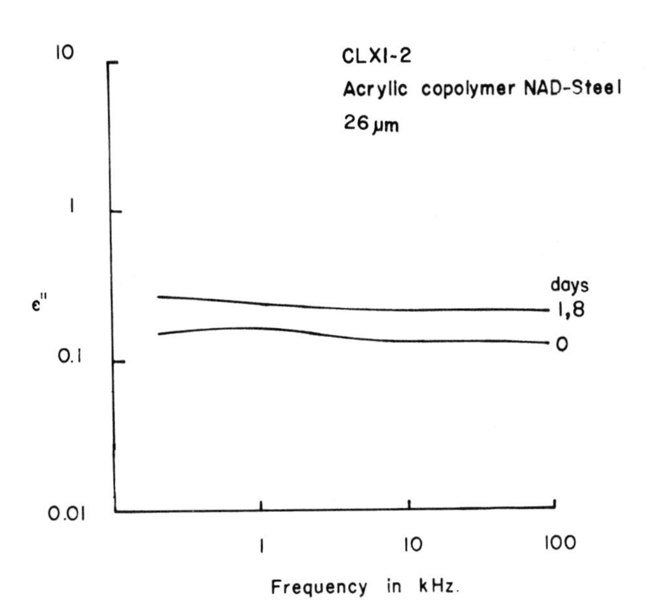

Fig. 9.—A plot of log permittivity loss vs. log frequency for a steel coated with an acrylic copolymer exposed for 0, 1 and 8 days to 0.52M NaCl at room temperature. Figure taken from Kendig and Leidheiser[19].

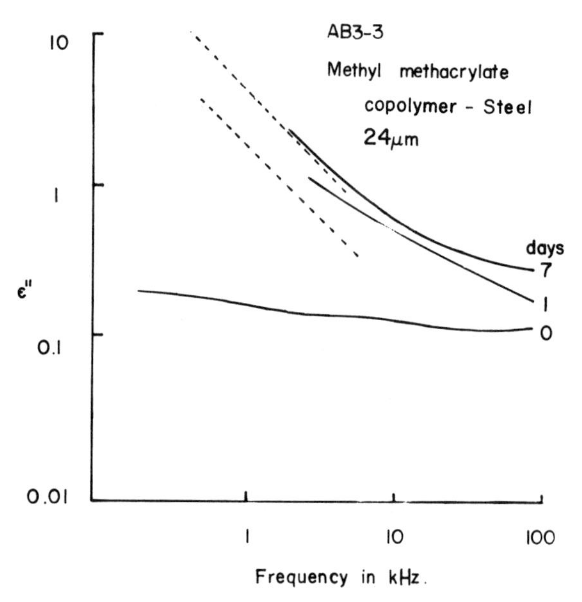

Fig. 10.—A plot of log permittivity loss vs. log frequency for a steel coated with a methylmethacrylate copolymer coating exposed for 0, 1 and 7 days to 0.52M NaCl at room temperature. The approach to a -1 slope at low frequencies is indicative of low resistance paths in the coating. Figure taken from Kendig and Leidheiser[19].

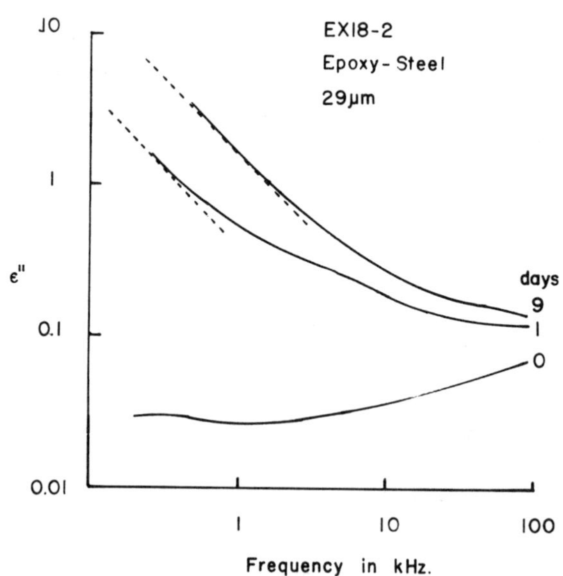

Fig. 11.—A plot of log permittivity loss vs. log frequency for an epoxy-coated steel exposed for 0, 1 and 9 days to 0.52M NaCl at room temperature. The approach to a -1 slope at low frequencies is indicative of low resistance paths in the coating. Figure taken from Kendig and Leidheiser[19].

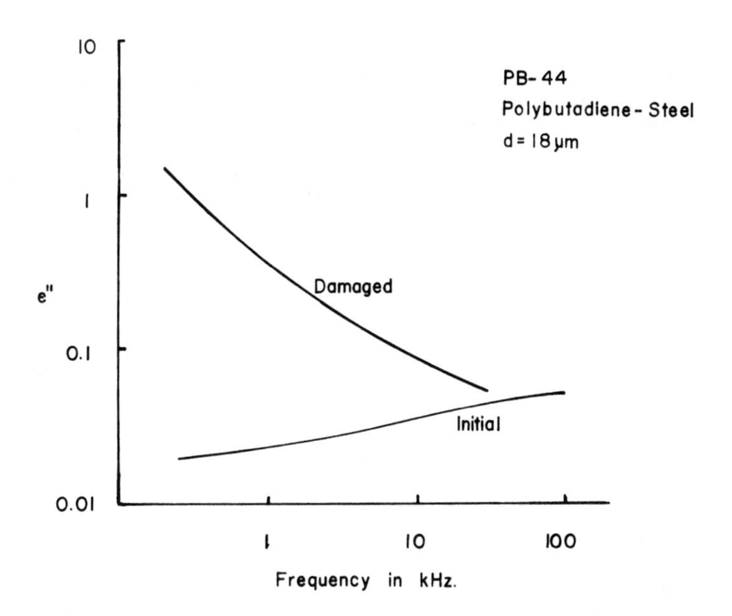

Fig. 12.—A plot of log permittivity loss vs. log frequency for a polybutadiene-coated steel that was intentionally pricked with a needle and was then exposed to 0.52M NaCl at room temperature. Note the approach to a -1 slope in the damaged sample. Figure taken from Kendig and Leidheiser[19].

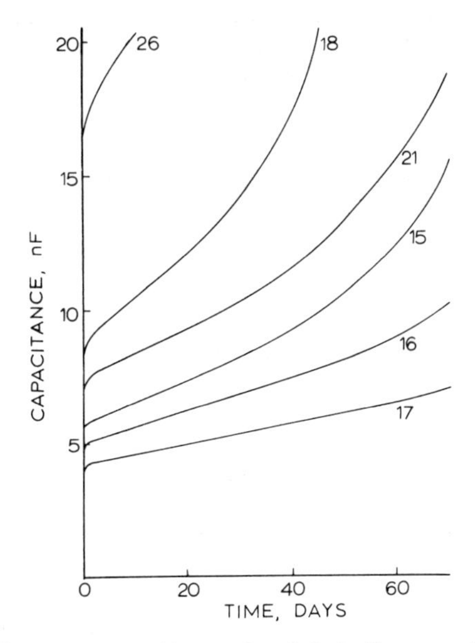

Fig. 13.—The capacitance at 3,000 Hz of polybutadiene coatings on steel substrates as a function of time of exposure to 3% NaCl at room temperature. Film thicknesses are: 26, 0.48; 18, 0.89; 21, 1.02; 15, 1.37; 16, 1.45; and 17, 1.70 x 10^{-3} cm. Figure taken from Touhsaent and Leidheiser[21].

gravimetric and capacitance measurements of water uptake are: (1) Assumptions must be made in developing a relationship between capacitance and water uptake. (2) Gravimetric measurements are subject to considerable error because non-penetrated water must be removed without affecting the penetrated water. Even with these difficulties, the agreement is satisfactory under conditions where water uptake is relatively small as shown for example in Figure 14. The agreement is less satisfactory in the case of large amounts of water uptake as shown by the work of Brasher and Kingsbury[22] in Figure 15.

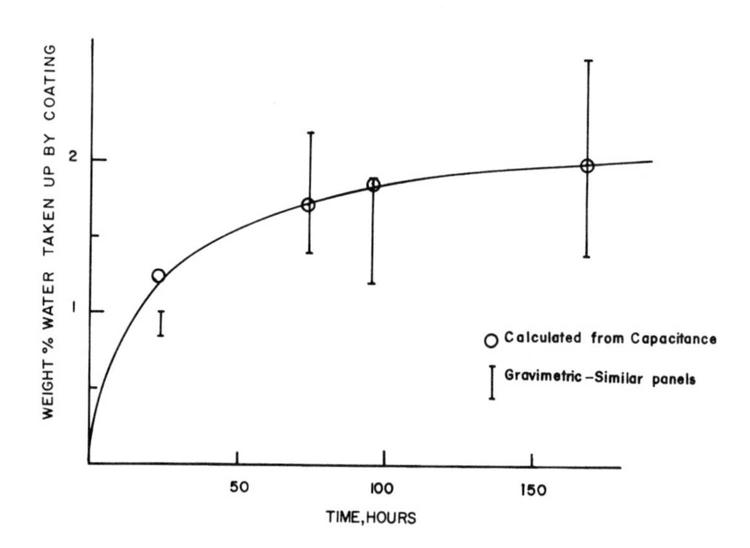

Fig. 14.—Wt. % water uptake by polybutadiene coatings approximately 2 x 10^{-3} cm thick, during exposure to 0.5M NaCl. The difficulty in making gravimetric measurements on thin coatings is indicated by the range of values obtained. Figure taken from Leidheiser and Kendig[24].

Brasher and Kingsbury calculated the water uptake of coatings from capacitance values under the assumption that the water within the paint film is random and uniform. They applied a formula given by Hartshorn, Megson, and Rushton[25].

$$K_m = K_p^{V_p/V} \cdot K_w^{V_w/V} \cdot K_a^{V_a/V} \tag{2}$$

where

K_m = Measured permittivity (dielectric constant)

K_p, K_w, K_a = Permittivity of paint, water, and air, respectively

V_p, V_w, V_a = Volume of paint, water, and air, respectively

V = Total volume = $V_p + V_w + V_a$

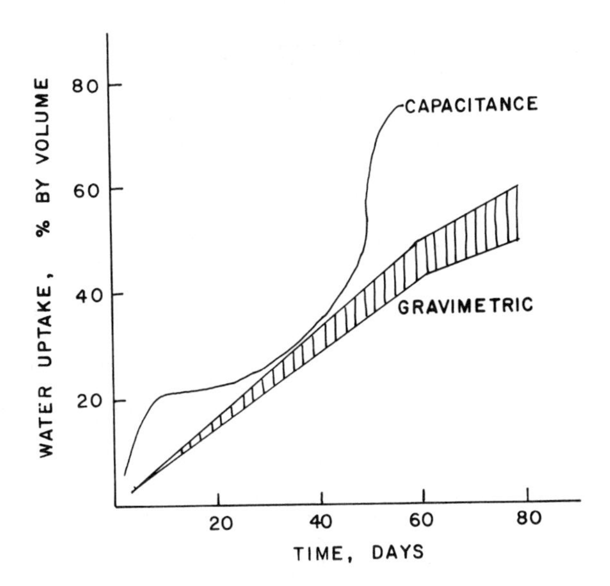

Fig. 15.—Comparison between water uptake by gravimetric and capacitance measurements as determined by Brasher and Nurse[27].

However, since $K_a = 1$

$$K_m = K_p^{V_p/V} \cdot K_w^{V_w/V}$$

Comparing K_m at any time t with K_m at time o,

$$K_m/K_{m,o} = K_w^{V_w/V} / K_w^{V_{w,o}/V}$$

because $K_p^{V_p/V} / K_p^{V_{p,o}/V} = 1$

But $V_{w,o} = 0$; therefore

$$K_w^{V_{w,o}/V} = 1$$

and

$$K_m/K_{m,o} = K_w^{V_w/V} = 80^x$$

where x = fraction by volume of water taken up.

But, if it is assumed that capacitance changes are proportional to permittivity changes, $K_m/K_{m,o} = C_m/C_{m,o}$, where C_m and $C_{m,o}$ are measured capacitances at time t and at time o respectively,

$$C_m/C_{m,o} = 80^x, \quad i.e., \quad x = \frac{\log(C_m/C_{m,o})}{\log 80}$$

$$X_v = \frac{100 \cdot log\left(C_m/C_{m,o}\right)}{log\ 80} \qquad (3)$$

where

X_v = percentage of volume of water taken up.

A significant feature of the coatings used by Brasher and Kingsbury is that in the first few hours X_v calculated from the capacitance values shows a much sharper rise than X_v measured gravimetrically. Brasher and Kingsbury interpreted this experimental finding by assuming that the initial distribution of water in their films was not random but has a tendency to take the form of columns normal to the surface. In other words, pores, capillaries, or pinholes normal to the surface are the first to be filled with water; the water then spreads out laterally to give a uniform random distribution. This concept is in marked contrast with the generally accepted theory of diffusion with an advancing front, a mechanism that would make the value of water uptake calculated from Equation (3) lower than that found gravimetrically. Reasonably good agreement between gravimetric and capacitance values was found for most samples, and the assumption of random distribution of water seems valid for these. If there is no replacement of a residual polar solvent in the coating or no bound water on the surface, the difference at all stages between capacitance and gravimetric measurements can be attributed to a concentration gradient normal to the surface, or pores, capillaries, or pinholes normal to the surface.

Brasher and Kingsbury conclude that capacitance measurements can be used instead of gravimetric methods to find water uptake where the type of paint used is such that capitance results agree with gravimetric results, i.e., where random distribution holds. Capacitance measurements may be continued to determine the effective lifetime of a coating. Comparisons between water uptake vs. time curves for capacitance techniques and for gravimetric techniques can give information on the mode of entry and distribution of water in the coating. The mode of entry and distribution can be very important. De and Kelkar[26] attempted to isolate the effect of the pigment in paint films by using Brasher and Kingsbury's technique on clear films. They found that both a good and a poor protective coating absorbed about the same amount of water measured gravimetrically. Therefore, they attributed the failure of the poor coating to increased ionic uptake and concluded that ionic uptake was the controlling factor in coating performance. However, this explanation may not be the whole picture since the mode of entry of water was not considered.

Brasher and Nurse[27] studied the effect of osmotic pressure and electrolyte concentration on protective coatings using the same capacitance-resistance method. They employed three types of solutions to separate the effects of ionic strength and osmotic pressure: (1) sucrose solution to study the effect of osmotic pressure in the absence of an electrolyte, (2) sodium chloride and other salts to study the effect of salts, and (3) sucrose and sodium chloride combined to study the effect of varying salt concentration at constant osmotic pressure. Results of measurements in these three types of solutions showed that in solution (1) water uptake was higher in low osmotic pressure solutions, in solution (2) the more dilute the salt solution the greater was the water uptake by the paint and the earlier the onset of paint breakdown, and in solution (3) the period before breakdown was shorter the higher the chloride content. In general, the water uptake in initial stages was controlled by the osmotic pressure, and in mixed solutions of constant osmotic pressure the initial water uptake was independent of the proportion of sucrose to sodium chloride. Breakdown occurred more quickly the more dilute the solution, showing that as long as a minimal amount of chloride ions was present, the primary controlling factor in the life of the paint was the amount of water taken up. When water uptake was large, the paint was uniformly less protective and general atteck occurred. For small values of water uptake, the points of breakdown were at chance weak spots resulting in localized corrosion. However, in solutions of constant osmotic pressure at any given water uptake, the rapidity of breakdown was a function of the activity of the chloride ion in the surrounding solution.

Conclusions about the mechanism of paint deterioration were drawn from these results, and comparisons between values of water uptake calculated by capacitance and gravimetric methods were made. In sodium chloride solutions a rise in capacitance preceded visible breakdown. Electro-endosmosis associated with the setting up of corrosion currents was rejected as an explanation for this finding because the rise in capacitance occurred before corrosion was noted. The gravimetric measurements support the rejection of electro-endosmosis since they show that no sudden increase in water uptake occurred as would happen with electro-endosmosis; rather, the uptake increased steadily with time. Brasher and Nurse concluded, therefore, that the capacitance rise was caused by a re-orientation in the mode of distribution of water within the paint film. It was hypothesized that at a certain stage water droplets linked up to form a network of capillaries largely normal to the surface of the paint, thus causing water uptake calculated from capacitance measurements to show a sharp rise while water uptake found gravimetrically did not show such a rise. This capillary formation quickly led to paint breakdown since it provided paths of greatly reduced electrical resistance for the passage of corrosion currents. Brasher and Nurse concluded that the probable sequence of events in the deterioration of paint is: (i) uptake of water randomly distributed, (ii) re-orientation of the mode of distribution of the water in the film, probably by a linking together of droplets to form continuous pores normal to the paint surface, (iii) establishment of equilibrium between the solution in the pores and the solution in which the paint is immersed, and (iv) onset of corrosion currents, whose intensity is determined by the electrical conductivity of the solution in the pores which is approximately equal to that of the surrounding solution.

It must be pointed out that the conclusions of this study are not necessarily valid for a type of polymer different than that used in the study or for the same polymer pigmented differently. The results are presented here to show how the method may be used to evaluate protective coatings and the mechanisms causing their failures. For instance, in another type of coating the water uptake calculated by the capacitance method may remain steadily below that found by the gravimetric method until the coating fails, suggesting that water is distributed parallel to the surface and that the coating fails when this layer finally extends to the surface, thus allowing corrosion currents to take place.

Gentles[28] used the same capacitance method to evaluate the differences in the modes of distribution in coatings pigmented differently. He presented the following chart to indicate how the water distribution may be determined.

Type of water distribution	Value of $\frac{X_V\ cap.}{X_V\ grav.}$
Distribution in pores of capillaries	>1
Random distribution	$=1$
Distribution in layers	<1

Indeed, he found that the value of X_V *cap.*/X_V *grav.* varied with the amount of pigment in the paint, the type of pigment, and the type of polymer.

The effect of the metal substrate cannot be neglected. Tomashov, Mikahilovski, and Leonov[29] isolated the effect of the metal substrate by using platinum as the substrate. They found that the electrochemical nature of the metal was one of the factors determining the initial deterioration of the insultaing film. Miller[30] found that a particular polymer may protect one metal well and may not protect another. The reactivity of the metal substrate and the type of product formed by corrosion have a great influence upon the life of a protective coating.

H. C. O'Brien[31] developed a new method using capacitance measurements to evaluate the performance of protective coatings. By comparing gravimetric and capacitance measurements as previously discussed, he found that for lightly pigmented bituminous coatings water was distributed in layers parallel to the coating surface. In these cases, while resistivity measurements may vary by several thousand percent because of pinholes in the coatings, the dielectric constant values are more indicative of the quantity of water absorbed because these pinholes do not effect the dielectric constant very much. O'Brien found it valuable, then, to relate the changes in capacitance to a calculated value of the depth of penetration of the water front into the sample. He took a series of dielectric constant values over time; when these values approached that of water, the sample was completely penetrated by water and had little, if any, protective value.

O'Brien's experimental set-up was somewhat different from those described above. A metal plate was coated evenly and its thickness measured. A one-pint can with the bottom removed was sealed to the plate and filled with water to form a parallel plate condensor with metal and water as the two plates. Immediately after the can was filled, the initial capacitance and resistance were measured. The specific inductive capacitance (K_o) for the whole coating was calculated from equations for circular parallel plate capacitors.

$$C_o = 0.0885\ K_o S/t$$

$$K_o = C_o t/0.0885\ S$$

where
- C_o = initial capacitance in mmfd.
- K_o = dielectric constant (81 for water was used by O'Brien)
- S = area of plate in sq. cm.
- t = coating thickness in cm.

By considering the coating to be composed to two layers, a water-saturated layer and a water-free layer, having capacitances of C_1 and C_2 respectively, one can estimate the approximate depth of water penetration, t_1. The capacitance of the water-saturated layer is:

$$C_1 = (0.0885)(81)S/t_1$$

The capacitance of the water-free layer with thickness t_2 is:

$$C_2 = 0.0885\ K_2 S/t_2$$

Substituting these two equations into the equation for equivalent capacitance:

$$1/C = 1/C_1 + 1/C_2$$

Noting that $t = t_1 + t_2$:

$$1/C = t_1/7.1685\ S + (t-t_1)/C_o t$$

and rearranging:

$$t_1 = 7.1685\ St(C_o-C)/(CC_o t-7.1685\ CS) \tag{3}$$

Using this equation and values between the initial value C_o and the final value C obtained when $t_1 = t$, graphs can be made in which capacitance is plotted vs. depth of penetration; from these graphs the depth of penetration for any capacitance can be found. Graphs of depth penetration vs. elapsed time can then be made by measuring the capacitance as a function of time. Thus, the rate of water absorption with time may be plotted, and the theoretical t_1, or penetration depth, may be used as an index of coating performance. A presumed good coating, after an initial increase, will establish a low t_1 with a very low rate of change. An evaluation of the penetration vs. elapsed time curve can furnish an index of ultimate performance and can be an aid in predicting the time of saturation failure. O'Brien found that a coating twice as thick as the t_1 established in the 'steady state' gave protective coatings of long lifetime. The prediction of coating life was made by finding the slope of the steady state (how fast the layer is advancing) and by calculating how long it would take for t_1 to equal t using this slope. This method permits projected life studies of coatings and enables the establishment of safe film thickness specifications. The method is fast, nondestructive, and sensitive enough to detect small changes in the coating composition.

In testing good protective coatings, however, it may be desirable to use accelerated tests to substantiate short-term tests; to prove the results of O'Brien's method without them may take years. Konecke[32] has studied various methods of acceleration. Among these were: (a) applied voltage, (b) increased temperature, (c) decreased film thickness, (d) increased electrolyte severity, and (e) aeration. He found best results with (a), (b), and (d). Impressed currents (a) were used to screen large numbers of polymers to determine which ones would make good protective coatings. However, this method may cause some coatings to fail although they would not under service conditions. The mechanism of paint failure may change when using an impressed current or increased electrolyte severity. Therefore, the preferred method to achieve acceleration is by increasing the temperature.

Miller[30] used a variation of O'Brien's technique to study acrylic and polyimide coatings. Capacitance cells were made by sealing the open ends of 1½-inch diameter polystyrene containers against the surfaces of coated panels. Disc electrodes were mounted 1½ inches above the panel surfaces by supporting screws attached to the ends of the containers. The specific inductive capacitance was calculated from equations for circular parallel plate capacitors:

$$K = \frac{C_o t_o}{0.0885\ S}$$

where

K = dielectric constant of the film
C_o = initial capacitance in mmfd.
S = area of plate in sq. cm.
t_o = film thickness

As the electrolyte penetrated the film, a portion of the dielectric layer became conductive and in effect formed a portion of the top plate of the condenser. The unpenetrated thickness, t, was inversely proportional to the capacitance at any given time.

$$t = \frac{K(0.0885)S}{C} \tag{4}$$

The film thickness penetrated $(t_o - t)$ was plotted vs. time to obtain curves similar to O Brien's. Miller also studied various metal substrates and found that a coating system should be considered a combination of a substrate, a surface, a primer or surface treatment, and a topcoat. He concluded that the capacitance measurement technique was an excellent method for determining the depth of penetration of electrolytes into organic films and that there was a correlation between the ionic permeability of films and their ability to protect metal from corrosion.

K. A. Holtzman[33] used capacitance measurements to calculate the permeability of paint films and obtained results similar to the gravimetric measurements. Holtzman's treatment was different from that of previous workers. Instead of using Equation (3) proposed by Brasher and Kingsbury to calculate the percent water uptake, Holtzman used an equation derived by Böttcher[34] for two-phase mixtures:

$$X_v = \frac{(K-K_2)(2K+K_1)}{3K(K_1-K_2)}$$

where K, K_1, and K_2 refer to the permittivities of the mixture, water, and film respectively, and K and K_2 are obtained from the relationship:

$$K = \frac{C_p L}{.06954\ x^2}$$

where the film thickness, L, and the diameter of the circular specimen, x, are given in centimeters and the parallel capacitance, C_p, is given in picofarads. The quantity of water absorbed, Q_t, was calculated from the equation:

$$Q_t = \frac{\rho\ X_v X_f}{1-X_v}$$

where ρ is the density of water at the temperature of the cell and V_f is the volume of the film in cm^3.

Holtzman then used a solution of Pick's second law derived by Carpenter[35]:

$$\log \frac{dQ_t}{dt} = \log \frac{8DQ_\infty}{L^2} - \frac{.434\pi^2 Dt}{L^2}$$

which relates the change in the quantity of absorbed water with time, dQ_t/dt, to the quantity of absorbed water at infinite time, Q_∞ to the diffusion constant, D, to the time, t, and to the film thickness, L. The diffusion constant, D, was obtained from the slope, G, of a plot of $\log dQ_t/dt$ vs. t.

$$D = -\frac{GL^2}{.434\pi^2}$$

The intercept, i, of this plot yielded, Q_∞, which in turn permitted the calculation of the solubility constant, S:

$$S = \frac{Q_\infty}{V_f(p-p_o)}$$

Finally, the permeability, P, was calculated from the definition,

$$P = DS$$

This study, as well as most others, makes it clear that the formula used by Brasher and Kingsbury (Equation (3)) does not give the same value for the amount of water in the coating as gravimetric measurements. Morozumi and Fujiyama[34], after studying the Brasher-Kingsbury method, concluded that the error must be small in most cases. However, there are some cases where the two methods do not agree. The causes for this disagreement may be (1) nonrandom distribution of water in the coating which occurs especially in the initial period of water abosrption, (2) water bound to the coating so that its dielectric constant is nearly that of ice, (3) changes due to swelling of the coating (4) accumulation of water in delaminated interfacial areas of multiple coatings, and (5) absorption of electrolytes along with water. Although all of these conditions may result in discrepancies between capacitance and gravimetric measurements, the capacitance values constitute a better indicator of coating performance. The correlation between capacitance values and performance is not determined by the concentration of water in the coating because in all cases there is enough water present to produce corrosion. Rather the controlling factor is the effect of the absorbed water on the electrical properties of the coating. It is these electrical properties that reflect the corrosion of the substrate. Thus the value of water uptake calculated by the capacitance method (Equation (2)) represents an "active" water concentration which is a better indicator of coating performance than gravimetric measurements of water absorption.

Corrosion Potential

The corrosion potential of a metal in an aqueous solution, in combination with other facts, can often provide information about the mechanism of the reaction and the rate controlling factor. The value of the corrosion potential by itself provides little information, but its change with time and its value in reference to other samples may be usefully applied. In the case of painted metals the area of metal exposed to the electrolyte is a function of the permeability of the paint to water and to ions and the integrity of the coating. Any electrical measurements must be made with negligible current flow or else the very fact of current flow may induce irreversible changes in the coating. For corrosion potential measurements a very high impedance voltmeter is an absolute necessity.

A review of the corrosion potential of painted metals covering the literature through 1970 has been prepared by Wolstenholme[37]. The early work of Burns and Haring[38], Haring and Gibney[39], Whitby[40] and Zahn[41] focused on the magnitude of the corrosion potential and how it changed with time. The summation of all these studies on steel substrates was that negative potentials were indicative of corrosion beneath the paint and positive potentials were indicative of the absence of corrosion. However, anomalous cases were noted in which the generalization did not hold. These very empirical measurements were followed by a more thorough study by Wormwell and Brasher[42]. They noted that the shapes of the potential/time curves during the first few hours or first days were quite misleading as a guide to the ultimate protective properties of the paint. A typical corrosion potential-time curve is superimposed on a weight loss curve in Figure 16 taken from Wormwell and Brasher. It will be noted that the corrosion potential moved in the active direction for a day or two, then moved in the noble direction

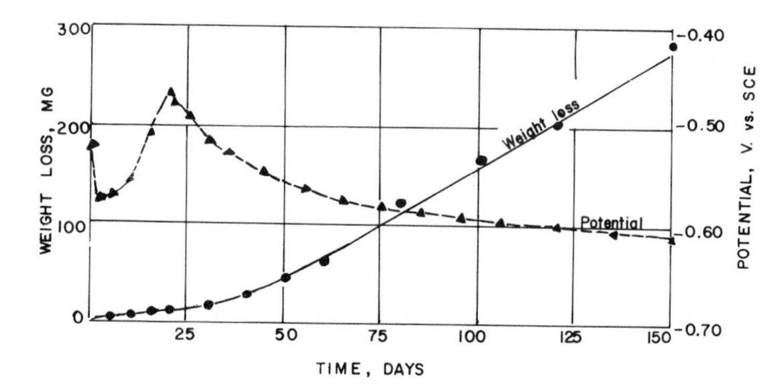

Fig. 16.—The relationship between the weight loss and time and the corrosion potential and time for steel coated with an anti-corrosive paint and exposed to artificial sea-water at pH 8.1. Figure taken from Wormwell and Brasher[42].

for about 20 days, and finally moved in the active direction until the end of the test at 150 days. A similar type of corrosion potential-time behavior has been noted by Kendig and Leidheiser[19] in the case of thin polybutadiene coatings on steel immersed in 0.5M NaCl as shown in Figure 17. Since polybutadiene coatings are transparent at the thickness used it was possible to appraise the degree of corrosion and degree of delamination non-destructively. It will be noted from Figure 17 that the degree of delamination, which is indicative of the area service as cathode, increased greatly during the first 10 days when the corrosion potential was moving in the noble direction. When the degree of corrosion increased abruptly at the end of 31 days, the corrosion potential dropped to the vicinity of -600 mv. vs. SCE. The explanation for these facts is given in Figure 18. During the first 31 days, the cathodic area increases greatly because of delamination of the coating and the cathodic polarization curve moves to the right. Since the surface area on which the anodic reaction is occurring does not increase at as rapid a rate, the anodic polarization curve moves only slightly to the right with the consequence that the intersection of the cathodic and anodic curves is at increasingly positive values. At the onset of severe corrosion at the end of 31 days, the anodic/cathodic surface area ratio increases greatly and the corrosion potential moves abruptly to a negative value.

As a generalization it can be concluded that movement of the corrosion potential in the noble direction is indicative of an increasing cathodic/anodic surface area ratio and is indicative that oxygen and water are penetrating the coating and arriving at the metal/coating interface. Movement of the corrosion potential in the active (more negative) direction is indicative that the anodic/cathodic surface area ratio is increasing and that the overall corrosion rate is becoming significant. Increasingly positive potentials with time suggests that alkaline conditions caused by the reaction, $2H_2O + \frac{1}{2}O_2 + 2e^- = 2OH^-$, are developing locally at the metal/coating interface and that delamination is of concern. Increasingly active potentials indicate rusting beneath the coating and represent the signal that the coating lifetime is limited.

Wormwell and Brasher have used corrosion potentials and their changes

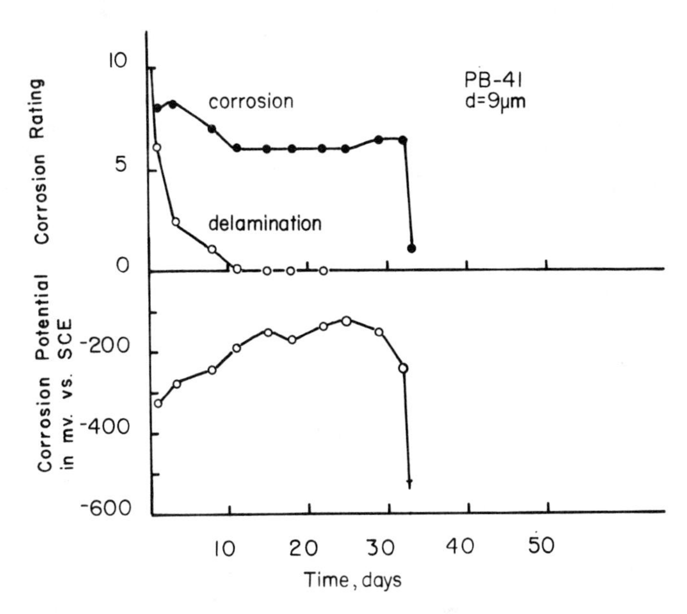

Fig. 17.—Upper curve is the ASTM D610 rating of corrosion and delamination vs. time of exposure to 0.52M NaCl for a steel coated with a 9 μm thick coating of polybutadiene. Lower curve is the corrosion potential determined during the experiment. Figure taken from Kendig and Leidheiser[19].

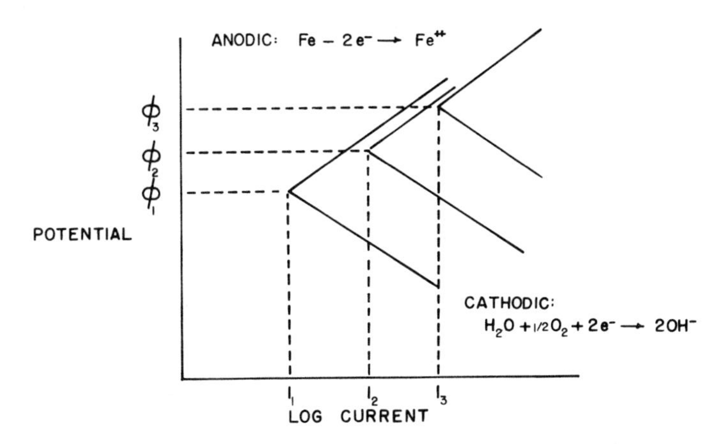

Fig. 18.—Schematic description of the change in corrosion potential with time for polybutadiene-coated steel exposed to aerated NaCl solution. The potential moves in the noble direction with time because the cathodic/anodic area ratio increases. Figure taken from Kendig and Leidheiser[19].

with time to investigate the following factors: thickness of the paint, surface finish on the steel, method of pretreatment before painting, and type of paint.

The important role of oxygen in determining the value of the corrosion potential is shown in Figure 19. In the absence of oxygen the corrosion potential wandered but generally had a value between -600 and -650 vs. SCE in 0.5M NaCl solution, whereas in the presence of oxygen the corrosion potential moved in the noble direction for reasons discussed above.

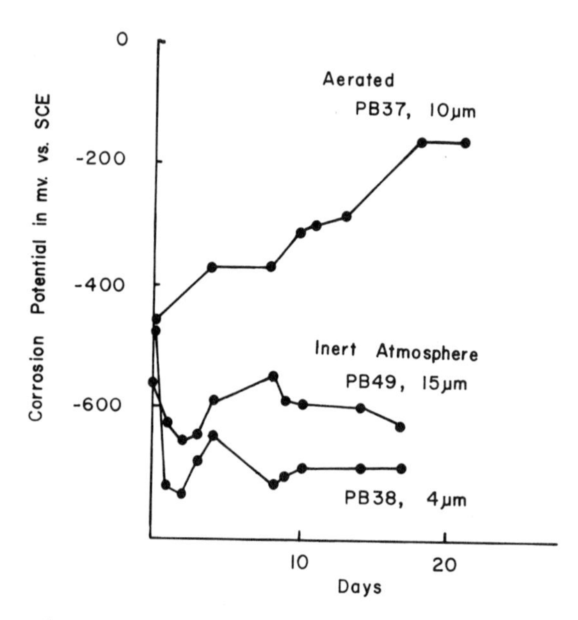

Fig. 19.—Comparison of typical corrosion potential vs. time curves for polybutadiene-coated steel exposed to deaerated and aerated NaCl solution. Figure taken from Kendig and Leidheiser[19].

Thin paint films (20-30 μm) with very low electrical resistances show active corrosion potentials which became more positive as the paint film is increased in thickness[43]. Wormwell and Brasher reached a somewhat similar conclusion in that they noted that the maximum noble potential was at -0.5 v vs. SCE for one thickness of paint coating, was at -0.4 v at two thicknesses, and was at -0.2 v for 3 thicknesses. The time to achieve the most noble value also increased with the thickness of paint. The relationship between corrosion potential and coating thickness is very different for polymeric, unpigmented coatings that are thin relative to paint films. An example is shown in Figure 20 where it will be noted that the potential immediately after immersion in the electrolyte became more active as the coating was increased in thickness. The explanation for this reverse effect from that observed by Yakubovitch et al[43] and Wormwell and Brasher[42] is that the thin polybutadiene coatings are very permeable to water and oxygen and the dominance of the area available for the cathodic reaction. The rate of supply of water and oxygen at the metal/polymer interface for the cathodic reaction immediately after immersion is a function of the coating thickness.

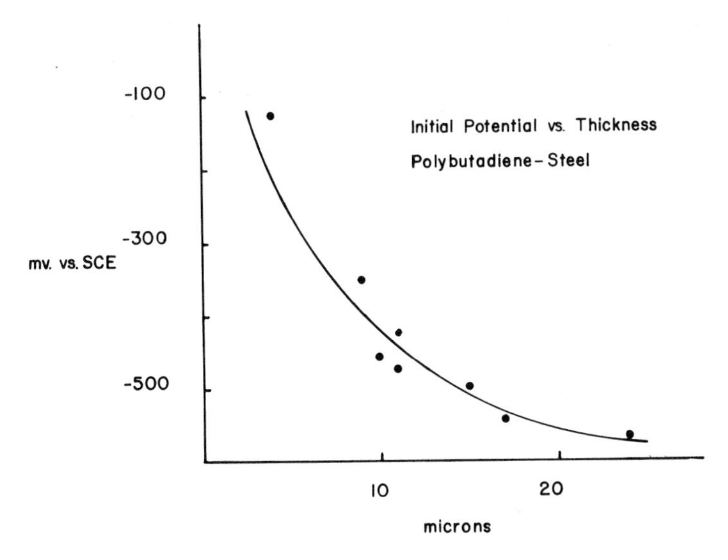

Fig. 20.—Extrapolated initial corrosion potential of polybutadiene-coated steels exposed to 0.52M NaCl at room temperature as a function of coating thickness. Figure taken from Kendig and Leidheiser[19].

Polarization Curves

An interesting application of polarization curves was made by Wiggle, Smith, and Petrocelli[8]. These workers were interested in the degree of porosity of phosphate conversion coatings. They determined the anodic polarization curve for untreated steel of known surface area and of steel that was pretreated with a phosphate solution. The percent of the sample surface area available for the anodic dissolution was then determined from the displacement of the polarization curves of the steel with and without the phosphate coating.

These same workers[8] have used polarization to understand better the corrosion mechanism at scribe marks on painted panels. They noted that anodic polarization at a current of 35 microamp. of scribe marks on SAE 1010 steel with a total length of 8-10 cm caused some undercutting on applied paints with no conversion coating. Undercutting was insignificant on anodic polarization when zinc phosphate or iron phosphate conversion coatings were applied between the metal and the paint. In a similar way they noted that severe undercutting occurred on cathodic polarization of steel with no conversion coating or with a zinc phosphate conversion coating. No significant undercutting occurred when an iron phosphate conversion coating was used.

Zurilla and Hospadaruk[44] have made an interesting application of applied potentials. They were interested in appraising the resistance to corrosion of phosphated steels serving as substrates for painted metal. They reasoned that the cathodic half reaction of the overall corrosion reaction that occurs adjoining a defect in the coating is controlled by the quality of the phosphate coating. Since the cathodic reaction develops a high pH, they chose NaOH of pH 12 as the test medium. They polarized the steel to -0.55 v. vs. SCE in the air-saturated NaOH solution and measured the oxygen reduction current. Similar panels were then coated with paint, scribed in the conventional manner and the time to failure, as determined by a 3 mm. undercutting away from the scribe mark, was measured. The results are shown in Figure 21. It is apparent that low activity of the phosphated steel for the cathodic half reaction coincides with long lifetime in the salt spray test.

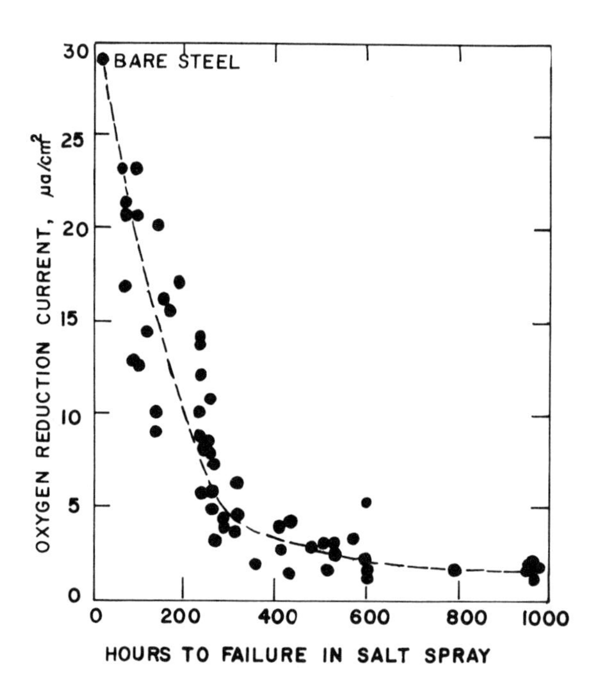

Fig. 21.—The oxygen reduction current measured for various phosphate steels at -0.55 v. (vs. SCE) in air-saturated NaOH, pH 12, as related to time to failure in salt spray for scribed and painted panels. Figure taken from Zurilla and Hospadaruk[44].

COMMENTARY

The survey made herein indicates that electrical measurements can be of great assistance in determining the rate of penetration of water and ions into organic coatings. D. C. resistance measurements and A. C. resistance measurements at frequencies below 10^3 Hz yield information on the combined effect of water and ions in the coating. The data summarized in Table 1 indicate that corrosion of the substrate is of concern when the resistance of the coating drops below 10^6 ohms/cm^2. Although service data are not available for quantitative assessment, work in our laboratories at Lehigh suggests that there is a direct relationship between the conductivity of the coating and the rate of corrosion. It is not immediately obvious why such a relationship should exist. Many metals, free of an organic coating, can be exposed to strongly conducting solutions without corrosion occurring at any significant rate. It is suggested that the occluded corrosion cell which forms at the base of any ionic conducting paths in the polymer coating provides good conductive means between the polymer and the metal substrate because of the ionic nature of the corrosion reactions:

$$Fe - 2e^- = Fe^{++} \qquad \text{(anodic reaction)}$$

$$H_2O + \tfrac{1}{2}O_2 + 2e^- = 2OH^- \qquad \text{(cathodic reaction)}$$

Without the availability of these ionic reactions, the passage of charge across the polymer/metal interface will not occur readily and ion motion through the water-penetrated polymer will be impeded by space charge effects. A low electrical resistance of the coating then requires (a) ion and water penetration in the coating, (b) ready motion of ions through the coating, and (c) ongoing ionic transfer reactions (corrosion) at the polymer/metal interface. Resistance measurements integrate these three effects but do not give information as to which is the rate controlling or limiting process.

Capacitance measurements have proven to be very useful because they can non-destructively determine the amount of water that penetrates the coating. The interpretation of experimental information as to the distribution and nature of water in the coating is not clearcut, but it does appear that capacitance measurements give appropriate values of the total water present in the coating. It is a generalization that when water uptake by the coating is great that the corrosion protection value of the coating is not great and conversely when the water uptake of the coating is low, the corrosion protection value of the coating is great. In fact Guruviah and Rajagopalan[12] find that the rate of water uptake in a laboratory experiment correlated rather well with the lifetime of 12 different coatings on steel upon outdoor exposure. Low water uptakes yielded the longest service exposure lifetimes and large water uptakes yielded the shortest service exposure lifetimes.

It is apparent that corrosion would not occur if steel could be coated with a polymer that was completely impervious to water, or to oxygen, since both water and oxygen are required for the corrosion reactions. Water and oxygen are reactants in the cathodic half-reaction and water is necessary as a recipient for iron ions generated by the anodic half-reaction. The practical reality of such a coating is not likely for two reasons. A perfectly impervious coating would still be subject to time-dependent chemical changes because of environmental exposure and to mechanical defects because of thermal shocks or impact. Other means of corrosion protection besides the impervious nature of the coating would necessarily have to be built into the material if the coating were to have wide usefulness.

Corrosion potential determinations provide useful information under some circumstances but the information generated by such measurements requires other facts before conclusions can be drawn. Changes in corrosion potential with time, by themselves, do not lead to unequivocal conclusions about the mechanism or rate controlling step. However, the change in corrosion potential with time does give information about the changes in the relative anodic/cathodic surface area. If the corrosion potential moves in the noble direction with time, an increasing cathodic/anodic surface area ratio is indicated and delamination at the metal/coating interface should be of concern. If the corrosion potential moves in the active direction with time, an increasing anodic/cathodic surface area is indicated and the loss of metal at the coating/metal substrate is of concern.

Measurements of polarization curves of polymer-coated metals is not recommended because the actual flow of current under the applied potential may, and usually does, affect the properties of the coating/metal interface.

Care must also be taken in measurements of corrosion potential because too large a current during the measurement may cause defects in the coating.

ACKNOWLEDGEMENT

Appreciation is expressed to Robert Touhsaent for permission to use information directly from his M.S. thesis at Lehigh University and to Christy Roysdon for her assistance in searching the recent literature.

Appreciation is also expressed to the editor (Dr. W. Funke) and the publisher (Elsevier Sequoia) of *Progress in Organic Coatings* for permission to reproduce a manuscript in press at the time of the Discussion.

REFERENCES

1. Mayne, J.E.O. and Mills, D.J., *Oil and Colour Chemists Assocn. J. 58, 155 (1975).*
2. Wirth, J.K., *Chem. Fabrik 11, 455 (1938).*
3. Wirth, J.K., *Korrosion u. Metalschutz 16, 69 (1940).*
4. Wirth, J.K., *Korrosion u. Metalschutz 16, 331 (1940).*
5. Wirth, J.K., *Angewandte Chemie 54, 369 (1941).*
6. Wirth, J.K., *Korrosion u. Metalschutz 18, 203 (1942).*
7. Bacon, R.C., Smith, J.J. and Rugg, F.M., *Ind. Eng. Chem. 40, 161 (1948).*
8. Wiggle, R.R., Smith, A.G. and Petrocelli, J.V., *J. Paint Technol. 40, 174 (1968).*
9. Weinmann, K., *Deutsche Farben-Zeitschrift 21, No. 1, 3 (1967); 21, No. 2, 59 (1967).*
10. Orzachovskij, M.L., *Lakokras. mater. i. ich primen No. 2, 40 (1968).*
11. Menges, G. and Schneider, W., *Kunststofftechnik 12, No. 12, 343 (1973).*
12. Guruviah, S. and Rajagoplan, K.S., "Proc. Semin. Electrochem. 14th," 1973, published 1974, p. 486.
13. Kinsella, E.M. and Mayne, J.E.O., *Br. Polym. J. 1, 173 (1969).*
14. Jedlicka, W.W. and Geschke, T.H., *J. Paint Technol. 41, 680 (1969).*
15. Riedel, G. and Voight, C., *Korrosion-Dresden 5, 13 (1974).*
16. Haruyama, S. and Tsuru, T., "Passivity and Its Breakdown on Iron and Iron Base Alloys," R.W. Staehle and H. Okada, Editors, Natl. Assocn., Corrosion Engrs., 1976, p. 41.
17. Menges, G. and Schneider, W., *Kunststofftechnik 12, No. 10, 265.*
18. Menges, G. and Schneider, W., *Kunststofftechnik 12, No. 11, 316.*
19. Kendig, M.W. and Leidheiser, H.Jr., *J. Electrochem. Soc. 123, 982 (1976).*
20. Much of the wording in this section is taken directly from the thesis of Robert E. Touhsaent, "A Capacitance-Resistance Study of Polybutadiene Coatings on Steel," submitted in candidacy for the Master of Science degree at Lehigh University, January 1972. The author of this report was the thesis director.
21. Touhsaent, R.E. and Leidheiser, H.Jr., *Corrosion 28, 435 (1972).*
22. Brasher, D.M. and Kingsbury, A.H., *J. Applied Chem. 4, 62 (1954).*
23. Leidheiser, H.Jr.,Kim, D.K. and Wang, W., unpublished results.
24. Leidheiser, H.Jr. and Kendig, M.W., *Corrosion 32, 69 (1976).*
25. Hartshorn, L., Megson, N.J.L. and Rushton, E., *J. Soc. Chem. Ind., London, 56, 266t (1937).*
26. De, C.P. and Kelkar, V.M., "First Intern. Congress on Metallic Corrosion, 1961," published 1962, p. 533.
27. Brasher, D.M. and Nurse, T.J., *J. Appl. Chem. 9, 96 (1959).*
28. Gentles, J.K., *Oil Colour Chemists Assocn. 46, 850 (1963).*
29. Tomashov, N.D., Mikahilovski, Yu.N., and Leonov, V.V., *Corrosion 20, 125 (1964).*
30. Miller, R.N., *Materials Protection 7, No. 11, 35 (1968).*
31. O'Brien, H.C., *Ind. Eng. Chem. 58, No. 6, 45 (1966).*
32. Koenecke, D.F., *Official Digest 32, 71 (1960).*
33. Holtzman, K.A., *J. Paint Techno. 43, No. 554, 47 (1971).*
34. See discussions by C.A.R. Pearce, *Brit. J. Appl. Phys. 6, 358 (1955)* and H. Looyenga, *Physica 31, 401 (1965).*
35. Carpenter, A.S., *Trans. Faraday Soc. 43, 529 (1947).*
36. Morozumi, T. and Fujiyama, C., *Shikazai Kyokaishi 44, No. 4, 161 (1971).*
37. Wolstenholme, J., *Corrosion Science 13, 521 (1973).*
38. Burns, R.M. and Haring, H.E., *Trans. Electrochem. Soc. 69, 169 (1936).*
39. Haring, H.E. and Gibney, R.B., *Trans. Electrochem. Soc. 76, 287 (1939).*
40. Whitby, L., Paint Research Assocn., Tech. Paper No. 125 (1939); Paper No. 96 (1938).
41. Zahn, H., *Corrosion 3, 233 (1947).*

42. Wormwell, F. and Brasher, D.M., *J. Iron Steel Inst. 164, 141 (1950)*.
43. Yakubovitch, S.V., Nitsberg, L.V. and Karyakina, M.I., "Proc. 6th Intern. Conf. Electrodeposition," 1964, p. 321.
44. Zurilla, R.W. and Hospadaruk, V., Tech. Paper No. 780187, Soc. Automotive Engrs. 1978, 6 pp; to be published in *Trans. SAE*.

IV. PRETREATMENT PROCESSES AND CONVERSION COATINGS

THE PROPERTIES OF PASSIVE FILMS ON METALS

J. B. Lumsden

Rockwell International Science Center
Thousand Oaks, CA

JESSE LUMSDEN

ABSTRACT

The properties of passive films on metals are discussed focussing on iron. Among the topics discussed are the thermodynamic criteria for the existence of protective films, their growth kinetics, and their composition. Changes in their kinetics and composition are correlated with changes in the environment.

INTRODUCTION

The corrosion resistance of metals depends on the stability of an insoluble product layer film. Since metals are virtually always in the film covered state, the properties of these films will affect in varying degrees any surface related process. In addition to corrosion, metallic surfaces are an important consideration in wear, lubrication, catalysis, adherence of coatings, stress corrosion cracking, erosion, optical properties, and electron emission. Thus in the broadest sense, understanding the nature of films on metal surfaces is valuable to all endeavors where metallic surfaces are important.

THERMODYNAMIC STABILITY

The existence of protective films on metals is broadly predictable from thermodynamic criteria for the stability of compounds relative to the pure metals. Thus an oxide film exists on nickel of the type NiO if

$$Ni + 1/2O_2 \rightarrow NiO \quad (1)$$

the

$$\Delta G < 0 \quad \text{for} \quad (2)$$

$$\Delta G = \Delta G^o + RT \log p(O_2). \quad (3)$$

At partial pressures below 10^{-10} atmospheres at 1000°C, the oxide is not stable. The stabilities of oxides formed in gaseous environments have been schematized graphically. Thus the relative stability can be portrayed in a Richardson and Jeffes [1] plot or as phase diagrams involving $p(O_2)$ and temperature.

Analogous to the gaseous case, a protective film will form on a metal exposed to an aqueous solution if the solubility product is exceeded for

$$Ni^{++} + H_2O = NiO + 2H^+. \quad (4)$$

The equilibrium concentration of Ni^{++} is related to the pH according to

$$\log [Ni^{++}] = 12.18 - pH \quad (5)$$

at room temperature. Here the film becomes unstable, i.e. higher $[Ni^{++}]$ as the pH is lowered. The film may also form by direct oxidation of the metal according to the half cell reaction

$$Ni + H_2O = NiO + 2H^+ + 2e \quad (6)$$

$$E_o = 0.110 - 0.060\ pH. \quad (7)$$

The equations (4-7) describe acid solubility. The nickel oxides are also soluble in caustic solutions according to the reaction

$$NiO + H_2O = HNiO_2^- + 3H^+ + 2e^- \quad (8)$$

where the $HNiO_2^-$ ion is stable. These equilibrium equations have been portrayed graphically by Pourbaix [2,3,4] showing the conditions under which corrosion (where the metal ions are the thermodynamically stable species), passivation (where the surface is covered by an insoluble film), and immunity (the film-free metal is thermodynamically stable) can be predicted.

It should be emphasized that the existence of thermodynamic stability does not, *a priori* imply protection; an insoluble product may be stable but not protective. There are other limitations associated with Pourbaix diagrams as well. For example, kinetic phenomena are not considered, also the lack of a stable phase, as calculated from bulk thermodynamics, does not preclude film formation.

PASSIVE FILM PROPERTIES

Figure 1 shows a schematic of the physical system and the processes which are of importance in passivity. At the metal-film interface a metallic solid is in contact with an ionic solid. The film consists of a distribution of alloying and environmental species. Experimental evidence suggests that the film is heterogeneous with an excess of metal ions near the substrate region and an anion excess region extending several atom layers from the film solution interface.

At constant applied potential, in the simplest case, the reaction current consists of two components, the current consumed by the growing film and the current associated with film dissolution. Thus the total current is given by

$$i_T(t) = i_F(t) + i_D(t). \quad (9)$$

In characterizing film growth kinetics the mechanistic picture of those events leading from submonolayer coverage to complete coverage and the onset of steady-state growth must be considered. However, transient conditions are not considered here.

Two mechanisms have been proposed to describe film growth after a continuous layer has formed on the surface: One proposed early by Cabrera and Mott [5] is a field assisted cation diffusion model. The rate determining

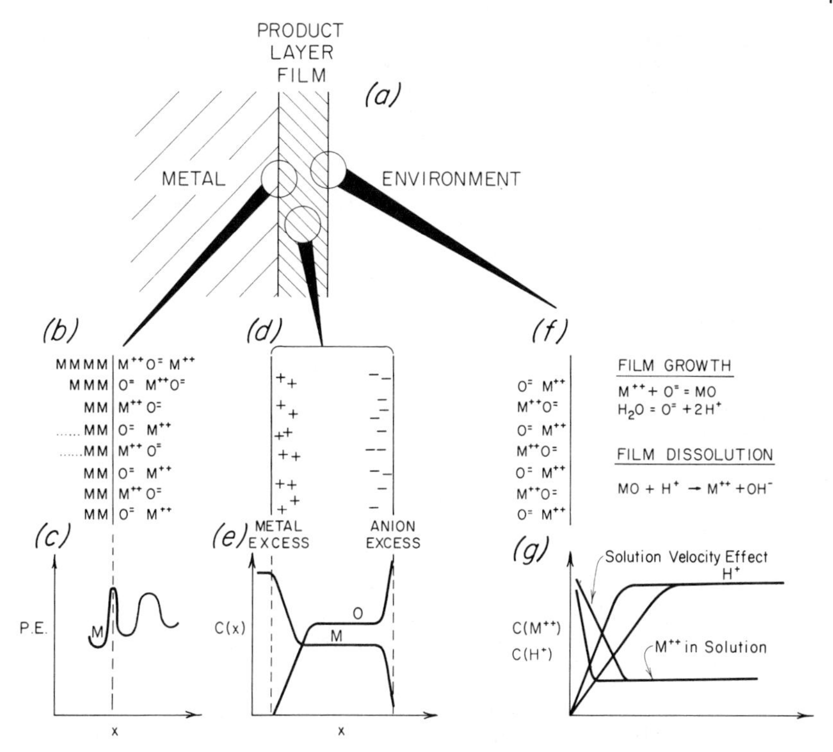

Fig. 1.—Schematic description of essential problems in considering a film-covered metal in aqueous solution; (b) shows arrangement of atoms and ions; (c) shows potential energy vs. distance plot for the ionization of a metal atom at the interface; (d) shows space charge distribution; (e) shows anion and cation distributions in film; (f) film dissolution at growth in solution; (g) mass transport processes for anion and cations.

step is considered to be the surmounting of the potential barrier at the metal-film interface by the metal ion. This model gives an ion current that varies exponentially with the field. The film thickness, x, follows inverse logarithmic kinetics

$$1/X = A - B \ln t. \tag{10}$$

A second model has been developed by Sato and Cohen [6] following a mechanism proposed by Trapnell [7]. This involves the simultaneous place exchange of all metal-oxygen pairs in a given row normal to the metal surface. The ion current varies exponentially with potential, while the film thickness varies logarithmically with time:

$$X = C + D \ln t. \tag{11}$$

Vermilyea, Young, Dewald, Sato, Cohen and others [6,8-11] have investigated these models electrochemically. In most cases it has been found that neither model uniquely describes all of the results.

Ellipsometric measurements by Kruger and Calvert [12] of iron exposed to a pH 8.4 borate-buffer solution have shown that growth kinetics can be described with an equal degree of confidence by both logarithmic and inverse logarithmic kinetics. Goswami and Staehle [13] obtained the same result for iron base alloys containing nickel and chromium.

Lukac, Lumsden, and Staehle [14] have found that the two rate laws also cannot be distinguished graphically at temperatures ranging from the near-freezing point to the near-boiling point of the borate buffer solution. It was shown that this arises because the total change in film thickness at a given potential is not very large (approximately 10%); in this case, simple relationships exist among the growth constants which hold as long as this change is small. The same investigators further found that neither model was completely in accordance with the observed temperature dependences of the growth constants.

The protective film grows at a decreasing rate (proportional to the $1/t$ for logarithmic kinetics) until a limiting thickness is reached. At this point, the dissolution rate is equal to the growth rate. The limiting thickness for the passive film on iron is approximately 50Å.

Seo, Lumsden, and Staehle [15,16] have used Auger electron spectroscopy to determine the composition of films formed on Fe exposed to the borate buffer solution. The composition depth profile was obtained by sputter-etching the film with argon ion bombardment followed by Auger analysis. Using this technique, the O/Fe composition ratio was determined as a function of depth. In every case, a region of constant composition was observed near the surface. Using the sputtered surfaces of stoichiometric oxides as standards, the film compositions corresponding to ratios in the shoulder regions are $Fe_2O_3 \cdot n\,H_2O$ where $n = 0.6\text{-}1.8$, increasing as the potential becomes more noble. Small amounts of boron were also detected in the surface and near-surface layers.

Shifts in energy were observed in the lines from certain electron transitions. A comparison of these spectra with those taken from stoichiometric oxides suggested that the iron in the outer layers is in the ferric state, while closer to the substrate it is either in a mixed ferric-ferrous state or is ferrous.

This type of heterogeneous oxide had been predicted earlier from electrochemical measurements [17,18]. Although in the above results this passivating film was formed in a borate buffer solution, its structure is not electrolyte dependent. The same type of film is formed in concentrated nitric acid [19] and one normal sulfuric acid.

EFFECT OF INORGANIC INHIBITORS

The passive film is a very protective oxide; however, it is highly susceptible to breakdown when exposed to water vapor or aqueous solutions. Breakdown can be hindered by the addition of inhibitors to the environment. Inhibitors may either be organic or inorganic, but only inorganic inhibitors will be discussed here.

There are two models for the action of inhibitors in suppressing corrosion. One is a purely electrochemical mechanism in which the only role ascribed to the inhibitor is to provide a mixed potential in the passive region [20]; the protective properties of nonoxidizing inhibitors are compatible with this idea since they require the presence of dissolved oxygen to be effective. A second idea assumes that inhibition is due to the formation of a protective layer, which acts as a physical barrier slowing the dissolution process [21].

Lumsden and Smialowska [22] have investigated the effects of inorganic inhibitors on iron. In most cases the ellipsometric results indicated film growth kinetics that were different from those characteristic of the passive film. Logarithmic kinetics were observed for iron exposed to chromate solutions; the film grew at a rate of 35Å/decade time to a limiting thickness of several hundred angstroms; this can be compared to a growth rate of 3Å/decade time and a limiting thickness of less than 50Å for the passive film. In a tungstate solution, the film thickened to approximately 300Å in 100 and then no longer thickened. The kinetics of film growth for iron exposed to a monohydrogen phosphate solution, pH 9.1, are complex. Film growth was logarithmic initially, thickening to approximately 75Å in one hundred minutes. Then the film grew rapidly to a thickness of several thousand angstroms in 200 minutes. The kinetics and optical constants which are characteristic of the passive film were observed for iron exposed to a pH 7 nitrite solution and a pH 12 phosphate solution.

The above results suggest that inhibitors promote film formation and that it is the protective properties of this film which determine the inhibitors' effectiveness. This was confirmed by Auger analysis. On exposure to chromate solutions, a film is formed that is very similar to the film which exists on stainless steel. An iron tungstate film and an iron phosphate film are formed in tungstate and phosphate solutions. The only inhibitor that operated exclusively by the electrochemical mechanism was the nitrite; the passive film was formed in this solution. Passive film formation was also observed in the pH 12 phosphate solution; however, this was not because inhibition resulted from the electrochemical mechanism, as explained below.

The inorganic inhibitors have one or more of the following properties in common: (1) they are salts of weak acids and thus have buffering capacity; (2) they are oxidizing agents; (3) they form insoluble salts with iron. Thus the picture that emerges is one in which the passive film is constantly being broken down and repaired. Breakdown can proceed by reductive dissolution (this will occur spontaneously in an aqueous environment unless an oxidizing agent is present), or at defect sites as a result of electrostatic and mechanical stresses. When breakdown occurs, the exposed metal surface undergoes anodic dissolution. Hydrogen ions are subsequently generated by hydrolysis of the metal ions causing a localized decrease in the pH, which will result in further breakdown and pit growth. Thus if an inhibitor has buffering ability it will control the pH and prevent the buildup of acid conditions. Its oxidizing ability (or the presence of dissolved oxygen) converts iron from the ferrous state to the ferric state. The ferric ions form insoluble salts blocking the defect site and preventing its propagation. The nitrite inhibitor oxidizes the exposed iron surface to form a protective oxide film. Corrosion is inhibited by the pH 12 sodium phosphate solution as a result of its good buffering properties, which prevent localized acidification, and as a result of the high pH. Ferrous hydroxide is very insoluble at this high pH. Thus any breakdown sites are quickly repaired. The precipitated ferrous hydroxide is then oxidized to Fe_2O_3 by dissolved oxygen.

REFERENCES

1. Richardson, E.D. and Jeffes, J.H.E., *J. Iron. St. Inst. 160, 261 (1948); 163, 397 (1949).*
2. Pourbaix, M., "Thermodynamics of Dilute Aqueous Solutions with Applications to Electrochemistry and Corrosion," Pergamon Press, London, 1949.
3. Pourbaix, M., "Atlas of Electrochemical Equilibria in Aqueous Solutions," Pergamon Press, London, 1966.
4. Pourbaix, M., "Lectures on Electrochemical Corrosion," Plenum Press, New York, 1973.
5. Cabrera, N. and Mott, N.F., *Rept. Progr. Phys. 12, 163 (1948).*
6. Sato, N. and Cohen, M., *J. Electrochem. Soc. 111, 512 (1964).*
7. Lanyon, A.H. and Trapnell, B.M.W., *Proc. Roy. Soc. 227A, 387 (1955).*
8. Dewald, J.F., *J. Electrochem. Soc. 102, 1 (1955).*
9. Bean, C.P., Fisher, J.C. and Vermilyea, D.A., *Phys. Rev. 101, 551 (1956).*
10. Diggle, J.W., "Oxides and Oxide Films," J.W. Diggle, Ed., 1973, Vol. 2, p. 281.
11. Bulman, G.M. and Tseung, A.C.C., *Corrosion Sci. 12, 415 (1972).*
12. Kruger, J. and Calvert, J.P., *J. Electrochem. Soc. 114, 43 (1967).*
13. Goswami, K.N. and Staehle, R.W., *Electrochima Acta 11, 1895 (1971).*
14. Lukac, C.N., Lumsden, J.B., Smialowska, Z. and Staehle, R.W., *J. Electrochem. Soc. 122, 1571 (1975).*
15. Seo, M., Lumsden, J.B. and Staehle, R.W., *Sur. Sci. 42, 337 (1974).*
16. Seo, M., Lumsden, J.B. and Staehle, R.W., *Corrosion Science 17, 209 (1977).*
17. Nagayama, M. and Cohen, M., *J. Electrochem. Soc. 109, 781 (1962).*
18. Nagayama, M. and Cohen, M., *J. Electrochem. Soc. 110, 670 (1963).*
19. Lumsden, J.B. and Staehle, R.W., *ASTM S.T.P. 596, 39 (1976).*
20. Stern, M., *J. Electrochem. Soc. 105, 638 (1958).*
21. Hoar, T.P. and Evans, U.R., *J. Chem. Soc. 2476 (1932).*
22. Lumsden, J.B. and Szklarska-Smialowska, Z., *Corrosion 34, 169 (1978).*
23. Cohen, M., *J. Electrochem. Soc. 121, 191C (1974).*

ANODIC COATINGS FOR ALUMINUM

William C. Cochran and
Donald O. Sprowls

Alcoa Laboratories,
Alcoa Center, PA

WILLIAM COCHRAN

ABSTRACT

Classifications and minimum requirements of porous type anodic coatings according to ASTM and military specifications are summarized. Properties (coverage, composition, resistance to dissolution, adhesion and brittleness) affecting the coatings' protective value are reviewed. Exemplary test data are cited indicating the generally good weathering performance and protective value of the coatings in industrial and seacoast atmospheres. Limitations in the coatings' protective abilities are craze-cracking, edge defects, and microflaws arising from the behavior of certain alloying constituents during anodizing. Edge defects and microflaws can be minimized by radiusing of edges and by avoidance of high current densities during the initial period of anodizing. The inherent brittleness of anodic coatings makes it difficult to avoid craze-cracking. Anodic coatings alone are not reliable for protection of susceptible alloys against stress corrosion cracking (SCC). Data are cited showing that anodic coatings can retard, accelerate, or have little effect upon resistance to SCC, depending upon conditions. Two approaches for possible improvement in the coatings' ductility and protective ability that appear worthy of further investigation are "eutectic anodizing" and pulsed current anodizing.

INTRODUCTION

Anodic oxide coatings have been used extensively for the protection of aluminum alloys since the 1920's when commercial anodizing processes were first developed. Many applications of anodic coatings have an aesthetic purpose of achieving and maintaining a desired decorative appearance over periods of many years. Obviously, to accomplish this, the coating must protect the substrate metal from unsightly corrosion. In other applications, anodic coatings serve purely engineering functions, imparting protection against corrosion, wear resistance, hardness, maintenance of optical properties or insulation to surfaces of aluminum components. The purpose of this paper is to discuss characteristics of the coatings affecting their use in the corrosion control of aluminum alloys. The performance of anodized aluminum in different environments is also reviewed.

CLASSIFICATION OF ANODIC COATINGS

Barrier layer or nonporous oxide films, formed in aqueous electrolytes having low solvent power for the oxide, are too thin (<1 μm) for effective corrosion control and are seldom used for this purpose. Barrier coatings are produced in borate or tartrate electrolytes and are used principally on high purity aluminum foil for electrodes in electrical capacitors.

Porous type anodic coatings in the 1-30 μm thickness range are normally used for corrosion protection and decorative finishes. They are produced in aqueous, acid electrolytes having moderate oxide solvent power, e.g., sulfuric, chromic, or oxalic acid, and in mixed acid electrolytes containing a major quantity of organic sulfonic acid with a minor quantity of sulfuric acid. The latter electrolytes are used for producing integrally colored, hard type coatings (2XXX series alloys excluded). Chromic acid electrolyte does not attack aluminum and is used to anodize aluminum components where there is danger of entrapment of electrolyte in crevices or joints. Chromic acid anodic coatings are limited to thicknesses up to 8μm, dependent upon alloy and anodizing conditions. Oxalic acid anodizing is little used commercially because the sulfuric acid process is less costly and more convenient to employ. Porous phosphoric acid type anodic coatings from 1-3 μm thick are sometimes used as basis for the electrodeposition of other metals on aluminum, or for adhesive bonding, but are not employed for protective purposes.

Porous anodic coatings always have a thin, nonporous barrier zone at the metal/oxide interface. Porosity can vary from a few to 25% of the volume of the coating. For best corrosion protective properties, the porous outer region of the coating is sealed by hydrothermal treatment in water, metal salt solutions, or steam (rarely).

Hard Coatings represent a special class of porous anodic coatings used principally for the good wear and scratch resisting properties they confer upon aluminum alloy surfaces. They are produced under conditions of lower solvent action by the electrolyte on the oxide, consequently are denser and can be made thicker (50-100 μm) than conventional porous anodic coatings. Hard coatings require higher forming voltages (25-100V) than conventional sulfuric acid coatings (15-20V).

ASTM *Classification* of anodic coatings recommended for various service conditions is summarized in Table 1. Coating thickness is the principal classification parameter.

Military Specification MIL-A-8625C classification and requirements for anodic coatings, for use by Departments and Agencies of the Department of Defense, are summarized in Table 2. Weight of the anodic coating and its protective ability in neutral salt spray are the principal criteria for Types I and II coatings. Thickness and resistance to wear are specified for Type III coatings. Coatings are further classified as dyed or not-dyed.

CHARACTERISTICS OF COATINGS AFFECTING THEIR PROTECTIVE QUALITIES

Coating Coverage. By its very nature the anodizing process has good throwing power, i.e., ability to develop coating uniformly over all surfaces of an aluminum component exposed to the anodizing electrolyte. As oxide coating forms on one area, it partially impedes the current flowing there, forcing it to flow to uncoated or more thinly coated surfaces where the film electrical resistance is lower. Thus, recessed areas and inside surfaces of

TABLE 1.—ANODIC COATING TYPES AND SERVICE CONDITIONS (adapted from ASTM B580)

	Type	Min. Coating Thickness µm	Electrolyte		Service Condition		Typical Applications
A	Engineering Hard Coat	50	H_2SO_4 or H_2SO_4 + oxalic acid	5.	Very Severe Prolonged atmospheric weathering, high wear conditions.	A,B	Unmaintained exterior architectural components. Machinery or marine parts.
B	Architectural Class I	18	H_2SO_4, or organic sulfonic acid +				
C	Architectural Class II	10	H_2SO_4	4.	Severe Resist scratching, abrasion, weathering and corrosion.	C	Maintained exterior architectural parts.
D	Automotive, Exterior	8	H_2SO_4			D	Exterior automotive trim.
E	Interior, Moderate Abrasion	5	H_2SO_4	3.	Moderate abrasion, occasional wetting.	E	Appliances, furniture, nameplates, reflectors.
F	Interior, Limited Abrasion	3	H_2SO_4	2.	Mild Indoors, minimum wear.	F	Automotive-interior, housewares, enclosed reflectors.
G	Chromic Acid	1	CrO_3	1.	Crevice Condition, humid, little or no abrasion.	G	Aircraft assemblies with lap joints. Base for paint.

TABLE 2.—MILITARY SPECIFICATION MIL-A-8625C REQUIREMENTS ANODIC COATINGS FOR ALUMINUM ALLOYS

	Type	Min. Coating Wt., g/m^2		Salt Spray Resistance 336 Hours	Max. Color Change Class 2 Coatings 200 hours - carbon arc light	Alloy Limitations*
		Class 1 (Not Dyed)	Class 2 (Dyed)			
I	Chromic acid, sealed	1.9	4.7	Max. of 15 isolated pits in 10 dm^2, or 5 in 2 dm^2, none >0.8 mm in diameter.	3E units	Max. Cu = 5% Max. Si = 7.5%
II	Sulfuric acid,	5.6	23,2†		3E units	--
III	Hard coatings‡ not sealed*	--	--	--	3E units	Max. Cu = 5% Max. Si = 8%

* Unless otherwise specified in contract.
† 13 g/m^2 for 2XXX alloys and certain casting alloys.
‡ Nominal thickness = 51 ±10% μm, unless otherwise specified. There is an abrasive wear test requirement.

holes on aluminum parts exhibit coating coverage. The good conductance (.05 S cm^{-1}) of sulfuric acid anodizing electrolyte contributes to its excellent throwing power.

Chemically, porous anodic coatings as formed are substantially amorphous, anhydrous aluminum oxide, but may contain up to a few percent of free plus combined water, and always incorporate anions from the electrolyte, typically 15% SO_4, 3% oxalate, 7% PO_4 and <0.1% Cr from the respective acid electrolytes. Small quantities of alloying elements in the aluminum alloy occur in the anodic coating, such as iron, magnesium, copper, manganese, chromium, zinc and silicon. They may be present as oxidation products or in unreacted intermetallic constituent particles. When sealed in hot nickel acetate solutions, the coatings incorporate a small amount of nickel hydroxide in the outer region, which does not adversely affect resistance to corrosion. Nickel salt solutions are used because they seal the porosity more efficiently than plain hot water. Advantage is sometimes taken of the coating's porosity to sorb chromates in the pores from hot, aqueous solutions. Incorporation of chromate inhibitor in anodic coatings is effective in preventing corrosion of the metal substrate in chloride environments so long as sufficient leachable hexavalent chromium is present.

Resistance to dissolution by dilute aqueous solutions is an important property of anodic coatings affecting their protective ability. The rate of dissolution of anodic coatings is pH dependent and is minimal in the range 4 to 8.5. The curve for the solubility of aluminum trihydroxide as a function of pH[1] (Fig. 1) indicates minimum solubility in the same pH range as for anodic coatings. Like aluminum hydroxides, the anodic oxide is amphoteric and dissolves readily in strong acids and bases. In aqueous solutions of intermediate pH, the solubility is low.

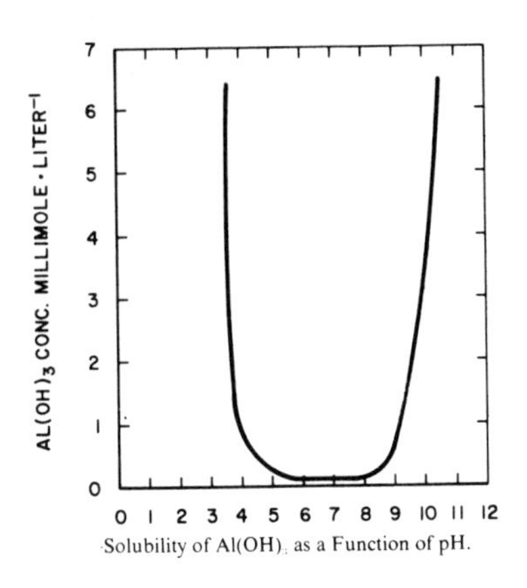

Fig. 1.—Solubility of Al(OH)$_3$ as a function of pH.

Adhesion of anodic coatings is excellent since the oxide coating is integral with the metal substrate from which it was formed. The coating will not normally spall off upon bending or impact of the anodized part, nor when it is subjected to thermal shock, by immersion in liquid nitrogen or oxygen, nor when exposed to temperatures up to the melting point of the aluminum alloy. Thick ($\geq$50 μm), hard coatings may exhibit spalling or flaking on the compression side of a part when severely bent.

Mechanical Properties. Like glass, anodic coatings are very hard, but brittle. They have the hardness of corundum and typical elongation values of only 0.3 to 0.4%. The coating will crack from deformation of the anodized part by bending, stretching, or by thermal expansion, since the linear coefficient of expansion of the oxide is lower than that of the basis metal. As a rule, the thicker the coating, the greater the tendency for cracking. These fine hairline fissures or craze-cracks act as stress risers and can lower the fatigue strength of anodized structural parts by as much as 35%, depending upon factors such as thickness and sealing of the coating. In severe corrosive environments, craze cracks can become sites for corrosive attack of the basis metal.

Corner or edge defects (Fig. 2) can occur on anodized parts because the coating grows perpendicular to the surface from which it is formed.

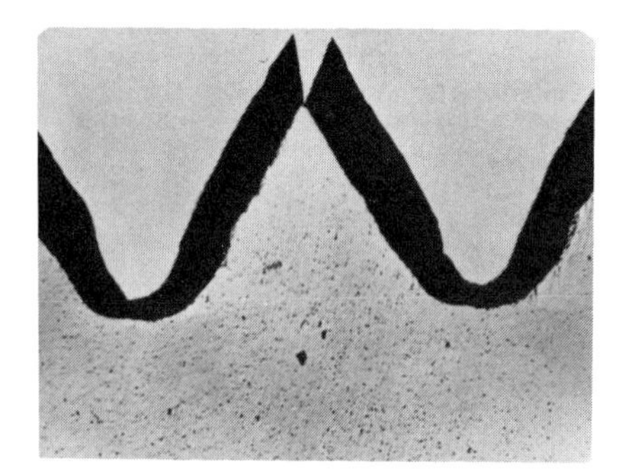

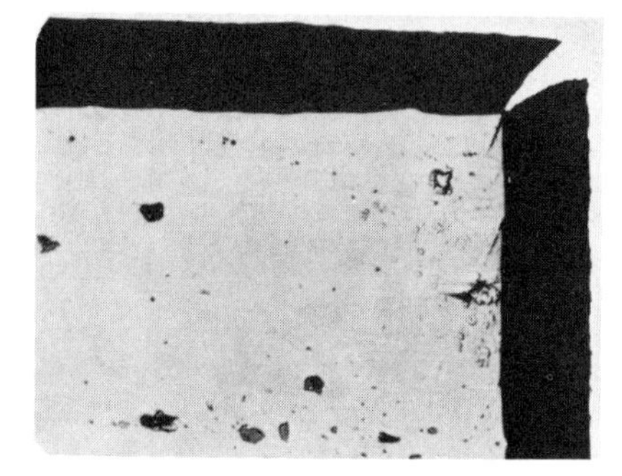

Fig. 2.—Voids in hard anodic coating at sharp corners and edges of parts.

A void occurs at sharp edges because the coating cannot fill the gap. By rounding the corners of sharp edges before anodizing, such voids are eliminated. For example, for a 25 μm thick coating, the minimum radius of an edge should be 0.8 mm to achieve a continuous coating over it.

SOURCES OF MICROFLAWS

Alloy Composition and Metallurgical Structure

Besides craze cracks and edge defects, the other source of flaws in anodic coatings arises from the behavior of alloying constituents during anodizing. High purity aluminum yields the most flaw-free anodic oxide films. However, many of the common wrought alloys of aluminum (1XXX, 3XXX, 4XXX, 5XXX, 6XXX series) yield anodic coatings having protective value similar to that achieved on high purity aluminum.

Of various alloying elements commonly added to aluminum, the most detrimental from the anodizing standpoint is copper, when present in amounts >1 wt% (2XXX and certain 7XXX series alloys). Copper, whether in solid solution or present as an intermetallic microconstituent, e.g., $CuAl_2$, Al_7Cu_2Fe, tends to dissolve during anodizing, resulting in lower coating efficiency and coatings of lower density and thickness and higher porosity. Examples of voids in hard coatings caused by Al_7Cu_2Fe particles in alloy 7075 are shown in the coating cross-section micrographs of Fig. 3[2]. Investigators at Boeing Commercial Airplane Company have recently developed an improved anodizing procedure to yield relatively pit-free, hard-type coatings on 7XXX-series alloys[3]. Hard coatings applied with full current density at the outset developed relatively large voids, or pits in the coating owing to "gas rupture" of the coating at sites of large Al_7Cu_2Fe constituent particles. "A major reduction in gas rupture void formation was obtained by utilizing a slow current buildup and a lower than normal current density."

Iron-bearing constituents, principally the aluminum-iron-silicides and $FeAl_3$, are attacked anodically during anodizing and may be partially or completely dissolved, resulting in voids in the oxide coating. Magnesium silicide (Mg_2Si) tends to dissolve during anodizing and if present as large platelets, can result in voids.

Size and distribution of alloy constituent particles are important. Large constituent particles, or strings and clusters of smaller particles, in the substrate matrix are undesirable for achieving void-free anodic coatings that provide continuous coverage of the basis metal with insulating oxide. As a general rule, when the maximum dimension of a constituent particle or cluster of particles is 1/3 or less of the final oxide coating thickness, the coating will likely be adequately continuous and protective.

Constituent Blocking and Mound Formation

Two principal ways in which constituent particles can disrupt the integrity of anodic coatings are illustrated schematically in Figures 4 and 5. Micrographs of coating cross-sections showing examples of constituent blocking and mound formation are shown in Figure 6. To minimize such microflaws, the current density during anodizing should be kept low initially then gradually increased to its normal value after the initial coating has formed.

PERFORMANCE IN DIFFERENT ENVIRONMENTS

Industrial Atmosphere

Properly sealed, porous type anodic coatings provide good protection to aluminum alloys. As shown in Fig. 7, the number of pits into basis metal decreases exponentially with increasing coating thickness[4]. The specimens in this test were of alloy 1100 sheet with conventional sulfuric acid anodic coatings, sealed in boiling water. They were exposed without maintenance for 8-1/2 years at 45° from vertical, facing south in the New Kensington, PA, industrial atmosphere. After exposure, the anodic coating on each specimen was dissolved off in stripping solution, without attacking the metal substrate, and the number of pits per cm^2 in the metal surface was determined. The pits were of pin point size and had penetrated less than 50 μm into the metal. Specimens with coatings of 22 μm or greater thickness were practically pit-free on the skyward facing surfaces. In another test, the benefit of an 11 μm thick coating on 6053 alloy in reducing the number of corrosion sites upon 16 years exposure is shown in the cross sectional views of Fig. 8[5].

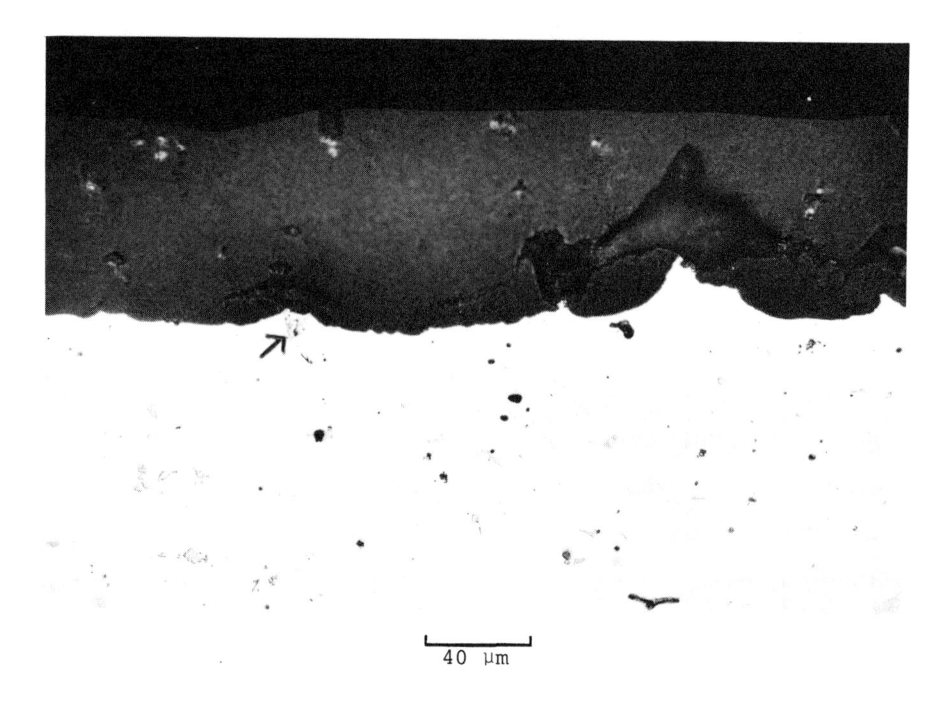

Fig. 3.—Gas rupture voids in hard anodic coatings on 7075-T6 forging caused by Al_7Cu_2Fe constituent particles 500X.

(a)

CONSTITUENT PARTICLE IN AS-FABRICATED SURFACE.

(b)

PARTICLE IN RELIEF AFTER CAUSTIC ETCH.

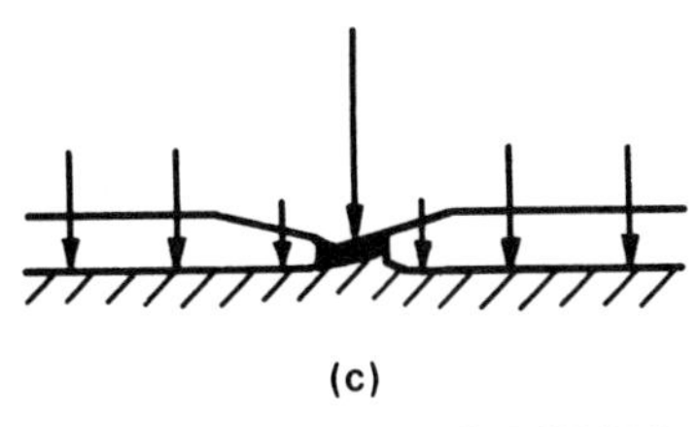

(c)

PARTICLE BLOCKING DURING COATING FORMATION. LENGTH OF ARROWS INDICATES MAGNITUDE OF CURRENT.

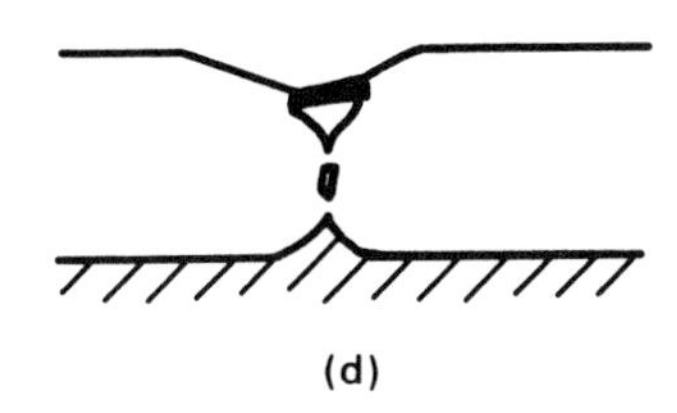

(d)

FULLY DEVELOPED COATING WITH OCCLUDED, UNDERCUT METALLIC PARTICLES CAPPED BY CONSTITUENT PARTICLE.

Fig. 4.—"Blocking" by constituent particle during anodizing.

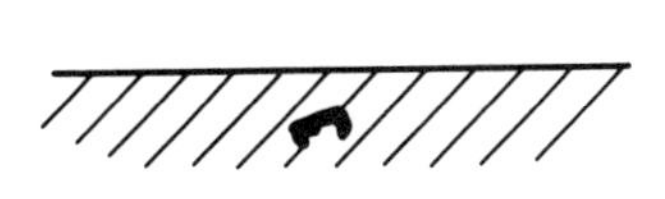

(a)

AS-FABRICATED SURFACE WITH INTERMETALLIC PARTICLE BELOW SURFACE.

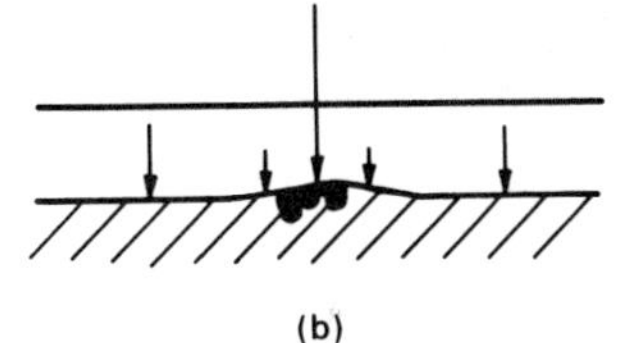

(b)

PARTICLE "ROBS" CURRENT FROM ADJACENT MATRIX. LENGTH OF ARROWS INDICATES MAGNITUDE OF CURRENT.

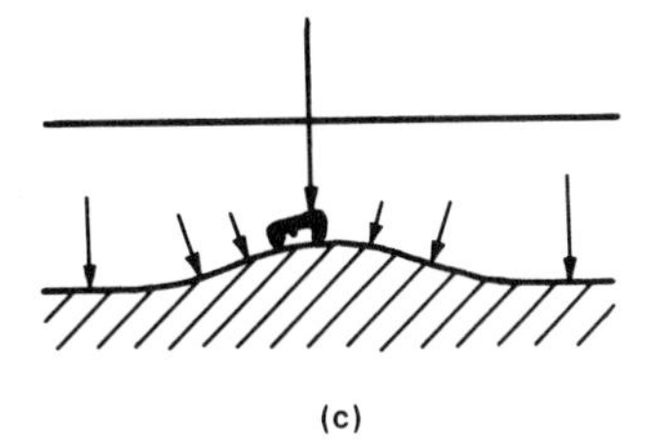

(c)

PARTICLE CONTINUES TO ROB CURRENT AND BLOCKS COATING FORMATION.

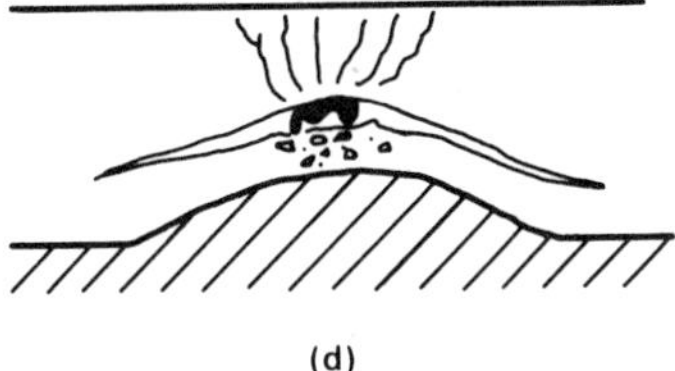

(d)

FULLY DEVELOPED MOUND WITH OCCLUDED METALLIC PARTICLES AND CRACK. SPONGY COATING ABOVE PARTICLE RESULTS FROM HIGHER, LOCALIZED TEMPERATURE.

Fig. 5.—Mound development.

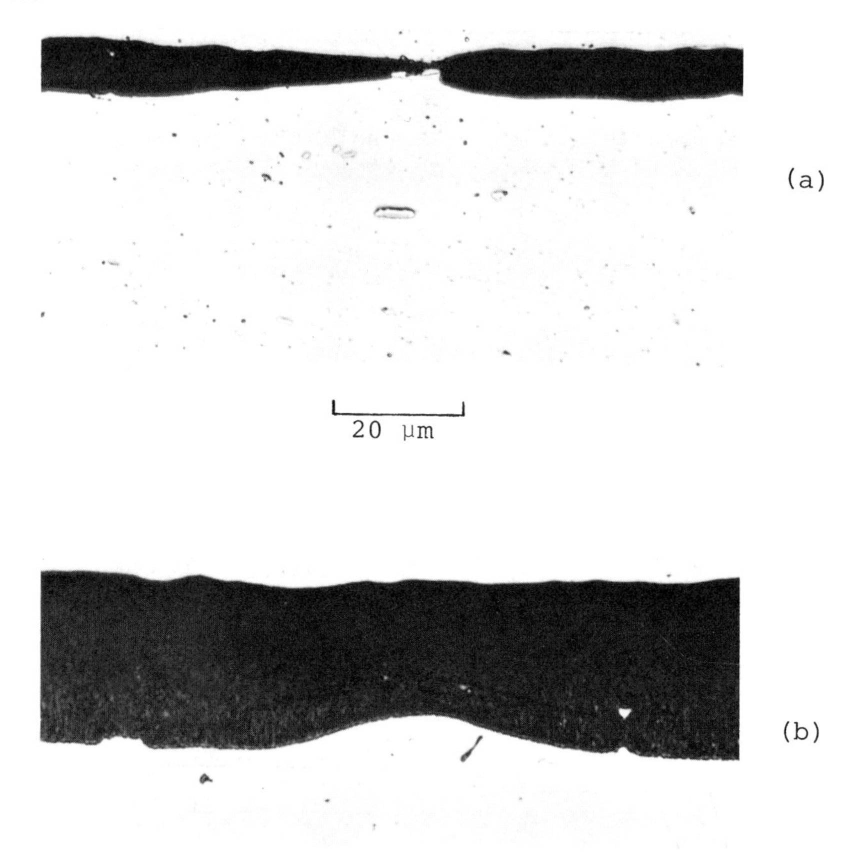

Fig. 6.—Cross-sections of anodic coatings showing microflaws, 1000X. (a) constituent blocking (b) mound formation.

The good resistance to erosion by weathering of sulfuric acid type anodic coatings on several aluminum alloys exposed up to 18 years in the New Kensington industrial atmosphere is shown in Fig. 9[4]. The data are plotted from an arbitrary base of 25 μm original coating thickness. Thickness changes were determined by microscopic measurement of cross sections of original and exposed coatings. The average erosion rate is only 0.33 μm/year.

Many examinations of cross-sections of weathered anodic coatings show that they erode uniformly. An example of the erosion tendency of an integrally colored, hard type coating 22 μm thick, produced in sulfophthalic (10 wt%) plus sulfuric acid (0.6 wt%) electrolyte is shown in Fig. 10[6]. The good appearance retention and freedom from pitting of similar coatings after 11 years' atmospheric exposure is illustrated in Fig. 11[6].

Seacoast exposure of 23 μm thick sulfuric acid coatings on various alloys for 3 years caused no visible surface pitting on most of the specimens, except for 7039 extrusion alloy, and along edges of 4 of the sheet alloys (Table 3)[7]. In this same test, it was observed that 51 μm thick

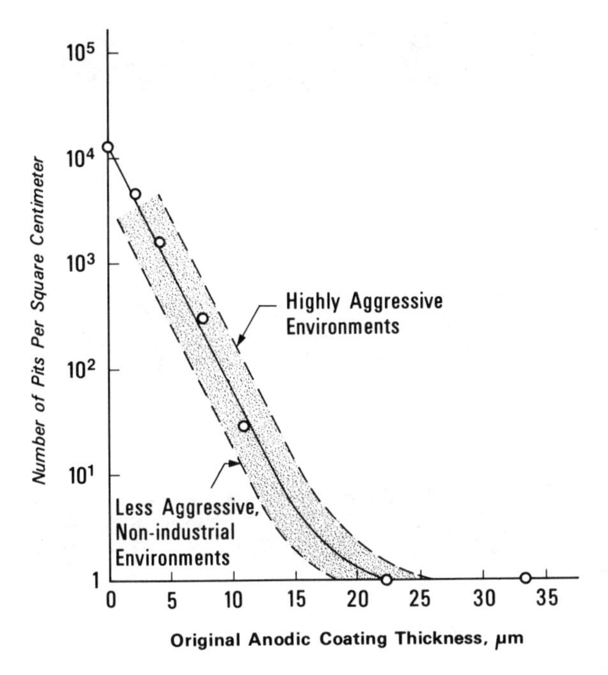

Fig. 7.—Pitting of anodized 1100 alloy as a function of coating thickness. 8.5 yr. exposure to industrial atmosphere.

coatings did not protect those alloys exhibiting pitting any better than did the 23 μm coatings. These results confirm a general observation that optimum protection against atmospheric corrosion is achieved in the 18-30 μm coating thickness range and that use of thicker coatings adds little.

Exterior Automotive Exposure, sometimes termed the "streetosphere" because of the presence of de-icing salts and other ingredients of road splash, to distinguish it from static exposures in fields or on buildings, is well resisted by the anodized aluminum alloys used for bright trim and bumpers. To retain brightness and image clarity on brightened parts, the anodic coatings employed are limited to 8-10 μm in thickness. These coatings perform well (Fig. 12)[5], partly because of the cleaning maintenance they receive from normal car washings. The development of a hazy or "bloomed" appearance of the coating is considered more of a problem than pitting in these applications. Blooming is a result of microroughening of the anodic coating surface causing light scattering, and can arise from inadequate sealing, or use of too harsh alkaline cleaners in car washing installations.

PROTECTION AGAINST STRESS CORROSION

Anodic coatings are generally not considered as a practical means of protection against stress-corrosion cracking (SCC) unless used as a part of a protective system that includes other methods such as shot peening and paint coatings. Test data can be cited to show that the presence of an anodic coating can either retard SCC, accelerate SCC, or have no appreciable

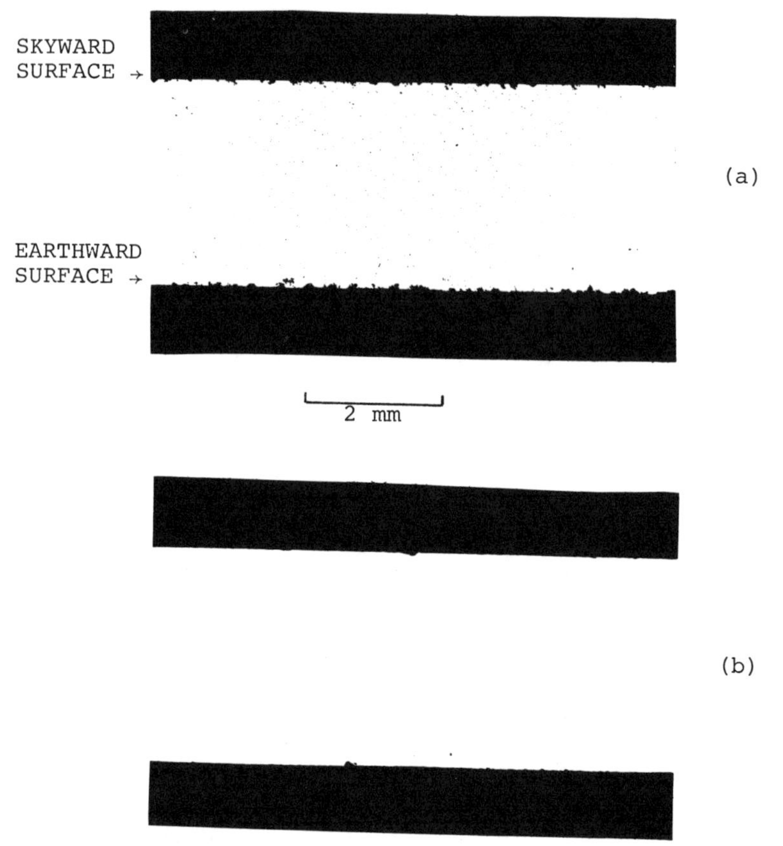

Fig. 8.—How anodic coating protects 6053 alloy from pitting during 16 year exposure to industrial atmosphere (a) bare specimen (b) anodized specimen. Cross-sections through specimens, 12X.

effect. There are several factors that determine which of these types of performance will be obtained, and examples of these effects are presented below.

Test data for 7075-T651 alloy plate stressed in the critical short transverse direction show that a 5 μm anodic coating can retard SCC. Test conditions involved a material with relatively high susceptibility under a low stress applied after anodizing. Strain at the surface of the specimens was low in the elastic range and did not crack the anodic coating during the stressing. The average length of time to failure was increased and so was the range of times and the standard deviation (Table 4). The role of the coating in preventing surface corrosion, thereby delaying the initiation of SCC, was indicated in this experiment by per cent losses in ultimate strength after 15 days' exposure of 11% for anodized compared to 21% for unprotected specimens exposed with no applied stress.

The test data in Table 5 show that a thicker oxide coating can be more protective than a thin one, as is the case in preventing general corrosion.

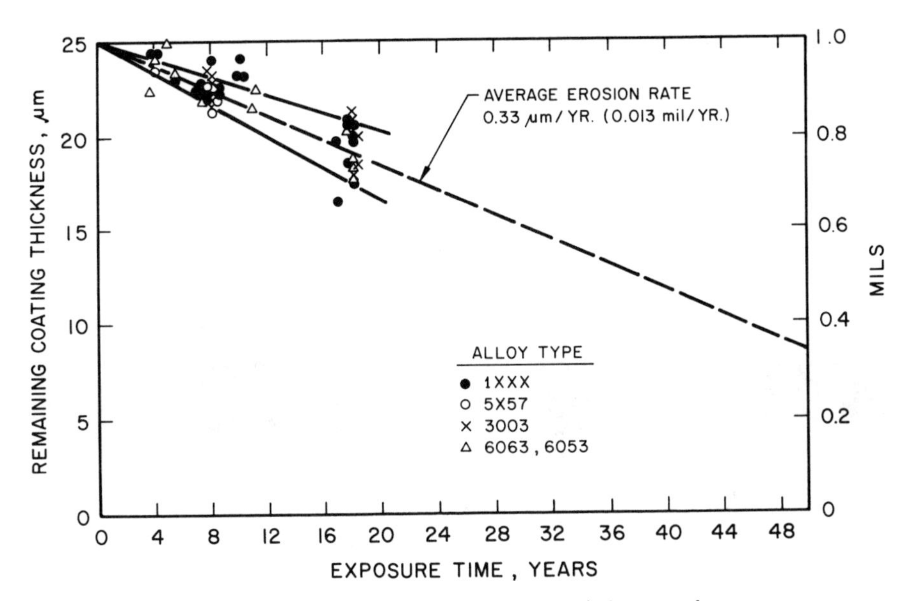

Fig. 9.—Weathering of anodic coatings in industrial atmosphere.

It must be noted that these test specimens had the advantage of being stressed before anodizing so that the oxide coatings were not cracked even though the strain was in the vicinity of the elastic limit of the metal.

A coating of the same thickness (5.1 μm) proven to be protective in the first two illustrations is shown by the data in Table 6 to accelerate the occurrence of SCC under less favorable conditions. These data were obtained on test materials that were stressed by plastically deforming the metal after anodizing, thus, cracking the oxide coatings. The specimens were dented to simulate a type of damage that could occur in service. The 5.1 μm coating on Alloy A accelerated the initiation of SCC in both the laboratory test and in the outdoor atmospheric exposure. Although a still thicker coating (11.3 μm) tended to be protective on alloy B in the accelerated test, it did not prevent the initiation of SCC in the atmospheric exposure.

Alumilite 205 sulfuric acid anodized coatings, which contain an impregnated chromate corrosion inhibitor, have been useful in commercial applications for preventing general corrosion. Therefore, this type of coating and a 50 μm thick hard coating, Alumilite 226, were included in a contract study for NASA (Marshall Space Flight Center) on the use of protective coatings for the prevention of SCC[8]. Interference stressed O-rings of stress corrosion susceptible materials were used in this program to obtain a relatively large metal surface area with a uniform high elastic strain. The rings were stressed before anodizing to represent the conditions often present in aerospace structures containing residual stresses. The results of atmospheric exposure tests (Table 7) show that neither type of coating was effective. Although the inhibited Alumilite 205 coating appeared to increase the time to failure slightly for alloy 2014-T651, the opposite effect occurred for alloy 7079-T651. The 50 μm thick hard coating had no

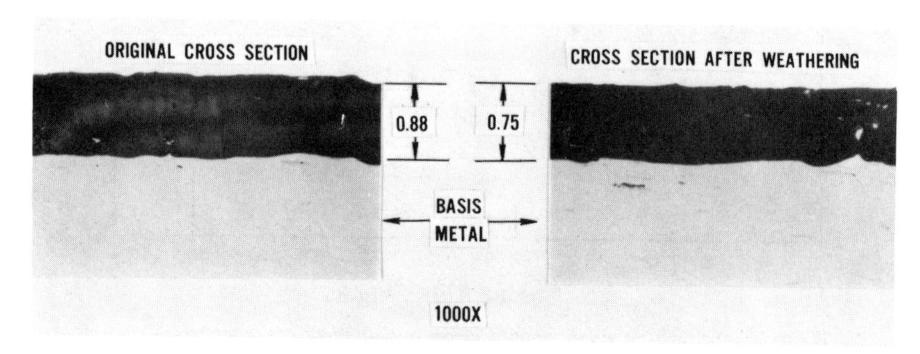

Fig. 10.—Weathering of integral bronze colored coating, 11 years' exposure in industrial atmosphere, average coating thicknesses in mils, color change: 4.5 e units, alloy 6063, anodized in sulfophthalic/sulfuric acid.

Fig. 11.—Alcoa duranodic 300 integral color finish, 11 years industrial atmosphere exposure, lower panels are unexposed controls.

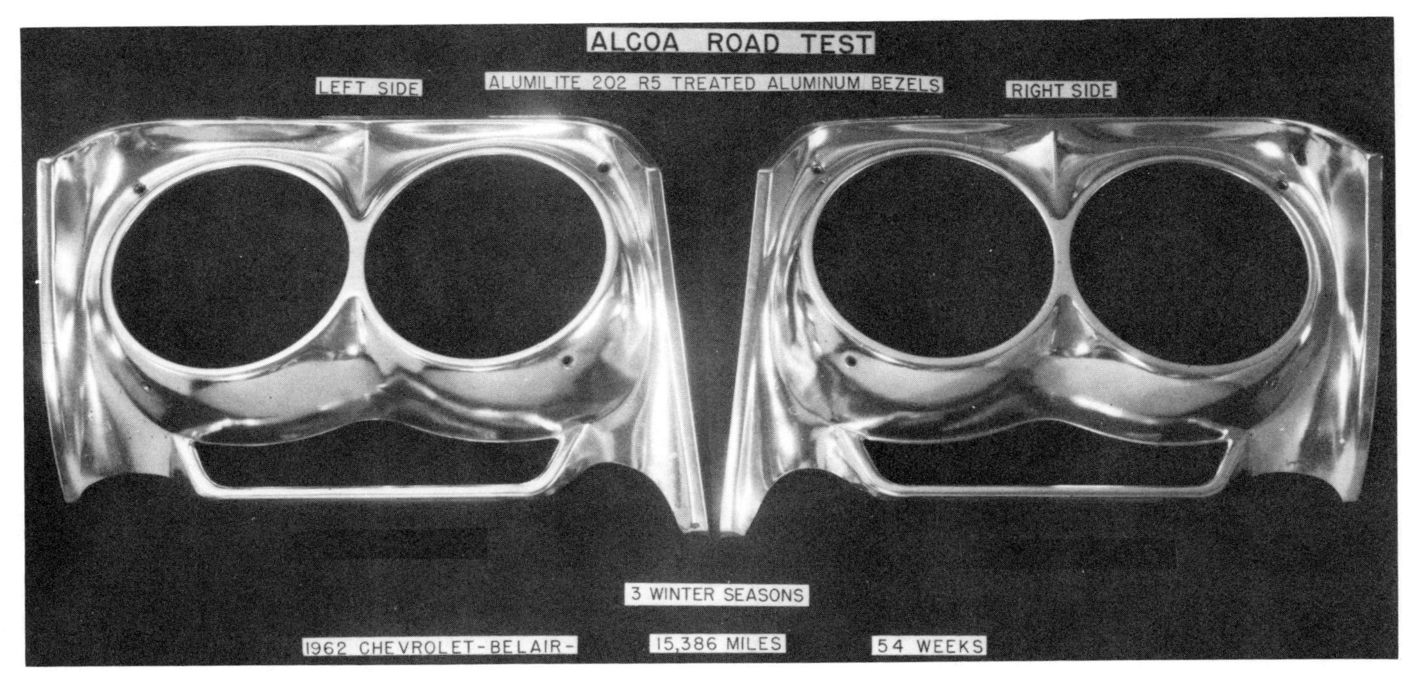

Fig. 12.—Shows good condition of bright anodized headlamp bezels after a 3 winter service exposure. 8 μm thick H_2SO_4 anodic coatings.

TABLE 3.—THREE YEAR SEACOAST EXPOSURE OF ANODIZED ALUMINUM ALLOYS
23 μm Thick H_2SO_4 Anodic Coatings, Sealed in Boiling Water
10 x 15 cm Test Panels

Alloy	Form	Results
1100	Sheet	No visible pitting
5086-H34	Sheet	No visible pitting
5456-H343	Sheet	No visible pitting
6061-T6	Sheet	No visible pitting
7039-T6	Sheet	No visible pitting
7075-T6	Sheet	Edge pitting only
7079-T6	Sheet	Edge pitting only
Alclad 2024-T3	Sheet	Edge pitting only
Alclad 7075-F	Sheet	Edge pitting only
6061-T6	Extrusion	No visible pitting
6063-T5	Extrusion	No visible pitting
6070-T6	Extrusion	No visible pitting
6351-T6	Extrusion	No visible pitting
7039-T6	Extrusion	Scattered, small pits

appreciable effect on the performance of either alloy. Even though the stress in these specimens was fairly high, the rings were generously radi-used to avoid coating defects at the edges and the coatings were applied to prestressed assemblies so that the coatings would not be cracked during stressing. The relatively small protective value of these particular anodic coatings compared to other protective systems included in this NASA contract study is shown graphically in Fig. 13. The more effective environmental barrier provided by the epoxy paint (representing a good organic coating system) is evident; and it also is evident that when that good barrier was damaged by scribing a narrow groove through the coating, the protection against SCC was almost completely lost. It is interesting that the performance of the anodic coatings was not affected by similar scribe damage.

In a subsequent contract with the Air Force to develop an improved stress corrosion inhibitive oxide coating, work was concentrated on anodic coatings produced in a low melting point, fused nitrate-nitrite eutectic mixture and impregnated with a sparingly soluble chromate inhibitor[9]. Limited accelerated SCC test data showed longer times to fail for 7075-T651 specimens with a eutectic anodized coating than those with a conventional sulfuric acid coating of the same thickness (Fig. 14). Also, in corrosion fatigue tests of 7075-T651 (Krause-type bending fatigue specimen), the experimental oxide coating had less of a deleterious effect on the number of cycles to failure. This combination of advantages is unusual because one of the inherent disadvantages of anodic oxide coatings is that they are brittle, and when they are thick enough to provide significant protection against corrosion, they also detract from the fatigue resistance of the substrate.

FUTURE INVESTIGATIONS

Although they perform well in most architectural and decorative trim applications, conventional anodic oxide coatings that are used for protection against corrosion have certain inherent characteristics that make them impractical for many structural applications involving high strength aluminum alloys. For example, their brittleness is deleterious to fatigue resis-

TABLE 4.—PROTECTION AGAINST SCC BY A 5 μm SULFURIC ACID COATING ON SPECIMENS STRESSED WITH MODERATE ELASTIC STRAIN *

	Days to Failure †	
	Degreased	Anodized
	4	5
	4	5
	4	8
	4	12
	4	12
	5	18
	5	29
	8	33
	8	36
	10	36
	15	38
	15	38
	17	43
	26	52
	26	66
Mean T_f	10.3 days	28.7 days
Range	22 days	61 days
Std. Deviation	7.8 days	18.2 days

NOTES: * Short transverse 3.2 mm dia. tension specimens machined from 50 mm thick 7075-T651 plate and stressed 32% YS (150 MPa).

† Exposed to 3.5% NaCl solution by alternate immersion (ASTM G44).

tance, especially the thicker coatings that offer greatest corrosion protection. Because of their brittleness, even relatively minor plastic deformation causes cracks that may interfere with protection against stress-corrosion cracking. Their effectiveness as a barrier to the environment to control pitting corrosion, as well as SCC, and corrosion fatigue, is impaired by microflaws in the coatings. Such flaws are related to composition of the substrate alloy, coating procedure and coating thickness. Some of the experimental work cited above, however, suggests that improvements in these characteristics of anodic coatings may be possible (and hopefully practical).

The exploratory work on anodization in molten baths of eutectic nitrate-nitrite mixtures [9,10] points toward a desirable combination of improved protection against both SCC and corrosion fatigue. Additional evaluation of the properties and characteristics of such coatings seem desirable.

Other novel approaches aimed at minimizing coating microflaws and increasing the coating's ductility should be considered. One such approach is the use of pulsed currents, or wave forms of current other than straight

TABLE 5.—EFFECT OF THICKNESS OF ANODIC COATINGS ON PROTECTION AGAINST SCC OF SPECIMENS STRESSED WITH HIGH ELASTIC STRAIN

7075-T6511 Extruded Bar, Transverse Stress 75% Y.S. (350 MPa) *

Specimen Preparation	Days to Failure †
As Machined, degreased	4, 4
Stressed; then anodized in CrO_3 (2.1 μm)	15, 16
Stressed; then anodized in H_2SO_4 (11.3 μm)	64, 365

NOTES: * C-ring test specimens 19 mm O.D. x 1.57 mm thick machined from a 50 mm square bar.

† Exposed to 3.5% NaCl solution by alternate immersion per ASTM G44 except that the tests were performed under ambient lab atmospheric conditions and the 3.5% NaCl solution was made from commercial salt and New Kensington, PA, tap water.

d.c., during anodizing. Some proprietary anodizing processes employing specific wave forms of current, or pulsed currents, are claimed to produce more ductile coatings with less tendency to crack. It is further claimed that less flawed anodic coatings were obtained on high strength alloys than is the case in conventional anodizing. Since there is little published data supporting these claims, it seems appropriate that coatings produced by this promising approach should be further investigated.

TABLE 6.—EFFECT OF THICKNESS OF ANODIC COATINGS ON PROTECTION AGAINST SCC OF SPECIMENS STRESSED BY PLASTIC DEFORMATION *

Coating Thickness † (Alloy) **	Depth of Indentation, mm	Time to Failure: Hours in Boiling 6% NaCl		Time to Failure: Days in Industrial Atmosphere	
		Bare	Anodized	Bare	Anodized
5.1 µm (Alloy A)	0.3	153-177	49- 52	OK 2255	OK 2255
	1.0	98-105	52- 55	2255 ‡	1460-1835
	2.5	98-105	26- 28	2255 ‡	900-1100
11.3 µm (Alloy B)	0.3	77- 96	OK 240	2108 ‡	2108 ‡
	1.0	24- 48	147-163	2108 ‡	2108 ‡
	1.0	24- 48	147-163	2108 ‡	2108 ‡

NOTES:

* Stressed after anodizing by pushing a 25 mm dia. ball into the metal surface; the deepest indentation is the maximum that the metal would take without rupture.

† Anodized in H_2SO_4 and sealed in boiling water.

** Extruded shapes 6.4 mm thick heat treated to T5 condition:
A - 0.09% Fe, 0.09% Si, 0.18% Cu, 0.00% Mn, 0.77% Mg, 6.75% Zn, 0.01% Cr
B - 0.06% Fe, 0.07% Si, 0.27% Cu, 0.00% Mn, 1.07% Mg, 4.38% Zn, 0.00% Cr

‡ Fine cracks detected with aid of 25X orthoscope.

TABLE 7.—PROTECTION AGAINST SCC BY AN INHIBITED OR A VERY THICK ANODIC COATING OF SPECIMENS, STRESSED WITH HIGH ELASTIC STRAIN * ‡

Alloy	Atmospheric Environment	Median Time to Failure Days **		
			Anodized	
		Bare	(5 μm) ‡	(50 μm)
2014-T651	Seacoast at Pt. Judith, RI	14	56	43
	Seacoast at Pt. Comfort, TX	35	130	35
	Industrial at New Kensington, PA	117	129	73
7079-T651	Seacoast at Pt. Judith, RI	43	14	43
	Seacoast at Pt. Comfort, TX	35	9	124
	Industrial at New Kensington, PA	54	28	59

NOTES: * Interference stressed O-ring specimens 3.18 mm thick x 13 mm wide x 51 mm O.D. machined from 64 mm rolled rod.

† Rings and stressing plugs of the same alloy anodized after stressing to 75% transverse Y.S.

** Specimens exposed in quintuplicate, with all specimens failing.

‡ Sealed with $K_2Cr_2O_7$ inhibitor.

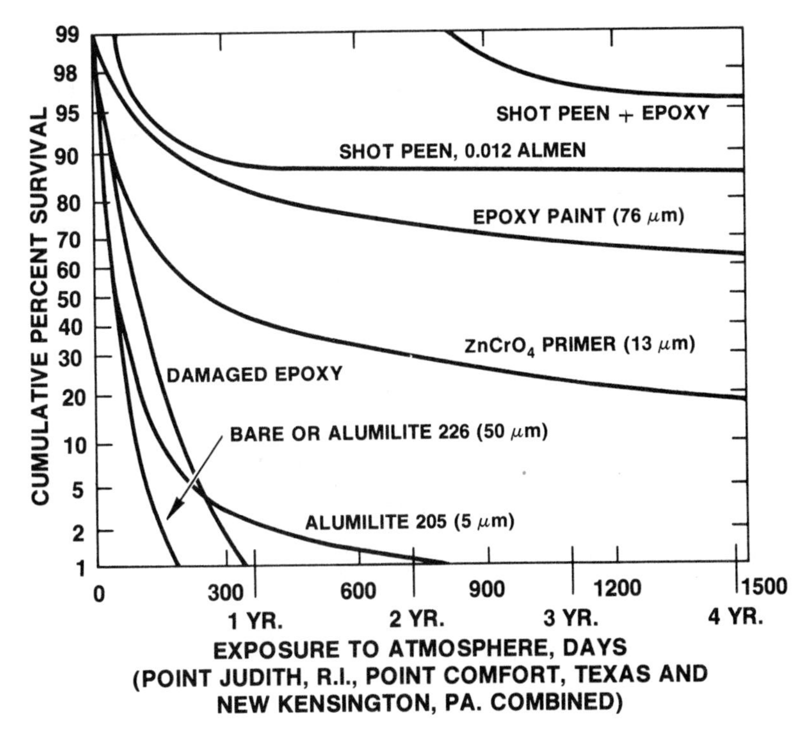

Fig. 13.—Relative protection against stress corrosion cracking of various protective systems, data are combined for alloys 2014-T651 and 7079-T651, 30 specimens per protective system.

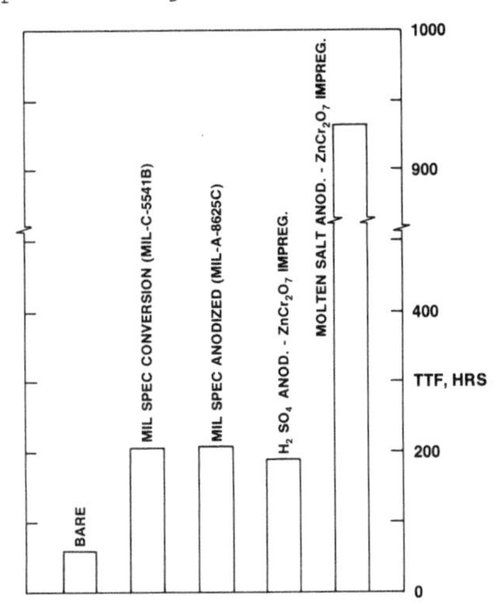

Fig. 14.—Al 7075-T651 alternate immersion tests in a 3.5 wt% NaCl solution (pH = 3), coated samples, scratched, stressed to 80% Y.S. Mitchell and Hurley, Tyco Laboratories, Inc.

REFERENCES

1. Wefers, K. and Bell, G.M., "Oxides and Hydroxides of Aluminum," Technical Paper No. 19, Aluminum Company of America, 1972.
2. Graff, W.R. and Powers, J.H., unpublished data, Alcoa Laboratories.
3. Cotton, W.L. and Thierry, R.P. "Pitting - A Problem with Hard Anodizing," 14th Annual Airlines Plating Forum, April 25-27, 1978, Tulsa, Okla. Published by the American Electroplaters' Society.
4. Mader, O.M., "A Critical Review of Aluminum in the Building Industry," *Metals and Materials, 303-307 (July 1972).*
5. Sowinski, G., unpublished data, Alcoa Laboratories.
6. Cochran, W.C., "Durable Integral Colored Aluminum Curtain Wall," Paper No. 4, Pittsburgh Chapter of Amer. Soc. for Metals, 1974.
7. Hale, L.A., unpublished data, Alcoa Laboratories.
8. Sprowls, D.O., Lifka, B.W., Vandenburgh, D.G.,Horst, R.L. and Shumaker, M.B., Contract NAS 8-5340, "Investigation of the Stress-Corrosion Cracking of High Strength Aluminum Alloys," Alcoa Laboratories' Final Report for the period May 6, 1963 to October 6, 1966.
9. Mitchell, C. and Hurley, G.F., Contract F33615-74-C-5016, "Materials and Approaches for Improved Stress Corrosion Inhibitive Coatings," Tyco Laboratories' Technical Report AFML-TR-72-191 Park III, Final Report for the period Nov. 1973 to June 1974.
10. Horne, D.H., "Eutectic Anodizing of Aluminum for Extended Service Life," Paper No. 203, presented at Corrosion/78, March 6-10, 1978, Houston, Texas.

CHARACTERIZATION OF CHROMIC-ACID ANODIZED ALUMINUM ALLOY SURFACE

William J. Russell and
Carolyn A. L. Westerdahl

Large Caliber Weapons System
Laboratory Arradcom,
Dover, NJ

WILLIAM RUSSELL

ABSTRACT

The topography of 2024-T3 aluminum-alloy surfaces has been characterized by using high-magnification transmission electron microscopy (TEM) at various stages during the chromic-acid anodizing procedure used to prepare aluminum surfaces for painting, adhesive bonding and improved corrosion resistance. The surface structure characteristic of the unsealed chromic-acid anodized alloy was found to form during the first 6 minutes of the 37½-minute anodizing cycle; the surface structure characteristic of the sealed anodized alloy was found to form during the last half of the 8-minute sealing cycle. The surface structure of FPL deoxidized and of unsealed anodized surfaces consists of a repeating pattern of close-packed depressions 200 to 800 angstrom units in diameter. The structure of the sealed surface consists of crevices 500 to 2,000 angstrom units wide between mounds of material formed during sealing.

The chemical composition of the surfaces was determined after anodizing and also after sealing the surface, using electron spectroscopy for chemical analysis and Auger electron spectroscopy. These surfaces were found to consist of aluminum, oxygen, carbon, and nitrogen with traces of silicon and manganese. Copper and magnesium, two of the major alloying elements in the alloy used, were not detected.

INTRODUCTION

The object of this investigation was to characterize a chromic-acid anodized aluminum surface. This study was limited to the top-most layer of the subject surface.

Aluminum, as received from the supplier or after fabrication, does not have a sufficiently corrosion resistant surface. Good corrosion resistance can be obtained only after the surface of the metal has been properly cleaned and prepared. The optimum procedures used to clean and prepare aluminum surfaces have been for the most part empirically determined. Optimization has been accomplished by preparing specimens in different ways and then testing to determine which combination of processing conditions produced the most durable surface. The chromic acid anodized surface is somewhat unique in that it is used for a corrosion resistant surface, for an excellent pretreatment prior to painting and as a process for preparing aluminum prior to adhesive bonding.

DISCUSSION

Study of Surface Topography

The chromic anodized surfaces studied in this work were made using the Bell Helicopter Company Process Specification BPS FW 4352 Rev E (1). This process is very similar to the process used to prepare surfaces meeting the requirements of MIL-A-8625 (2). The topography of the surface was examined with the transmission electron microscope after each step in the process. In this way the effect of each procedure upon the surface was noted. The chemical composition of the surface, after the anodizing and sealing operations were completed, was studied using Auger and electron spectroscopy for chemical analysis. These techniques determined the chemical makeup of the surface material directly in contact with the environment.

The most direct way to study a surface is to look at it under sufficient magnification to bring out the structures of interest. A magnification of approximately one hundred thousand diameters was required to see what was happening as the surfaces went through the various stages of processing. The required magnification was obtained using a transmission electron microscope. This instrument is unsuitable for study of the surface directly, since the specimens are opaque to the electron beam. Thus it was necessary to make very thin transparent copies of the surface (replicas). These were made by vacuum evaporating platinum onto the surface at a low angle to bring out fine details of the surface structure and increase the contrast. The surface was then coated with a uniform film of vacuum-evaporated carbon. The films of material were freed from the surface by etching the metal out from underneath, after which they floated free. The replicas were then rinsed and air dried. Replicas made in this manner are called single-stage replicas and are ready for study with the transmission electron microscope.

Degreasing

The metal when received from the supplier is contaminated with oil, dirt, and marking ink. It is also covered with a continuous film of aluminum oxide approximately 165 angstrom units thick (3). The oxide film is formed during rolling and heat treating operations when sheets of the required thickness and strength are fabricated. The first step in the chromic-acid anodizing process is to vapor degrease the surface by flushing the oil and dirt from the specimen with solvent, the vapors of which condense to a liquid on the cool metal. Figures 1 and 2 are TEM micrographs of an acetone-degreased surface. Figure 1 is at relatively low magnification and shows the structure of the oxide on the surface when it is received from the supplier.

Alkaline Cleaning

After degreasing to remove all solvent soluble contaminants, the metal is immersed in alkaline cleaner to remove all water soluble and saponifiable soils. This cleaner can be either inhibited or noninhibited. The noninhibited cleaner etches the surface of the metal and removes the oxide layer as well as the water soluble surface contaminants. The inhibited cleaner contains silicates or other additives that either prevent or minimize the etching of the surface by the cleaner. The noninhibited cleaner was used

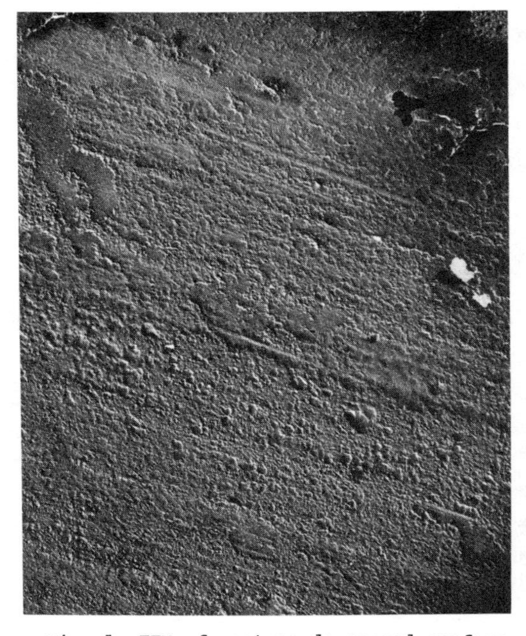

Fig. 1.—TEM of acetone-degreased surface (6,750 X).

Fig. 2.—TEM of acetone-degreased surface (100,000 X).

for this work.

When some aluminum alloys are alkaline etched, a film of black powdery material or smut is formed on the surface of the metal. The smut does not rinse off and has to be removed before a true replica of the underlying surface can be made. Removal was accomplished by making a series of casts of the dried surface using solvent-softened plastic replicating tape. After the solvent had evaporated and the casts or plastic replicas had hardened, they were stripped off the surface. This technique removed the dried powdery material without damaging or changing the surface of the metal. The procedure was repeated until all the foreign material was removed and the surface was clean. A final careful plastic cast of the cleaned surface was then made. This cast was used to make a two-stage replica of the surface.

When the replicas were studied, the alkaline-cleaned surface was found to be reasonably smooth. All traces of the roll marks on the degreased surface had been removed by the cleaner. Crevices noted in the surface were probably sites where ridges of oxide had become entrapped or embossed into the surface of the metal during fabrication. When the oxide was dissolved out during alkaline cleaning, a void was left. Figure 3 is a low-magnification micrograph showing the voids. When studied at a higher magnification, the surface appears to have a wave-like pattern with random shallow pits approximately 200 angstrom units in diameter (Fig. 4).

Deoxidizing

After the metal was alkaline cleaned, it was immersed in a deoxidizing bath consisting of a hot solution of sulfuric acid and sodium dichromate in water. When the specimens covered with the black smut were placed in the deoxidizing bath, the smut dissolved and normal etching proceeded. Chemical desmutting is a normal production procedure since it removes the smut without excessive mechanical work. Figure 5 shows the surface pattern that developed after 9 minutes in the deoxidizer. The pattern consists of etched-out depressions (concave) about 2 to 3 microns in diameter. When the structure of the surface within these etched out depressions is examined at higher magnification (Fig. 6), a dimpled concave sub-structure is seen. These irregular shaped dimples are 400 to 800 angstroms in diameter. The surface structures shown in Figures 5 and 6 are typical of 2024-T3 alloy surfaces etched in the sulfochromate deoxidizer whether the surface has been alkaline etched or only solvent degreased prior to processing (4). When specimens which have been degreased but not alkaline cleaned are immersed in the deoxidizer, the oxide layer dissolves away within the first 1/2 minute of processing. After this the metal is etched, developing a surface structure typical of the etchant-alloy combination (5). The typical surface structure is the same as that shown in Figures 5 and 6. Thus it appears that treatment prior to deoxidizing has no visible effect on the surface structure developed in the deoxidizing solution.

Anodizing

The anodizing cycle as described in the reference specification is a complex procedure requiring specified increases in the anodizing voltage during a given time sequence. Specimens were removed at various times during the cycle to determine if changes in the voltage and time of anodizing had any effect on the surface structure.

Previous transmission electron microscopy (TEM) studies of anodized coatings used specimens of the coating obtained by dissolving the aluminum substrate underneath the coating, leaving the oxide layer intact (6). The

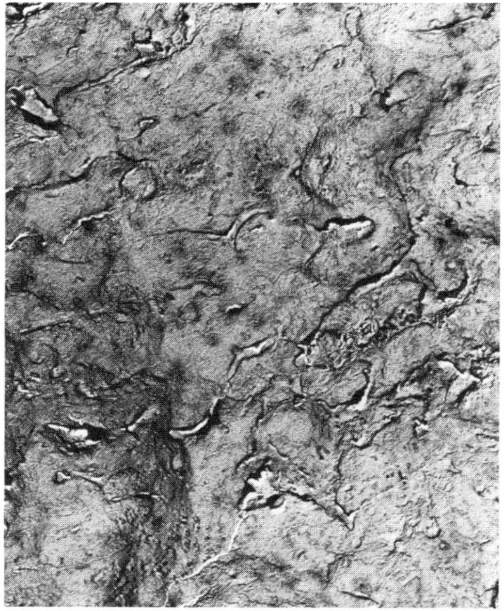

Fig. 3.—*TEM* of Alkaline-cleaned surface (6,750 X).

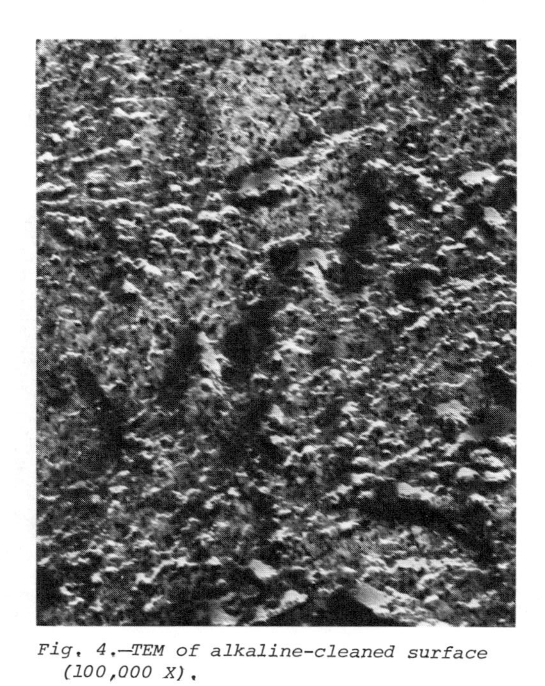

Fig. 4.—*TEM* of alkaline-cleaned surface (100,000 X).

Fig. 5.—TEM of (FPL) deoxidized surface (6,750 X).

Fig. 6.—TEM of (FPL) deoxidized surface (100,000 X).

oxide was sufficiently thin to be used as a specimen. This procedure had two disadvantages for our purposes. It revealed the internal structure of the coating, not the surface, and there was no assurance that the technique used to dissolve the metal did not alter the sample surface. Thus, for this study, anodized surfaces were replicated using the single-stage process previously described. This procedure enabled detailed study of the surface at maximum resolution and also assured that minimal changes occurred as a result of the replicating process.

Figure 7 is a high magnification TEM micrograph showing the structure of the surface after the first two minutes (at 7-1/2 volts) of anodizing. This figure shows a surface structure that is in transition between the structure developed in the deoxidizer (Fig 6) and the fully developed anodized structure. The structure of the deoxidized surface shown in Figure 6 is becoming obscure or indistinct, and the structure typical of a chromic acid anodized surface is beginning to emerge.

Figure 8 is a high magnification TEM micrograph of the structure of the surface after 2-1/2 minutes (at 7-1/2 volts) of anodizing. The structure in this figure has become almost typical of the unsealed chromic-acid anodized surface. Figure 9 shows the surface after 5-1/2 minutes of anodizing (22-1/2 volts max). All traces of the surface structure typical of the deoxidized surface have disappeared. No additional change appears to occur during the remainder of the anodizing cycle other than a possible deepening of the pores. Figure 10 shows the surface after anodizing has been completed (35 minutes at 40 volts). A careful comparison of Figure 8 (2-1/2 minutes), Figure 9 (5-1/2 minutes), and Figure 10 (35 minutes) shows that the surface structure during the latter part of the anodizing cycle is essentially the same even to the size of the pores. The pores appear as dimples or conical depressions in the surface. The only change that can be detected during the latter part of the anodizing cycle is a possible increase in the depth of the pores. This change is difficult to measure, since data were accumulated the limiting resolution of the microscope with carbon platinum replicas. Since it was impossible to determine the structure of the interior of the pores using the single-stage replicating process (Fig. 10), a two-stage replica was made. A cast of the surface was taken using solvent-softened plastic replicating tape. This cast was the inverse of the surface. It was convex where the anodized surface was concave and concave where it was convex. Thus, the pores in the anodic coating were reproduced as spikes protruding from the surface of the plastic cast. When the cast was replicated and examined with the TEM, the pores were found to be truncated cone-like hollows in the anodic coating. These were narrow at the base, opening up as they approached the surface. Figure 11 is a micrograph showing the structure of the pores using a two-stage replica.

Sealing

After anodizing has been completed, the surface is frequently immersed in a hot dilute solution of chromic acid to seal the pores and improve the corrosion resistance of the coating. The mechanism of the sealing process is described by Barkman as a swelling of the aluminum oxide on the surface of the coating (6). Corrosive attack upon the underlying metal is thus prevented. Figure 12 shows the surface of a sealed anodized specimen. This is a micrograph of a surface that has undergone the complete anodizing and sealing procedures. As can be seen, the surface is radically different from an unsealed anodized surface. The pore structure shown in Figure 10 is completely covered by a material that appears to have been extruded from the pores, forming a pattern of hills and ridges. Thus the sealing operation not only seals and closes the pore structure, it actually develops a

Fig. 7.—TEM of surface chromic-acid anodized for 2 min. at 7½ volts (100,000 X).

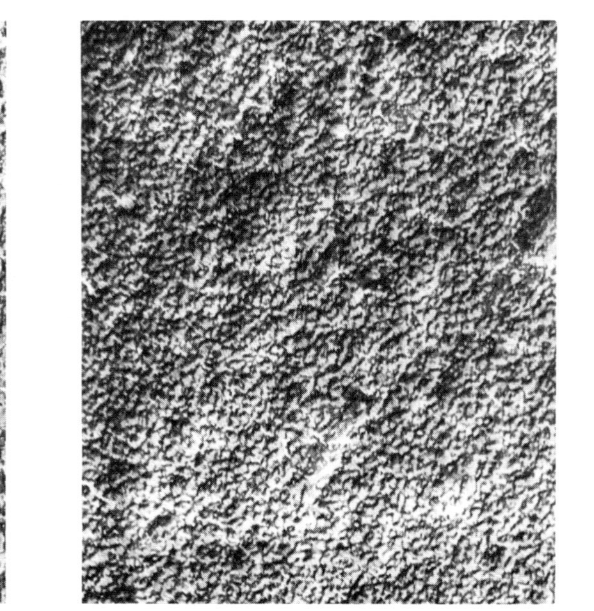

Fig. 8.—Surface chromic-acid anodized for 2½ min. at 7½ volts (100,000 X).

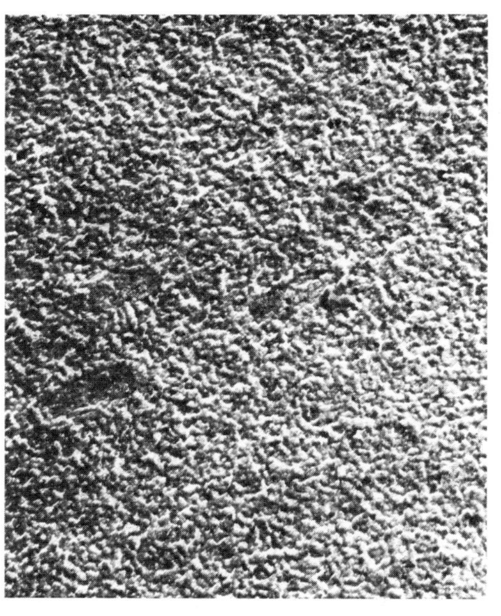

Fig. 9.—Surface chromic-acid anodized for 5½ minutes (100,000 X).

Fig. 10.—Surface chromic-acid anodized for full cycle (100,000 X).

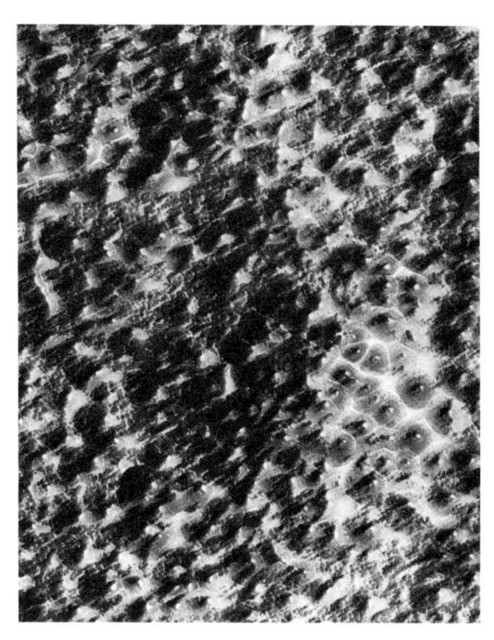

Fig. 11.—Two-stage replica of chromic-acid anodized surface showing internal structural pores (100,000 X).

definite surface structure distinctly different from the unsealed anodized surface.

Figure 13 is a micrograph of the sealed surface prepared in a slightly different manner. The area was selected to show the structure of the extruded material. The sphere in the lower left corner is a 0.3-micrometer-diameter latex particle which was used to interpret surface structure and determine surface profile.

When specimens that had been anodized and sealed for one-half the normal 8-minute cycle were examined, the surface structure was found to be very similar to that of an unsealed specimen. When a two-stage replica was made and examined, the surface of the specimen sealed for one-half the normal cycle was again found to be similar to the unsealed specimen. The internal structure of the pore was apparently unaltered. Thus it appears that the material found on the surface of the sealed anodic coating forms during the last half of the sealing cycle. Strict control of the sealing solution, composition, temperature, and time of processing are critical. Slight changes may have a significant effect.

The differences that occur in the surface structure of the specimens as a result of the deoxidizing, anodizing, and sealing operations are evident at only very high magnifications. At lower magnification the structure of these surfaces is very similar to that shown in Figure 5.

Surface Profile

There is difficulty in interpreting the topography of a surface with

Fig. 12.—Sealed chromic-acid anodized surface (70,000 X).

Fig. 13.—Sealed chromic-acid anodized surface (33,000 X).

TEM. Optical illusions are frequently encountered, and it is impossible at times to determine whether a structure is concave or convex. As an aid for this determination, the specimens under study were sprayed with a latex solution prior to replicating. Since the spheres of necessity had to be convex, it was possible to interpret the surface by using the spheres as a reference. With the spheres as a guide, profiles of five of the surfaces were roughly drawn to scale (Fig. 14).

The profiles of the first two surfaces are notable in that they are not repetitive. The surface structures, although distinctive, are random. The surface of the degreased specimen is fairly smooth with gradual changes in slope over distances of a half micron or more. The surface of the alkaline-etched specimen was also non-repetitive, although there was evidence of a wave-like pattern. The surface profile of the deoxidized specimen was drawn from Figure 6. This surface consists of a repetitive pattern of small cup-like depressions or dimples. The cups, although random in shape, are of a fairly consistent size (from 200 to 400 angstroms) in diameter and are closely packed. The profile of the unsealed anodic coating was drawn after a careful study of Figures 10 and 11. As can be seen from the profile, this surface also consists of a repetitive pattern of depressions. These are small pores in the coating whose diameters at the surface are about 200 to 300 angstroms with a secondary pattern of larger pores whose diameters at the surface are 400 to 800 angstroms.

The profile of the sealed anodic coating was sketched from Figure 13. The structures on this surface consist of a series of piles or lumps of material that appear to have come out of the pores. It is again noted that these structures have a diameter of 400 to 1000 angstroms. In this case however, the structures are convex.

Chemical Composition of the Surface

The chemical reaction occurring between a surface and a corroding environment are strongly affected by the chemical composition of the surface. The material immediately below the surface is not in contact with the corrodant and thus has little interaction with it. Therefore, an effort was made to determine the composition of the top-most layer. When the chemical composition of a surface is analyzed, the results obtained are frequently far different from the results expected. Small quantities of material absorbed or reacted onto a surface can cover a large percentage of the total area.

ESCA Studies

Surfaces of sealed and unsealed chromic-acid anodized 2024-T4 aluminum specimens were studied using electron spectroscopy for chemical analysis (ESCA). This technique, which is also known as X-ray photo electron spectroscopy, yields information about the chemical composition of surfaces. X-rays incident to the surface being studied are used to knock out inner shell electrons from the surface atoms. Those electrons which are close enough to the surface to escape (0 to 50 angstroms) (7) are collected and their kinetic energy is analyzed. This energy is the difference between the energy imparted by the X-ray and the energy which binds the electron in the inner shell of the atom. The energy differences detected are unique for each element and thus provide a means of chemical identifiaction. Since the kinetic energy of the electrons depends on the chemical bonding state of the element from which they come, this technique also yields some information on the type of chemical bonding to which the surface elements are subjected.

Table 1 is a tabulation of the results of studies showing the binding energy at which signal peaks were obtained, their relative intensities and

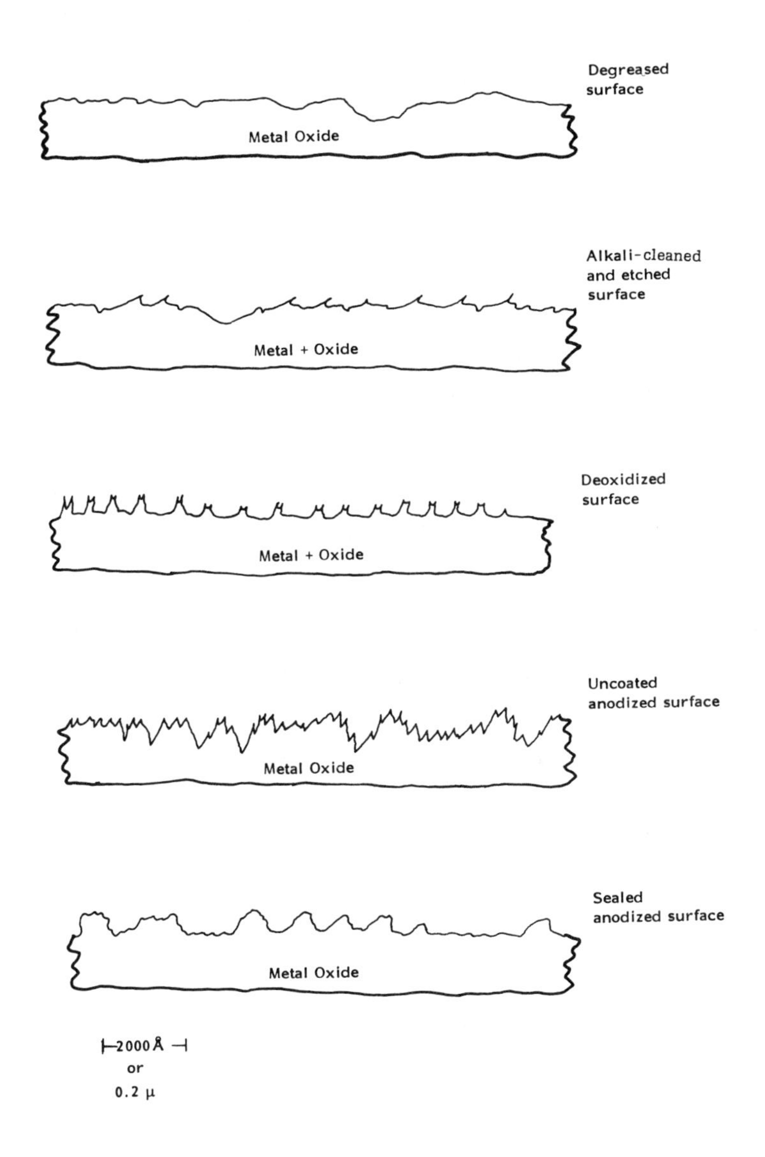

Fig. 14.—Sketched profiles of treated surfaces (to scale).

TABLE 1.—RESULTS OF ESCA STUDIES

Binding peak energy (EV)	Element	Intensity* Unsealed specimen	Sealed specimen
645	Manganese $2P_3$	None detected	VVV weak
586	Chromium $2P_1$	Moderate	Moderate
577	Chromium $2P_3$	Weak	V weak
533	Oxygen 1S	VV strong	VVV strong
399	Nitrogen 1S	VV weak	Strong
285	Carbon 1S	Moderately strong	Strong
154	Silicon 2S	V weak	Weak
120	Aluminum 2S	Strong	V strong
80	Aluminum 2P	V strong	VV strong
28	Oxygen 2P	V strong	VV strong

*V = very

the elements from which they came. It will be noted that the intensities are reported as *very very very* strong to *very very very* weak. No numerical values are given because the common methods of surface analysis (ESCA, Auger, SIMS, ISS) do not yield truly quantitative data (7). Table 2 is a listing of the bulk elemental composition of 2024 aluminum alloy, the material used in these experiments.

TABLE 2.—CHEMICAL COMPOSITION OF 2024 ALLOY

Element	%
Cu	3.8-4.9
Mg	1.2-1.8
Mn	0.3-0.9
Si	0.5 Max
Fe	0.1 Max
Cr	0.25 Max
Zn	0.05 Max
Other elements 0.15	
Al	Remainder to make 100%

When the data in Tables 1 and 2 are compared, the effects of the anodizing and sealing process on the chemicsty of the surface begin to emerge.

As can be expected, the ESCA spectrum shows strong signal peaks for both aluminum and oxygen. These peaks obviously result from the aluminum oxide that is formed during the anodizing process. Signal peaks are shown for chromium and are much greater than expected if the source of the chromium is only from the material from which the specimen in the base was made. The chromium peaks obtained are probably due to the presence of chromium absorbed from the anodizing bath. The slight decrease in the chromium content of the sealed specimen (2 P_3 peak) suggests that the soaking in the hot sealing solution removed some of the chromium from the surface. The weak peak shown for silicon (which is probably of a higher concentration than chromium in the bulk alloy and is essentially insoluble in the processing solutions) contributes to the proposition that much of the chromium detected was absorbed from the electrolyte during the anodizing process. Copper was not detected in the surface. This result is surprising since a fairly high concentration (4 to 5%) of this material is present in the aluminum alloy. The failure to detect copper indicates that the copper is dissolved out of the surface during the cleaning and anodizing process. The same is true for magnesium. No trace of this material was detected even though it was present (1 to 2%) in the aluminum alloy. Only a very slight trace of manganese was detected in the sealed specimen; none in the unsealed specimen. This material was probably of a higher concentration than silicon in the alloy; thus, like copper and magnesium, it must have been selectively etched out of the surface during processing. Iron and zinc were not detected.

Two elements (which were not present in the base alloy and not part of the final processing solutions) were strongly detected. The strongest, carbon, was present on the surface of both unsealed and sealed specimens. The structure of the peak suggests that much of the carbon detected was in the form of carbonate, although hydrocarbon was also probably present. The source of the carbonate was probably carbon dioxide from the air. The second element found was nitrogen. The structure of the nitrogen peak did not rule out the presence of amine compound on the surface. The presence of an amine could also account for some hydrocarbon contamination. The source of the nitrogen was probably amine contamination of the deionized water which was obtained from a commercial resin bed dionizer, and used to rinse and seal the finished specimen. This supposition was supported by the great increase in the nitrogen signal obtained when the specimen was sealed by immersing it in a very dilute hot solution of chromic acid in deionized water.

When all the peak signal intensities for the sealed and unsealed specimens were compared, the intensities of the sealed specimens were always greater (except for chromium). The cause of this effect has not been determined although this effect is consistent enough to be significant.

Auger Electron Spectroscopy

Anodized specimens, both sealed and unsealed, were submitted to The Center for Surface and Coatings Research at Lehigh University for an Auger electron spectroscopic analysis. After analysis, it was reported that there was no major difference detected in the surface composition of the sealed and unsealed specimens. The surface layer (5 to 25 angstroms deep) consisted of aluminum, oxygen, chromium, and carbon. A copy of The Center's report is included as Appendix I.

These results confirmed the presence of the above named elements, but did not detect the presence of nitrogen, silicon, and manganese which were detected as being present at lower concentrations by ESCA.

CONCLUSIONS

1. The surface topography developed during each of the processing steps used to chromic-acid anodize and seal a surface is different when studied with the transmission electron microscope at 100,000 times magnification.

2. The topography of the unsealed chromic-acid anodized surface develops during the first 6 minutes of anodizing. No additional changes in surface structure were noted during the remainder of the anodizing cycle.

3. The topography characteristic of a sealed anodized surface forms during the latter half of the sealing cycle. No observable changes occurred until after the first half of the cycle.

4. The chemical composition of chromic-acid anodized sealed and unsealed surfaces consists of aluminum, oxygen, chromium, carbon, and nitrogen with traces of silicon and manganese. Copper and magnesium, two of the major alloying elements in the alloy used, were not detected.

APPENDIX 1

AUGER ELECTRON SPECTROSCOPIC ANALYSIS OF ANODIZED ALUMINUM

Prepared

by

G.W. Simmons

Center for Surface and Coatings Research
Lehigh University, Bethlehem, Pennsylvania

Objective of Analysis

The principal objective of the analysis was to compare the surface composition of two anodized aluminum specimens. One specimen was in the as anodized condition and the other was anodized and then sealed in a chromate bath.

Experimental Procedure

The specimens were mounted in the spectrometer directly as received from Mr. Russell of Picatinny Arsenal. Although the aluminum oxide surface is dielectric, no interference due to surface charging was encountered. The area of the specimen surface represented by the spectra is determined by the 0.5 mm diameter of the incident electron beam. The sensitivity of the Auger electron spectroscopy technique is on the order of one atomic percent, and the escape depth of Auger electrons varies from 5 to 25 Angstroms over the energy range of 0-2000 eV.

Results of Analysis

Spectra taken of the two specimens have been included in the report for reference. No major differences were observed for the surface composition of the anodized and anodized-sealed specimens. Carbon was the only element found in addition to oxygen and aluminum expected for the oxide. The amount of carbon was approximately the same for both specimens.

It is interesting to note that if chromium is present on the surface of the sealed specimen, the concentration is too low to be resolved in the spectrum. Unfortunately, the sensitivity for chromium in the presence of oxygen is reduced owing to the overlap in energy of the chromium and oxygen Auger electron transitions.

The specimens may be different in their degree of hydration, but it is not possible to make conclusions of this fact from these Auger electron spectra.

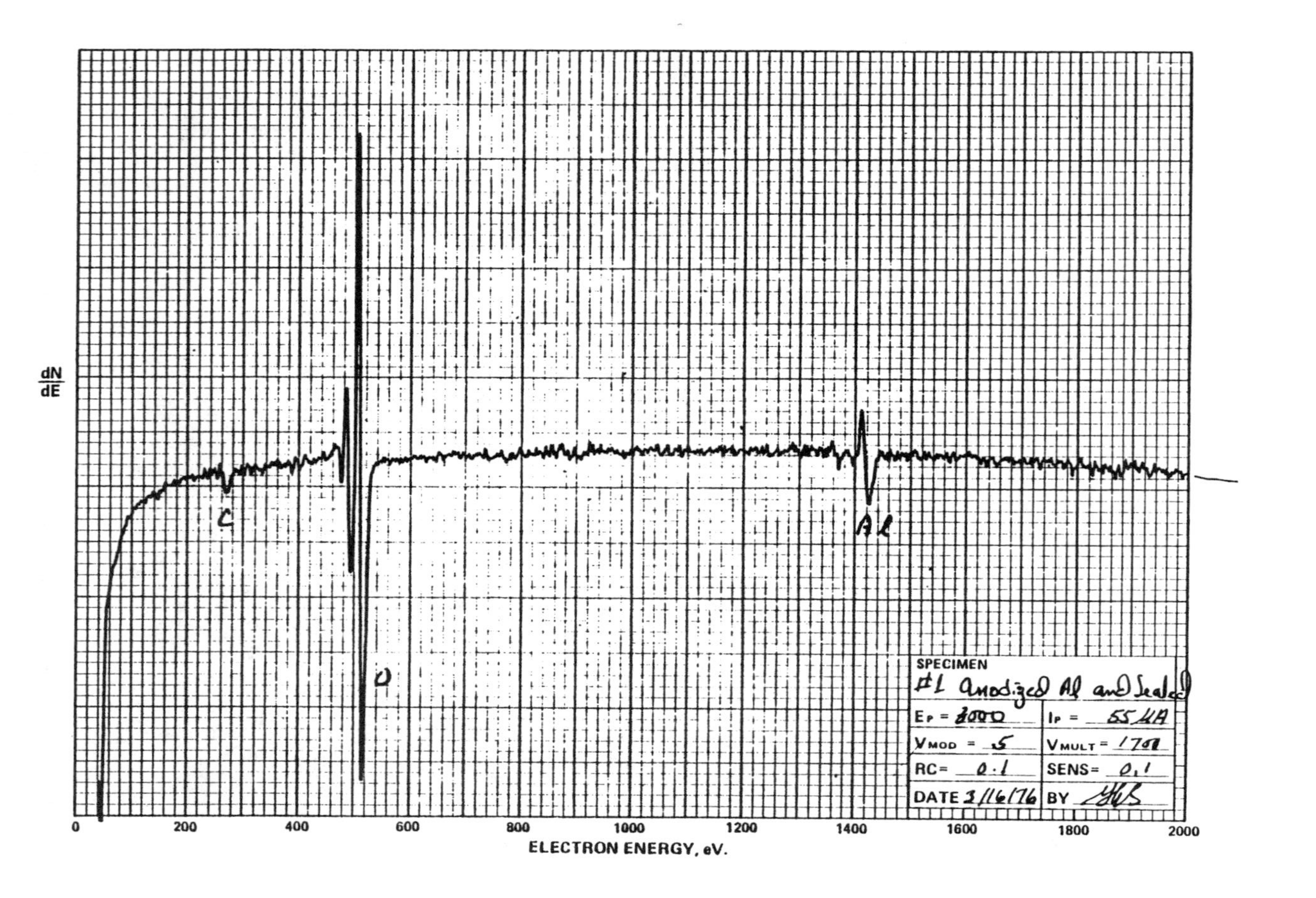

dN/dE
C
O
Al
SPECIMEN
#1 Anodized Al and Sealed
Ep = 3000
Ip = 55 μA
Vmod = 5
Vmult = 1700
RC = 0.1
SENS = 0.1
DATE 3/16/76
BY
0
200
400
600
800
1000
1200
1400
1600
1800
2000
ELECTRON ENERGY, eV.

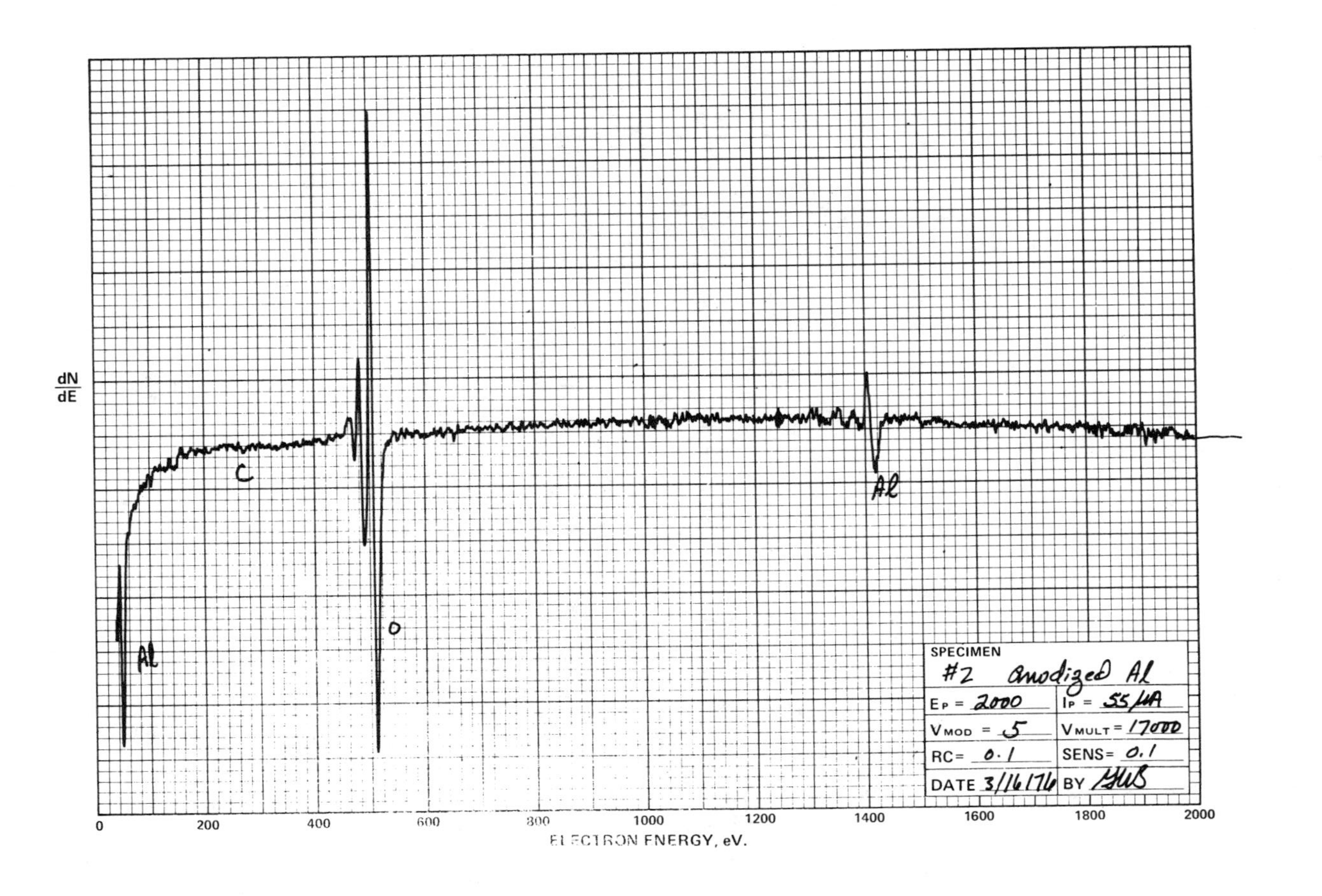

dN/dE
Al
C
O
Al
SPECIMEN
#2 Anodized Al
EP = 2000
IP = 55 μA
VMOD = 5
VMULT = 17000
RC = 0.1
SENS = 0.1
DATE 3/16/76
0
200
400
600
800
1000
1200
1400
1600
1800
2000
ELECTRON ENERGY, eV.

APPENDIX 2

EXPERIMENTAL PROCEDURES

MATERIALS

Aluminum

The specimens were cut from 1.7-mm-thick 2024-T3 aluminum alloy sheet into 25 by 100 mm strips. These were processed and used for the TEM and Auger electron spectroscopy. Since the ESCA instrument required a cylindrical specimen 11 mm in diameter and 19 mm long, these were machined from 13 mm cylindrical stock 2024-T4 alloy. The T3 temper material was not available in this size.

Chemicals

All chemicals used were of the chemically pure grade with the exception of the sodium dichromate and chromic acid, which were technical grade, and the alkaline cleaner, which was a proprietary product. The alkaline cleaning material was non-inhibited and etched the alloy during the cleaning cycle.

SURFACE PREPARATION OF SPECIMENS

Degreasing

The surfaces of the specimens were wiped with a clean acetone-saturated paper towel until they were free of all oil, grease, and marking ink.

Alkaline Cleaning

The degreased specimens were immersed in the alkaline cleaner at 11 g per liter and 60°C for 5 minutes. The specimens were then rinsed with flowing tap water to remove all traces of cleaning solution. Specimens for TEM study were flushed with deionized water, then blown dry with a jet of dry filtered compressed air.

Deoxidizing

The alkaline-cleaned specimens were immersed in a solution (FPL) consisting of 33 1/3 g of sodium dichromate ($Na_2 Cr_2 O_7 \cdot 5 H_2O$), 181 mg. of concentrated sulfuric acid (*Sg 1.84*), and one liter of deionized water at 65°C for 9 to 11 minutes. The specimens were then flushed with flowing tap water to remove all traces of deoxidizer. Specimens for TEM study were flushed with deionized water, then blown dry with a jet of dry filtered compressed air.

Anodizing

The deoxidized specimens were made the anode in a 10% solution of chromic acid ($Cr O_3$) in deionized water at 35 ± 1°C. The anodizing voltage sequence was (in order):

7-1/2 volts for the first 2-1/2 minutes

11-1/2 volts for one minute

16-1/2 volts for one minute

23 volts for one minute

30 volts for one minute

35 volts for one minute

40 volts for 30 minutes

When anodizing was terminated, the specimens were promptly removed from the anodizing solution and flushed with flowing tap water, then with deionized water. Specimens for TEM, ESCA, and Auger electron spectroscopic study were blown dry with a jet of dry filtered compressed air.

Sealing

The anodized specimens were immersed in a bath (100 ppm chromic acid in deionized water) at 82°C for 8 minutes. The specimens were flushed with deionized water to remove all traces of sealing solution and blown dry with a jet of filtered compressed air.

TRANSMISSION ELECTRON MICROSCOPY

Replication of Specimens

All surfaces were replicated within 24 hours after processing was completed. The specimens were stored in kraft paper envelopes until replicated. Unless stated otherwise, the specimen was placed in a vacuum evaporator and a 1/2-inch length of 8-mil platinum palladium wire was evaporated onto the surface from an approximate distance of 15 centimeters at an angle of about 30°. The specimen was then coated with carbon at an angle of 90° to the surface using a rotating stage. The surface of the coated specimen was scored in 2 x 2 mm squares and the replicas were freed from the surface by etching. A solution of one part by volume of Kellers Etch Conc., (Table 3) two parts by volume of concentrated nitric acid, and four parts deionized water was used to dissolve the metal underneath the evaporated platinum carbon replica. When the replicas floated free, they were rinsed and mounted on TEM grids. These were single-stage replicas.

Two-stage replicas were made as follows: An impression of the surface

TABLE 3.—COMPOSITION OF KELLER'S ETCH CONCENTRATE

Amount	Component
10 ml	48% Hydrofluric acid conc
15 ml	37% Hydrochloric acid conc
25 ml	70% Nitric acid conc
50 ml	De-ionized water

was obtained using polyvinyl acetate replicating tape softened with acetone. The softened tape was pressed against the surface and allowed to dry. It was then removed and placed in a vacuum evaporator where it was coated with platinum palladium and carbon, as described above. The coated plastic was cut into 2 x 2 mm squares and the plastic was removed by washing with acetone. The two-stage replicas were then mounted on TEM grids.

Specimens for ESCA and Auger

The specimens for spectroscopic analysis were machined to fit the instruments, then anodized and sealed as previously described.

REFERENCES

1. Bell Process Specification, "BPS FW 4352 Rev E," Bell Helicopter Company, Fort Worth, Texas, October 1968.
2. Military Specification, "Anodic Coatings for Aluminum and Aluminum Alloys," MIL-A-8625C.
3. Pattnaik, A. and Meakin, J.D., "Characterization of Aluminum Adherend Surfaces," Technical Report 4699, Picatinny Arsenal, Dover, N.J., July, 1974.
4. Work currently in progress at this laboratory.
5. Russell, W.J. and Garnis, E.A., "A Study of the FPL Etching Process Used for Preparing Aluminum Surfaces for Adhesive Bonding," *SAMPE Quarterly (April 1976)*.
6. Barkman, E.F., "Sealing and Post Anodic Treatments, Anodized Aluminum," *ASTM STP 388 (1965)*.
7. Robinson, A.L., "Surface Analysis: Multiple Techniques for Monolayers," *Science 191 (4233), 1253 (1976)*.

CONVERSION COATINGS—CHROMATE AND NON-CHROMATE TYPES

Nelson J. Newhard, Jr.

Metalworking Chemicals Division,
Amchem Products, Inc.,
Ambler, PA

NELSON NEWHARD, JR.

ABSTRACT

Three types of conversion coatings are described, i.e., chromium phosphate, chromium chromate, and non-chromium type. The utility of each is discussed, along with compositions of the baths and probable compositions of the coatings. The findings of a considerable number of researchers are explored in an attempt to relate their results with the conclusions reached by earlier investigators. The use of sophisticated instrumentation such as ESCA, EPA, XPS, SEM, Auger, etc., is mandatory today in order to isolate and define certain component species of the coatings. Several unresolved areas of conversion coatings are listed as possible targets for future research.

INTRODUCTION

Chemical conversion coatings on aluminum can be considered to be attempts by the environment or by man to convert the surface of the metal to an insoluble inorganic compound. Nature wants this insoluble compound to be an aluminum oxide, via corrosion products, and to return the metal to its natural oxide state. Because man has found the metal to be extremely useful to him in his everyday life, he wants the conversion coating to resist this transformation to the oxide, i.e., to prevent corrosion from occurring. We will be concerned with the latter step in the presentation of this paper and will concentrate on three main conversion coatings, i.e., chromium phosphate, chromium chromate, and a non-chromium-containing coating.

Robert Ayres[1] lists five minimum requirements necessary to delineate a true reaction system for conversion coatings. These have the overall effect of establishing conditions for an irreversible physiocochemical process.

1. The metal must be surrounded by a fluid such as a mild acid which will render its surface reactive.

2. The medium must contain soluble anions or anion-formers (phosphate, chromate, ferrocyanide, dissolved oxygen, etc.).

3. The character of the medium at the interface must be so changed by

the anion-cation interaction that it does not dissolve the reaction product.

4. The insoluble chemical compound which forms must precipitate sufficiently close to the metal surface so that it remains an integral part of the surface.

5. A transition layer must develop across which the binding forces holding the coating to the metal can be exerted.

The formation of oxide coatings on aluminum by electrochemical means, in which aluminum is the anode, is another form of conversion coating but[1] "anodizing processes are examples of surface conversion systems in which the driving forces are provided by externally applied electric currents... they may best be called electro conversion coatings rather than chemical conversion coatings".

The non-chromium conversion coatings are still undergoing considerable research activity but have already made a considerable impact in the container industry where chromium in plant effluents is no longer acceptable by either EPA or by local waste-treating facilities.

DESCRIPTION OF THREE TYPES OF CONVERSION COATINGS

Chromium Phosphate

The conversion coating with the greatest versatility and longest use is the one known as chromium phosphate, or amorphous phosphate. The latter term was used to distinguish this type of coating from the commonly used crystalline phosphate coatings normally applied to steel. This particular chemical film, introduced in 1945, can be produced in weights that are so low that their presence becomes difficult to measure even though their performance is excellent. They can also be produced in weights that exceed 500 mg/sq. ft., producing a color which is dark emerald-green in appearance. Any weight or color between these two extremes can be produced on a continuous basis from baths which are maintained by the proper addition of needed replenishing chemicals.

The basic components of the chromium phosphate bath are:

1. Chromium Trioxide - CrO_3
2. Phosphoric Acid - H_3PO_4
3. Hydrofluoric Acid - HF

This combination is the simplest form of the chemical components that can be employed but many variations on this composition have been utilized during the past 33 years. The chromium source must be hexavalent but can be introduced as a chromate or dichromate as long as sufficient acidity is supplied by the remaining two ingredients to enable the solution to react with the surface of the aluminum. The phosphate portion can be supplied in various forms also, such as the acid shown above, or salts thereof. However, again the total system must be sufficiently acid to enable a reaction with the aluminum to take place. The fluoride contribution is the key to the continuous production of the particular coating and weight that is desired. Introduction as the acid, or as salts thereof are the most common methods employed, but complex fluoacids or salts have also been used.

When salts of the acids are introduced into the bath, precipitation with dissolved aluminum (from used baths) will occur forming such salts as cryolite (Na_3AlF_6) or elpasolite (NaK_2AlF_6) depending upon the cations introduced. Advantage has been taken of the formation of this precipitate in production baths where architectural/decorative coatings are desired. By keeping the bath free of aluminum, via precipitation, the decorative coating can be maintained indefinitely, whereas if the cations were not available, the build up of aluminum in the bath would result in coatings ranging from emerald green to whitish green to a white powder, depending upon the amount of dissolved aluminum present.

The chromium phosphate coating can be heated to temperatures up to the melting point of the aluminum without adversely affecting the properties of the coating. This is not true of most of the other types of conversion coatings, however. Originally, the chromium phosphate coating was intended as a substitute for anodic processes and was introduced into aircraft plants for that purpose. Because it had many unique properties compared to the anodic processes, a new military specification had to be written around it in order for it to be used for aircraft/military applications. This specification -- AN-C-170 -- was the forerunner of the present specification MIL-C-5541B, which covers conversion coatings applied to aluminum for use in aircraft or naval applications.

Interestingly enough, the material around which the first such specification was written is no longer on the Qualified Product List because of the advent of the chromium chromate coatings and the introduction of more stringent tests. The chromium chromate coatings provided considerably more in the way of unpainted corrosion resistance than did the chromium phosphate coating, and the latter was relegated to other uses.

Paint-bonding capabilities, decorative uses, and corrosion resistance for beverage containers, have been the main reasons for the continued popularity of this coating. Many square miles of unpainted insect screening, miles of cyclone fencing, and quite a few industrial buildings have shown the utility of this weathered-copper look that reduces glare and blends in with the natural surroundings. The Echo satellite was another use for this coating where thermal control was required to extend the satellite's useful life in outer space.

Paint bonding applications have centered mostly around such areas as extrusions, siding, mobile homes, etc., where the weight of the coating is generally in the 35-80 mg/sq. ft. range. Many paint manufacturers are readily inclined to guarantee their products over this type of pretreatment, for 20 to 30 years, indicating the benefit that they expect to gain from properly treated metal. The thickness of 100 mg/sq. ft.^{2} coating has been calculated as approximately 4800Å. Many millions of beer/beverage containers have also been protected by this kind of coating which resists dome staining during pasteurization of beer, and also helps to bond the interior lacquer and external decoration. It is a very versatile conversion coating for aluminum.

Chromium Chromate

The conversion coating commonly known as chromium chromate was introduced to industry approximately 25 years ago. This chemical film can be produced in coating weights ranging from approximately 5 mg/sq. ft. to approximately 200 mg/sq. ft. (accelerated bath) with colors ranging from colorless to dark brown. The color can be mostly eliminated from fresh coatings by leaching them in hot water, when this is a requirement. There may be a slight loss of corrosion resistance due to the leaching, however, and care must be taken to insure an adequate supply of high purity water.

Leachable ions from the coating quickly saturate the hot water bath, preventing additional leaching from taking place. Leaching of aged or oven dried coatings is not very practical or successful, however.

The basic components of the chromium chromate baths are:

Accelerated Bath (b)	1. Chromium Trioxide - CrO_3	Unaccelerated Bath (a)
	2. Hydrofluoric Acid - HF	
	3. Potassium Ferricyanide - $K_3Fe(CN)_6$	

The above represents the simplest forms of (a), the Unaccelerated bath, and (b), the Accelerated bath. The latter indicates the addition of ferricyanide ion to the bath, which then becomes an additive to the coating composition, considerably increasing the amount of total coating weight. Again, the components can be added in forms other than the above, i.e., chromates or dichromates, simple or complex fluorides, and acid when needed, usually as nitric acid, for pH adjustment.

When the chromium chromate coatings are heated to temperatures in excess of approximately 70°C, the resistance of the coating to corrosion in salt spray drops off proportional to the rise in temperature and the time at temperature. Depending upon the temperature involved, degradation of the coating can be slight at temperatures between 65°C and 95°C up to complete degradation of the accelerated coating at temperatures of approximately 300-320°C.[2] The question that is always unanswered in this case, however, is how much of this drop in corrosion resistance is due to the degradation of the coating with heat, and how much is due to a breakdown in the metallurgical structure, due to heating. The time at temperature is very important, of course, and we are concerned about times of 30 minutes to 2 hours, sufficient to permit changes to occur in both the metal and the coating.

Because of the excellent corrosion resistance of the chromium chromate coatings, both the unaccelerated and accelerated types have been approved for use under military specification MIL-C-5541B. This specification requires an essentially corrosion-free surface after panels are exposed to salt fog for 168 hours. Another part of the specification deals with electrical resistance properties of the film along with corrosion resistance, and for this application the unaccelerated chromium chromate coating is usually preferred.

The main applications for the chromate coatings are aircraft and military use, continuous coil for paint-bonding purposes, and general purpose parts lines. They offer excellent paint bonding properties, and for this purpose the coating weights are usually of the 25 to 40 mg/sq. ft. range. The minimum coating weight requirement for military use is 40 mg/sq. ft. The thickness of a 100 mg/sq. ft. coating has been calculated as approximately 6000Å.

Again where siding manufacturers employ the chromium chromate conversion coatings on their continuous coil lines, the major paint suppliers have shown no reluctance to warranting their paints for 20 to 30 years, because field experience has shown excellent results. The chromium chromate coatings have filled a very vital need during the past 25 years enabling such items as siding, mobile homes and many other commercial items to become economically feasible in the market place.

Non-Chromium

The newest of the conversion coatings that has achieved a place in industry is the non-chromium type, which by definition, is a conversion coating that contains no chromium in the processing bath. When we are restricted in the use of, or have to eliminate entirely those chemicals that have always been part and parcel of the metal treating industry, the search for suitable replacements can become a lengthy one. The difficulty is especially acute when the replacement chemical must also be free of those properties that eliminated their predecessors.

Accordingly, the search for non-toxic chemicals that can perform well in the commonly used metal treating baths today, is one that is receiving wide attention and is still in its initial phases. Where requirements are not too stringent, as in the container field, the non-chromium coatings are performing satisfactorily in providing resistance to dome staining during pasteurization and in promoting the adhesion of inks, varnishes, basecoats, and interior lacquers. However, the non-chrome processes are much more sensitive to cleaning, rinsing, contamination, etc., than their predecessors which contained chromium. At present there are no non-chromium processes that can compete on a general performance basis with either the chromium phosphate or chromium chromate processes.

The present useful non-chromium coatings are colorless and have weights that are generally less than 10 mg/sq. ft. Attempts to produce coatings heavier than that have been unsuccessful for the most part. It is interesting to note that in the early research days of developing conversion coatings for aluminum, the present non-chromium coatings would have been overlooked or rejected because of their lack of both color and weight.

One of the earliest practical non-chromium processes developed for use in the container industry, was that originated by a gentleman who received his doctorate from Lehigh University in 1973.[3] His dissertation on "The Reaction of Aluminum with H_2O and D_2O" was an excellent background for his introduction as an R & D chemist into an industry that was searching for answers in this area. His name is Narayan Das.

His experimentation with non-chromium coating baths for treating aluminum containers culminated in patent 3,964,936.[4] The essential components of the coating bath described in this patent:

1. Boric Acid
2. Source of Fluoride
3. Source of Zirconium
4. Nitric acid for pH adjustment

The colorless coating obtained from the above will not meet the requirements of MIL-C-5541B either on the basis of weight -- 40 mg/sq. ft. -- or on the basis of corrosion resistance. However, it does perform satisfactorily in meeting the requirements of the container industry.

Other formulations of the non-chromium type are still in the research and development stages. One such formulation includes the addition of phosphate to the bath, which helps to provide a means of identifying that a coating is present on the can, even though it is invisible. This is by way of the "muffle test" in which a treated can is subjected to a temperature of 1000°F in a muffle furnace for a 5 minute interval. The phosphate in the coating and magnesium in the alloy form a complex which has a gold to tan to dark brown coloration after muffling, if the can has been properly treated. The magnesium of the alloy migrates to the surface during heating, permitting the complexing with phosphate to take place at the surface. The heavier the apparent coating, the darker the coloration after muffling. Any further meaning or interpretation from the color exhibited, as to weight of coating, expected performance, etc., is not as yet fully understood.

COMPOSITION OF CONVERSION COATINGS

Chromium Phosphate

Early analysis of the chromium phosphate conversion coating involved wet analysis of stripped coatings.[2] The postulated composition of this coating based on the wet analysis was:

$$Al_2O_3 \cdot 2CrPO_4 \cdot 8H_2O$$

TGA curves on this coating were obtained to determine the moisture content, which agreed very well with the amount lacking after the chromium, aluminum, and phosphate were determined.

Nimon and Korpi[5] studied the composition of the amorphous phosphate conversion coating and determined that a fresh coating consisted essentially of "hydrated chromium phosphate". When the coating was dried, both "hydrated chromium and aluminum phosphate are present in the same relative amounts". They further stated that "it is possible that the coating contains Al_2O_3 but only in trace amounts". There appears to be a conflict here in that if the fresh coating consisted essentially of hydrated chromium phosphate, but after heating consisted of half chromium phosphate and half aluminum phosphate, then half of the chromium would have had to have been lost by some means because no additional phosphate would have been available for the aluminum.

Further, fresh coatings upon wet analysis show a definite amount of aluminum which could be due possibly to the stripping techniques employed. However, used baths produce coatings which show a considerably greater amount of aluminum upon wet analysis than do fresh baths. The change in the color of the coating from an emerald green to a whitish green, as the bath increases in its amount of dissolved aluminum, further indicates a change in coating composition involving a contribution by the aluminum in the bath. The apparent differences in the composition of coatings obtained from fresh vs. used baths complicates formulating an exact composition for this type of coating.

A paper by J.A. Treverton and N.C. Davies[6] shows agreement with Nimon and Korpi's first conclusion, i.e., that hydrated chromium phosphate is the major component of a fresh coating. They suggested also that chromic oxide was also a part of the film that was closest to the metal surface. Figure 1 shows an ESCA spectrum listing the atomic percentage of the elements in this coating. Figure 2 shows a simulated cross-sectioned structure of the coating on the metal. As can be seen, the chromium phosphate is in the outer layer of the coating with the chromic oxide being concentrated towards the metal itself, but with several possible forms of aluminum interaction products at the metal/coating interface. The findings of several types of aluminum salts at the coating/metal interface could explain the aluminum found in early wet analysis determination.

Chromium Chromate

Some of the early analytical work involving chromium chromate coatings using wet analysis, TGA and IR spectral analysis indicated an empirical formula of[2]:

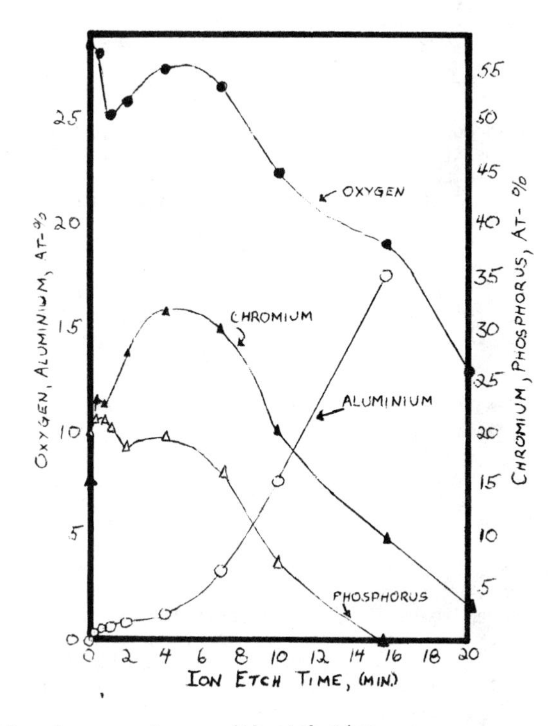

Fig. 1.—Variation in percentages with etch time.

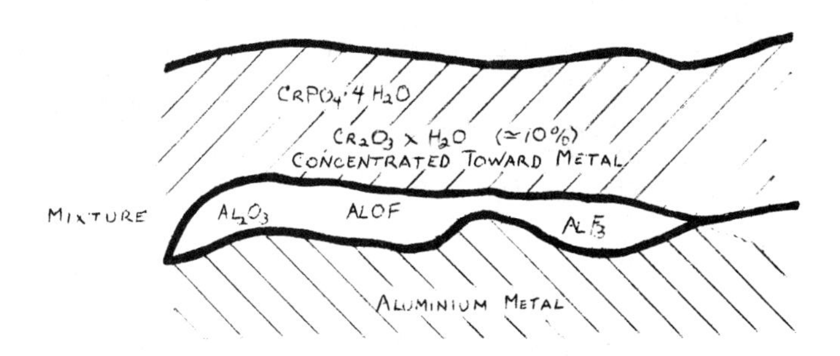

Fig. 2.—Structure of chromium phosphate film.

$$Cr(OH)_2HCrO_4 \cdot Al(OH)_3 \cdot 2H_2O,$$

for the unaccelerated coating, and:

$$CrFe(CN)_6 \cdot 6Cr(OH)_3 \cdot H_2CrO_4 \cdot 4Al_2O_3 \cdot 8H_2O,$$

for the accelerated coating. Sufficient sophisticated instrumentation was lacking at that time to carry the analysis further, but it was determined that hexavalent chromium was present in the fresh coating, but its apparent absence after the coating aged or was oven dried was unexplained.

W.G. Zelley[7] concluded (1) "chromate conversion coatings...provide a substantial protection to 2219 alloy in salt spray exposure through the presence of an anodic inhibitor, hexavalent chromium, in the coating". Also, (2) "Thermal treatments decrease the corrosion resistance...by rendering the hexavalent chromium less soluble and thus preventing it from being leached in a sufficient amount to serve as an effective inhibitor".

In a paper presented by Sunderland[8] on the Analysis of Chromate Conversion Coatings on Aluminum, he stated "it is commonly believed that an amorphous, insoluble gel-like film is formed on the surface, and that this film is impregnated with water of hydration and contains traces of substrate metal, some processing solutions, and hexavalent and trivalent chromium. The protective function of the film is believed to be twofold: the insoluble gel acts as a barrier to a corrosive medium and the hexavalent chromium in some manner inhibits corrosion, should the gel be penetrated." After an investigation involving the use of Auger electron analysis, secondary ion mass analysis, and an ion scattering spectrometer, Sunderland concluded that "throughout the region of variation, the aluminum oxide is largely converted to the chromium oxide Cr_2O_3. This conclusion is supported by the photoelectron spectrum...which shows chromium to be in the +3 oxidation state". Other sources such as Shapiro[9] and Gallacio[10] believe that hexavalent chromium is present in the fresh coating and is available for leaching into scratched or damaged areas. However, "heating above 50°C insolubilizes the chromates in chromate conversion coatings. At higher temperatures the insolubilization is more complete".

Thus, it appears that there is some conflict as to whether hexavalent chromium is present in the conversion coatings we classify as 'chromium chromate'. Treverton and Davies[6] concluded that "after 24 hours, hexavalent chromium was no longer leachable from the pretreated surface" and "no hexavalent chromium was found". The wet analysis of freshly prepared films performed very early in this work, in which hexavalent chromium was determined to be present, may have been due to the fact that "the chromate ions may not have completed their reaction with the metal surface".

In summation of the findings regarding the presence or absence of hexavalent chromium in the chromate conversion coatings, it appears that most of the researchers conclude that if hexavalent chromium is present in fresh films, it is either not present or unavailable after 24 hours of aging. For all practical purposes, therefore, the chromate conversion coatings can be considered to contain no detectable hexavalent chromium. For the balance of the composition, researchers[2,6,7] agree that ferricyanide is in the accelerated coating, probably as chromium ferricyanide, and that[6,8] "the aluminum oxide is largely converted to the chromium oxide throughout the region of variation".

Figure 3[6] shows an ESCA spectrum in terms of atomic % of the components of the accelerated chromate coating. Figure 4[6] shows a simulated cross-

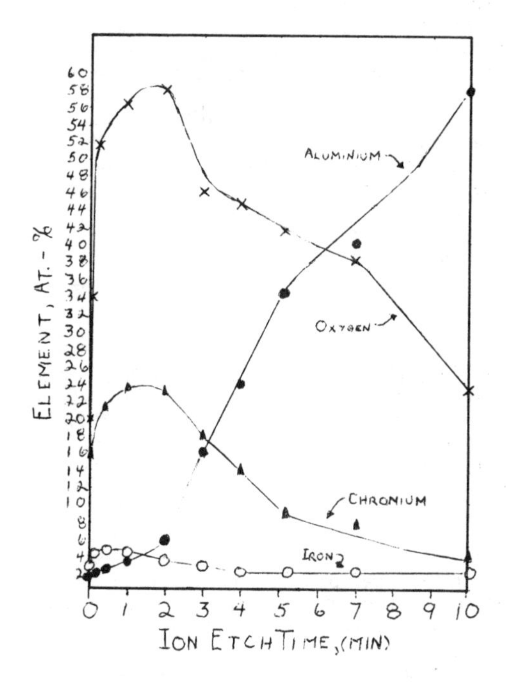

Fig. 3.—Variations in percentages with etch time.

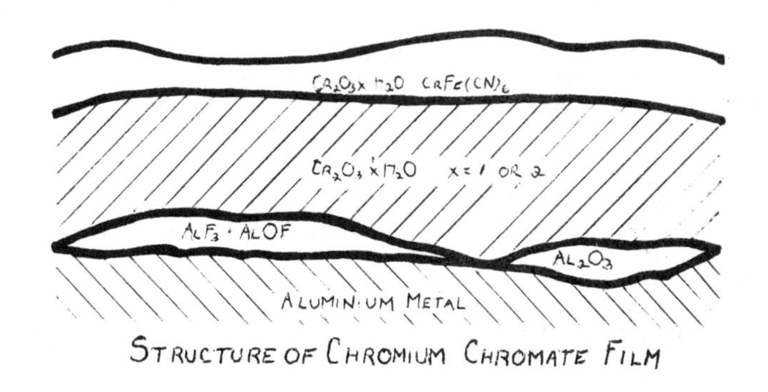

Fig. 4.—Structure of chromium chromate film.

sectioned structure of the coating on the metal substrate. Treverton and Davies found the chromium ferricyanide to be in the outer layer of the coating along with some chromic oxide. The bulk of the coating inward toward the base metal appears to be chromic oxide and again, adjacent to the metal/coating interface, are the reaction products AlF_3, $AlOF$, and Al_2O_3.

The unaccelerated coating is expected to be similar, but with no chromium ferricyanide present. The complexity of the chemistry involved in the conversion coating processes has prevented our learning much about these coatings up until the advent of such research instruments as ESCA, SEM, Auger, XPS, EPA, TEM, etc. With these research tools at our command, much has been learned over the past five years. It is expected that a great deal more will be known and understood after another 5 years of investigative research.

Non-Chromium

As indicated earlier, the essential ingredients of one of the initial non-chromium containing formulations which found immediate acceptance in the container industry, included boric acid, a source of fluoride, a source of zirconium, and nitric acid for pH adjustment. The coating composition is more difficult to determine because of its extremely low coating weight, and only through the use of such sophisticated research instruments as ESCA, EPA, XRES and Auger, can it be proven that a conversion coating actually exists.

The mechanism that produces pseudo boehmite films on aluminum via boiling DI water, can be considered as being similarly involved in the reaction of aluminum with the above chemical baths. An anhydrous layer of Al_2O_3 is believed to be adjacent to the base metal (in the case of the pseudo boehmite films) with an aluminum oxide/hydroxide layer next to it as in Figure 5[3]. The formation of a porous $Al(OH)_3$ layer at the outer edge of the film is believed to occur due to the hydration of the $AlOOH$ layer. This outer layer of randomly oriented, loosely connected platelets has poor corrosion resistance and adhesive properties. The introduction of zirconium and fluoride into the film by way of the coating bath is pictured in Figure 6[4].

When zirconium is bonded to the hydrous alumina film, the formation on the surface, of hydrated oxide of aluminum is prevented. Auger Electron Spectrum Analysis of the coating has shown the presence of both Zr and F in the coating.

An Electron Probe Microanalysis[11] method for determining zirconium was developed by using a polished massive ZrO_2 standard, the emissions from which were then compared to emissions from the test specimens themselves. The zirconium content of the test specimens -- the non-chromium coatings -- was determined to be approximately 4.1 mg/sq. ft. for a percentage of the total coating weight of approximately 36%. The major constituents of the coating were found to be zirconium, aluminum, oxygen, and fluorine. The total weight of the coating, therefore, could be calculated as being approximately 11.4 mg/sq. ft.

If a pictorial arrangement of the layers of the coating on the metal were made as determined by ESCA it would probably look like the drawing shown in Figure 7.

The thickness of the coating was measured from transmission electron micrographs of ultra thin microtomed sections and found to be approximately[11] 200-250 Å (Figure 8). This is much thinner than the normal chromium-

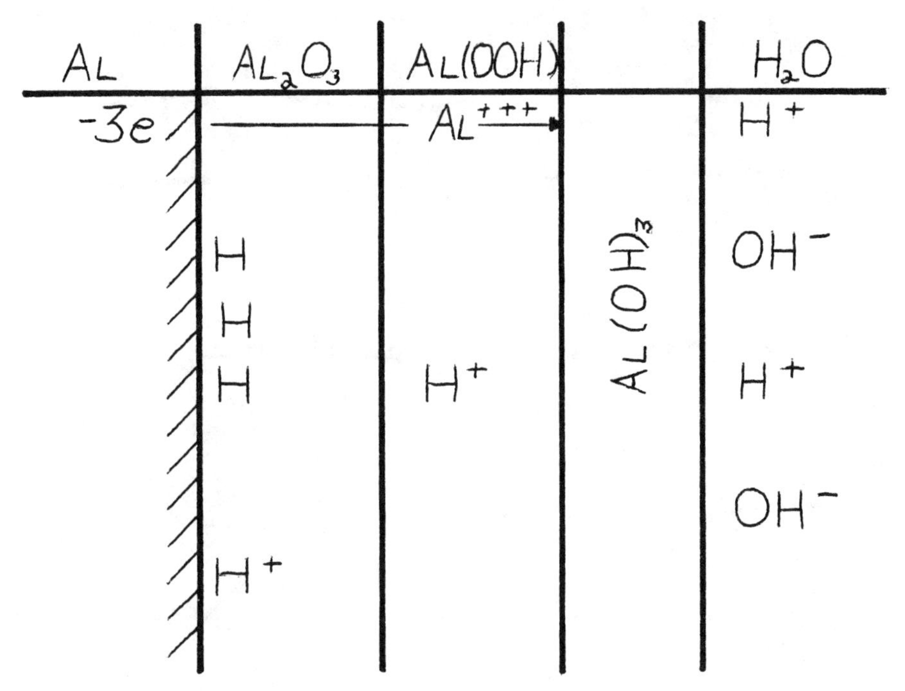

Fig. 5.—Reaction of aluminum with water.

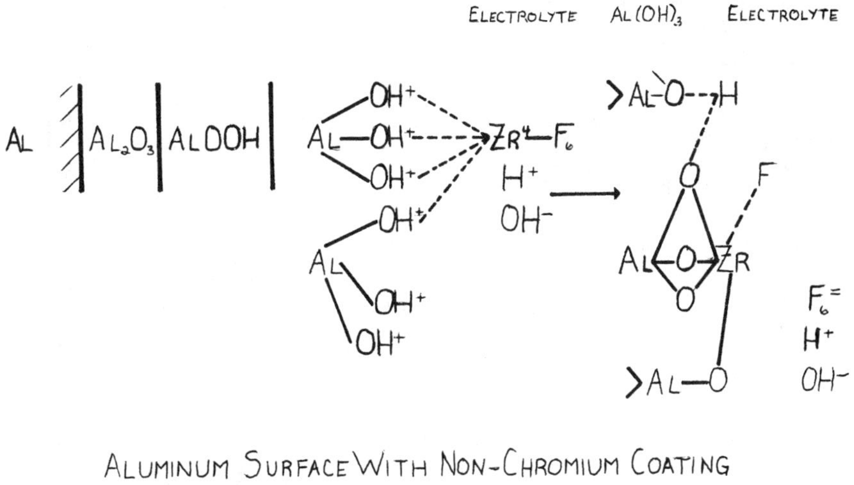

ALUMINUM SURFACE WITH NON-CHROMIUM COATING

Fig. 6.—Aluminum surface with non-chromium coating.

containing coatings described earlier, and increases in concentration, time, or temperature do not show appropriate increases in coating weights.

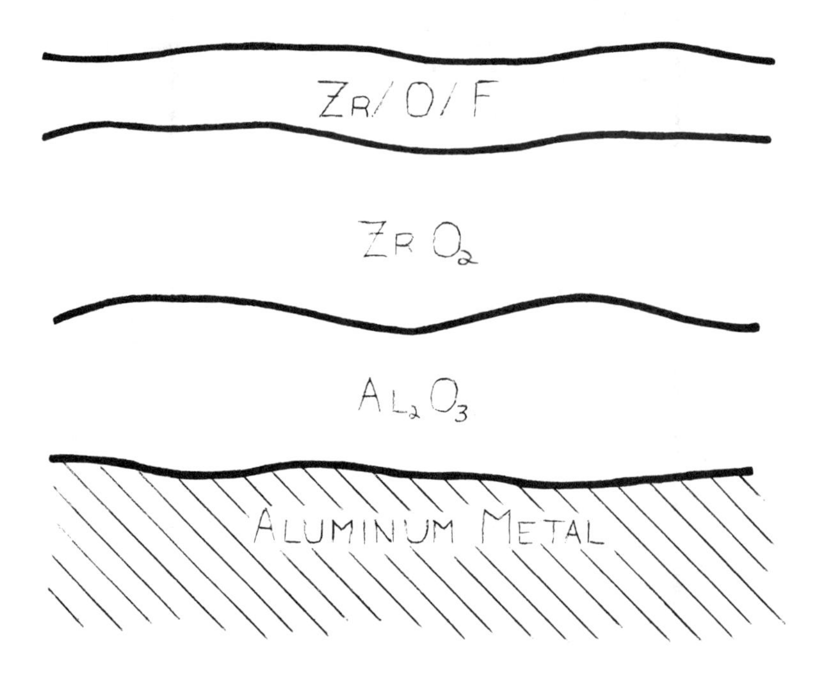

Fig. 7.—Structure of non-chromium film.

UNRESOLVED TECHNICAL CONSIDERATIONS

Researchers involved in the chemistry of treating metals, particularly aluminum, learn very quickly that providing solutions to one or two problems, usually generates at least twice as many questions as were originally posed. Some of the questions that I have found to be particularly perplexing are:

1. The part played by the metal itself in providing adequate corrosion resistance. Ketcham and Brown[12] have concluded that "corrosion resistance is also influenced by the composition and heat treatment of the aluminum alloys as well as the mill practices employed in their production". The question of greatest importance here is how to tell when the one is influenced by the other.

2. The effect of each of the steps involved in producing the conversion coating. Again Ketcham and Brown have found that "performance of a chromate conversion coating...is heavily influenced by each step in the processing used - cleaning, deoxidizing, chromating, rinsing, and drying". Again, the question of greatest importance is how to tell which of the processing steps is at fault, if the part looks excellent before exposure to salt spray, but fails as a result of that exposure.

3. The ability to produce a true boehmite film on aluminum by chemical means at low temperature, rather than by electrochemical means or by the

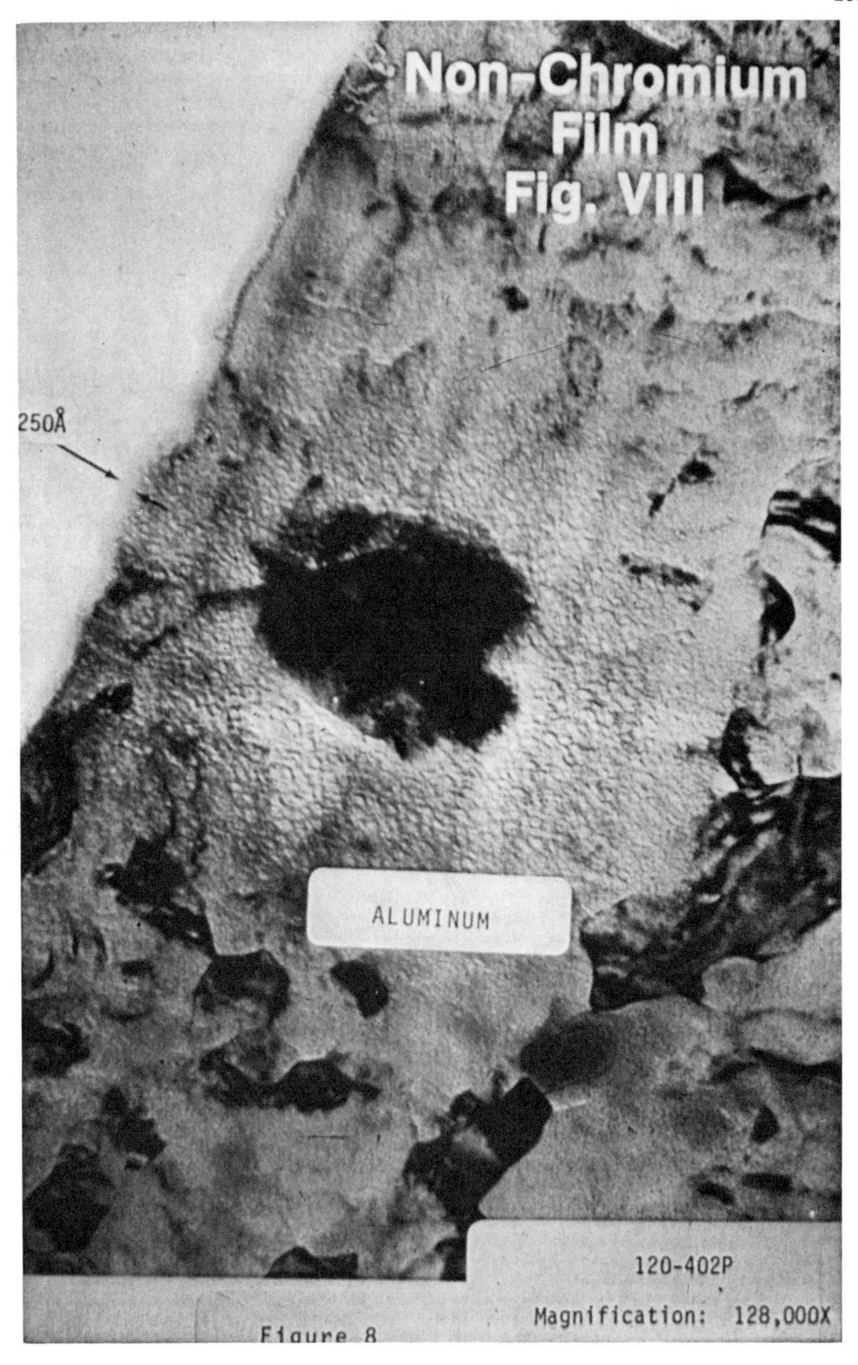

Fig. 8.—Non-chromium film.

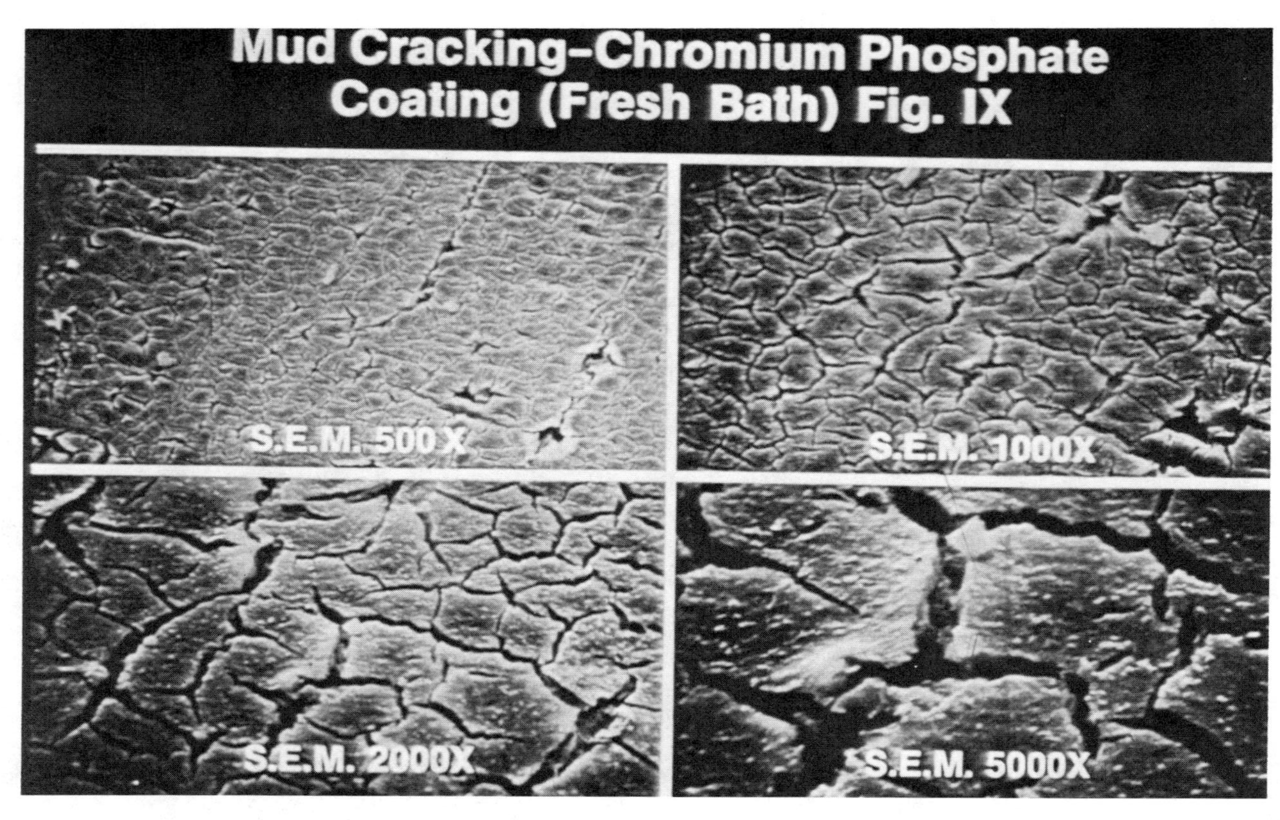

Fig. 9.—Mud cracking - chromium phosphate coating (fresh bath).

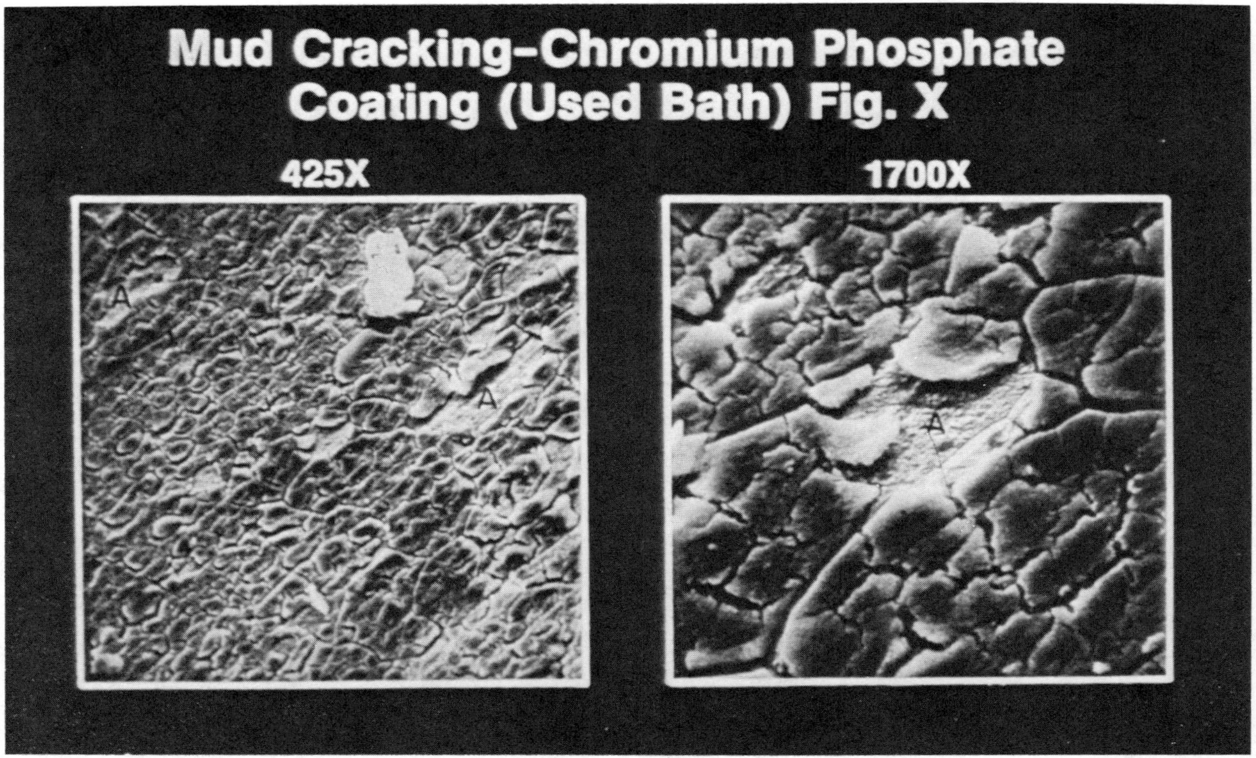

Fig. 10.—Mud cracking - chromium phosphate coating (used bath).

reaction of the metal with boiling DI water. This problem appears to be solvable, but the right combination of oxidizing, accelerating, activating, and passivating agents has not yet been found.

These are only a few of the many areas in which a considerable research effort could be made. Another one in which I will go into some depth is the matter of mudcracking of the dried film. This effect appears in both the chromium chromate and chromium phosphate coatings. Figure 9 shows the appearance of a chromium phosphate coating at SEM magnifications of from 500X to 5000X. The weight of the film was approximately 500 mg/sq. ft., a very heavy coating, which probably contributed to its ease of fracture.

Figure 10 represents the same type of coating but produced from a used bath. This coating weight was approximately 180 mg/sq. ft., and by examining the center surface area carefully, it is apparent that pieces of the "mud" are missing. This was true in other areas of the coating, at magnifications of 425X to 1800X, and an Auger analysis of the area "A", where a "mud" piece was missing, showed that no chromium or phosphate was present. Aluminum was the major component found (probably base metal), with possibly a trace of fluorine. All of the adjacent areas showed the normally expected elements, *Cr*, *P*, *Al*, *F*.

Of interest is the fact that the specimens which showed pieces of mud cracking removed, performed poorly under paint when tested in an accelerated cycling acetic acid salt fog. We believe the absence of a continuous film contributed to that failure.

Mudcracking appears to be a normal function of these kinds of conversion coatings. It would be interesting to study this phenomenon further to determine:

1. Does the cracking go clear through to the base metal?

2. If so, does this occur in coatings produced both from fresh and used baths?

3. What is the effect of aluminum build-up in the bath on the degree of mudcracking?

4. Does the manner in which the coatings are dried affect the patterning of the mudcracking?

5. Can mudcracking be eliminated, and if so, will the coating show improved corrosion resistance?

In this paper I've tried to present a brief description of the most commonly used conversion coatings on aluminum; a bibliography of how other researchers view these coatings; and posed some questions that need to be answered in order for us to better understand these coatings. It has been interesting studying the many opinions and papers referenced. I hope that some of the questions posed will be of sufficient interest to warrant additional investigation in the future.

REFERENCES

1. Ayres, R.F., *Metal Progress 78(9),101 (1960)*.
2. Newhard, N.J., *Metal Finishing 70(7), 49 (1972); 70 (8), 66 (1972)*.
3. Das, N., "The Reaction of Aluminum with H_2O and D_2O," Doctoral Dissertation, Lehigh University, 1973.
4. Das, N., U.S. Patent 3,964,936.
5. Nimon, L.A. and Korpi, G.K., *Plating 59(5), 421 (1972)*.
6. Treverton, J.A. and Davies, N.C., *Metals Technol. 4, 480 (1977). London*.
7. Zelley, W.G., "Development of Improved Conversion Coatings for Aluminum Alloys," Final Report NA S8-11226, July 1966.
8. Sunderland, R.J., "Proc. Int. Vac. Cong., 6th,"p. 347, 1974.
9. Shapiro, H., "Restrictions on the Use of Chromate," *Products Finishing (1968)*.
10. Gallacio, A., Pearlstein, F. and D'Ambrosio, M., *Metal Finishing 64(8), 50 (1966)*.
11. Special report by outside consultant.
12. Ketcham, S.J. and Brown, S.R., *Metal Finishing 74(11), 37 (1976)*.

IMPORTANT CORRELATIONS ARISING FROM THE SCRIBE TEST FOR APPRAISING PAINT/PRETREATMENT/SUBSTRATE SYSTEMS

Ronald W. Zurilla

Engineering and Research Staff,
Ford Motor Company,
Dearborn, MI

R. W. ZURILLA

ABSTRACT

The principal factors that influence the delamination of paint systems in a scribe/salt spray test are reviewed. The effects of steel surface carbonaceous deposits and the porosity level of zinc phosphate coversion coatings on the rate of paint delamination are discussed. Methods for determining the amount of surface carbon contamination on commercial steels and for measuring relative phosphate coating porosity levels are described. The electrochemical reduction of oxygen at the steel/ phosphate/ paint interface is shown to produce an alkaline solution which causes loss of paint adhesion. The importance of resin composition for inhibiting paint undercutting is illustrated. Additional information required to give a more complete explanation of the paint delamination phenomenon is presented. The basic electrochemistry of painted steel corrosion and paint delamination is outlined.

INTRODUCTION

Paint coatings have been used for many years to provide protection and decoration for articles fabricated of steel. It is generally recognized that for a paint system to exhibit good corrosion protection, it is essential to use well-cleaned steel, high quality phosphate conversion coatings and high quality paints. The need for improved paint systems of consistently high quality has become more critical because of the recently increased severity of the automobile environment caused by the greater usage of large quantities of road deicing salts. As part of continuing programs in the automotive and related industries to satisfy this need, the salt spray test has proven beneficial for appraising paint systems that prevent and retard delamination. In this paper the current status of our understanding of the paint delamination phenomenon in the scribe salt/spray test is reviewed and the relationship of delamination to properties of the paint/phosphate/steel substrate system is discussed. The basic electrochemistry of painted steel corrosion and delamination is briefly outlined.

ELECTROCHEMISTRY OF PAINTED STEEL CORROSION

The fundamental aspects of the corrosion of metals and the influence of various environments on the rate and form of corrosion have been reviewed in detail[1-5]. This prior work shows that the corrosion of steel in the presence of oxygen and water is an electrochemical process in which the oxidation and reduction reactions involve charge transfer across a solid-liquid interface. The oxidation reaction (anode) is the dissolution of iron to form soluble ferrous ions:

$$Fe \rightarrow Fe^{+2} + 2e^- \tag{1}$$

The excess electrons generated at the anode site are conducted through the metal and are consumed at the cathode site by the reduction reactions. In a neutral aqueous solution the principal reactions at the cathode site are the reduction of oxygen and the evolution of hydrogen:

$$O_2 + 2H_2O + 4e^- \rightarrow 4OH^- \tag{2}$$

$$2H_2O + 2e^- \rightarrow H_2 + 2OH^- \tag{3}$$

Both reactions produce a local increase in solution pH, and as the pH increases, the reduction of oxygen becomes the predominant cathodic process. The total rates of the anodic and cathodic reactions are identical since a net excess of electrons cannot be produced.

In the absence of dissolved salts the anodic and cathodic reactions occur at relatively close atom sites. The electrical circuit is completed by the diffusion of the Fe^{+2} and OH^- ions which react to form ferrous hydroxide. The hydroxide is oxidized by oxygen to form insoluble hydrated ferric oxide (rust) which slows the rate of corrosion by restricting the movement of ions.

When dissolved ionized salts are present, such as sodium chloride solutions, the anodic and cathodic reactions are the same as described above, but the electrical circuit is completed by the diffusion of Na^+ and Cl^- ions. As a consequence, $NaOH$ is formed at the cathode site and $FeCl_2$ at the anode site. The ferrous chloride can react with the sodium hydroxide to form ferrous hydroxide (or a basic salt) which is subsequently oxidized to form rust. Sodium chloride is also regenerated according to the reactions:

$$FeCl_2 + 2NaOH \rightarrow Fe(OH)_2 + 2NaCl \tag{4}$$

$$4Fe(OH)_2 + O_2 \rightarrow 2Fe_2O_3 \cdot H_2O + 2H_2O \tag{5}$$

Alternatively, the ferrous chloride can be hydrolyzed at the anode site to ferrous hydroxide and hydrochloric acid as follows:

$$FeCl_2 + 2H_2O \rightarrow Fe(OH)_2 + 2HCl \tag{6}$$

The ferrous hydroxide also can be oxidized to rust as shown in Equation 5. The formation of rust can restrict the transport of oxygen to the corroding anode and shift the cathode reaction to sites just outside the rust deposits. Thus, a differential oxygen concentration cell is produced which can accelerate the corrosion process.

Undoubtedly, the chemical reactions are more complex than illustrated above. However, the overall result of the electrochemical and chemical reactions is to produce rust, set up oxygen and electrolyte concentration gradients, cause a wide separation of the anode and cathode sites and accelerate corrosion. The rust deposits generally form complex mixtures of hydrated ferrous and ferric oxides, and the electrolyte within the rust becomes acidic due to hydrolysis reactions. The cathodic sites are shifted to the periphery of the rust where the electrolyte becomes alkaline. As will be discussed shortly, this local increase in pH is an important factor in the loss of paint adhesion.

For painted steel, the corrosion process can be considered as occurring in stages consisting of initiation, paint delamination and active steel corrosion. The initiation of paint corrosion can be influenced by many variables, particularly the environmental conditions, the nature of the paint polymer, and the presence of imperfections and impurities. Reviews[6-10] of the initiation of corrosion emphasize that paint films are not totally impermeable to the corrosive elements. Water and oxygen are readily absorbed by all paints and diffuse through the coatings at relatively high rates. Ionic species also penetrate the films, but at a much lower rate than those of water and oxygen. Eventually, sufficient oxygen, water and ionic species (*NaCl*) diffuse to the steel substrate and set up a corrosion cell. With time the corrosion products (rust and *NaOH*) can produce a local loss of adhesion and a macroscopic break in the paint. Once a break occurs in the paint film by either underfilm corrosion or mechanical damage, paint delamination can occur at an accelerated rate. Figure 1 schematically shows the location of the corrosion reactions at a damaged or cracked paint film.

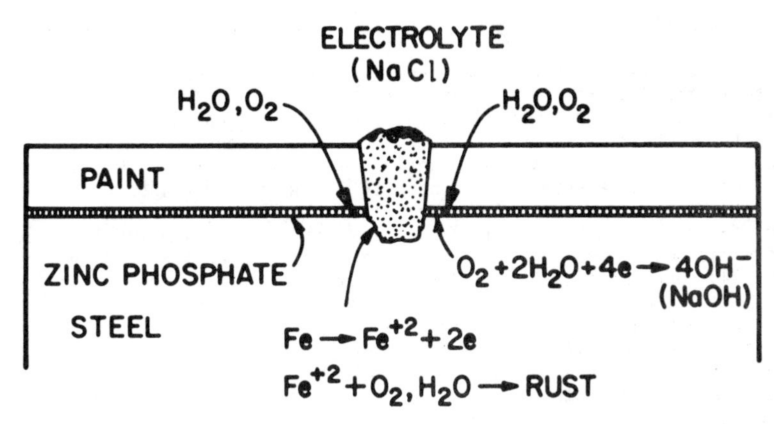

Fig. 1.—Schematic representation of the corrosion reactions at a damaged or cracked paint film.

The anodic dissolution of steel occurs beneath the rust layers and the cathodic reduction of oxygen is shifted to the zinc phosphate/steel interface, since oxygen transport to the anode is slowed by the rust deposits. With time a highly alkaline solution, with a pH greater than 12, can be formed which produces extensive delamination of the paint film by destroying the conversion coating[11] and by causing saponification of the paint[12].

The steel beneath the undermined paint film is protected from corrosion as long as a high pH solution is maintained. If the paint film is cracked, the alkaline solution can be diluted by the ingress of water. As the electrolyte pH beneath the paint approaches neutral values, the unprotected steel can begin to corrode and produce extensive pitting and penetration of the substrate. Hence, the process of cathodically induced paint delamination is an important precursor to extensive corrosion of the metal. The process of localized corrosion is discussed by many authors[1-5].

SCRIBE/SALT SPRAY TEST

The salt spray test was originally developed as a method of evaluating materials for protecting steel in seacoast environments. With time the test has been adopted by the aircraft, automotive, and structural industries to appraise coatings exposed to deicing salts. Currently, the method is commonly used by many organizations as an acceptance test for materials, and as a means to investigate the corrosion protection provided by paint systems. The ASTM salt spray (fog) test[13] consists primarily of exposing coated samples to a mist of sodium chloride solution for an arbitrary time period in an enclosed chamber maintained at an elevated temperature. The ASTM designation specifies a 5% sodium chloride solution and a temperature of 35°C. The pH of the collected salt solution is controlled in the range of 6.5 to 7.2. ASTM recommends that the samples be positioned in the chamber at an angle of 15 to 30 degrees from the vertical and parallel to the principal direction of horizontal flow of salt spray through the chamber. Free settling of salt mist also should occur on all specimens. Additional details concerning the equipment and procedure are given in the test specification.

For the evaluation of paint systems in a salt spray test, it is common practice to scribe the painted samples with a sharp tool prior to testing. The scribing procedure serves as a means of exposing the underlying metal substrate in order to simulate damaged areas or paint ruptures caused by underfilm corrosion. After testing, the scribed specimens are rinsed and dried; after which tape is used to remove any weakly adherent paint. The test duration and the acceptable amount of paint undercutting are arbitrary choices; the values selected will generally depend on experience and the quality of the paint system. For specification purposes, the time and the maximum permitted amount of delamination are often mutually established by the organizations concerned with the evaluation of a specific product, for example, 240 hours of testing and 3 mm maximum undercutting for an automotive primer on phosphated steel. Alternatively, the time required to produce a certain amount of delamination can be used as a means of evaluating paint systems in the salt spray test. Figure 2 illustrates the type and the different amounts of paint delamination that can occur between two painted samples in a salt spray test for 240 hours. The paint system is an epoxy ester-melamine primer applied to zinc phosphated steel. The sample designated as "pass" showed minimal paint undercutting from the scribe whereas the sample designated as "fail" exhibited paint delamination much greater than 3 mm.

When paint systems exhibit delamination in a scribe/salt spray test as shown in Figure 2, two important observations can be made. In the area of undercut paint, the steel is essentially uncorroded, and the thin film of solution is alkaline, often in excess of pH 12. These observations are in agreement with the paint delamination mechanism described in the prior section. Despite the criticism[6,14-16] that the test does not assess many of the modes of paint degradation which can occur in service, this agreement indicates that the scribe/salt spray test is a valid means of evaluating the ability of a paint system to prevent and retard the lateral spread of corrosion when initiated. The following sections review the influence of the steel substrate, phosphate conversion coating and paint properties on the rate of paint delamination in a scribe/salt spray test.

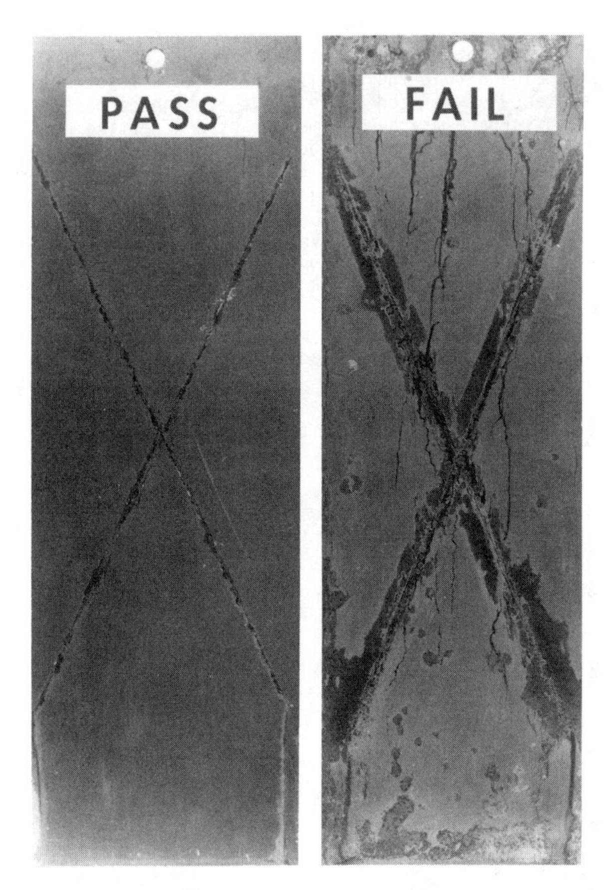

Fig. 2.—Paint delamination observed for phosphated and painted steel samples in a 240 hour scribe/salt spray test. Pass: <3mm undercutting of paint; Fail: >3mm.

ROLE OF STEEL SURFACE IN PAINT DELAMINATION

The variability in the corrosion performance of phosphated and painted commercial steel has been observed for many years. Wirshing and McMasters[17] showed that cold-rolled steel samples, obtained from either one or many sources, gave variable salt spray results after the samples were phosphated and painted under controlled conditions. Attempts to identify the cause and to clean the steel by conventional commercial methods were unsuccessful. Improved salt spray performance was obtained only when the steel was ground or sand blasted before painting. Although no direct experimental evidence was presented, Grossman[18] suggested that the variability of painted commercial steel was due to carbonaceous residues on the steel surface. It was postulated that these residues formed from rolling lubricants during annealing of the steel. It was confirmed that grinding of the steel surface would produce good corrosion performance after phosphating and painting.

Bronder and Funke[19] also reported variable salt spray results for painted steels that were obtained from either the same or a number of manufacturers. They concluded that the steel surface had a major influence on the corrosion resistance of painted parts and that high carbon concentrations near or on the surface may be the main factor affecting performance. Recently, Hospadaruk et al.[20] demonstrated that carbonaceous deposits on the surface of steel are responsible for the variable salt spray performance of painted commercial steel. After cleaning the steel samples with a hot alkaline cleaner, carbon was identified as the major surface contaminant by Auger electron spectroscopy. The relative amount of surface carbon was measured by determining the ratio of the peak-to-peak heights for carbon and iron as a function of sputtering time. A good correlation was found between surface carbon and corrosion test results. As shown in Figure 3, a high level of surface carbon contamination on steel causes early failure in a scribe/salt spray test for zinc phosphated, epoxy ester-melamine painted steel samples.

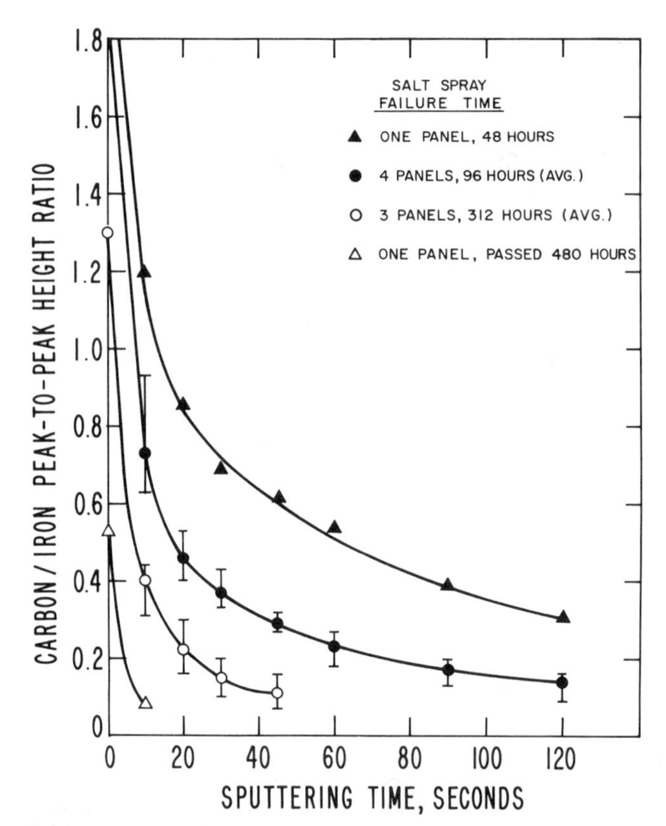

Fig. 3.—Relative amount of surface carbon detected by Auger electron spectroscopy for steels having different salt spray performance.[20]

Failure was defined as the time required to produce 3 mm of paint delamination. No relationship was found between the salt spray performance and any of the other surface impurities detected by Auger analysis. Wojtkowiak and Bender[21] also recently found a good correlation between salt spray corrosion resistance and the amount of carbon on the surface of steel measured by Auger electron spectroscopy and ESCA. The salt spray tests in this work determined the amount of paint undercutting for a paint system that consisted of an alkyd primer, acrylic sealer and acrylic lacquer on phosphated steel.

Auger spectroscopy and other surface analysis methods examine only a very small surface area and require a relatively long analysis time using sophisticated, expensive equipment. Hospadaruk et al.[20] describe a fairly rapid method to measure the amount of carbonaceous deposits on one-square-foot steel samples. The proecdure consists of first cleaning the steel with either a vapor degreasing treatment followed by a spray of hot alkaline cleaner, or only the alkaline cleaner treatment. After use of either cleaning method, the samples are hot water rinsed, dried, and wiped repeatedly with a glass fiber pad saturated with 1:1 hydrochloric acid solution to collect all of the residual surface carbon deposits. The pads then are oven-dried and analyzed for carbon in a combustion furnace equipped with a CO_2 infrared detector. This analysis provides average surface carbon levels and makes no distinction between various forms of carbon that may be present on the surface. With the use of this analytical procedure a reasonably good correlation was obtained between the amount of surface carbon and the salt spray performance of phosphated and painted commercial steel, as shown in Figure 4.

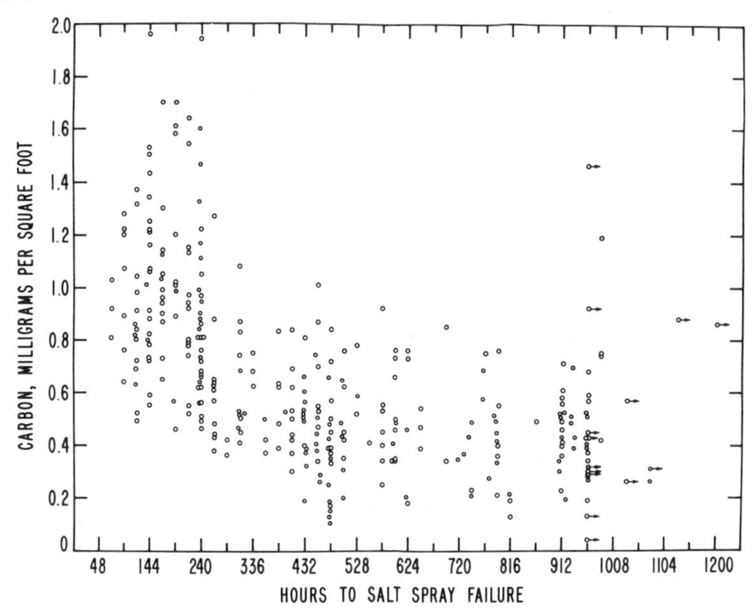

Fig. 4.—Relationship of surface carbon contamination to the salt spray performance of phosphated and painted commercial steel.[20]

For this work an epoxy ester-melamine primer was used, and the amount of surface carbon was measured on adjacent panels that were prepared by spray alkaline cleaning only. The results in Figure 4 indicate that steels having less than 0.4 mg/ft^2 of surface carbon deposits have a high probability of passing a 240 hour salt spray test for the paint system used. The considerable scattering of the results was attributed to incomplete removal of surface oils, non-uniform deposits on the steel, an average analytical carbon level for a relatively large area, and differences in the contamination levels between the analytical and salt spray samples.

When the steel samples were prepared for surface carbon analysis by vapor degreasing and alkaline cleaning, the carbon levels were generally lower than those obtained with only the alkaline cleaner preparation, and were related to salt spray performance in a similar manner to that given in Figure 4. This cleaning method is useful for identifying steels that have a high probability of failing a 240 hour salt spray test, since most of the steels with surface carbon deposits in excess of 0.8 mg/ft^2 failed within 240 hours. Despite the scatter of the data in Figure 4, the reported salt spray results show the wide variability in corrosion performance associated with commercial steel. The evidence presented demonstrates that the variability is caused primarily by carbonaceous deposits on the surface of commercial steel. These surface deposits are likely a result of current steelmaking practices.

ROLE OF CONVERSION COATING IN PAINT DELAMINATION

Zinc phosphate conversion coatings are applied to steel surfaces to promote the adhesion of paint and to inhibit paint delamination in a corrosive environment. The mechanism of deposition and the technology for producing good quality coatings have been extensively studied and recently reviewed[22,23]. As a quality control check, the phosphating industry has relied upon the measurement of coating weights[22,24]. However, it is generally agreed that coating weight, except at extreme values, does not directly relate to the salt spray performance of a scribed, phosphated and painted steel sample.

For phosphated commercial steel, Cheever[25] developed a rapid test, called Ferrotest, to measure coating density. This test consists of contacting a phosphated steel sample with a filter paper wetted with sodium chloride-potassium ferricyanide solution. After one minute, the amount of blue color produced is visually compared to color standards. Good agreement was reported between the Ferrotest ratings and two-week salt spray results for phosphated and painted samples. Essentially the same correlation was recently reported by Wojtkowiak and Bender[21]. This work indicated that phosphate coating porosity may have a strong influence on paint delamination that occurs during salt spray testing. Recently, Zurilla and Hospadaruk[26] showed that relative phosphate porosity values obtained from oxygen reduction curves consistently correlated with salt spray performance. Typical current-potential curves for oxygen reduction on phosphated steels having different salt spray performance are given in Figure 5. The salt spray results are based on the time required to produce 3 mm delamination from a scribe for phosphated and epoxy ester-melamine painted steel. The oxygen reduction curves were measured at a voltage scan rate of 2 mV/sec. In this test, a glass cell with O-ring exposed 21 cm^2 of phosphated steel to an unstirred, air-saturated, pH 12, sodium hydroxide solution. A graphite rod and a saturated calomel electrode were used as a counter and reference electrode, respectively. The commercial steels investigated had a wide range of surface carbon contamination as measured by the acid-wipe method described previously[20], and were phosphated in a 6-stage laboratory process.

The polarization curves in Figure 5 show that the oxygen reduction current increases as the salt spray performance decreases, which indicates an increase in the electro-chemical active area (or porosity) since oxygen is

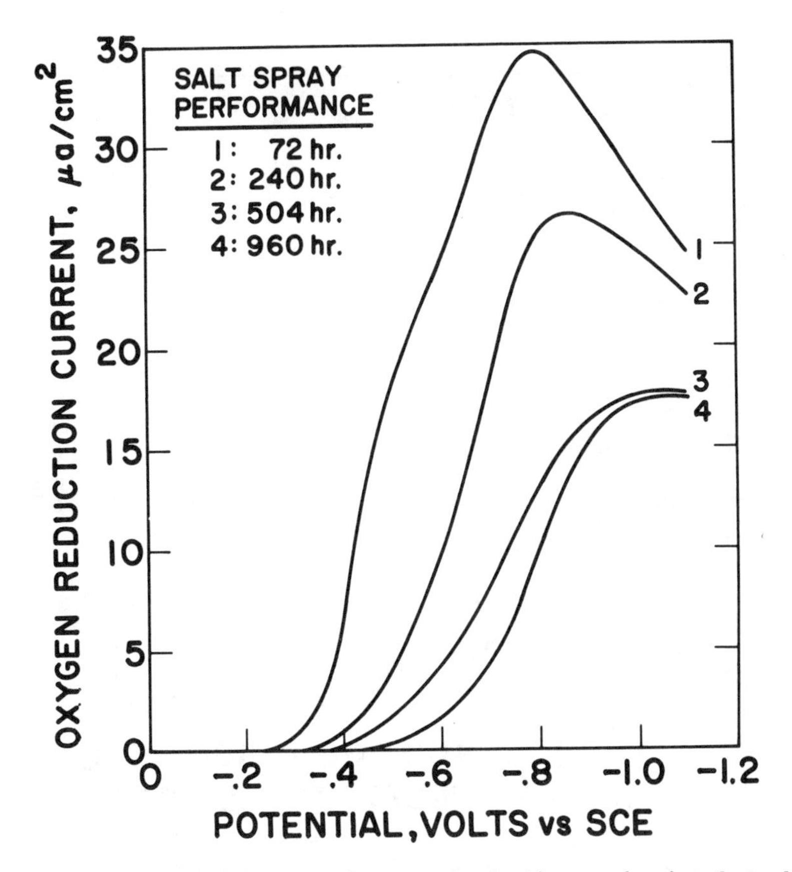

Fig. 5.—Current-potential curves for oxygen reduction on phosphated steels having different salt spray performance after painting.[26]

not reduced on the zinc phosphate crystals. The complex shape of the polarization curves at large negative potentials may be influenced by the size and shape of the pores in the coating and by reactions involving hydrogen peroxide which can be formed as an intermediate in the reduction of oxygen[27]. Current-potential curves obtained for many additional phosphated steels with known salt spray performance were in excellent agreement with those given in Figure 5.

The relationship of the oxygen reduction current measured at -0.55V to the salt spray performance for a large number of samples is given in Figure 6 where each circle represents a phosphated sample from a different coil of steel. A value for bare steel is included for comparison. The relationship shown in Figure 6 is approximately hyperbolic and indicates that the greatest change in phosphate coating porosity occurs for steels that fail within 72 to approximately 300 hours in salt spray. Similar results were found by using currents measured at other potentials in the range of -0.4 to -0.7 V. Within this range the oxygen reduction currents are not influenced by use of an anodic inhibitor in the final rinse of the phosphating process. At more positive potentials, the currents can be influenced by anodic reactions. At

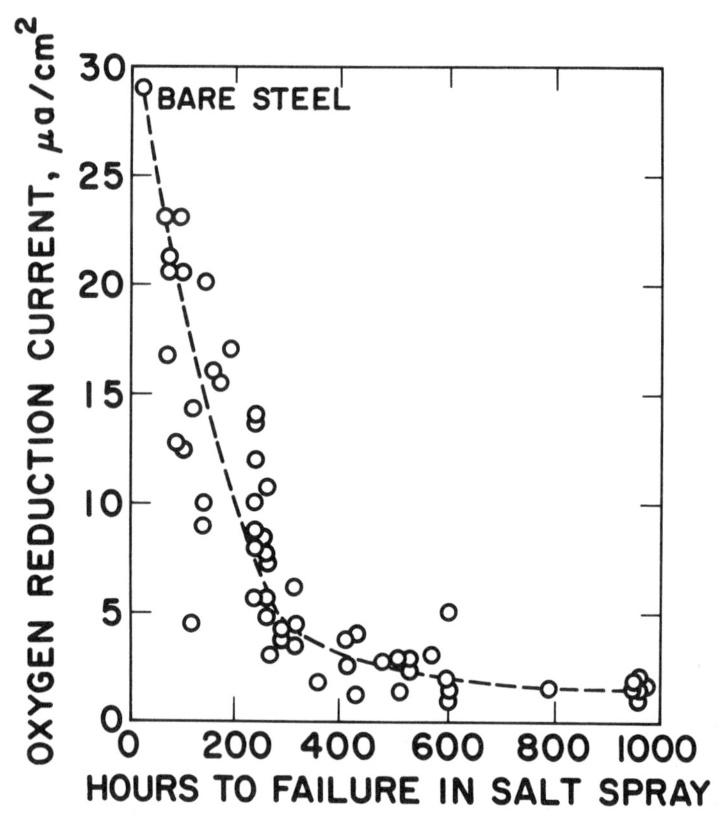

Fig. 6.—Relationship between oxygen reduction currents measured at -0.55V for phosphated steels in air-saturated, pH 12 NaOH and the salt spray performance after painting.[26]

more negative voltages the electrochemical reactions are complex and the currents are strongly influenced by oxygen diffusion. Scatter of the results in Figure 6 may be due partly to the difficulty of determining the time of failure in a salt spray test and to the variability of commercial steel over short distances. Similar relationships of oxygen reduction current and salt spray performance would be expected for primers other than the epoxy ester-melamine primer used in the work, with the curve shifted to higher or lower salt spray values depending upon the paint quality.

The differences among the current-potential curves measured for phosphated commercial steel can be attributed primarily to surface carbon contamination[26]. The effect of surface carbon contamination on the oxygen reduction current, measured at -0.55 V for the phosphated steels, is shown in Figure 7. Surface carbon contamination was determined by the method described previously[20] for the bare commercial steels cleaned by vapor degreasing and alkaline treatment. Although there is considerable scattering of data, the results indicate that an increase in surface carbon contamination produces an increase in coating porosity and a decrease in salt spray performance. The scatter of results may be due to variable carbon deposits among samples, and to the analytical procedure which determines average con-

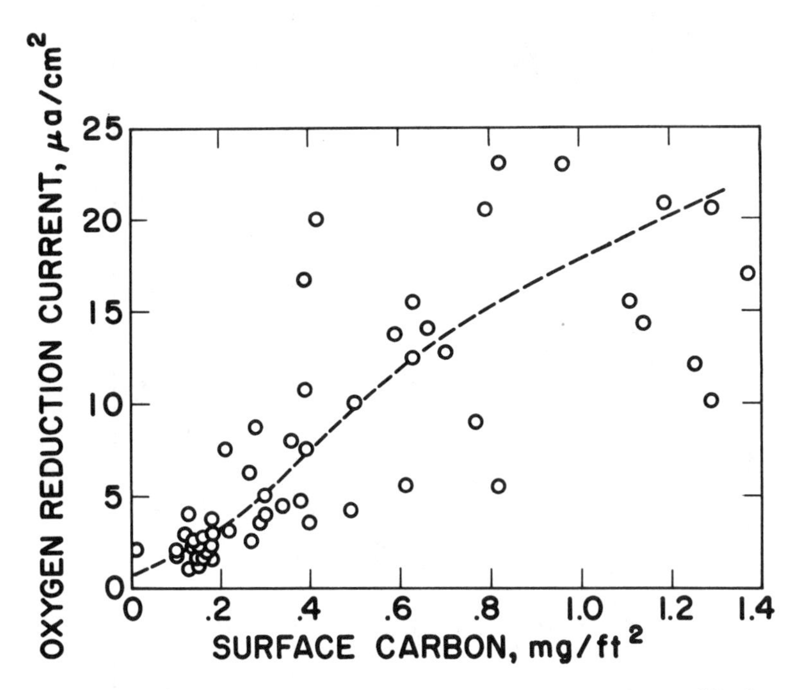

Fig. 7.—Relationship of oxygen reduction current measured at -.55V for phosphated steels in air-saturated, pH 12 NaOH to the steel surface carbon contamination level.[26]

tamination levels over a relatively large area but does not identify thick segregated deposits that can produce poor phosphate coverage. Wojtkowiak and Bender[21] also have concluded that surface carbon deposits on commercial steel interfere with the formation of a high quality phosphate coating.

Based on principles of electrochemical kinetics, the porosity of a phosphated steel would be related to the ratio of currents measured for a phosphated sample and a bare steel sample of the same geometric area, provided that the oxygen reduction reaction is solely kinetically controlled and anodic reactions are negligible. Conventional Tafel plots of potential versus the logarithm of current were not linear which indicated that the currents were not solely kinetically controlled. Since the reduction of dissolved oxygen is often controlled by both diffusion and charge transfer[28], diffusion-corrected Tafel plots[29] were constructed. Typical plots are given in Figure 8 for phosphated and for bare steel. In these plots i_D corresponds to the diffusion-limited current density for oxygen reduction, and the values were chosen by the following procedure. For bare steel the current-potential curves often exhibited a well-defined current plateau over a voltage range of several hundred millivolts, and the current plateau value was used as i_D. Because diffusion-limited currents were not observed for the phosphated steels (as shown in Figure 5), i_D was selected as the value which yielded a linear plot of potential versus the diffusion-corrected currents with a slope of approximately 0.12 V. The values of phosphate coating porosity were calculated from the ratio of diffusion-corrected currents for phosphated and bare steel. The results indicate that an increase

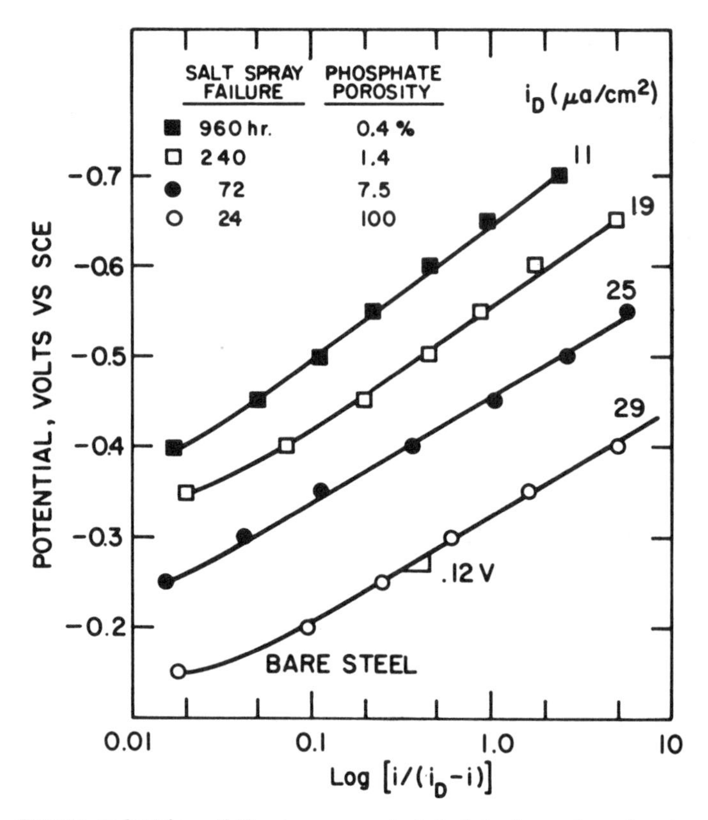

Fig. 8.—Oxygen reduction diffusion-corrected Tafel plots for phosphated and bare steel in air-saturated, pH 12 NaOH. Salt spray performances noted were based on time to produce 3 mm undercutting for painted samples.

in coating porosity causes a decrease in the salt performance. The fact that the values of i_D required to give linear Tafel plots are lower for phosphated steel than for uncoated steel suggests that the coating porosity (or total electrochemically active area) is small, as found by the calculations. Also, the linear kinetic plots with a slope of 0.12 V, as shown in Figure 8, indicates that the first electron-transfer step is rate controlling for oxygen reduction on steel in an alkaline solution, which is in agreement with results found for many other metals[28].

Phosphate porosity levels were calculated for additional phosphated steels according to the procedure described above. Ranges of calculated levels are listed in Table I for ranges of steel surface carbon contamination and the corresponding salt spray performance. Though the indicated ranges are approximate, the results indicate that porosity levels of only a few percent can significantly reduce the effectiveness of phosphate coatings for retarding paint delamination, and that levels of 0.5% or less are required for excellent corrosion performance. The calculated values of porosity listed in Table I are in good agreement with levels of 0.5 to 1.5% reported as typical[22]. The values are considered as estimates because of the method and assumptions used for the calculations. The porosity levels also may be "apparent" values since the electrochemical reduction of oxygen may occur at surfaces having a thin coating that is rapidly dissolved by the alkaline electrolyte as well as at uncoated areas.

TABLE I.—ESTIMATES OF PHOSPHATE COATING POROSITY FOR RANGES OF STEEL SURFACE CARBON CONTAMINATION AND THE CORRESPONDING SALT SPRAY PERFORMANCE.[26]

Steel Surface Carbon (mg/ft^2)	Salt Spray Failure Time (hours)	% Coating Porosity
> 0.8	< 120	> 2
0.3-0.8	300-120	0.5-2
0-0.3	> 300	< 0.5

The good correlation of oxygen reduction currents (or porosity levels) measured for phosphated steels in an alkaline solution and the salt spray performance of painted and phosphated steels is attributed[26] to the fact that the undercutting of paint, which occurs in environments containing oxygen, water and sodium chloride, is caused by the accumulation of alkali from the reduction of oxygen beneath the paint film. The rate of reduction of oxygen and the buildup of alkali should be greatest for the most porous phosphate coatings and result in the earliest time to exhibit a large amount of paint delamination.

ROLE OF PAINTS IN DELAMINATION

A considerable number of paint systems have been evaluated in a scribe/ salt spray test. While this work has shown that paints of different formulations exhibit different relative amounts of delamination, usually the information provided is insufficient to relate the loss of paint adhesion to the specific properties of the resin, pigment and other additives. Concerning the overall mechanism of paint delamination, the early investigations[30-32] showed that loss of paint adhesion in a salt environment is caused by the accumulation of alkali from the reduction of oxygen at the paint-metal interface. Evans and Taylor[12] indicated that the generated alkali could produce resin saponification or attack of the pigment. For paint applied to phosphated steel, Wiggle et al.[11] showed that delamination was due to dissolution of the phosphate conversion coating by cathodically produced hydroxide.

The effect of resin composition on delamination was investigated by Smith and Dickie[33]. They found, that when coatings with no inhibitive pigments were applied to unphosphated steel, that polyester-melamine and maleinized oil primers exhibited a greater amount of paint delamination in a 24-hour salt spray test than polybutadienephenolic and epoxy ester based primers. The least amount of paint undercutting was observed for an epoxy resin cathodic-electrodeposited primer. The relative differences in delamination were attributed to the relative susceptibility to attack of the resin by alkali, since the same ranking was found when the coatings were cathodically polarized in an oxygen saturated $NaCl$ solution as that found in the salt spray tests.

Corrosion inhibitive pigments can provide improved corrosion protection for paints in a corrosive service environment[6,7]. These materials can inhibit the initiation of corrosion and the undercutting of the paint film by slowing the anodic dissolution of the steel substrate, but their effectiveness depends strongly on the specific paint formulation and the service environment. Also, certain pigments can function by cathodically protecting the steel, by slowing the permeability of oxygen, water and/or electrolyte, or by forming a passive film on the steel surface. Though differences in salt spray performance have been demonstrated for various pigmented paints, the role of pigments in specific resins and the mechanism for retarding delamination have not been adequately shown.

As discussed previously, the accumulation of alkali that produces delamination is generated by the electrochemical reduction of oxygen. Since the reduction of oxygen occurs at the paint/steel or paint/phosphate/steel interface, it is necessary for sodium ions to diffuse to the cathodic sites to maintain electrical neutrality. The transport of sodium ions can occur either by diffusion through the paint film or by ingress at the exposed interface at the scribe, or by both paths. Paints have been shown to be permeable to *NaCl*[34] and that the rate of diffusion of ions is much smaller than that of water and oxygen[7]. As discussed in reviews[7-9] of the transport properties of paint, the sodium chloride diffuses through the film as ions and not as ion pairs, and the rate depends upon such factors as the resin type, pores, water uptake and cross-linking density. The effect of these variables on the loss of paint adhesion in a salt spray test has not been adequately investigated. Also, the nature of the bonding of the paint to a phosphate coating and the location of paint delamination have received little attention.

CONCLUSIONS

The experimental evidence indicates that paint delamination in a scribe-salt spray test is caused primarily by the accumulation of an alkaline solution beneath the paint film and that the alkali is generated by the electrochemical reduction of oxygen. Alkali-resistant resins in paints are effective for inhibiting the loss of paint adhesion. Commercial steels were shown to have surface carbonaceous deposits that interfere with the formation of a low porosity zinc phosphate conversion coating. Low phosphate coating porosities are required to inhibit effectively the reduction of oxygen and to slow the rate of paint delamination.

Additional work is needed to determine the role of pigments and other additives for retarding the loss of paint adhesion. Also, the rate and path of diffusion of sodium ions to the steel/phosphate/paint interface and the location of bond cleavage needs to be investigated to provide a greater understanding of paint delamination.

REFERENCES

1. Evans, U.R., "The Corrosion and Oxidation of Metals," St. Martin's Press, Inc., New York, 1960.
2. Uhlig, H.H., "Corrosion and Corrosion Control," 2nd ed., John Wiley & Sons, Inc., New York, 1971.
3. Fontana, M.G. and Greene, N.D., "Corrosion Engineering," 2nd ed., McGraw-Hill Book Co., New York, 1978.
4. Tomashov, N.D., "Theory of Corrosion and Protection of Metals," The MacMillan Co., New York, 1966.
5. Shrier, L.L., Ed., "Corrosion," Vol. 1 and 2, 2nd ed., Newnes-Butterworths, London, 1976.
6. Burns, R.M. and Bradley, W.W., "Protective Coatings for Metals," 3rd ed., Reinhold Publishing Corp., New York, 1967.
7. Mayne, J.E.O., "Corrosion," Vol. 2, 2nd ed., L.L. Shrier, Ed., Newnes-Butterworths, London, 1976, p. 15:24-37.
8. Jullien, H., *Prog. Org. Coat. 2, 99 (1973-74).*
9. Koehler, E.L., "Localized Corrosion, " NACE-3, R.W. Staehle, B.F. Brown, J. Kruger, A. Agrawal, Ed., National Association Corrosion Engineers, Houston, 1974, pp. 117-133.
10. Smith, A.G. and Dickie, R.A., "Fundamentals of Corrosion Protection," Paper 780913, SAE Conference:Designing for Automotive Corrosion Prevention, Nov. 8-10, 1978.
11. Wiggle, R.R., Smith, A.G. and Petrocelli, J.V., *J. Paint Tech. 40, 174 (1968).*
12. Evans, U.R. and Taylor, C.A., *Trans. Inst. Met. Finishing 39, 188 (1962).*
13. ASTM Designation B117-73, "Standard Method of Salt Spray Testing," Annual Book of ASTM Standards, Part 27, 1976.
14. Valentine, L., *J. Oil and Colour Chem. Assoc. 46, 674 (1963).*
15. LaQue, F.L., *Materials and Methods 35, 77 (1952).*
16. Ailor, W.H., Ed., "Handbook on Corrosion Testing and Evaluation," John Wiley and Sons, Inc., New York, 1971.
17. Wirshing, R.J. and McMasters, W.D., *Paint and Varnish Production 46, Sept., 1956.*
18. Grossman, G.W., *Paint Mag., Sept. 1961, pp. 7-10.*
19. Bronder, E. and Funke, D., *Trans. Inst. Metal Finishing 51, 201 (1973).*
20. Hospadaruk, V., Huff, J., Zurilla, R.W. and Greenwood, H.T., "Paint Failure, Steel Surface Quality and Accelerated Corrosion Testing," Paper 780186, Society Automotive Engineers National Meeting, Detroit, Mich., March, 1978.
21. Wojtkowiak, J.J. and Bender, H.S., *J. Coating Tech. 50, 86 (1978).*
22. Lorin, G., "Phosphating of Metals," Finishing Publications Ltd., Middlesex, England, 1974.
23. Biestek, T. and Weber, J., "Electrolytic and Chemical Conversion Coatings," Portcullis Press Ltd., Redhill, Surrey, 1976.
24. Spring, S., "Preparation of Metals for Painting," Reinhold Publishing Corp., New York, 1965.
25. Cheever, G.D., *J. Paint Technol. 41, 259 (1969).*
26. Zurilla, R.W. and Hospadaruk, V., "Quantative Test for Zinc Phosphate Coating Quality," Paper 780187, Society Automotive Engineers National Meeting, Detroit, March 1978.
27. Evans, U.R., "The Corrosion and Oxidation of Metals: First Supplementary Volume," St. Martin's Press, New York, 1968, p. 348.
28. Hoare, J.P., "The Electrochemistry of Oxygen," Interscience Publishers, New York, 1968.
29. P. Delahay, "Double Layer and Electrode Kinetics," Interscience Publishers, New York, 1965, pp. 153-177.
30. Evans, U.R., *Trans. Electrochem. Soc. 55, 243 (1929).*

31. Gay, P.J., *J. Oil Colour Chem. Assoc. 32*, 488 (1949).
32. Anderton, W.A., *Official Digest 36*, 1210 (1964).
33. Smith, A.G. and Dickie, R.A., *Ind. Eng. Chem. Prod. Res. Dev. 17*(1), 42 (1978).
34. Kittleberger, W.W. and Elm, A.C., *Ind. Eng. Chem. 44*, 326 (1952).

ASPECTS OF SUBSTRATE SURFACES INFLUENCING INDUSTRIAL COATING SYSTEMS CORROSION PERFORMANCE

Ralph E. Pike

Consultant
Media, PA

RALPH E. PIKE

ABSTRACT

Steel surface quality and/or phosphate process operation are most frequently identified as the contributing cause of the variability, and often catastrophic failure, in the salt spray corrosion performance of organic coating systems. This exploratory study represents one effort to identify and measure substrate surface attributes considered most likely to influence systems corrosion behavior. Techniques for measurement of steel surface cleanliness, topography, reactivity, and contaminant composition were evaluated, standardized, and applied to a spectrum of commercial cold rolled steel samples representing either controlled variation in the steel manufacturing process or predetermined abnormal corrosion behavior.

High concentrations of carbon, retained on the steel surface after alkaline power wash, appear as the most probable poor corrosion performance determinent. Other observations suggest that the presence of carbon may be an indirect manifestation of several possible causes of early system failures.

INTRODUCTION

In early 1976 a project was initiated at the Du Pont Experimental Station by the Fabrics and Finishes Department addressing the complex problem of corrosion control in organic coating systems.

This project had as its broad long-term objective to assimilate, analyze and focus the state-of-the-art knowledge of the diverse facets of coatings systems performance that most significantly influence the service life of fabricated steel products. The experimental program focused on four areas of previously identified technical need:

- Basic principles of design, operation, and quality control of commercial metal conditioning processes;

---

*The data base for this paper was developed in 1977 by the author then employed as a Research Fellow in the Fabrics and Finishes Department of E. I. du Pont de Nemours and Co., Inc., Wilmington, De.

- Ability to measure the "quality" of substrate surfaces before and after conditioning and to correlate these measurements with appearance paint adherence, and systems corrosivity;

- Understanding of the mechanisms for corrosion immunity at the substrate/coating interface during exposure to corrosive environments, specifically
  wet adhesion,
  inhibition, and

- Ability to correlate accelerated testing methods for coating with field performance.

The scope of experimental investigation, for practical purposes, would be initially limited to autobody type cold rolled steel substrates. The finishing system would employ a standard commercial alkaline cleaning and zinc phosphate conditioning process, followed by an alkyd type high bake dip primer. This system has been used internally in Du Pont for laboratory systems corrosion evaluation for three decades. Corrosion performance would be determined as scribe creepage after ASTM B-117 salt spray exposure. Extreme care would be exercised in experimental panel preparation and salt spray testing using an individual predesigned series plan to ensure maximum aseptic handling conditions and consistent time intervals in processing or measurement. Corrosion would be interpreted as rank order within a given series using the scale and guide lines for individual panel ratings as shown in Figure 1.

This paper will attempt to provide an overview and interpretation of some of the findings uncovered in the initial phase of the investigation against these objectives and experimental plan.

It should be noted that this investigative effort was independently carried out during a period when a number of parallel programs were exceedingly active in various steel or automotive manufacturing laboratories, and other research facilities as a consequence of a high public focus on automotive corrosion. The conclusions only reflect the interpreted results of this predetermined experimental plan employed on five series of steel samples procured from various sources. The substrate samples were obtained with neither a fully characterized metallurgical and age history or an established record of field behavior. It is not known whether or not the reported results confirm or refute similar findings by other laboratories or whether the spectrum of samples tested in any way represent the pattern of commercial steel qualities traded in the market place. Extrapolation of the results beyond the experimental framework could be misleading.

RATIONALE

One of the most elusive targets of coatings research has been the development of industrial systems for finishing fabricated steel products which do not rust in service. This situation exists in spite of the fact that corrosion costs are reputed to represent as much as 3.5% of the gross national product. Nor is it a consequence of a lack of investigative effort. Massive volumes of literature have been published on many aspects of corrosion theory, mechanisms, and correlation studies.

The nature of the problem itself has defied well organized attempts to

SALT SPRAY CORROSION PERFORMANCE RATING SYSTEM

Rating #	Description	Hours Exposure	Points, 64th of inch creep
10	Excellent	168	0
		336	0-1
		504	0-1
		Max. Total	2
8	Good	168	0-1
		336	4 or less
		504	8 or less
		Max. Total	13
6	Fair	168	0-4
		336	8 or less
		504	16 or less
		Max. Total	28
4	Poor	168	0-8
		336	16 or less
		504	32 or less
		Max. Total	56
2	Very Poor	168	0-16
		336	32 or less
		504	64 or less
		Max. Total	112
0	Catastrophic	Max. Total	>152

Fig. 1.—Rating scale - creep corrosion.

gain appropriate scientific insight. There is, for example, substantial information on the electrochemical nature of the corrosion process at steel surfaces, but little information on the environment at the interface between the steel substrate and the applied coating when exposed to atmospheric corrosive conditions.

There are in existence millions of information bits relating to observed system failures as a function of metal conditioning, coatings application, and/or coatings composition. Yet few of these enable sound generalizations on basic principles for surface pretreatment, for methods or conditions of preferred coatings application, for rust inhibitive coatings formulation, or even projection to the coating system probability of field failure.

There are in existence hundreds of chemical compounds with demonstrated ability to retard corrosion reactions, as ingredients of coating composition, as metal pretreatment, or as additives to the corrodent medium. These additives reportedly may function (1) as inhibitors of anodic or cathodic corrosion reactions, (2) to control pH of the corrodent, (3) to create an insulating interfacial boundary layer, (4) to alter chemically the substrate surface to one of lower entropy, or (5) merely improve the physical properties of an adherent coating. Their utility must always be empirically established.

There must be an explanation why the research investment in corrosion and in corrosion resistant coating systems development has failed to solve the persistent corrosion dilemma. The most likely causes center around the compartmentalized approaches historically taken toward the problem resolution and the basic problems of measurement intrinsic to a complex non-homogeneous system.

Whether or not a finished fabricated steel product will fail to show adequate corrosion resistant service life is dependent on a number of interrelated factors:

- product design and specification
- steel processing history
- method of fabrication
- surface preparation
    - prefabrication
    - post-fabrication
- finishing process
- finishing materials

If one looks critically at past history, it is easy to recognize failures within given industry segments to consider fully the impact of their manufacturing and development activities on the overall corrosion problem.

In product design and specification, the automotive industry failed to recognize the impact of stainless steel trim, of gravel impingement, of underbody entrapment recesses in design, or ventilation of box members. The domestic steel industry has failed to ensure the manufacture of a more uniformly paintable substrate. The movement toward unitized welded autobody construction failed to identify fully the significance of the lack of precoating in joined sections. The chemical pretreatment industry fell short in developing or selling a process that ensures cleaning and conversion processes which adequately function under abnormal operating conditions. Primer application processes historically have not provided uniform, defect-free, coverage of edges and recessed areas. The primer manufacturer has often failed to guide adequately the end user in establishing realistic correlatable performance test requirements.

The practical solution of the problem, therefore, demands an analysis and weighting of the options offered, and integration of the best technology within some or all of the contributing segments -- or in other words, a fully multi-discipline approach. Substantial progress has been made in many of these areas. Certainly the designer and specifier now have an increased awareness of corrosion control responsibilities. The steel industry has responded significantly to improvement of the paintability aspects of commercial sheet products in surface texture and cleanliness control. Electrodeposition has offered processes and materials which trend to overcome prior coating deficiencies.

The principal technical void, clearly evident in our literature study, deals with the basic technical understanding of the nature of the steel and coatings interface. This project, therefore, was oriented to the examination of techniques for measuring properties of surfaces most likely to influence coating systems behavior, with particular emphasis on wet adhesion in corrosive environment exposure.

The experimental plan was initially structured on the following basic outline:

SURFACE ATTRIBUTE	METHOD
• Cleanliness	
Ease of cleaning	Surface Energy
Degree of cleaning	
Rate of deterioration	
• Topography	
Surface area	
Peak density, amplitude, sharpness	Profilometry
Phosphate nucleation	Electron Microscopy
• Surface Reactivity	
Acid response	Pickle Lag
Corrosivity	Anodic Polarization
• Surface Chemistry	
Carbon bonding	Electron Scanning
State of oxidation	Chemical Analysis
Crystallinity	X-Ray Back Scatter

EXPERIMENTAL DETAIL

The test procedures employed in this project were essentially based on common laboratory or published techniques employed by others in similar characterization efforts. As the program progressed, experience was acquired and results were interpreted and many of these procedures were modified. A general recap of this experience and related observations may be of benefit in result evaluation or to others engaged in similar explorations on this unproven experimental ground.

Since a variety of test procedures were to be applied to a wide spectrum of steel samples, received in varied stages of preservation, it was necessary to develop rigidly standardized procedures to ensure that measurements would be carried out on as uniformly common surface as was practical to achieve. During a preliminary study of steel surface energetics, three laboratory cleaning procedures were developed and standardized with carefully controlled time, temperature, and handling sequences. Method A represented a six stage combination Perclene ® solvent and alkaline cleaning process. Method B represented a four stage alkaline cleaning process. Method C represented an eight stage modification of Method A which included wet Scotchbrite ® polishing. Although all three procedures were applied in measurements on ease of cleaning, Method A was exclusively applied to all steel surfaces which were subjected to characterization and analytical test procedures. Refer to Figure 2.

Steel samples in the as received and as cleaned condition were wrapped in kraft or envelope paper, placed in polyethylene bags and stored in deep freeze conditions (-30°C). The surface energetics investigation had demonstrated this storage method enabled maintenance of a relatively constant surface energy, a criteria presumed to represent a stable surface state. To ensure maximum correlatable conditions of test, all procedures were carried out on a single 4" x 12" test panel aseptically segmented, coded, treated and commonly stored and handled. Replication was employed wherever practicable. Each series was controlled with panels of a common pretested lot of Parker unpolished autobody steel panels. Figure 3 describes the structure of the five individual series from which the observations and conclusions of this paper were drawn.

STANDARDIZED CLEANING PROCEDURES

CLEANING METHOD A:

(1) Dipped 5 minutes in slow boiling Perclene®.

(2) Dried at ambient laboratory temperature (∿22-24°C).

(3) Dipped 5 minutes in hot, well-stirred 2.34% aqueous Parcocleaner 2351 (∿70°C).

(4) Rinsed 30 seconds in hot tap water (∿60°C).

(5) Rinsed 60 seconds in distilled water at ambient laboratory temperature.

(6) Dried with nitrogen gas blow-off at ambient laboratory temperature (Time = ∿60 seconds).

CLEANING METHOD B:

(1) Dipped 5 minutes in hot, well-stirred 2.34% aqueous Parcocleaner 2351 (∿70°C).

(2) Rinsed 30 seconds in hot tap water (∿60°C).

(3) Rinsed 60 seconds in distilled water at ambient laboratory temperature.

(4) Dried with nitrogen gas blow-off at ambient laboratory temperature (Time = ∿60 seconds).

CLEANING METHOD C:

(1) Dipped 5 minutes in slow boiling Perclene®.

(2) Dried at ambient laboratory temperature (∿22-24°C).

(3) Dipped 5 minutes in hot, well-stirred 2.34% aqueous Parcocleaner 2351 (∿70°C).

(4) Rinsed 30 seconds in hot tap water (∿60°C).

(5) Wet polished at ambient laboratory temperature for 60 seconds with Scotchbrite® UF pads attached to an orbital sander. Tap water used as lubricant.

(6) Rinsed 30 seconds in hot tap water (∿60°C).

(7) Rinsed 60 seconds in distilled water at ambient laboratory temperature.

(8) Dried with nitrogen gas blow-off at ambient laboratory temperature (Time = ∿60 seconds).

Fig. 2.—Cleaning procedures.

DESCRIPTION OF STEELS EVALUATED

Series	Panel No.	Source	Description	Type	Quality	Rockwell Hardness, R_b	Du Pont Corrosion Rating
I	1	"A" Company	Coil 1	Rimmed	CQ	50	9
	2		2 Sequentially			50	9
	3		3 temper-rolled			50	9
	4		4 on Roll 1			50	9
	5		5			56	8
	11	"A" Company	1	Rimmed	CQ	44	9
	12		3			53	10
	13		5 Sequentially			53	9
	14		7 temper-rolled			53	10
	15		9 on Roll 2			59	10
	16		11			50	10
	17		13			49	10
II	21	Parker-Control	Unpolished CRS	Rimmed	Half-Hard	76	10
	22	"B" Company	Unacceptable Roughness		CQ	58	10
	23	"C" Company	Laboratory Stock		DQ	40	9
	31	"A" Company	Good Substrate	AK		46	8
	32	"A" Company	Poor Substrate			46	4
	33	"A" Company	Poor Substrate			46	3
III	60-1	"B" Company	Unacceptable Texture	Rimmed	CQ	56	6
	60-2	Parker-Control	Unpolished CRS		Half-Hard	78	9
	61	"X" Company	Electrocleaned	Not Known	Full-Hard	94	6
	62	"X" Company	Temper-Rolled w/o oil		CQ	58	3
	63	"X" Company	Oiled & Temper-Rolled		DQ	40	6
	64	"D" Company	Good Substrate	AK		46	7
	65	"D" Company	Poor Substrate			46	0
IV	70	Parker-Control	Unpolished CRS	Rimmed	Half-Hard	77	7
	71	Q-Panel-Control	Polished CRS		Qtr.-Hard	68	7
	72	"E" Company	Plant 1 Production		DQ	37	6
	73					48	8
	74		Rolled - Plant 1			45	7
	75		Annealed - Plant 2			40	8
	76					45	8
	77		Rolled - Plant 2			48	8
	78		Annealed - Plant 2			48	7
	79		Rolled - Plant 3		CQ	58	0
	80		Annealed - Plant 2			61	1
V	90	Parker-Control	Unpolished CRS	Rimmed	Half-Hard	77	9
	91	Q-Panel-Control	Polished CRS		Qtr.-Hard	64	5
	92	"F" Company	Designated "clean"		DQ	42	7
	93					49	6
	94					44	6
	95					45	9
	96					42	9
	97					42	10
	98		Designated "dirty"			43	2
	99					42	2
	100					45	10
	101					46	9
	102					42	3
	103					41	2

Fig. 3.—Series composition.

CLEANLINESS

The principal findings of this project investigation of methods for measurement of steel surface cleanliness and the relationship of these measurements with corrosion of coatings systems has been recently reported by Dr. K. R. Buser(1) at the Substrate Conditioning Conference sponsored by the Association for Finishing Processes of the Society of Manufacturing Engineers. His work demonstrated that steel surface cleanliness, ease of cleaning, and rate of surface quality deterioration can be quantified by surface energy determination. This measurement, however, is practically limited to the range of 22 to 72 dynes/cm by the availability of liquids with suitable wetting tension.

Figure 4 provides representative data obtained on samples of coil coating stock representing a classical case of good and bad production experience. Panel 31 represents normal satisfactory experience. Panels 32 and 33 represent a steel coil stock which provided varying degrees of finished system poor corrosion performance. The data clearly reflect differences between these samples in (1) ability to generate high surface energy immediately after

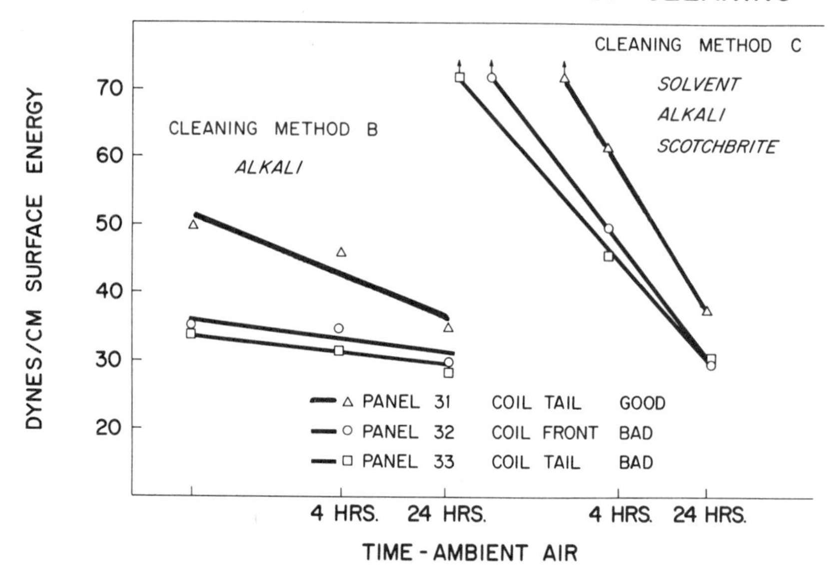

Fig. 4.—Surface energy of steel surfaces as a function of cleaning method, steel quality, and time after cleaning.

power alkaline wash, and (2) a rate of deterioration which essentially reduces to a common level after 24 hours atmospheric storage. Further demonstrated is the exceptionally higher surface energy values attained when mechanical abrasion is incorporated in the cleaning process.

TOPOGRAPHY

It has been commonly hypothesized that surface profiles should favorably influence phosphate nucleation and coatings adhesion. In cooperation with a major steel supplier, a series of panels was procured representing a continum of production from two finishing lines between temper roll changes. In each run final roll cut off was made when production quality no longer provided surface texture quality meeting internal control standards. Stylus profilometry was selected for characterization of surface texture.

The test procedure consisted of measurement of Roughness Average and Peak counts using a Bendix QHD profilometer, Model QHD. It measures the R_a in microinches at optional roughness width cutoffs of 0.010, 0.030 and 0.100 inches as well as carries out peak counting for peaks larger than 5, 10, 25, 50, 100 and 500 microinches PTV (*P*eak-*T*o-*V*alley distance). For this study the Net Peak Density (NPD) was defined as $NPD = PPI_{50} - 2\ PPI_{250}$ *where* PPI_{50} *and* PPI_{250} *are the number of peaks per inch for peaks larger than 50 and 250 microinches, respectively.*

A representative profilometer curve is shown in Figure 5. This procedure was applied to all samples in this study. Representative data, as

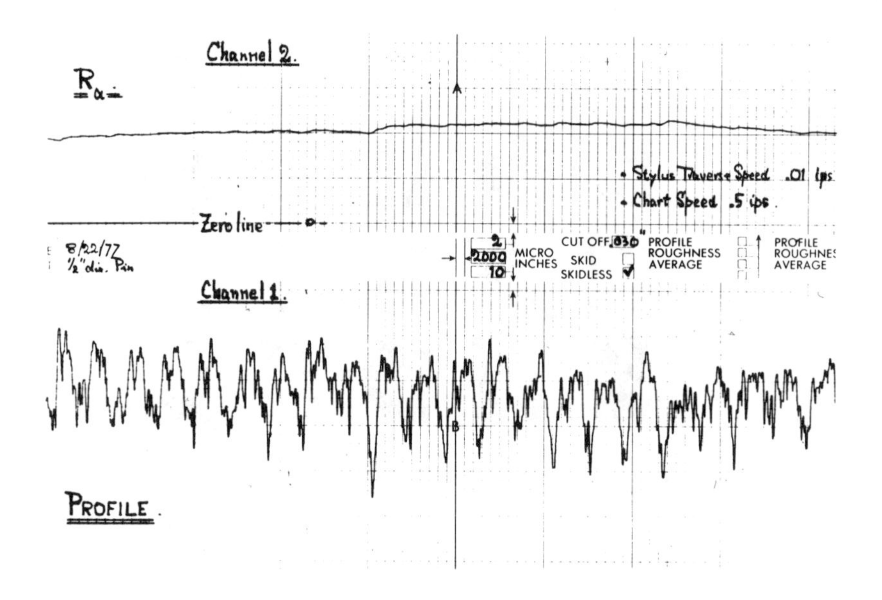

Fig. 5.—Chart output from Bendix QHD Profilometer.

shown in Figure 6, as well as the complete data mass gave no evidence that a correlation exists between the surface texture as described by profile measurements and the corrosive behavior of coated systems. There appears to be a qualitative agreement between some aspects of surface appearance photographed with scanning electron microscopy and profilometer data, but little correlation with phosphate quality.

There is no indication that profilometry, in the way it is presently applied, would be able to predict corrosive behavior. The fundamental reason for this simply is that a profilometer explores only the external surface characteristics on a macroscale while corrosive action is controlled much more by the microstructural aspects of the surface. To indicate those micro aspects that, to a large extent, determine the corrosivity of the sheet steel, tests other than profilometry and peak counting must be established. It should be clear that a profilometer with a diamond stylus of 0.0005" radius is not going to resolve microfeatures (cavities, cracks, fissures) on a surface in the order of 10 microns or less.

SURFACE REACTIVITY

Pickle Lag

Several years ago, American Can[2] developed the rapid "Pickle Lag Test" for correlation with performance of tinned steel in food containers. A standard size specimen, 8x65 mm, is immersed in *6N HCl* at 90°C. The pressure generated by the hydrogen evolved is recorded as a function of time. The initial rate of hydrogen evolution increases with time, but then becomes constant. By extrapolating the slope of the constant rate period backward to

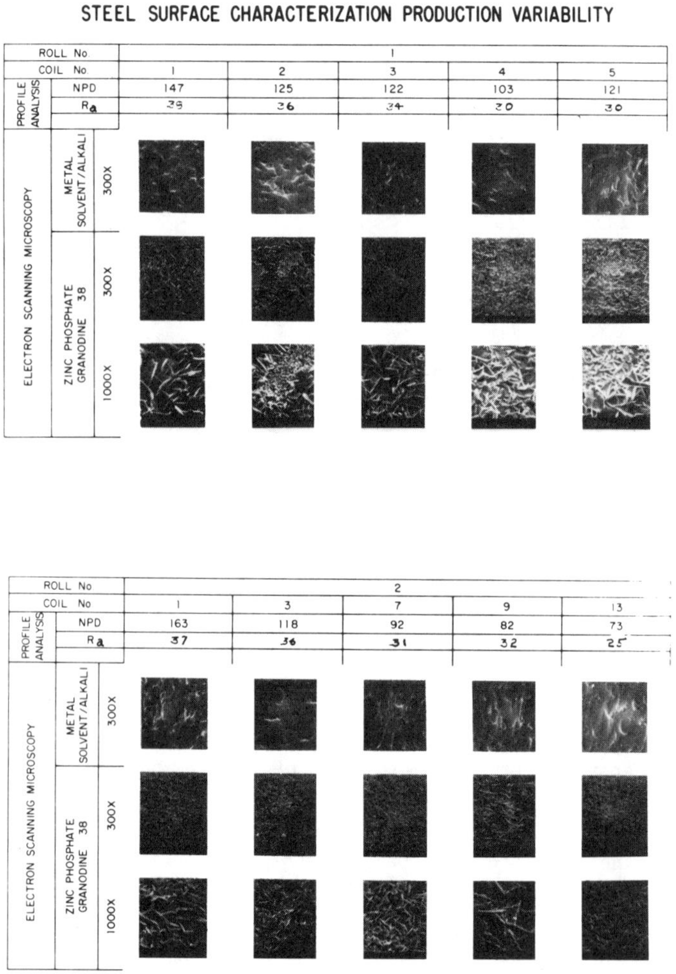

STEEL SURFACE CHARACTERIZATION PRODUCTION VARIABILITY

ROLL No.		1				
COIL No.		1	2	3	4	5
PROFILE ANALYSIS	NPD	147	125	122	103	121
	R_a	39	36	34	30	30
ELECTRON SCANNING MICROSCOPY	METAL SOLVENT/ALKALI 300X					
	ZINC PHOSPHATE GRANODINE 38 300X					
	ZINC PHOSPHATE GRANODINE 38 1000X					

ROLL No		2				
COIL No		1	3	7	9	13
PROFILE ANALYSIS	NPD	163	118	92	82	73
	R_a	37	36	31	32	25
ELECTRON SCANNING MICROSCOPY	METAL SOLVENT/ALKALI 300X					
	ZINC PHOSPHATE GRANODINE 38 300X					
	ZINC PHOSPHATE GRANODINE 38 1000X					

Fig. 6.—Series I - sequential coil variation in surface texture.

zero pressure, a time in seconds, the "pickle lag," is determined. A short pickle lag on steel sheet intended for food containers has been correlated with long service life.

Figure 8 presents representative results of "pickle lag" testing on Series IV. This series, which offered the widest spread in corrosion quality of production samples procured, shows little evidence that "pickle lag" testing can be effectively applied to corrosion performance prediction in this situation. Wide differences in surface reaction rates and subsequent etch patterns suggest that modification of these techniques, e.g., use of phosphoric acid, might well be expected to generate some quantitative relationship between phosphatability of surfaces and acid response behavior.

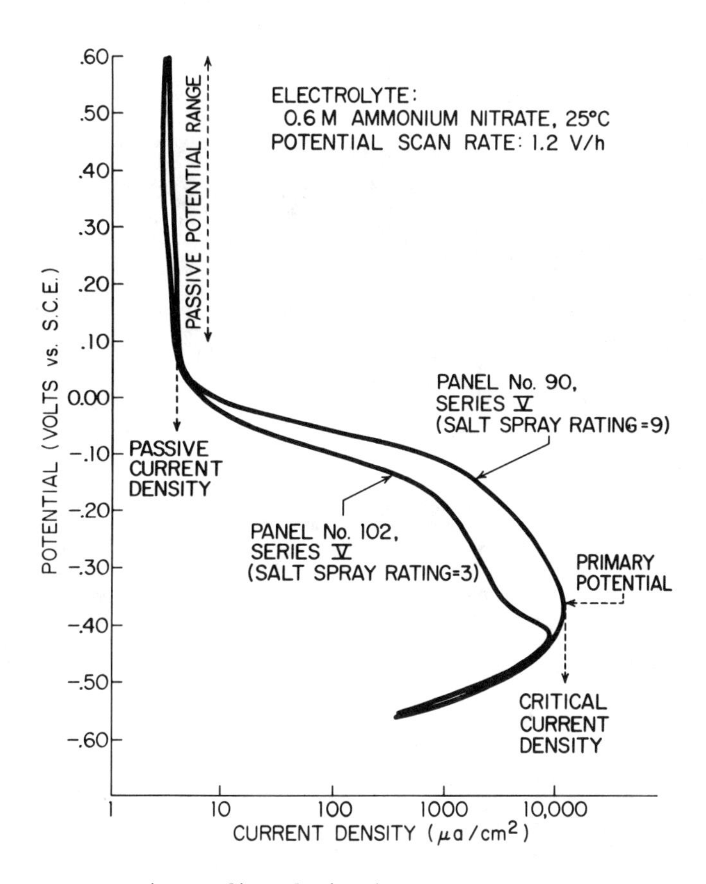

Fig. 7.—Representative anodic polarization curves.

Anodic Polarization Curves

Based on anodic polarization procedures reported by R. C. Chance[3] of General Motors Research Laboratories, characteristic curves of corrosion current density as a function of applied potential were developed for the full series. In application considerable scatter of results was encountered, principally attributed to external factors and the wide deviation between the environment in the controlled electrochemical procedure and conditions for salt spray test.

SERIES IV   SURFACE REACTIVITY

			70	71	72	73	74	75	76	77	78	79	80
PANEL No.			70	71	72	73	74	75	76	77	78	79	80
DESCRIPTION			PARKER	Q-P	29568 1F	30817 1F	1B	2F	2B	65362 1F	1B	47554 1F	1B
PICKLE LAG TEST 6N HCL AT 90°C	SECS-INDUCTION		36	30	70	34	53	43	46	7	4	43	17
	TO LEVEL B		55	54	123	137	159	149	145	31	32	78	58
	TO LEVEL A		32	27	67	55	65	62	60	12	13	42	24
	AB-PROPAGATION RATE		23	27	56	82	94	87	85	19	19	36	34
	PICKLE LAG, SECS		23	21	44	16	25	20	20	4	2	26	9
ACID ETCH	SEM 500x					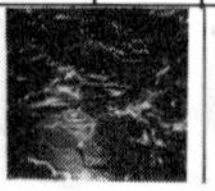		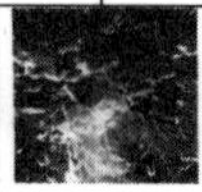		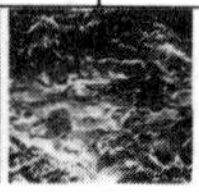		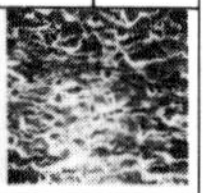	
SURFACE ENERGY	120"	DYNES/cm	46	36	36	36	36	47	41	36	36	36	36
SALT SPRAY	DUPONT SCRIBE 1/64"	168	1	0	7	0	1	1	1	1	1	11	3
			3	1	7	1	1	3	1	1	1	27	7
		336	3	3	7	3	1	FAIL	1	3	31	FAIL	FAIL
			19	3	7	15	5	11	1	3	3	FAIL	FAIL
		500	15	23	7	7	7	11	7	7	31	FAIL	FAIL
			19	7	11	15	5	23	3	3	3		
	RANK ORDER	MFR			5	8	7	10	9	3	6	4	2
		DUPONT	7	7	6	8	7	8	8	8	7	0	1

Fig. 8.—Pickle lag test results for a series of rimmed steel production panels.

A parameter that best represented the differences in salt spray corrosion results was the current density at a potential between the active corrosion peak and the passive region (specifically - 0.3 volts vs standard calomel electrode). Figure 7 provides an example of differences observed between "good" and "poor" corrosion examples. A significant correlation at the 95% confidence level of this parameter with salt spray results was demonstrated, further reinforcing the basic supposition that high surface reactivity, per se, plays a beneficial role in systems behavior.

SURFACE CHEMISTRY

ESCA (Electron Spectroscopy for Chemical Analysis)

This more recent analytical method has provided the researcher with an opportunity to analyze a very thin surface layer less than 100Å thick. It has been described in the literature extensively. A more recent and pertinent review article is found in Reference 4.

Figure 9 typifies the result obtained from a steel surface. This figure shows a so-called "wide scan," extending over the entire span of elements that can be analyzed. Since the sensitivity changes over the length of the scan, the instrument had to be adjusted twice, as is evident from the figure. Prominent are the signals for carbon and oxygen, more so than the iron signal. This just re-emphasizes the fact that the steel surface, even if cleaned thoroughly by establishing cleaning methods, consists only to a small degree of iron atoms while other elements like oxygen and carbon predominate. Because ESCA does not provide information about analytical concentration in the definition of number of atoms per volume unit, it is preferable to present ESCA data as ratios of signals. This is not "quantitative" information, but excellent for the comparison from sample to sample. Instrument factors, sample size and roughness and the like are eliminated where ratios are computed. While the light elements such as carbon and oxygen give single

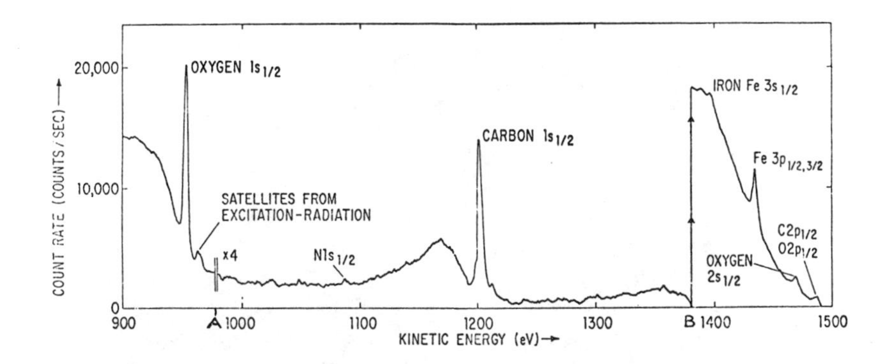

Fig. 9.—ESCA - "wide scan" from kinetic energy 900 - 1487 eV (binding energy 586-0 eV). All elements are represented in this energy range by at least one peak.

and straightforward signals, medium and heavy elements like iron and molybdenum yield multiple, more complicated signals of less prominence, as is also apparent from Figure 9. We elected to use the ratios of carbon over iron signals and carbon over oxygen as most important ESCA information. The signals involved are indicated as carbon $1S_{1/2}$, Fe $3_{p1/2/3/2}$ and oxygen $1S_{1/2}$, which defines these by way of the atomic level from which they originate. The respective binding energies are 284, 56 and 532 eV, respectively. With Al radiation (1486 eV) exciting these elements, kinetic energies of 1202, 1430 and 954 eV result.

X-Ray Diffraction (XRD)

XRD is the interaction of x-rays with the periodic arrangement of atoms in solids. The variant of XRD applied in this study seeks information about the size, integrity and arrangement of the iron crystallites in the steel. The arrangement of Fe atoms themselves in the crystallites is well known. *(Cubic body-centered; length of the elementary cube a = 2.8665Å; 1Å = 10^{-8}cm - 0.1 nm = 4 x 10^{-9} inches.)* These "crystallites," i.e. small crystals, can be subdivisions of a grain or can be individual grains themselves.

Choice of the monochromatic x-radiation for the tests is of considerable importance because this selection can enhance or subdue effects of the crystallite features on the pattern. All XRD tests have to be carried out in back reflection because of the need to nondestructively test the samples. The cone-back reflection method was used because it yields the theoretical maximum of the interference pattern obtainable from a sample in nondestructive arrangement. X-radiation from a cobalt target *(Wavelengths λ:$K\alpha_1$ = 1.788965Å; $K\alpha_2$ = 1.79285Å)* was selected as the analyzing radiation because it produces an interference pattern favorable for the test purpose. Refer to Figure 10.

An *ad hoc* single number term, K, was selected to represent the relative "fineness" or "coarseness" of crystalline structure as interpreted from the semi-circular densitometer track patterns obtained. Early experimentation, not statistically verified in the full series, suggested high values of K

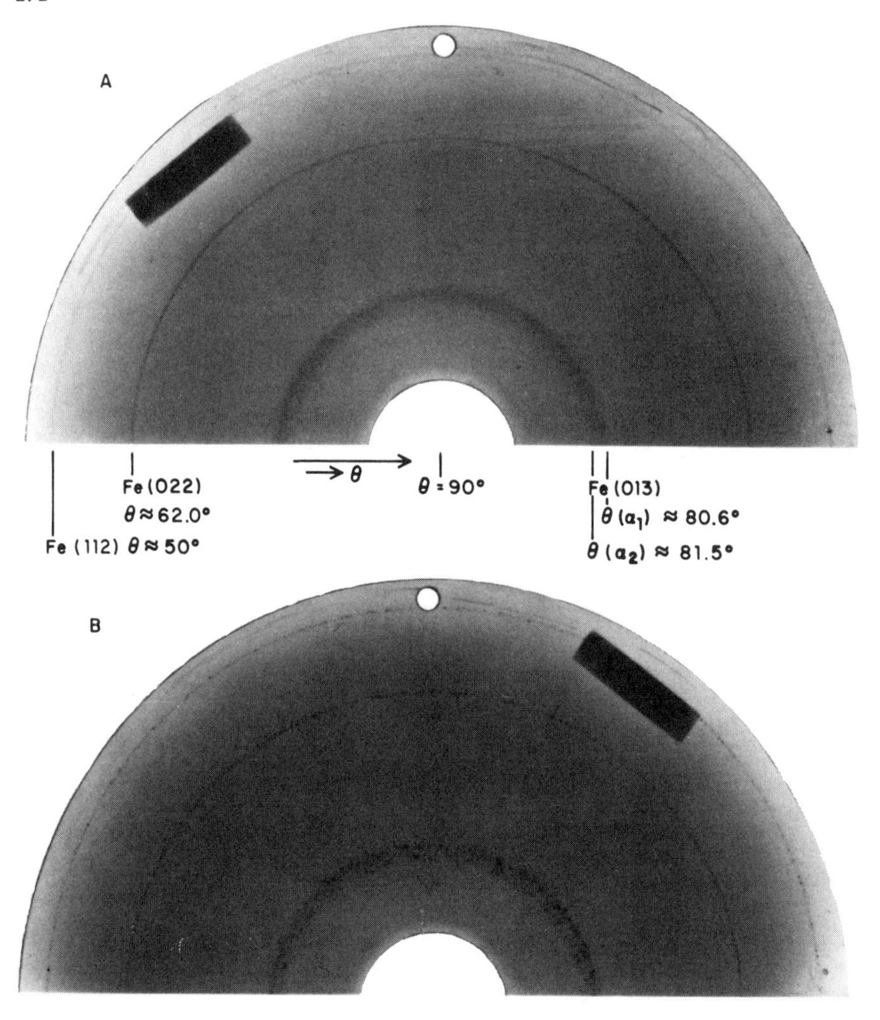

Fig. 10—Back reflection diagrams of steel samples showing good (A) and poor (B) salt spray test results.

which represent a fine uniformly dispersed crystalline structure gave superior corrosion performance. Substantial statistical efforts to establish a correlation using other techniques for diffraction pattern interpretation were made without the ability to accommodate gross disparities on samples with catastrophic corrosion failure.

Representative results of the surface analysis efforts are shown in Figure 11. They include measurements of surface carbon as determined by chemical analysis by the steel supplier on parallel samples. High surface carbon levels, whether analyzed spectroscopically or chemically, consistently provides the signal for high probability of poor systems corrosion performance.

SERIES V — SURFACE ANALYSIS

		90	91	92	93	94	95	96	97	98	99	100	101	102	103
PANEL No.		90	91	92	93	94	95	96	97	98	99	100	101	102	103
DESCRIPTION		PARKER Q-P		123-A		103-A		111-B		101-B		108-A		120-A	
ESCA PEAK RATIOS	C/O	1.5	1.3	1.1	1.1	0.8	1.4	1.1	1.7	1.6	1.4	1.1	1.3	1.6	1.8
	$C/Fe_{56}=D$	20	21	18	16	12	17	17	22	24	24	16	18	29	33
	C-H/C=O	13	10	12	12	9	12	12	14	14	14	15	11	11	10
	Fe_{56}/Fe_{47}	>100	2	7	5	7	>10	9	>100	>10	>10	4	6	7	4
CARBON mg/FT^2				0.20	0.11	0.10	0.14	0.12	0.15	1.08	1.05	0.08	0.08	0.67	0.75
X-RAY CRYSTALLITES	K	63	44	39	14	13	14	16	13	10	14	10	12	8	9
	K/D	3.2	2.1	2.1	0.8	1.1	0.8	1.0	0.6	0.4	0.6	0.6	0.7	0.3	0.3
SEM 10000×															
SURFACE ENERGY 120' DYNES/cm		46	36	34	38	34	36	27	38	36	34	27	28	34	41
SALT SPRAY — DU PONT SCRIBE 1/64"	168 HRS	1/3	5/1	3/1	3/1	3/1	1/1	1/3	0/1	11/27	11/31	0/7	0/1	3/19	11/27
	336 HRS	2/7	11/3	7/3	11/3	11/3	1/3	1/3	1/3	31/FAIL	23/FAIL	0/19	0/3	3/FAIL	27/FAIL
	500 HRS	3/11	15/5	9/5	11/5	15/3	3/3	3/5	1/3	FAIL/FAIL	FAIL/FAIL	1/19	3/7	FAIL/FAIL	FAIL/FAIL
SALT SPRAY — RANK ORDER	MFR														
	DU PONT	9	5	7	6	6	9	9	10	2	2	10	9	3	2

Fig. 11—Surface analysis determination for a series of rimmed steel panels showing good surface carbon/salt spray creep correlation.

Electron Probe Microanalysis

A few selected samples, representing extremes in salt spray systems corrosion behavior, were subjected to microprobe analysis. The results are shown in Figure 12. Carbon appears to exist in two modes: first as a fine deposit uniformly spread over relatively smooth surfaces, and secondly in high concentration in micro pits, crevices and cavities along with high concentrations of sodium, chlorine, sulfur and other contaminant residues from prior processing conditions. A model micropit configuration is shown in Figure 13.

CONCLUSIONS

This project stage has principally served to develop and refine procedures for characterization of the attributes of steel surfaces considered most likely to influence corrosion performance of coating systems. Experimental execution demanded skillful and experienced manipulation and interpretation. This fact would probably preclude routine application of these methods for control or specification purposes. Correlation analysis was limited by the nature of sample pattern and the wide spectrum of sample source variables encountered. Important correlations, however, would be expected to emerge if these refined procedures were similarly applied to a strategic sample set with a common steel composition, manufacturing procedure, thermal and storage history.

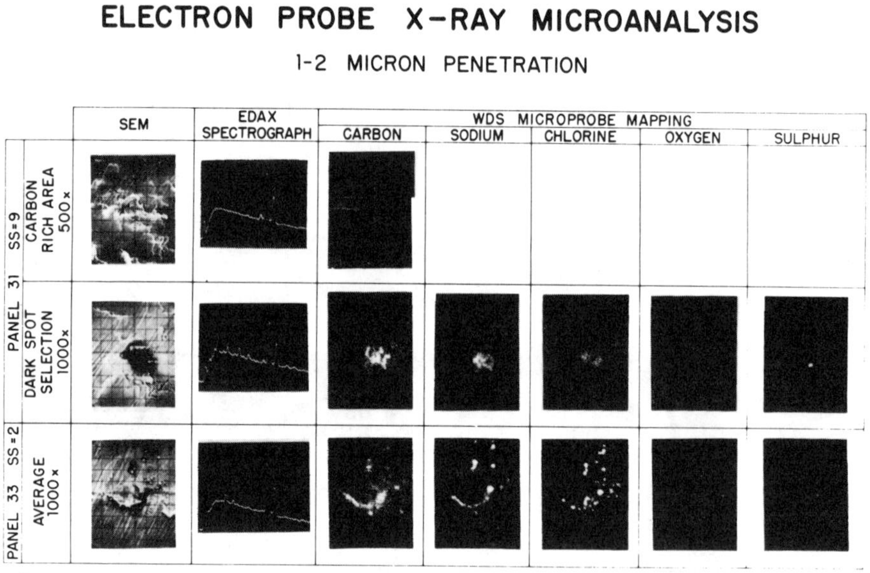

ELECTRON PROBE X-RAY MICROANALYSIS
1-2 MICRON PENETRATION
SEM
EDAX SPECTROGRAPH
WDS MICROPROBE MAPPING
CARBON
SODIUM
CHLORINE
OXYGEN
SULPHUR
PANEL 31 SS=9
CARBON RICH AREA 500x
DARK SPOT SELECTION 1000x
PANEL 33 SS=2
AVERAGE 1000x

Fig. 12.—Micrographs showing surface carbon on smooth areas and high contaminant concentration in micropits.

COLD ROLLED STEEL SURFACE
180Å
10⁴Å 1.0μ
FeOOH, Fe₂O₃
FeOOH, Fe₂O₃
Fe
UNIT CELL = BODY CENTERED CUBIC
ADSORBED LAYER [O₂, H₂O]
SURFACE CONTAMINANTS [C, Na, B, K, S, Ca, Si]
INCLUSIONS [Fe₃C, MnO, Al₂O₃, SiO₂, AlN]

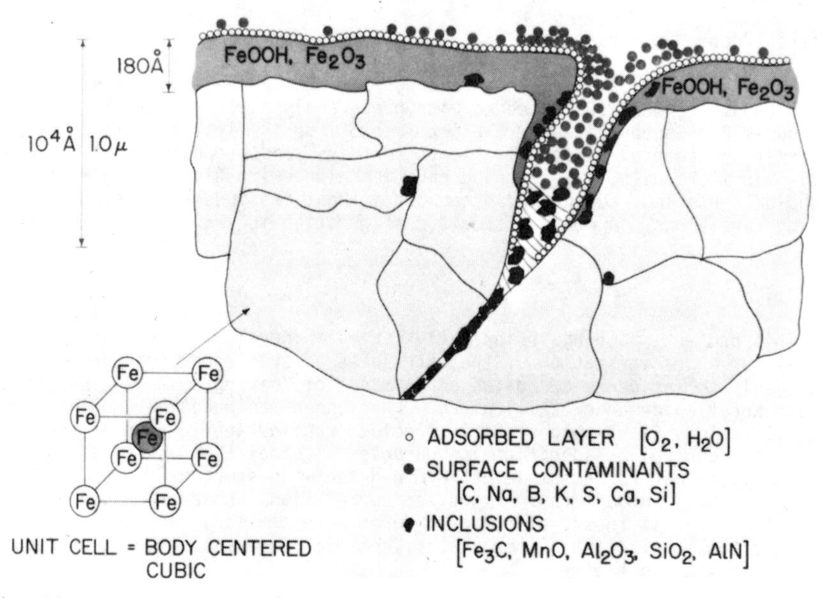

Fig. 13.—Cold rolled steel surface model concept.

The Entdecker Computer was utilized for statistical examination of the ESCA and SRD information developed. This conglomerate of data storing, computing, and display equipment enabled study of three dimensional relations between variables and data interrogation. The primary conclusion was that surface carbon as measured by ESCA and corrosion tendency as measured by salt spray creep correlates strongly with each other above the 99% confidence limit. Not proven, but strongly suggested in this statistical study, is the possibility that some measure of polycrystallinity has the potential for correlation interaction with the presence of surface carbon and consequently corrosion tendency. This is interpreted to imply that micropit or microcrevice occurrence and concentration may be a consequence of the thermal and roll history of the sample.

The pattern of corrosion failure demonstrates beyond reasonable doubt that random purchase of cold rolled steel holds a high probability for occasional weak to catastrophic failure in coatings system performance. Cleaning processes which employ mechanical abrasion prior to finishing can significantly reduce the failure probability. Examination by SEM, microprobe, ESCA or similar micro analytical techniques are essential to discrimination of "good" and "poor" substrate surfaces.

Efforts to determine the mechanism for failure induced by surface carbon presence have not been conclusive. No procedures were uncovered which adequately quantify the physical and chemical aspects of phosphate crystal quality for correlation study. It is our conjecture, however, that surface carbon present as a simple consequence of oil reduction during annealing reduces phosphate adhesion. Surface carbon detected at micropit sources, on the other hand, may only reflect the existence of companion soluble ion pockets which create very active electrochemical reaction sites. The former may lead to a general weakening of corrosion protection standards. The latter may be regionally catastrophic in nature. The potential for simultaneous existence of both carbon mechanisms may account for the sporadic disparities encountered in efforts of others to establish a statistically correlatable salt spray/surface carbon relationship.

ACKNOWLEDGEMENTS

The paper represents a progress report on the initial phase of a Du Pont Fabrics and Finishes Department program chartered to investigate aspects of paintability of steel surfaces. It is an interpretive compendium of the planning, rationalization, and experimental efforts of a group of highly qualified individual experts, comprising:

Dr. K. R. Buser, F. & F. Dept., Surface Energy
Mr. G. C. Bell, F. & F. Dept., Electrochemistry
Dr. H. K. Herglotz, Engineering Dept., ESCA and XRD
Dr. M. A. Streicher, Engineering Dept., Corrosion
Dr. R. P. Steijn, Engineering Dept., Profilometry

The excellent interactions, the responsiveness, the creativeness, and the technical contributions of this group are gratefully acknowledged.

REFERENCES

1. Buser, K.R., "The Energetics of Clean Surfaces," Society of Manufacturing Engineers, Technical Paper No. FC78-574, Oct. 1978.
2. Willey, A.R., Krickl, J.L., and Hartwell, R.R., "Steel Surface Properties Effect Internal Corrosion Performance of Tin Plate Containers," *Corrosion 12, 433 (1956)*.
3. Chance, R.L. and France, W.D.Jr., *Corrosion 25, 329 (1969)*; see also Chance, R.L., *Corrosion 33, 108 (1977)*.
4. Herglotz, H.K., "The Role of ESCA in Surface Characterization," Society of Manufacturing Engineers, Technical Paper No. FC77-673, Oct. 1977.

V. INHIBITORS IN COATINGS

MECHANISMS OF CORROSION CONTROL BY INHIBITORS

E. McCafferty

Naval Research Laboratory,
Washington, D.C.

EDWARD MC CAFFERTY

ABSTRACT

The adsorption of corrosion inhibitors is an important consideration in the protection of metals. It is particularly so in acidic solutions where air-formed oxide films are dissolved away, so that the inhibitor can chemisorb onto the cleaned metal surface. Various factors which affect the efficiency of chemisorption-type organic inhibitors in acid solutions are: the molecular structure of the inhibitor, the structure of the electrical double layer, the existence of adsorbed water molecules at the interface, competitive or cooperative adsorption with surface active anions, and the degree of hydrophobicity of the aliphatic tail or aromatic ring. Several of these factors are treated in some detail in case studies on organic diamines and on partially fluorinated aliphatic amines. Inhibition mechanisms in nearly neutral solutions are complicated by the retention of air-formed oxide films and by the participation of dissolved oxygen in the electrode processes. Adsorption of inhibitors is also an important consideration in nearly neutral solutions, although one of several types of subsequent reactions is usually required for protection. The role of adsorption processes is discussed with reference to the use of chelate inhibitors and the inhibition of localized corrosion in nearly neutral solutions.

INTRODUCTION

An organic coating is not a perfect barrier separating a metal substrate from its environment, but instead is like a membrane. Water, oxygen, and various ions are able to penetrate into such coatings [1-3]. When conductive paths develop within the coating, anodic dissolution and oxygen reduction processes occur at different areas on the metal surface within a confined volume of electrolyte [3,4]. Localized corrosion within such an "occluded corrosion cell" produces a decrease in local pH [5], while consumption of oxygen in the cathodic reaction generates an increased pH, due to formation of hydroxyl ions. Both the low and high pH's can be injurious to an organic coating [1,3].

Thus, soluble inhibitors incorporated into coatings should be those effective at low and/or high pH's. The large amount of work done on corrosion inhibition in bulk electrolytes can serve as a scaled-up model for the more localized processes within the coating. Of course there are differences in detail between the two situations. Ohmic resistance and concentration effects will be more important in the localized process, but nevertheless, principles of corrosion inhibition for bulk electrolytes should provide guidance for inhibition in localized geometries.

The mechanisms of corrosion inhibition are different for acidic and neutral solutions. In acid solutions, natural surface oxide films initially present are easily dissolved away so that inhibitors then interact directly with the metal surface. In neutral solutions, the system is more complex because the metal surface is oxide-covered and because dissolved oxygen participates in the electrode reactions.

Fundamental principles of corrosion inhibition in both types of solutions will be discussed in this paper. The discussion will be limited to aqueous corrosion inhibitors. (Vapor phase inhibitors will not be considered although their importance is recognized.)

Throughout this paper, special attention will be given to the role of adsorption processes in corrosion inhibition. Adsorption of corrosion inhibitors is important in both acidic and neutral media, although not with all inhibitors. Before getting into detail, then, it is well to consider first the main classes of inhibitors.

TYPES OF INHIBITORS

Inhibitors can be classified in several different ways. The two main types are (1) adsorption inhibitors and (2) film-forming inhibitors.

Adsorption inhibitors form a chemisorptive bond with the metal surface so that ongoing electrochemical reactions are impeded. Most organic inhibitors are the adsorption type. For example, aliphatic organic amines adsorb via the surface active $-NH_2$ group, which contains a pair of unshared electrons available for donation to the metal to form a chemisorptive bond. The hydrocarbon tails orient away from the interface toward the solution, so that further protection is provided by the formation of a hydrophobic network which excludes water and aggressive ions from the metal surface.

There are two types of film-forming inhibitors: passivating inhibitors and precipitation inhibitors. As implied by their name, the first type promotes the formation of passive films. There are two classes of passivating inhibitors: oxidizing and non-oxidizing. Chromate is a classical oxidizing inhibitor. With ferrous metals, the chromate ion is reduced to Cr_2O_3 or $Cr(OH)_3$ on the iron surface to produce a protective mixed oxide film of chromium and iron oxides(6). Adsorption processes are also important with such inhibitors, as a monolayer of chromate is first rapidly adsorbed before being more slowly incorporated into the oxide film as Cr^{+3}(7,8). Non-oxidizing passivators like benzoate, azelate, and phosphate also first adsorb onto the oxide-covered surface(9) before acting to induce or maintain passivity.

With precipitation inhibitors, a precipitation reaction between the corroding metal and the inhibitor deposits a three-dimensional barrier film on the metal surface. Phosphates and silicates are precipitation inhibitors(10). These types of inhibitors do not involve adsorption on the metal (or oxide-covered metal) surface and will not be considered in this paper.

From another point of view, inhibitors can be classified as anodic, cathodic, or mixed, depending on which partial electrochemical reaction is affected. As examples, chromate is an anodic inhibitor, some phosphates are cathodic inhibitors, and most organic compounds are mixed inhibitors.

ACIDIC SOLUTIONS

General Background

Much of the detailed information about corrosion inhibition mechanisms has evolved from the study of acidic solutions. The reason is that the dissolution of air-formed surface oxides allows focus on the direct interaction between the inhibitor and the clean metal surface. Another simplification often made in mechanistic studies is that the solution is deaerated, so that oxygen will not participate in the electrode reactions. Actually, the effect of dissolved oxygen does not appear important until the pH is as high as 2 to 4(11).

Despite these simplifications, the general results have considerable practical application. Corrosion inhibitors are used in a variety of acid cleaning operations, ranging from commercial pickling processes to special laboratory problems. A special problem of interest to the Naval Research Laboratory, for example, involves the cleaning of corrosion products from rusted fracture surfaces without affecting the underlying microscopic topographical features which must be preserved for fractographic examination and analysis(12). In addition, corrosion inhibition in deaerated acidic solutions can serve as a scaled-up model for localized corrosion processes, such as crevice corrosion, where the local electrolyte is both acidic and oxygen depleted.

The most important act of the inhibitor in acid solutions is to chemisorb on the metal surface. After a brief discussion of chemisorption, various factors which affect the chemisorption of inhibitors will be considered. These factors include the role of water molecules, the effect of the electrical double layer, competitive adsorption, and the effect of molecular structure of the inhibitor.

Chemisorption of Inhibitors

According to the adsorption theory of corrosion inhibition(13), the primary step in the action of the inhibitor is chemisorption onto the metal surface. The adsorption process is a general one occurring on both anodic and cathodic types of electrochemical sites. The type of electronic interaction involves an actual charge transfer or charge sharing between the inhibitor and the metal surface atom.

By interacting with surface metal atoms, the chemisorbed inhibitor interferes with ongoing electrode processes. The simplest picture is one of a "blockage" of surface sites, but this picture is not quite accurate. Like all adsorbed species, chemisorbed inhibitor molecules have a certain residence time at the surface. Thus, the inhibitor must play a dynamic role in interfering with the dissolution reaction by participating in a number of adsorption-desorption steps.

The main characteristics of the chemisorption process are its high heat of adsorpiton, persistence, and specificity. Table I compares the process of chemisorption with physical adsorption, a more general type of adsorption but one involving weaker interactions.

Chemisorption usually occurs through donation of electrons from a N, S, or O functional atom to the metal surface. As mentioned earlier, aliphatic organic amines adsorb via the surface active $-NH_2$ group which contains a pair of unshared electrons which are available for donation to the metal surface. See Figure 1.

Correlations between the molecular structure of inhibitors and their efficiency first led to the idea that chemisorption was important in corrosion inhibition. For example, in a homologous series of organic compounds differing in the functional donor atom, the order of corrosion inhibition is usually:

$$Se > S > N > O,$$

which is the reverse order of electronegativity. Sulfur compounds are better corrosion inhibitors than their nitrogen analogs (other factors being equal) because the S atom, being less electronegative than N, draws electrons the lesser to itself, and is thus the more efficient electron donor in forming the chemisorptive bond.

TABLE I.—CHARACTERISTICS OF ADSORPTION BONDS. FROM REF. (14).

	In Physical Adsorption	In Chemisorption
Type of electronic interaction	Van der Waals' or electrostatic forces.	Actual charge transfer or charge sharing.
Reversibility	Adsorbed species readily removed by solvent washing.	Adsorption is irreversible, more persistent.
Energetics	Low heat of adsorption, in H_2O < 10 kcal.	Higher heat of adsorption, in H_2O > 10 kcal.
Kinetics	Rapid adsorption, relatively independent of temperature.	Slow, with activation energies in excess of 6 kcal.
Specificity	Adsorbed species relatively indifferent to nature of adsorbent.	Specific interaction, strong dependence on nature of adsorbent.

The nature of substituents on aromatic rings influences the availability of electrons for formation of chemisorptive bonds. For example, the order of inhibitor effectiveness with substituted pyridines is(15):

pyridine < 3-methylpyridine (CH_3) < 4-methylpyridine (CH_3)

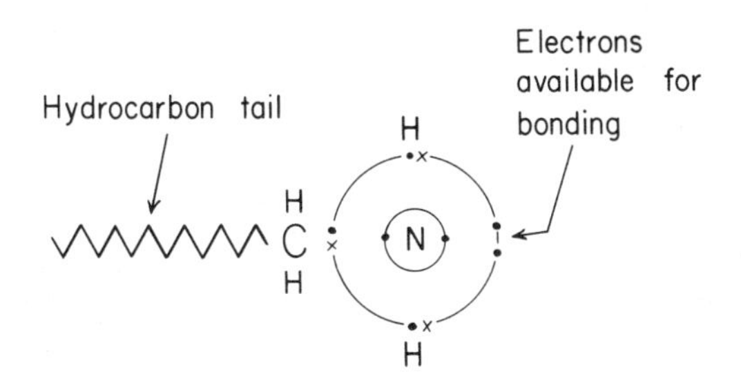

ATOMIC NUMBER OF N = 7

FILLED SHELLS: 2, 8, 18, ...

Fig. 1.—Chemisorption of amines by electron donation.

The methyl group is an electron-releasing (nucleophilic) substituent, and more effectively so in the para- than meta-position. Thus, the order of corrosion inhibition parallels the availability of bonding electrons.

More direct evidence as to chemisorption of inhibitors has been provided by studies which monitor the uptake of inhibitors onto the metal surface. These include solution depletion measurements[16,17], radiotracer techniques[18,19], infra-red studies[20,21], changes in the capacitance of the electrical double layer[22,23], and applications of the newer surface analytical techniques such as x-ray photoelectron spectroscopy (XPS)[24-26] and Auger analysis[27].

A measure of the bond strengths between various amines and the iron surface is provided by the calorimetry studies of Yao[28], who measured the heats of adsorption of various aliphatic amines from the vapor phase onto reduced iron powders. The heats of adsorption at initial coverages were 30 to 40 kcal/mole, and were consistent with values calculated on the basis of covalent bond formation.

Involvement of Water

When a metal is immersed into an aqueous solution, most of the surface is covered with adsorbed water molecules. The adsorption of organic molecules is thus a replacement reaction[18]:

$$Org_{(soln)} + nH_2O_{(ads)} \longrightarrow Org_{(ads)} + nH_2O_{(soln)}$$

where n is the number of adsorbed water molecules which must be desorbed to accommodate the organic molecule. If an aliphatic amine adsorbs in the vertical configuration, for example, two water molecules must first be removed because the cross-sectional areas are about 10Å^2 for water and 24Å^2 for the hydrocarbon chain.

The energy of interaction between a water molecule and iron has been calculated to be 15.4 kcal/mole[18]. Although substantial, this value is still less than those determined by Yao[28] for various amines on iron, so adsorption of the latter is favored. The overall free energy of adsorption for the process involves both organic and water molecules:

$$\Delta G_{ads} = \Delta G_{ads}^{Org} - n\,\Delta G_{ads}^{H_2O} \qquad (1)$$

Free energies (and heats) of adsorption are less for electrode processes than in vapor phase adsorption because of this involvement of pre-adsorbed water[29]. The two types of adsorption are related by appropriate thermodynamic cycles:

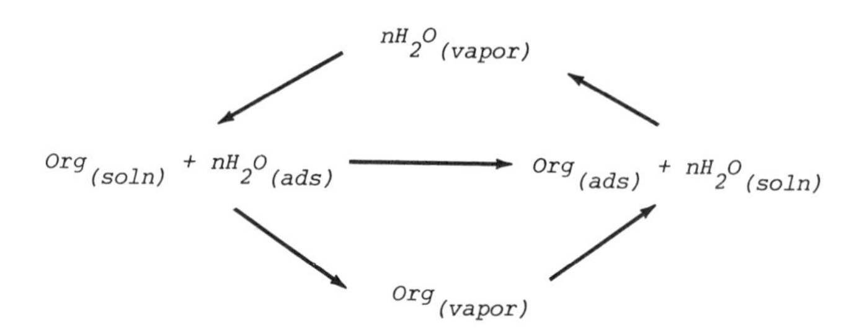

Heats of adsorption in electrochemical processes are difficult to measure directly, but sometimes can be calculated from such cycles, or from the fitting of inhibition data to adsorption isotherms based on these cycles[30].

Water molecules are also involved in the charge-transfer, i.e. corrosion reaction, as well as in the adsorption process. In the absence of adsorbable anions, the dissolution of iron in acidic solutions proceeds through a series of steps in which the chemisorbed water molecule participates as follows[31]:

$$Fe + H_2O \rightleftharpoons Fe\text{-}H_2O_{ads} \qquad (2)$$

$$Fe\text{-}H_2O_{ads} \rightleftharpoons FeOH^-_{ads} + H^+ \qquad (3)$$

$$FeOH^-_{ads} \rightleftharpoons FeOH_{ads} + e^- \qquad (4)$$

$$FeOH_{ads} \rightleftharpoons FeOH^+ + e^- \qquad (5)$$

$$FeOH^+ + H^+ \rightleftharpoons Fe^{++}(aq) + H_2O \qquad (6)$$

In acidic chloride solutions, it is the species $FeCl^-_{ads}$ which is the precursor to active dissolution[32,33].

Without going into detail here, organic adsorption inhibitors play a dynamic role at the interface rather than one of merely blocking active sites. As mentioned earlier, the inhibitor participates in a number of adsorption-desorption steps so as to impede the process of dissolution reactions such as in Equations 2 to 6. The effect of the benzoate anion on the above reaction scheme has been examined in detail by Kelly[34].

Competitive and Cooperative Adsorption

When the solution contains adsorbable aggressive ions, such as Cl^-, organic inhibitors must compete with the Cl^- ions for sites on the water-covered surface. The heat of adsorption of a chloride ion on iron has been calculated to be 45 kcal/mole[35], which is not too different from the energy of the metal-amine bond. However, the hydrocarbon tails of adsorbed aliphatic compounds or the hydrocarbon rings of adsorbed aromatic compounds serve to protect the surface from subsequent attack by Cl^- ions. This process of competitive adsorption is shown schematically in Figure 2(a). In this type of adsorption, the protonated inhibitor loses its associated proton(s) in entering the electrical double layer and chemisorbs by donating electrons to the metal, as discussed earlier.

In addition, the inhibitor can adsorb in another manner, as shown in Figure 2(b). The protonated inhibitor can electrostatically adsorb onto the halide-covered surface through its hydrogen ion. Evidence for this latter type of adsorption is provided by cases in which amines are more effective inhibitors in HCl than $HClO_4$[36]. Similar cases involving cooperative adsorption between halide ions and organic inhibitors have been reported[37,38].

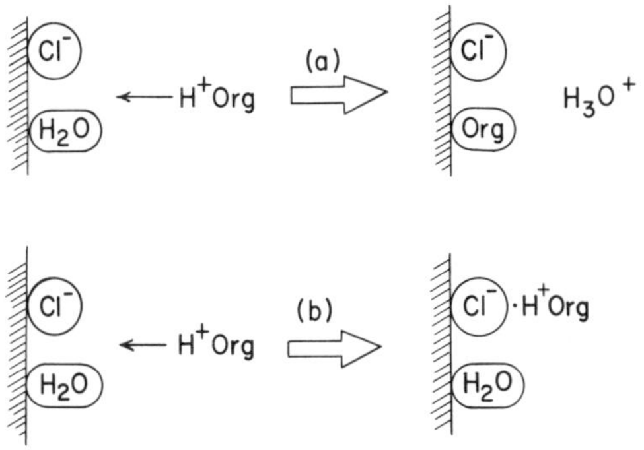

Fig. 2.—(a) Competitive adsorption; (b) Cooperative adsorption

The type of adsorption which occurs depends in part on the structure of the electrical double layer.

The Electrical Double Layer

The array of charges distributed across the metal/solution interface is called the electrical double layer, although there is often really a triple layer.

The first layer is a sheet of charges at the metal surface due to either an excess or deficiency of free electrons.

The second layer is formed on the solution side of the interface by specially adsorbed ions. The locus of centers of these charges form the *inner Helmholtz plane* (I.H.P.) of the "double" layer. (See Figure 3.) These ions, shown in Figure 3 as anions, lose their water sheaths, displace adsorbed water molecules from the metal surface, and in turn are adsorbed onto portions of the bare metal surface.

These charges are balanced in part by hydrated ions of opposite charge located in the *outer Helmholtz plane* (O.H.P.), which is the last of the three layers. The net ionic charge located between the surface and the O.H.P. is balanced by the array of charges in the *diffuse* layer, which extends out into the solution.

Intrusion of organic molecules into the electrical double layer changes its composition and structure. Thus, measurement of the capacitance of the double layer, C_{dl}, before and after addition of the inhibitor can be used to monitor its adsorption. In measuring C_{dl}, the faradaic processes, e.g., iron corrosion reactions, must be separated from double layer charging. This is done in the single pulse method, in which the double layer is charged quickly before the response due to the slower faradaic processes is observed. More detail is given in Figures 4 and 5.

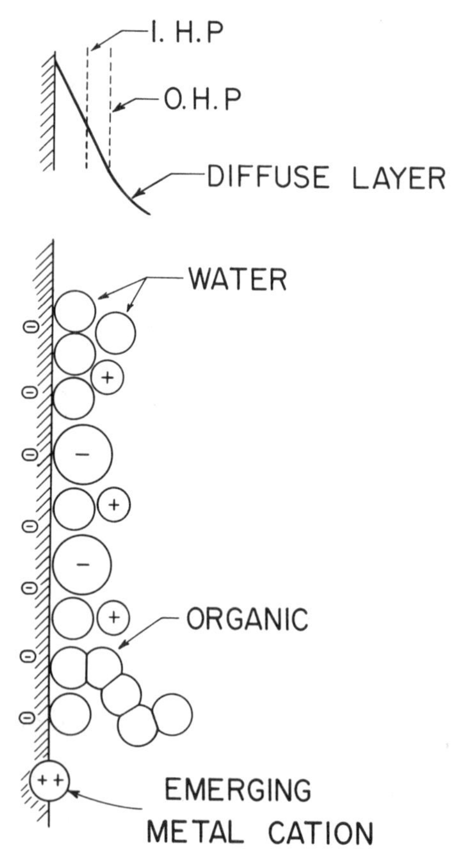

Fig. 3.—The electrical double layer at a corroding surface.

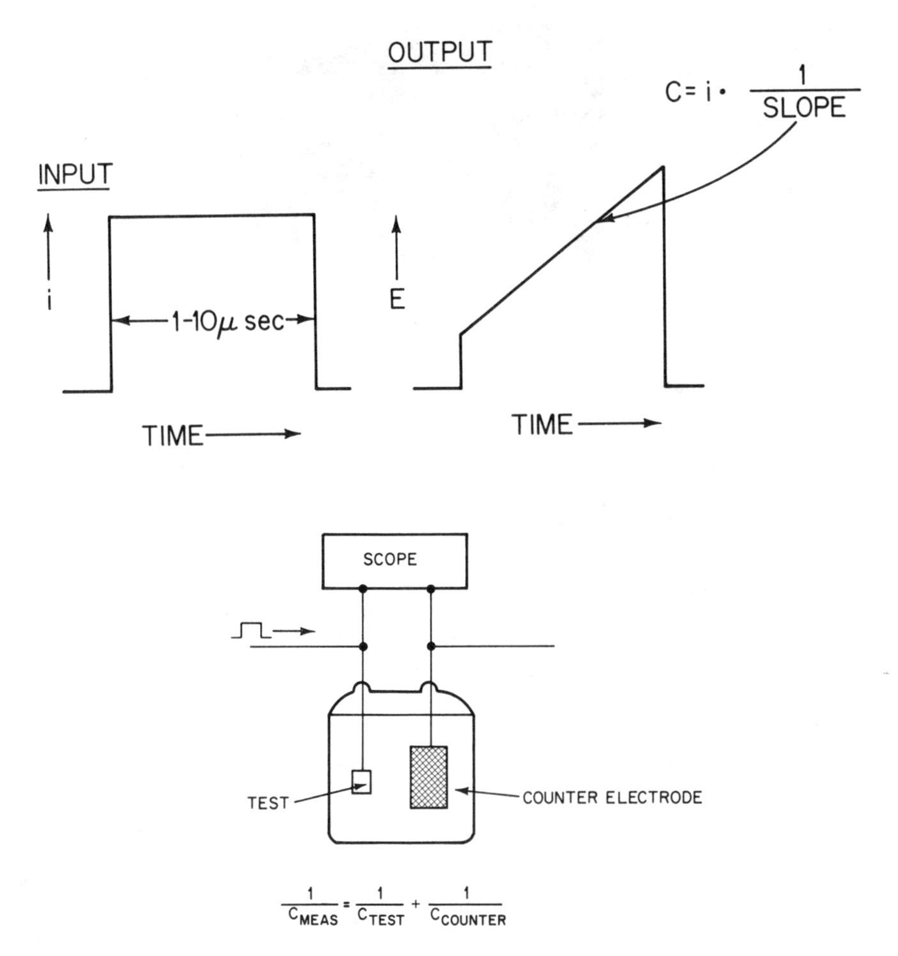

Figs. 4 and 5.—Measurement of double layer capacitances by the single pulse method. If the area of the test electrode is much smaller than that of the counter electrode, $C_{TEST} << C_{COUNTER}$ *so that* $C_{MEAS} = C_{TEST}$.

Figure 6 shows a typical double layer charging curve for iron, and Figure 7 shows double layer capacitance-potential curves for iron in *6N HCl* with and without added NH_2-$(CH_2)_{12}$-NH_2 inhibitor[23,39]. The decrease in C_{dl} shows that the diamine is adsorbed. Moreover, addition of the inhibitor leads to a general decrease in C_{dl} for potentials both anodic and cathodic to the open-circuit potential. The diamine is adsorbed on both types of sites, as first proposed by Hackerman and Makrides[13]. At cathodic potentials, C_{dl} rises due to H^+ adsorption; at anodic potentials due to the adsorption of Cl^- ions.

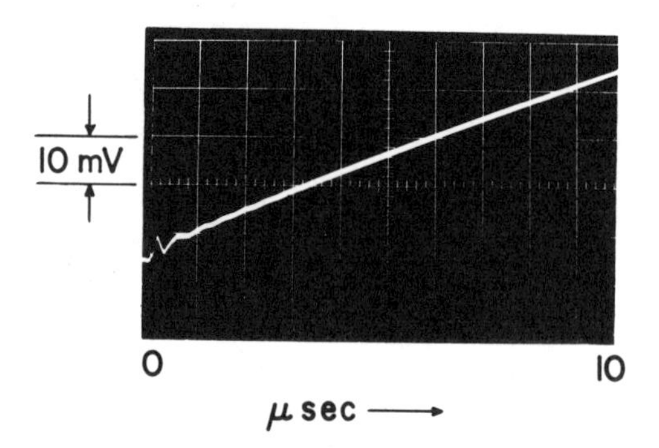

Fig. 6.—Typical double layer charging curve for iron in acid solutions.

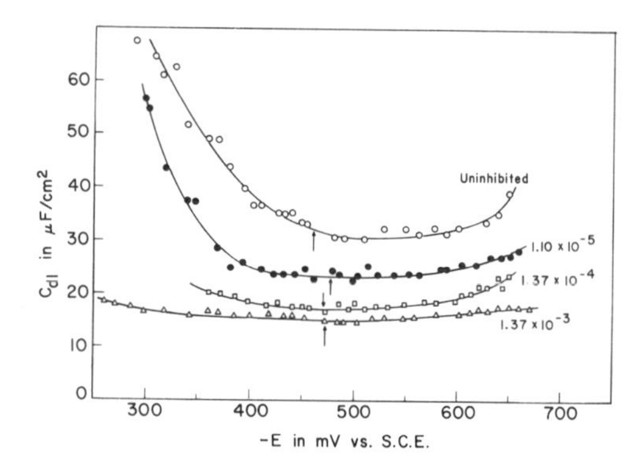

Fig. 7.—Double layer capacitance of iron in 6N HCl with and without added $NH_2-(CH_2)_{12}-NH_2$. The arrows indicate the open circuit potentials.

As mentioned earlier, the metal side of the interface can have an excess or deficiency of free electrons. That is, the metal surface can have a negative or positive charge. In the former instance, positive charges will be attracted to the surface, and in the latter instance, negative charges will be favored. The potential at which the metal has no net charge is called the point of zero charge (p.z.c.). The closer a metal is to its p.z.c., the easier for neutral organic molecules to adsorb at the metal surface. This is especially true for organic molecules which do not chemisorb strongly and would not be favored in competitive adsorption with other surface active solution species, such as chloride ions.

Antropov[40] introduced a scale of potentials called the ϕ-potential, which is a measure of the departure from the p.z.c.:

$$\phi = E - E_{p.z.c.} \tag{7}$$

From Figure 7, the open-circuit potential of iron in the solutions containing the diamine is about -0.46 V vs S.C.E., or -0.22 V vs N.H.E. Reported values for the point of zero charge of iron range from -0.33 V to -0.4 vs N.H.E.[41]

Using an average value of -0.37 V gives: $\phi\ (Fe) = -0.22 - (-0.37) = +0.15$. The positive sign means that the metal side of the interface is positively charged, or that a preponderance of negative ions (probably Cl^-) is favored on the solution side of the interface. Thus, cooperative adsorption by electrostatic attraction of the protonated inhibitor is possible.

To summarize the example considered, analysis of C_{dl} data shows that diamines adsorb at the iron/acid interface by both competitive and cooperative adsorption.

Effect of Molecular Structure

The molecular structure of organic molecules influences the adsorbability, and correspondingly, the effectiveness of corrosion inhibitors. Some general aspects of this topic have already been considered earlier.

The substituents in ring compounds are able to modify the effectiveness of would-be inhibitors because of their inductive effect on electron-donating centers, such as the $-NH_2$ group. Electron-withdrawing substituents, like $-Cl$, will decrease the charge density of electrons at the N atom resulting in decreased adsorption and reduced protection. On the other hand, electron-donating substituents like $-CH_3$ will increase the availability of donor electrons at the adsorption center and should result in increased protection.

Donahue and Nobe[42] have proposed a quantitative relationship between inhibitor effectiveness and the Hammet σ constant. The latter is a measure of the ability of the substituted group to provide electrons (negative σ) or withdraw electrons (positive σ) from a ring structure. These relationships have been verified experimentally for ring-substituted benzoic acids[43] and benzotriazoles[44], with increased inhibition systematically accompanying the ability of the substituent group to provide electrons to the ring structure.

There is another way in which changes in molecular structure can change the inhibitor effectiveness: changes in spatial configuration, rather than in electron configuration. Studies with polymer amines have shown that very short chains of adsorbable repetitive groups improve inhibitor effectiveness[45]. In a series of polyethylene amines:

$$- \overset{H}{N} - (CH_2)_2 - \overset{H}{N} -$$

the order of inhibitor effectiveness per nitrogen atom follows the order:

4 units > 3 units > 2 units > monomer.

Short chains of poly (4-vinylpyridine):

$$\left[-CH_2-CH(C_5H_4N)- \right]_n$$

were more effective than the monomer, but there was no added advantage in going from $n = 4$ to $n = 67$.

The effect of ring strain was shown in studies[46-48] which compare secondary amines and cyclic imines:

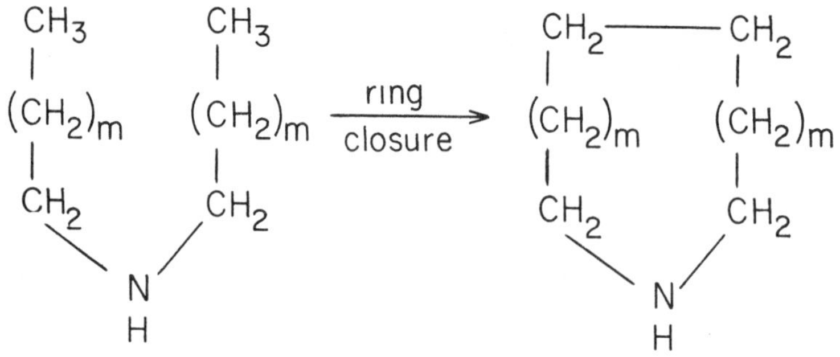

CH_3 CH_3
$(CH_2)_m$ $(CH_2)_m$ ring closure → CH_2—CH_2
CH_2 CH_2 $(CH_2)_m$ $(CH_2)_m$
N H CH_2 CH_2
N H

Both sets of compounds are secondary amines, have the same molecular weights, and same molecular areas. Yet, the closed ring compounds are much more effective inhibitors than their branched chain analogs. This structure effect is most pronounced for rings with 9 to 12 carbons. This high efficiency is due to the large strain in these rings, which is relieved with the formation of a strong chemisorptive bond.

Organic Diamines - A Case Study

The purpose of this section is to apply some of the foregoing considerations to a specific case study.

Organic amines were of interest due to the possibility of adsorption at the metal solution interface in several different configurations[23,39]. These configurations include the vertical position in which only one of the two end groups is adsorbed with the hydrocarbon chain extending outward (Figure 8c), and the flat position in which both end groups are adsorbed with the hydrocarbon network parallel to the surface (Figure 8a). In addition, if the molecule is long enough to be flexible, the hydrocarbon chain can be buckled between the two adsorbed end groups, as in Figure 8b, so as to present a hydrophobic "hump" to the solution.

Figure 9 shows typical polarization curves for iron in *6N HCl* with and without added diamines. Corrosion rates determined from the intersections of anodic and cathodic Tafel slopes are shown in Figure 10. The minimum corrosion rate for all diamines was 150 to 200 $\mu A/cm^2$. This value corresponds to 90% inhibition, where percent inhibition is defined as:

$$\%I = \frac{i_u - i}{i_u}, \qquad (8)$$

with i_u and i the uninhibited and inhibited rates, respectively.

In comparing a homologous series of adsorption type inhibitors, account must be taken of the following molecular parameters: electron donating ability of the adsorbing inhibitor, molecular area, and solubility.

The electron donating ability of the amines can be estimated from their acid dissociation constants, pK_a, which are related to base strengths. Larger values of pK_a imply stronger bases and hence better electron donors.

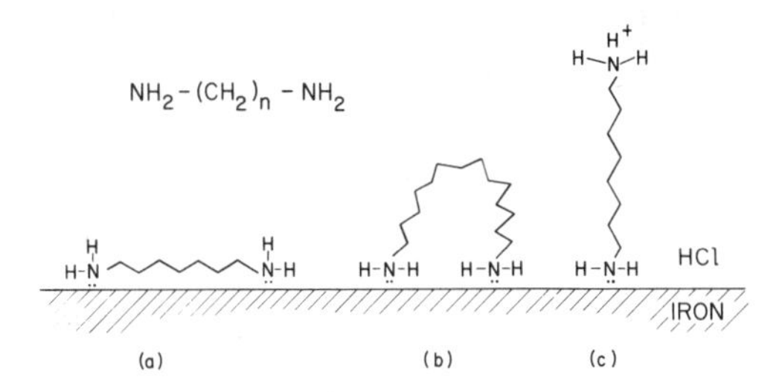

Fig. 8.—Possible configurations of aliphatic diamines upon adsorption.

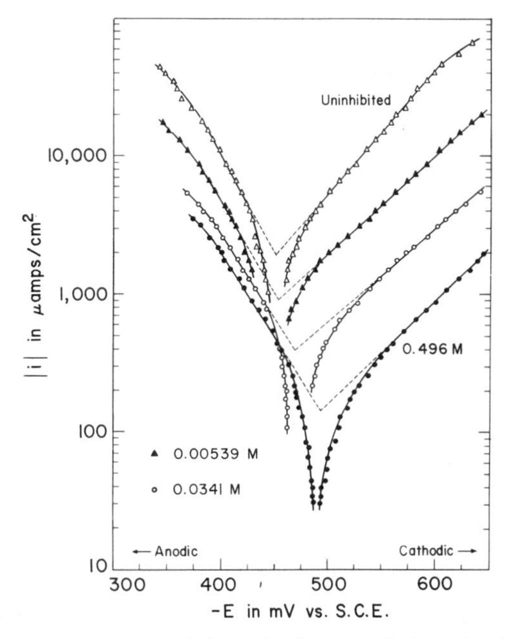

Fig. 9.—Polarization curves of iron in deaerated 6N HCl with $NH_2-(CH_2)_3-NH_2$ inhibitor.

As seen in Figure 10, ethylenediamine is the poorest of the inhibitors. This lack of effectiveness is explained by the fact that it is the weakest base, as seen from the pK_a data given in Figure 11.

The corrosion rates are the same for the C_3- through C_8- diamines. These diamines are all readily soluble and have essentially comparable base strengths (at least for C_5- through C_8-). Thus, the similar corrosion rates for the same concentration of these different diamines implies that the C_3- through C_8- diamines must all adsorb in the same configuration.

The longer chain diamines are more efficient than the C_3- through C_8- diamines for a given concentration of additive. However, this improvement

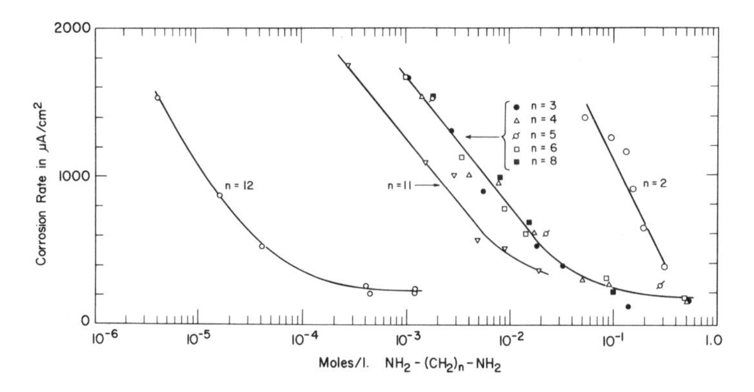

Fig. 10.—Corrosion rates of iron in 6N HCl for the homologous series of NH_2—$(CH_2)_n$—NH_2 inhibitors.

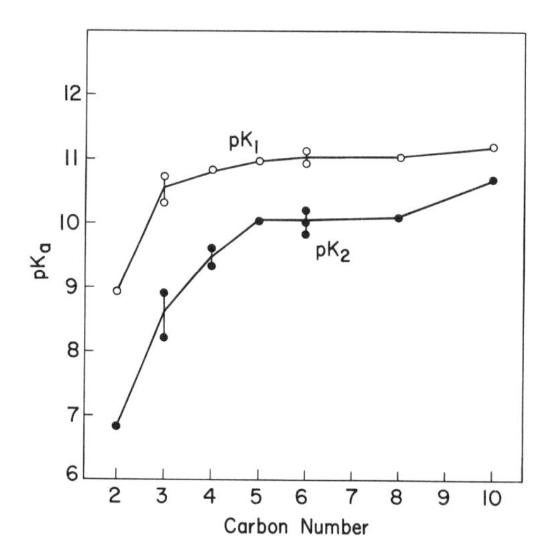

Fig. 11.—Acid dissociation constants pK_a for the homologous series NH_2—$(CH_2)_n$—NH_2.

is due to the decreased solubility of the longer chain diamines. As seen in Figure 12, the C_6- diamine is more efficient than the C_{12}- diamine on a relative concentration basis.

From Figure 11, the pK_a values of the C_6- diamine are similar to pK_a's for longer diamines. We have already taken differences in solubility into account. Thus, we can look for differences in molecular configurations. First, we need to add the experimental findings from double layer capacitance measurements.

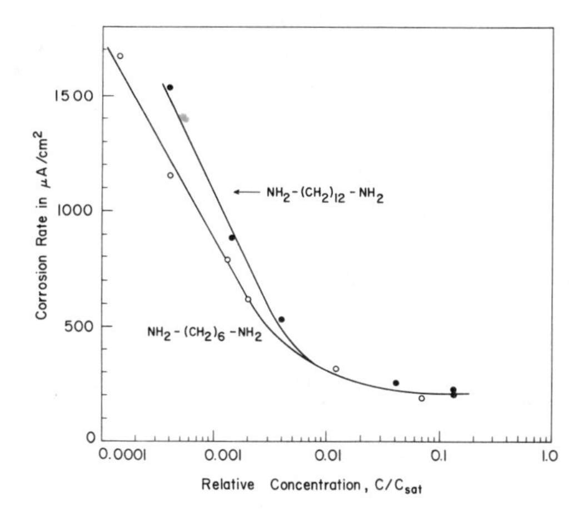

Fig. 12.—Comparison of C_6*- and* C_{12}*- diamines on a relative concentration basis.*

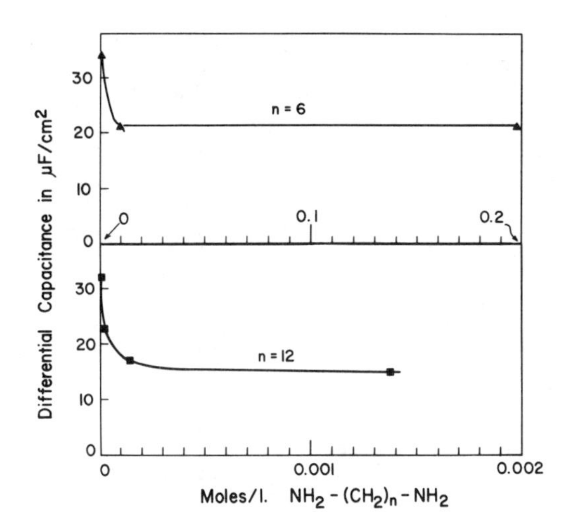

Fig. 13.—Limiting double layer capacitances for the C_6*- and* C_{12}*- diamines in 6N HCl.*

Figure 13 compares C_{dl} at the open-circuit potential for C_6- and C_{12}-diamines. Both additives reach limiting values of the double layer capacitance, as listed in Table II.

TABLE II.—LIMITING VALUES OF THE DOUBLE LAYER CAPACITANCE FOR IRON IN 6N HCl WITH TWO DIFFERENT DIAMINES

$H_2N-(CH_2)_6-NH_2$	22 μ F/cm^2
$H_2N-(CH_2)_{12}-NH_2$	14 μ F/cm^2

Five different possible combinations of molecular configurations can now be considered. These are listed in Table III. The first four cases listed there can be rejected as explained below.

Case I. C_6- and C_{12}- both adsorbed vertically. This configuration predicts that the double layer capacitance given by:

$$C_{dl} = \varepsilon/l \tag{9}$$

would be less for the C_{12}- diamine because the length l of the double layer would be longer for the longer diamine. The dielectric constant ε for the interface is that at infinite frequency, which is essentially constant for many homologous series of compounds[49]. Experimental results for C_{dl} agree with the order predicted by the model. However, with the vertical configurations, the corrosion rates should be the same because cross-sectional molecular areas would be the same (24 Å^2) in each case. This prediction is not observed experimentally, so Case I can be rejected.

Case II. C_6- and C_{12}- both adsorbed flat. Here C_{12}- would cover the larger area per molecule and be more effective than C_6-. However, Figure 12 shows the reverse is true. Moreover, C_{dl} should be the same for each case. The length l would be the same and so would the local dielectric constant ε, which would be determined mostly by the hydrocarbon chains located along the surface. Thus, this possibility can be rejected.

Case III. C_6- vertical and C_{12}- flat. By the line of reasoning above, the corrosion rates should be in the order C_{12}- < C_6-, and the double layer capacitances should be: C_6- < C_{12}-. Neither trend is observed experimentally.

Case IV. C_6- flat and C_{12} vertical. This arrangement predicts the corrosion rates to be C_6- < C_{12}- and the double layer capacitances to be C_{12}- < C_6-. Both trends are in agreement with experiment. However, if the C_6- diamine does not adsorb vertically, it is doubtful that the longer chain C_{12}- diamine will adsorb vertically. The humped configuration, as in Case V, is more likely.

Case V. C_6- flat and C_{12}- humped. This arrangement predicts the correct order of both corrosion rates and double layer capacitances. Hard-sphere molecular models show that the hydrocarbon chain of the C_{12}- diamine, but not the C_6- diamine, can bend over as in Figure 8b.

Confirming evidence as to these configurations is available from the field of surface chemistry. This evidence is summarized below:

1. Contact angle measurements of water on monolayers of mono- and dicarboxylic acids supported on a platinum substrate show that the two types of organic films behave similarly for twelve or more methylene groups[50]. See Figure 14. For the longer chain di-acids to terminate in a hydrocarbon group, as per the corresponding mono-acids, the di-acid molecule must buckle

over in the "inchworm" configuration.

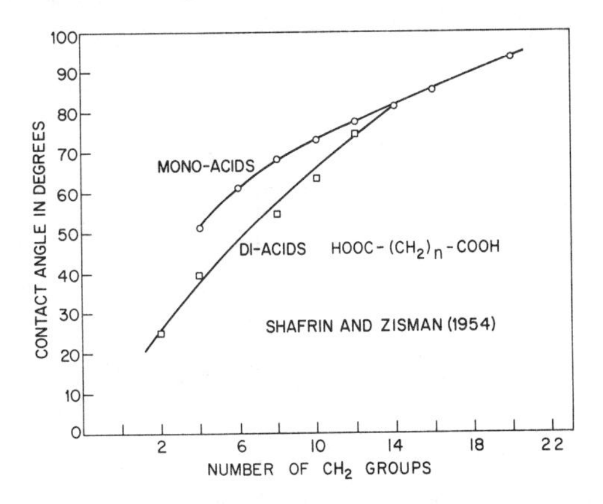

Fig. 14.—Contact angle measurements provide information on the configuration of adsorbed molecules.

2. From surface tension data for the C_4- and C_{10}- diamines at the water/chloroform interface(51), the limiting cross-sectional areas molecular areas are calculated from the Gibbs adsorption equation (see Figure 15) to be 151 $Å^2$ for the C_4- diamine and 58 $Å^2$ for the C_{10}- diamine. The smaller area of the C_{10}- diamine must mean that the hydrocarbon chain is buckled in the organic phase whereas the C_4- diamine lies flat at the interface.

$$\Gamma = -\frac{c}{RT} \cdot \frac{d\gamma}{dc}$$

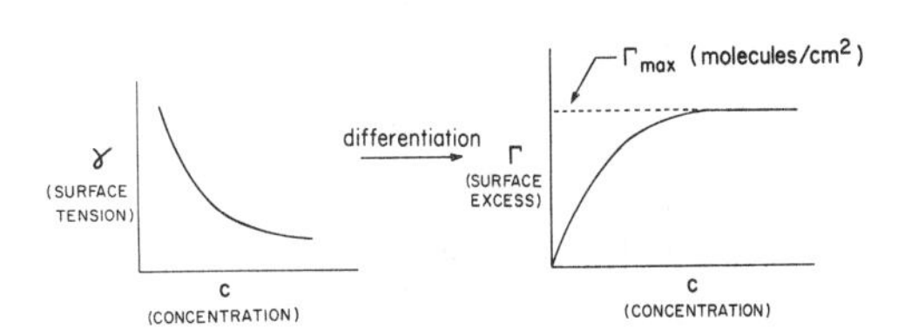

Fig. 15.—Interfacial tension data also provide information on molecular configuration at interfaces.

3. There is considerable evidence that short chain difunctional compounds adsorb in the flat configuration in various systems. At the air/water interface, the surface tensions of short chain difunctional acids, alcohols, and amines are lower than the corresponding monofunctional compounds(52,53) due to interaction of both polar end groups with the surface. At the metal/solution interface, the flat difunctional molecules exhibit the higher double layer capacitances due to a decreased distance between the capacitor plates(52-54).

To summarize the example considered, medium-sized diamines NH_2-$(CH_2)_n$-NH_2 *(n = 3 to 8)* are more efficient inhibitors than the longer chain compounds. These medium-sized inhibitors chemisorb through both end groups in a flat configuration, and are thus better inhibitors than the corresponding monofunctional amines(55).

TABLE III.—POSSIBLE CONFIGURATIONS OF C_6- AND C_{12}- DIAMINES AT THE IRON/ACID INTERFACE

	Corrosion Rates		C_{dl}	
	Predicted	Observed?	Predicted	Observed?
I 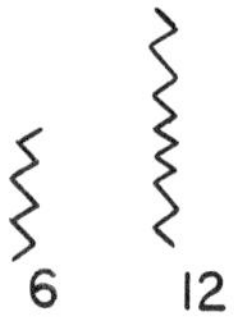 (6, 12)	$C_6 = C_{12}$	NO	$C_{12} < C_6$	YES
II 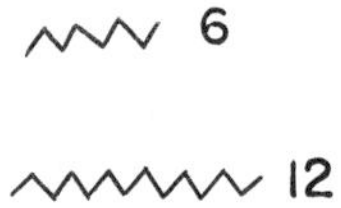 (6, 12)	$C_{12} < C_6$	NO	$C_{12} = C_6$	NO
III 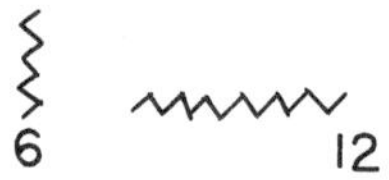 (6, 12)	$C_{12} < C_6$	NO	$C_6 < C_{12}$	NO
IV 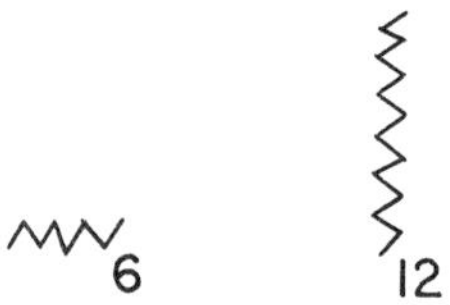 (6, 12); V (6, 12)	$C_6 < C_{12}$	YES	$C_{12} < C_6$	YES

Further Insights from Surface Chemistry

Using fundamental principles of corrosion inhibition, it should be possible to "tailor make" an inhibitor for a given application. Much useful information toward this goal is available from the field of surface chemistry.

For example, carboxylic acids are surface active agents which adsorb at metal surfaces by donation of electrons from the oxygen in the *-COOH* group to the metal. When adsorbed from solutions of non-electrolytes, the hydrocarbon tails are extended outward from the surface to form a close-packed hydrophobic monolayer. It is well known that surfaces terminating in fluorocarbon groups are less energetic and are less wettable than are surfaces terminating in hydrocarbon groups.

Figure 16 compares wettability data for films of fluorinated and conventional carboxylic acids preadsorbed onto a platinum or glass substrate[56]. The wetting liquid is methylene iodide, an organic liquid commonly used to rate surfaces. The higher the contact angle, the less wettable and the lower the energy of the surface. Figure 16 shows that fluorine substitution produces a "Teflon-like" surface with as little as five carbon atoms in a chain.

Thus, the idea of using fluorinated molecules as corrosion inhibitors is intriguing because of the possibility of in-situ formation of a low-energy "Teflon-like" surface. Such compounds should be effective corrosion inhibitors if a complete chemisorbed hydrophobic monolayer can be formed by adsorption from the corrosive medium. Of course, the corrosion environment is complicated by the presence of other surface active agents, such as water molecules and perhaps chloride ions, which compete with the inhibitor for the metal surface. However, one favorable aspect is that the actual configuration of the adsorbed inhibitor would not seem important. Either a flat or vertical configuration would present low-energy surface groups to the solution, as shown in Figure 17.

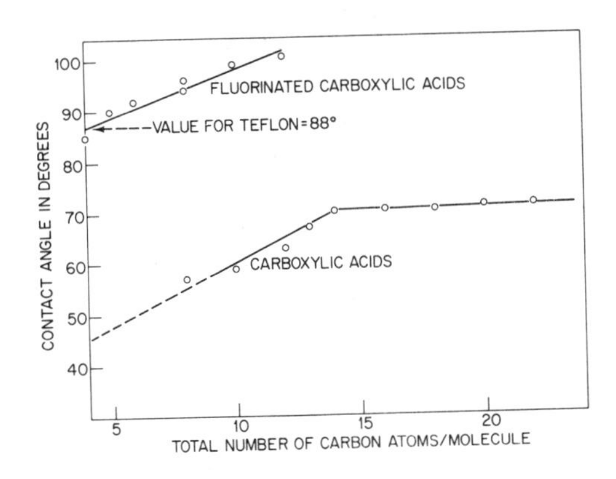

Fig. 16.—Adsorbed fluorocarbon films are lower energy surfaces than the corresponding hydrocarbon films.

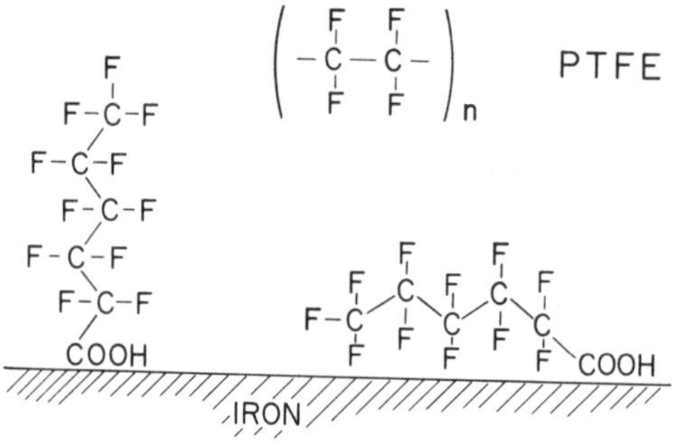

Fig. 17.—Can an adsorbed fluorinated carboxylic acid form a "Teflon-like" surface?

To test these ideas, corrosion rates of high-purity iron were determined in deaerated *3N HCl* with and without added butyric acids. Details have been given elsewhere[57]. Figure 18 shows that perfluorobutyric acid $CF_3(CF_2)_2\,COOH$ is a poorer - not better - inhibitor than its conventional analog, $CH_3(CH_2)_2COOH$.

The reason is that the electronegativity effect of F atoms adjacent to the carboxyl group is predominant over the potential increase in hydrophobic character. That is, the strongly electronegative F atoms withdraw electrons from the *-COOH* group to make it a stronger acid (and weaker base) so as to decrease the density of electrons available for bonding with the iron surface. The effect is schematically illustrated in Figure 19, which also lists ionization constants. Thus, the weaker bond strength with the totally fluorinated acid results in its poorer inhibitive effect.

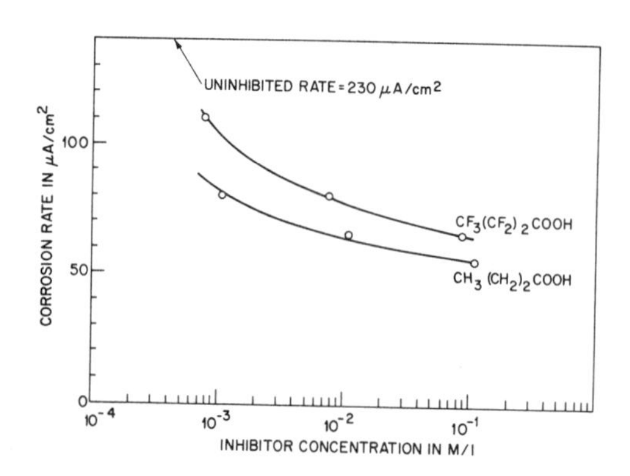

Fig. 18.—Corrosion rates of butyric acids in 3N HCl.

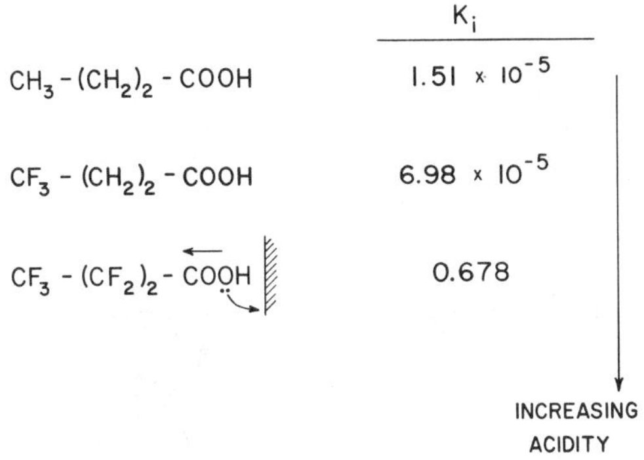

	K_i
$CH_3-(CH_2)_2-COOH$	1.51×10^{-5}
$CF_3-(CH_2)_2-COOH$	6.98×10^{-5}
$CF_3-(CF_2)_2-COOH$	0.678

INCREASING ACIDITY ↓

Fig. 19.—Effect of fluorine substitution on the ionization constant of butyric acid.

However, Figure 19 shows that the presence of methylene groups interposed between the surface active carboxyl group and fluorocarbon tail shields the *-COOH* group from the inductive effect of the fluorine atoms.

These observations lead us to predict that the most effective fluorinated compounds should be only partially fluorinated, as shown in Figure 20. The chain should feature several "insulating" $-CH_2-$ segments which shield the surface active *-COOH* group from the fluorine atoms in the hydrophobic tail.

Additional work on such compounds has been directed toward amines, rather than carboxylic acids, because the former class of compounds is usually the more effective as inhibitors. Figure 21 shows corrosion-time curves for iron in deaerated *3N HCl* with and without added propyl amine, $CH_3CH_2CH_2NH_2$, and pentafluoropropyl amine, $CF_3CF_2CH_2NH_2$ (58). It can be seen that the partially fluorinated amine is a better inhibitor than its conventional analog.

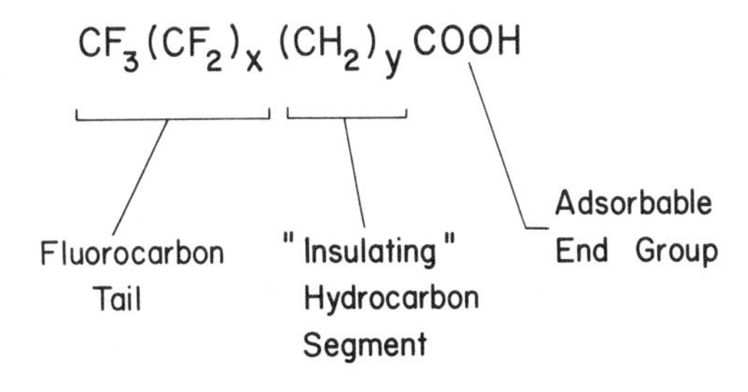

Fig. 20.—A "tailor-made" inhibitor.

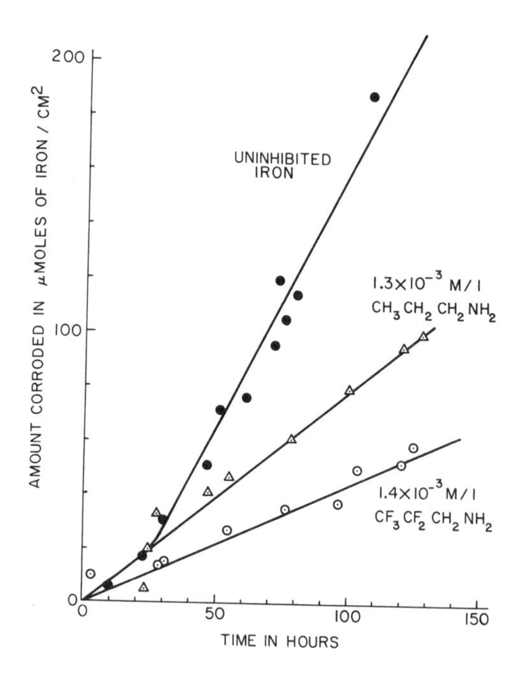

Fig. 21.—The partially fluorinated amine is the better inhibitor for iron in 3N HCl.

It is interesting to note that pentafluoropropyl amine is a better inhibitor than propylamine despite the fact that the latter is the stronger base. The acid ionization constants, pK_a, for the two compounds are given in Table IV. Thus, the beneficial aspects of an improved hydrophobic tail overcome the detrimental aspects of a weaker electron donor.

TABLE IV.—DISSOCIATION CONSTANTS FOR THE TWO AMINES

COMPOUND	pK_a*	Percent Inhibition†
$CH_3CH_2CH_2NH_2$	10.53	59
$CF_3CF_2CH_2NH_2$	5.7	77

*The higher pK_a indicates the stronger base and hence the stronger electron donor.
†From Figure 21.

Adsorption Isotherms

The adsorption isotherm is the link between the amount of inhibitor in solution and the amount taken up by the surface. However, direct experimental evidence on the adsorption of organic inhibitors during the actual corrosion process is rather scanty. One reason is that adsorption isotherms from solution are difficult to determine with the small and changeable surface areas involved. Applicable techniques, such as radiotracer measurements, require that the corrosion process be interrupted and the electrode removed from solution to monitor adsorption.

Adsorption isotherms can be inferred from changes in the double layer capacitance and in the corrosion rate, because both parameters are influenced by the surface coverage of inhibitor, θ_{org}. Relative, but not absolute, adsorption isotherms can be determined.

The usual model of the electrode surface is parallel capacitors for the inhibitor-bare and inhibitor-covered portions added on a coverage basis:

$$C = C_o (1-\theta_{org}) + C_1 \theta_{org} \tag{10}$$

where C_o and C_1 are the capacitances (per unit area) of the inhibitor-free and inhibitor-covered surfaces, respectively, and C is the composite total capacitance for any intermediate coverage θ_{org}. The values for C_1 are those limiting capacitances, as in Figure 13, for which addition of inhibitor caused no further decrease in capacitance. While there is a maximum adsorption, there may not be complete monolayer formation. Thus, θ_{org} is a relative coverage given by Γ/Γ_{max}, where Γ_{max} is the true surface excess (not usually determined) corresponding to capacitance C_1. From Equation 10:

$$\theta_{org} = \frac{C_o-C}{C_o-C_1} \tag{11}$$

Relative adsorption isotherms can also be determined from corrosion rates by considering the total rate i is given by the sum of two parallel reactions of rate i_o for the inhibitor-bare surface and i_{sat} for the inhibitor-covered surface. That is:

$$i = i_o (1-\theta_{org}) + i_{sat} \theta_{org} \tag{12}$$

or

$$\theta_{org} = \frac{i_o-i}{i_o-i_{sat}} \tag{13}$$

In the diamines example considered earlier, i_{sat} is the minimum corrosion rate in Figure 10, i.e. about 200 $\mu A/cm^2$. Again, θ_{org} obtained from Equation 13 is a relative coverage because Equation 13 implies $\theta_{org} = 1$ when $i = i_{sat}$.

Adsorption isotherms obtained from Equations 11 and 13 are shown in Figure 22 for $NH_2-(CH_2)_{12}-NH_2$ in concentrated HCl[39]. A similar shaped isotherm was also obtained for the C_{11}-diamine. These are Temkin isotherms, in which θ_{org} is linear in log concentration for intermediate coverages. These isotherms take into account decreases in the heat of adsorption with coverage, as shown in Figure 23. Temkin isotherms are reported frequently for corrosion inhibitors, and not surprisingly so in view of the intrinsic

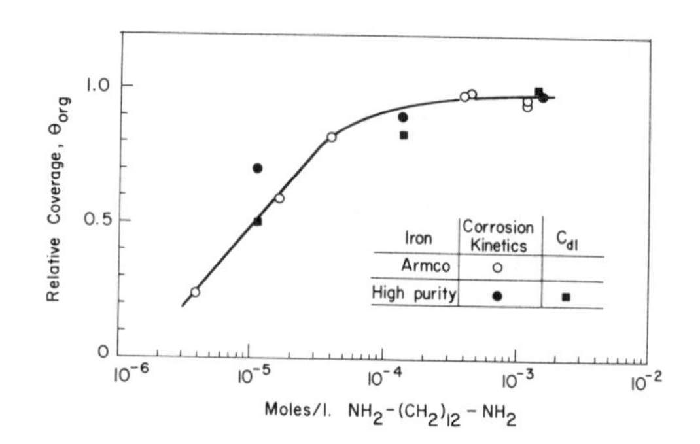

Fig. 22.—Adsorption isotherm for C_{12}-diamine in 6N HCl.

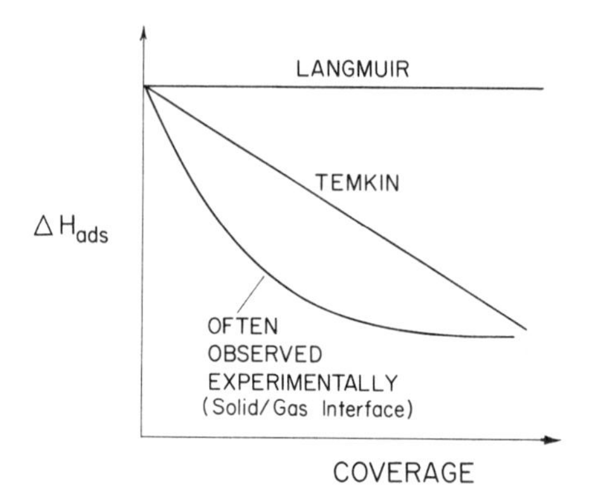

Fig. 23.

surface hetereogeneity which must exist for a corroding metal. In addition, lateral interactions must be considerable for the complicated interface which exists under the mixed corrosion potential. Both of these factors act to decrease the heat of adsorption with coverage, as required by the model.

More detailed information on adsorption isotherms in regard to corrosion inhibition has been given by Riggs[59]. This topic will be taken up again later in regard to inhibition of localized corrosion in neutral solutions.

NEARLY NEUTRAL SOLUTIONS

Introduction

The corrosion of metals in nearly neutral solutions differs from that in acid solutions for two reasons. First, in acid solutions, the metal surface is oxide-free but in neutral solutions, the surface is oxide-covered. Secondly, in acid solutions the main cathodic reaction is hydrogen evolution, but in air-saturated neutral solutions the cathodic reaction is oxygen reduction.

In keeping with the theme of this review, the main emphasis is on adsorption processes in regard to corrosion inhibition. First the role of oxide films will be considered, followed by two topics of recent interest. The first is the use of chelating compounds in neutral media. The second involves localized corrosion, which is a common problem in neutral solutions.

Effect of Oxide Films

Corrosion inhibitors can interact strongly with a surface even if it is oxide-covered. For example, heat of adsorption (from the vapor phase) of various aliphatic amines is actually slightly higher for oxide-covered iron than for reduced iron(28). This strong interaction was attributed to either one or both of the following two processes: (1) the peripheral dipole of the amine interacts with the surface Fe^{+2} or Fe^{+3} ions, and (2) the chemisorbed amine forms a coordination bond with the surface Fe^{+2} or Fe^{+3} ions.

Studies on oxide-covered iron(28) and on bulk iron oxides(60) show that water is also strongly adsorbed onto oxide surfaces. Oxide surfaces invariably terminate in $-OH$ groups due to the interaction with water vapor in the ambient atmosphere(61), as shown in Figure 24. In aqueous solutions, water molecules are hydrogen bonded onto these surface hydroxyl groups. The hydrogen bonded network can be quite extensive involving several molecular layers, so that the overall effect is a considerable attraction of water to the oxide film.

Thus, when the surface is oxide-covered, the displacement of adsorbed water by chemisorbed inhibitor molecules is again an important consideration, just as in the case of oxide-free surfaces. However, the nature of surface-adsorbate interactions differs in the two cases. With oxide-covered metal, solution species interact with surface oxide anions or cations rather than with metal atoms, as in acidic environments.

Because the metal surface is oxide-covered, the role of the inhibitor in neutral solutions is to make the oxide film protective or to maintain it so. The mechanism involves both adsorption and reaction steps. The type of reaction depends on the nature of the inhibitor. Chromate is adsorbed rapidly to form a monolayer, followed by additional slow uptake(6,7). The $CrO_4^{=}$ is incorporated into the mixed oxide film as Cr^{+3} according to the reaction:

$$2Fe + 2CrO_{4\,(aq)}^{=} + 2H_2O \longrightarrow Fe_2O_3 + Cr_2O_3 + 4OH^{-}_{(aq)} \quad (14)$$

With non-oxidizing inhibitors, such chemical reactions are not possible. Phosphate, for example, is adsorbed onto oxide-covered iron at about 0.7 monolayers at the active/passive transition(62). The phosphate ion is then incorporated into the oxide film by exchange with surface oxide (or hydroxide) ions(9). Other inhibitors, such as sodium azelate (sodium salt of $HOOC-(CH_2)_7-COOH$) are not taken up as uniformly or as extensively, but are believed to repair pores in oxide films by reaction there with dissolving cations to form

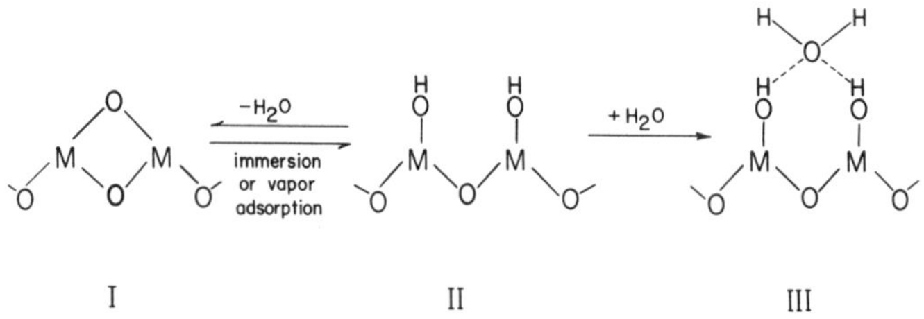

Fig. 24.—A metal oxide surface (I) exposed to ambient environments forms a hydroxyl layer (II), which provides additional sites for water molecules via hydrogen bonding (III).

insoluble local precipitates[63]. More detailed reviews on these processes have been given by Evans[64,65] and Thomas[66].

A number of studies have shown that the concentration of inhibitors required for protection increases with the concentration of aggressive ion present in solution[67-69]. This competitive effect is especially important in localized corrosion reactions, where attack by chloride leads to pitting on open surfaces and to breakdown of passivity within crevices. The role of chromate in inhibiting both types of attacks is discussed in a later section.

The presence of an oxide film on a metal adds another factor not usually considered - the semiconductor nature of the oxide film. Vijh[70,71] has recently correlated the corrosion potential of various metals in oxygenated chloride solutions with the semiconductivity of the corrosion film. The corrosion potential was more negative the greater the band gap width of the inorganic oxide or chloride. This relationship was interpreted in terms of n-type corrosion products, with the increased band gap width decreasing the rate of the partial anodic reaction but not affecting the cathodic reaction, so that the resulting mixed potential was shifted to more negative values.

Vijh[72] has pointed out that the semiconductivity of surface oxides is often masked by additional factors. For instance, the oxide film may be thin enough to sustain the electrode reaction by electron tunneling, or the electrode reaction may occur at active sites on the oxide surface or within pores or flaws in the oxide film. Nevertheless, more work needs to be done to clarify the effect of the semiconductivity of oxides on corrosion and its inhibition.

Chelating Compounds as Corrosion Inhibitors

Recently there has been an increased interest in chelating agents as corrosion inhibitors. Chelating compounds have been recently studied as corrosion inhibitors for steel in industrial cooling waters[73,74] and for zinc[75] and aluminum[76,77] in various environments. Much of this work has been prompted by the search to replace chromate inhibitors with compounds which are less toxic and less polluting.

Chelating agents are organic molecules with at least two polar functional groups capable of ring closure with a metal cation. The functional groups may either be basic groups, such as *-NH₂*, which can form bonds by electron donation, or acidic groups, such as *-COOH*, which can coordinate after loss of the proton. An example of a surface chelate is shown in Figure 25.

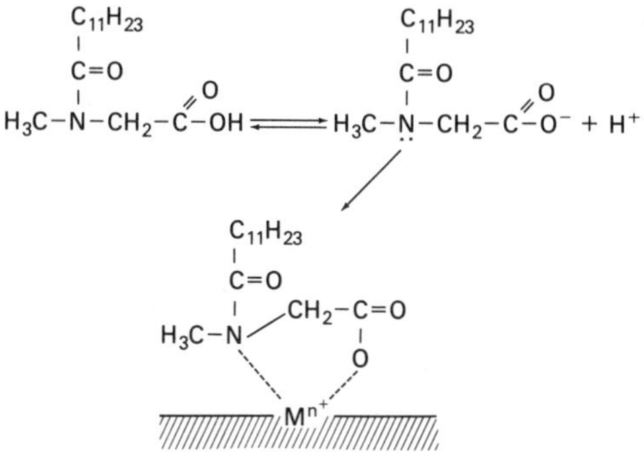

$$H_3C-N(C(=O)C_{11}H_{23})-CH_2-C(=O)-OH \rightleftharpoons H_3C-\ddot{N}(C(=O)C_{11}H_{23})-CH_2-C(=O)-O^- + H^+$$

$$\downarrow$$

$$H_3C-N(C(=O)C_{11}H_{23})-CH_2-C(=O)-O \quad (N \cdots M^{n+} \cdots O)$$

Fig. 25.—A sarcosine-type surface chelate. From ref. (73).

Chelating agents are well known in the fields of analytical and inorganic chemistry. The reactions occur in the bulk solution, however; and much less detail is known about the formation of chelates with a metal surface.

Table V lists some examples of surface chelate corrosion inhibitors. As noted by Zecher(74), the concept of using surface chelants as corrosion inhibitors is not new, as a number of patents describe the use, if not the mechanism, of such additives. One of the first organized studies was made twenty years ago by Hayakawa and Ida(78) on the use of ortho-hydroxyazo compounds on the corrosion of aluminum. These authors showed that there was inhibition only in basic solutions where chelate formation could occur.

Two mechanisms can be responsible for metal stabilization by surface chelation. The chelating molecule may interact with metal ions while they are still bound in the metal lattice (or with metal ions in the surface oxide film). In the second mechanism, surface metal ions dissolve and immediately react with the chelating agent near the surface, forming a complex of high molecular weight and low solubility which precipitates on the metal surface, forming a barrier film. Results in the literature support elements of both mechanisms.

Experimental evidence for the adsorption mechanism is summarized below:

1. Increases in adsorbability of homologous chelants result in increases in inhibitor effectiveness. For example, α-mercaptocarboxylic acids:

$$C_nH_{(2n+1)} - \underset{SH}{\overset{H}{C}} - COOH$$

TABLE V.—SOME EXAMPLES OF SURFACE CHELATING AGENTS AS CORROSION INHIBITORS IN NEARLY NEUTRAL SOLUTIONS

Compound	Formula (and probable surface bonding)	Application
Sarcosine type compounds	See Fig. 25	Mild steel in synthetic cooling water (73,74)
α-mercaptoacetic acids and esters	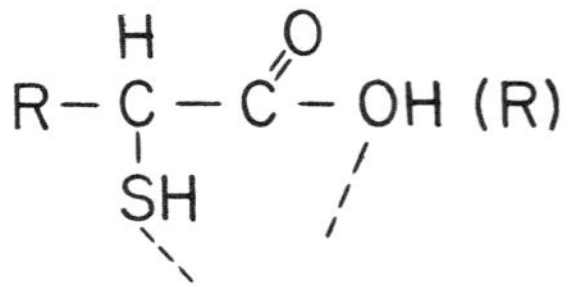 $R-CH(SH)-C(=O)-OH\ (R)$	Mild steel in synthetic cooling water (74); resistance to white rust staining of galvanized steel (75).
Alizarin	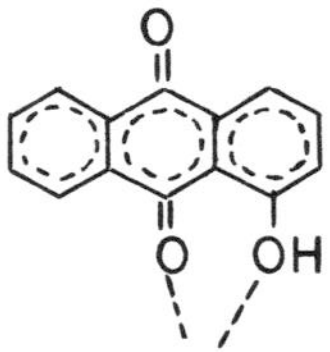 O, O, OH	Improvement in the paint adherence and corrosion resistance of aluminum alloys (76)
8-hydroxyquinoline	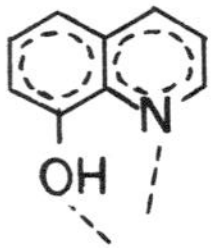 N, OH	Mild steel in synthetic cooling water (73)

were more effective inhibitors that the corresponding α-hydroxycarboxylic acids[74]:

$$C_nH_{(2n+1)} - \underset{\displaystyle OH}{\overset{\displaystyle H}{C}} - COOH$$

This improvement can be explained by the increased absorbability of the *-SH* group as compared to the *-OH* group[13].

In another example, corrosion rates increased when the $-CH_3$ group in the sarcosines in Figure 25 was replaced with *-H*. According to Weisstuch and co-workers[73], the inductive effect of the $-CH_3$ group increases the availability of electron pairs at the nitrogen atom for improved surface bonding, i.e., adsorption.

2. Steric requirements for surface chelation are more restrictive than those in solution. Not all compounds which are capable of forming chelates in the bulk solution are effective inhibitors. As an example, 2,2'-bipyridine and 1,10-phenanthroline are classical chelating agents for ferrous ions in solution[79]. However, in cooling water studies, the former provided no protection to iron while the latter did[73]. The explanation given was that the preferred trans type spatial orientation of 2,2'-bipyridine restricts the probability of surface chelation, but the rigid structure of 1,10-phenanthroline is favorable for surface chelation. See Figure 26.

It appears that the precipitation mechanism can operate if surface chelation is not favored. Visible films have been observed in several systems where there was only minimal protection. X-ray analysis of the films formed on iron in cupferron (3.5% inhibition) and zinc in 8-hydroxyquinoline (51% inhibition) showed that each film was the same complex as was formed in the bulk solution[80]. Ushenina and Klyuchnikov[81] have generalized that this deposition mechanism can provide only a slight inhibiting effect on corrosion and that the surface chelate affords the more effective protection.

Alteration in the molecular structure of surface chelants can improve their performance in neutral media, just as is the case for adsorption-type inhibitors in acid solutions. It is well known that in bulk solution, 5-membered chelate rings are the most stable, other factors being the same[79]. Recent results indicate that this structure effort carries over to surface chelates as well[73,74]. Weisstuch and coworkers[73] suggest that surface chelates should possess large hydrophobic substituent groups. These groups would decrease the solubility, so as to promote adsorption; and once adsorbed, would improve the hydrophobic barrier to electrolyte penetration.

Future work on surface chelating inhibitors should involve the use of surface analytical techniques such as Auger analysis, XPS, or infra-red to gather evidence as to the nature of these surface species.

Inhibition of Localized Corrosion

Crevice corrosion and pitting are often problems in neutral solutions. Crevice corrosion is a form of localized attack that occurs in narrow clearances, such as between overlapping or stacked metal sheets, under gaskets, within screw threads, or under marine deposits or paint films. Corrosion within the crevice is initiated by a differential aeration cell, the metal within the crevice being in contact with electrolyte devoid in oxygen

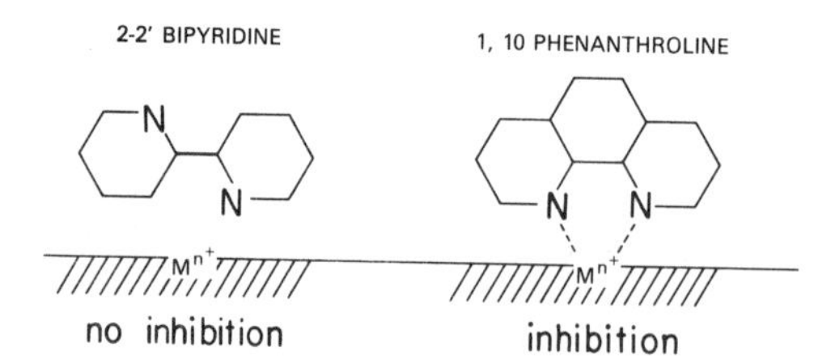

Fig. 26.—Steric effects in surface chelation. From ref. (73).

compared to metal outside the crevice. With open surfaces, localized breakdown of passivity in the form of pitting corrosion often occurs. Both types of localized corrosion propagate due to the development of a local internal acidity resulting from the restricted exchange of bulk and local electrolytes(5). In crevice corrosion (and stress-corrosion cracking), the long narrow diffusion path limits the access to bulk electrolyte. In pitting corrosion, the porous cap of corrosion products above a pit is the barrier.

The mechanism of inhibition of localized corrosion of iron in chloride solution has been studied recently using chromate as an inhibitor(82,83). Both crevice corrosion and pitting of iron can be inhibited by $CrO_4^=$. The explanation in both cases involves competitive adsorption between the aggressive Cl^- ion and the inhibitive $CrO_4^=$ ion. If adsorption of the former predominates, pitting occurs on open surfaces, and breakdown of passivity occurs within creviced surfaces.

Figure 27 shows the effect of chromate on the crevice corrosion current of iron in 0.6M *NaCl*. Details concerning the crevice assembly and the measurements are given elsewhere(82). It can be seen that crevice corrosion is inhibited by adding a sufficient amount of chromate.

For more dilute chloride solutions, a smaller amount of chromate is required for protection. See Figure 28, which summarizes the results of many different combinations of $CrO_4^=$ and Cl^- concentrations. In all instances, a critical minimum concentration of $CrO_4^=$ is required for inhibition, given a certain concentration of Cl^-. When the data in Figure 28 are replotted on the basis of ion activities, rather than concentrations, a straight line results, as in Figure 29. The borderline separating the regions of crevice corrosion and inhibition is given by:

$$\frac{d \log a_{Cl^-}}{d \log a_{CrO_4^=}} = const \qquad (15)$$

Figure 30 shows crevice potential-time curves for 0.6M Cl^- corresponding to the current-time curves in Figure 27. The electrode potential is that measured *within* the crevice, using a micro-reference electrode inserted into the crevice. For the chromate concentrations where crevice corrosion occurred (0.003, 0.01, and 0.03 M/l), the steady-state electrode potential of the corroding iron was -0.620 to -0.660 mV *vs. Ag/AgCl*. In fact, these potentials were observed for all solutions where crevice corrosion occurred, regardless

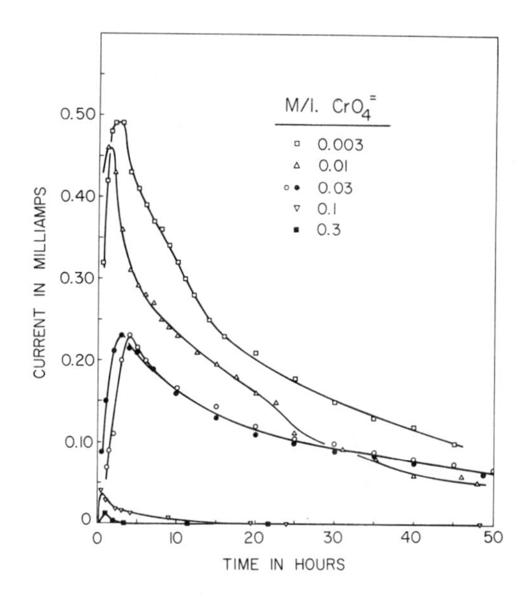

Fig. 27.—Crevice corrosion of iron (7.9 cm² area) in a 10 mil crevice in 0.6 M NaCl with various amounts of added chromate.

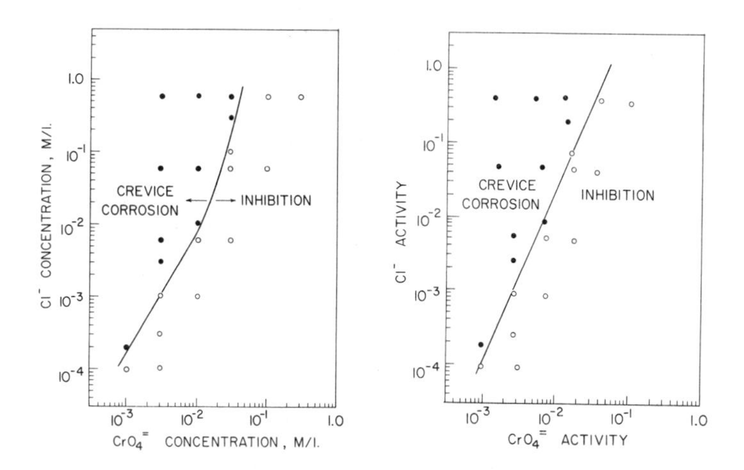

Figs. 28 and 29.—The minimum amount of chromate required to inhibit crevice corrosion of iron depends on the amount of chloride that is present.

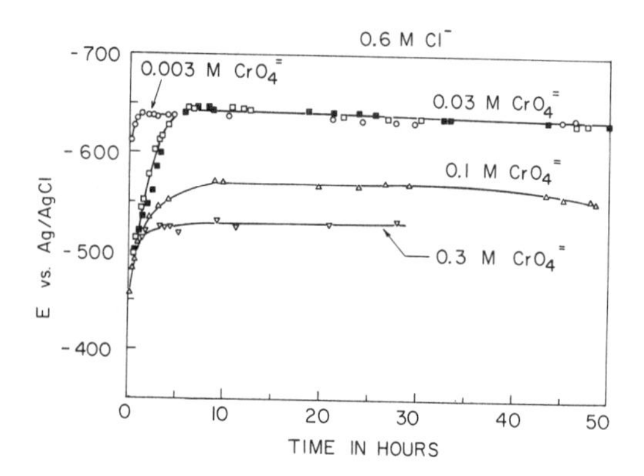

Fig. 30.—Electrode potential of iron in crevices, corresponding to the current-time curves in Figure 27.

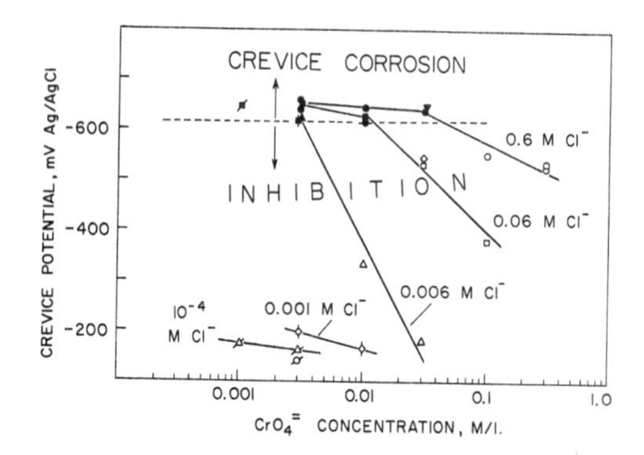

Fig. 31.—Inhibition of crevice corrosion always occurred when the electrode potential of iron (measured within the crevice) was more positive than -0.62 mV.

of the bulk concentration of electrolyte. See Figure 31.

Results analogous to those in Figure 29 were first obtained with iron by Matsuda and Uhlig[84] for the case of pitting corrosion. This same general relationship has been observed in free electrolytes containing various aggressive and inhibitive ions, including $Cl^-/CrO_4^=$ mixtures[67-69]. Matsuda and Uhlig have interpreted the linear log/log plots obtained for pitting on the basis of competitive adsorption between inhibitive and aggressive ions. This same explanation can also be proposed for the case of crevice corrosion because of the similarities between these two forms of localized corrosion.

Matsuda and Uhlig[84] described the adsorption process in terms of the empirical Freundlich isotherm, which has no theoretical basis. More recently, Strehblow and Titze[85] have applied the Langmuir isotherm to the competitive adsorption between inhibitive and aggressive anions in the case of pitting corrosion. For adsorption of the aggressive ion:

$$\frac{\theta_A}{1-\theta_A-\theta_I} = K_A' \, a_{A^-} e^{\Delta H_{ads,A}/RT} e^{\gamma_A z_A FE/RT} \tag{16}$$

where θ_A and θ_I are the surface coverages of aggressive and inhibitive ions, respectively, a_{A^-} the bulk activity of aggressive anion, $\Delta H_{ads,A}$ the heat of adsorption of the aggressive ion, and E the electrode potential across the interface. The factor γ_A (not to be confused with activity coefficients) takes into account the fact that the specifically adsorbed anions move through only part of the potential drop across the interface[85].

If the heat of adsorption ΔH_{ads} is independent of coverage, Equation 16 yields the usual Langmuir isotherm for charged intermediates[86], modified to account for the co-adsorption of two species. More likely, however, the heat of adsorption decreases with coverage due to: (a) *a priori* heterogeneity of surface sites having a distribution of energies, and (b) lateral repulsive interactions between neighboring adsorbed anions.

A reasonable approach to take these factors into account is provided through use of the Temkin isotherm, which assumes that the heat of adsorption decreases linearly with coverage, as discussed earlier.

If only one adsorbed species is involved,

$$\Delta H_{ads} = \Delta H^o - r\theta \tag{17}$$

where ΔH^o is the initial heat of adsorption and r is the Temkin parameter. If the chloride ion adsorbs in patches or clusters, then the adsorption of chloride and of chromate will be independent of each other for most coverages. That is:

$$\Delta H_{ads,A} = \Delta H_A^{\,o} - r_A\theta_A \tag{18}$$

and

$$\Delta H_{ads,I} = \Delta H_I^{\,o} - r_I\theta_I \tag{19}$$

Use of Equation 18 in Equation 16 gives:

$$\frac{\theta_A}{1-\theta_A-\theta_I} = K_A \, a_{A^-} e^{-r_A\theta_A/RT} e^{\gamma_A z_A/RT} \tag{20}$$

where

$$K_A = K_A' \, e^{\Delta H_A^{\,o}/RT}$$

Taking logarithms and using the usual approxiamtion that at intermediate coverages $ln\left[\theta_A/(1-\theta_A-\theta_I)\right] << r_A\theta_A/RT$ gives the Temkin isotherm:

$$\theta_A = \frac{2.303\ RT}{r_A} log\ K_A + \frac{2.303\ RT}{r_A} log\ a_{A^-} + \frac{\gamma_A z_A\ FE}{r_A} \tag{21}$$

Similarly

$$\theta_I = \frac{2.303\ RT}{r_I} \log K_I + \frac{2.303\ RT}{r_I} \log a_{I^-} + \frac{\gamma_I z_I\ FE}{r_I} \qquad (22)$$

As discussed earlier, the electrode potential for crevice corrosion is approximately -620 mV *(Ag/AgCl)* in all instances. With $E = E_{active}$, Equations 21 and 22 combine to give:

$$\frac{1}{r_A} \log a_{A^-} - \left(\frac{\theta_A}{\theta_I}\right) \frac{1}{r_I} \log a_{I^-}$$

$$= \left(\frac{\theta_A}{\theta_I}\right) \frac{1}{r_I} \log K_I - \frac{1}{r_A} \log K_A$$

$$+ \left(\frac{\theta_A}{\theta_I} \cdot \frac{\gamma_I z_I}{r_I} - \frac{\gamma_A z_A}{r_A}\right) \frac{FE_{active}}{2.303\ RT} \qquad (23)$$

Uhlig and co-workers[69,84] have proposed that pitting initiates when there is a critical ratio in the surface coverages of aggressive to inhibitive ions. This same effect can result in passive film breakdown within crevices (although the critical ratios are probably different in the two forms of localized corrosion). In the patchwise adsorption model, breakdown will occur at certain specific sites within the crevice and then spread to other areas as crevice acidification occurs. If there is a critical ratio in the coverages of adsorbed chloride to adsorbed chromate which leads to passive film breakdown,

$$\frac{\theta_A}{\theta_I} = \theta_{crit} \qquad (24)$$

then Equation 23 gives:

$$\frac{d \log a_{A^-}}{d \log a_{I^-}} = \theta_{crit} \frac{r_A}{r_I} \qquad (25)$$

which has the form of Equation 15.

The value of $E_{active} \cong -0.620$ *mV vs. Ag/AgCl* is not too different from the thermodynamic value of -0.590 V calculated by Pourbaix[87]. Pourbaix's value is based on the assumption that the internal contents of a stress-corrosion crack, corroding crevice, or corrosion pit on iron consists of magnetite-covered iron in contact with the local electrolyte, a saturated solution of ferrous chloride.

In the case of pitting corrosion on open surfaces, the characteristic electrode potential is the pitting potential, which is measured above the open surface. At potentials below (less positive than) the pitting potential, pits do not initiate; but above the pitting potential, pits grow. Figure 32 shows anodic polarization curves for iron in 0.003 M $CrO_4^=$ with varying amounts of chloride ion. With increased chloride concentration, pitting occurs more easily, i.e., at a lower positive potential.

Figure 33 shows that the relationship between the pitting potential and chloride ion activity is:

$$\left(\frac{d\ E_{pit}}{d\ log\ a_{Cl^-}}\right)_{a_{CrO_4^=}} = const' \qquad (26)$$

This relationship can also be explained by the competitive adsorption model in which each ion adsorbs according to the Temkin isotherm. In this case, E_{active} in Equation 23 should be replaced by E_{pit}, which is a mixed potential between the potentials of the developing pit and the otherwise passive surface. With this modification, Equation 23 gives:

$$\left(\frac{d\ E_{pit}}{d\ log\ a_{A^-}}\right)_{a_{I^-}} = \frac{2.303\ RT}{F} \cdot \frac{1}{\theta_{crit}\left(\frac{r_A}{r_I}\right)\gamma_I z_I - \gamma_A z_A}, \qquad (27)$$

which has the form of Equation 26.

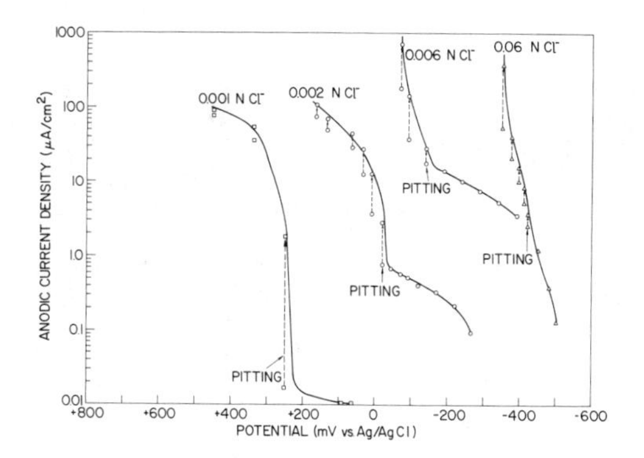

Fig. 32.—Anodic polarization curves for open iron in 0.003 M Na_2CrO_4. Pitting occurs with marked increases in current.

To summarize this section, quantitative relationships which hold for two different forms of localized corrosion can be explained on the basis of competitive adsorption of aggressive and inhibitive ions. The first relationship is that a certain minimum activity of chromate is required to inhibit crevice corrosion for a given activity of chloride. The second is that the pitting potential on open surfaces is related in a well-defined manner to the activity of aggresive ion.

SUMMARY

The adsorption of corrosion inhibitors is an important process in the protection of metals in both acidic and neutral environments. Mechanisms of inhibition differ in the two types of solutions. In acidic solutions, air-formed oxide films dissolve away so that the inhibitor interacts directly with the cleaned metal surface; whereas in neutral solutions, air-formed oxide films are retained.

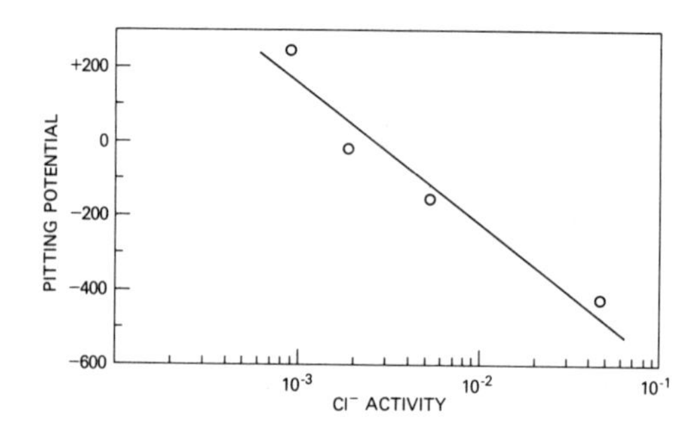

Fig. 33.—The pitting potentials in Figure 32 are related in a regular manner to the chloride activity.

With many organic inhibitors in acid solutions, the chemisorption step is all that is required to provide corrosion inhibition. Among the factors affecting chemisorption of inhibitors from acid solutions are the molecular structure of the inhibitor, the inhibitor solubility, the ability of the inhibitor to displace adsorbed water from the interface, competitive effects of other surface active species, and the structure of the electrical double layer. A detailed case study of organic diamines showed that the most effective members of the series were medium-sized diamines which adsorb via both end groups in the flat configuration. Studies with fluorinated aliphatic compounds showed that a partially fluorinated chain is needed to balance the effects of increased hydrophobicity and decreased electron donation.

Adsorption of inhibitors onto oxide-covered surfaces in neutral solutions is usually followed by additional necessary steps. These may be chemical reactions, as with oxidizing inhibitors such as chromate, or surface exchange reactions, as with non-oxidizing inhibitors. The role of adsorption has been discussed regarding the use of surface chelating inhibitors and in the inhibition of localized corrosion. In the latter case, quantitative relationships have been developed to explain the role of chromate in inhibiting the crevice corrosion and pitting of iron on the basis of competitive adsorption between chromate and chloride ions.

ACKNOWLEDGEMENT

Part of this work was supported by the Office of Naval Research, to whom the author is grateful. The skill and patience of Mrs. Sylvia V. Smith in the preparation of this manuscript is especially appreciated.

REFERENCES

1. Koehler, E.L., "Localized Corrosion," NACE, Houston, Texas, 1974, p. 117.
2. Svoboda, M. and Mleziva, J., *Progr. in Org. Coatings 2, 207 (1973/74).*
3. Leidheiser, H. Jr. and Kendig, M.W., *Corrosion 32, 69 (1976).*
4. Kendig, M.W. and Leidheiser, H. Jr., *J. Electrochem. Soc. 123, 982 (1976).*
5. Brown, B.F., *Corrosion 26, 249 (1970).*
6. Kubachewski, O. and Brasher, D.M., *Trans. Faraday Soc. 55, 1200 (1959).*
7. Brasher, D.M. and Kingsbury, A.H., *Trans. Faraday Soc. 54, 1214 (1958).*
8. Powers, R.A. and Hackerman, N., *J. Electrochem. Soc. 100, 314 (1953).*
9. Thomas, J.G.N., *Br. Corrosion J. 5, 41 (1970).*
10. Hackerman, N. and Snavely, E.S., "NACE Basic Corrosion Course," Chapt. 9, NACE, Houston, 1969.
11. Zembura, Z., Ziolkowska, W. and Kolny, H., *Bull. Acad. Polon. Sci., Ser. sci. chim. 13, 487 (1965).*
12. Dahlberg, E.P., Scanning Electron Microscopy/1974," IIT Research Institute, Chicago, 1974, p. 911.
13. Hackerman, N. and Makrides, A.C., *Ind. Eng. Chem. 46, 523 (1954).*
14. Hackerman, N., *Corrosion 18, 332t (1962).*
15. Ayres, R.C. Jr. and Hackerman, N., *J. Electrochem. Soc. 110, 507 (1963).*
16. Hackerman, N. and Roebuck, A.H., *Ind. Eng. Chem. 46, 1481 (1954).*
17. Makrides, A.C. and Hackerman, N., *Ind. Eng. Chem. 47, 1773 (1955).*
18. Bockris, J. O'M. and Swinkels, D.A.J., *J. Electrochem. Soc. 111, 736 (1964).*
19. Newman, D.S., McCarthy, J. and Heckaman, M., *J. Electrochem. Soc. 118, 541 (1971).*
20. Poling, G.W., *J. Electrochem. Soc. 114, 1209 (1967).*
21. Friberg, S. and Muller, H., "5th Scandinavian Corrosion Congress," Chapt. 5, Copenhagen, 1968.
22. Szklarska-Smialowska, Z. and Dus, B., *Corrosion 23, 131 (1967).*
23. McCafferty, E. and Hackerman, N., *J. Electrochem. Soc. 119, 146 (1972).*
24. Koudelkova, M., Augustynski, J. and Berthou, H., *J. Electrochem. Soc. 124, 1165 (1977).*
25. Wood, J.D., *Corrosion 34, 70 (1978).*
26. Bailey, R. and Castle, J.E., *J. Materials Sci. 12, 2049 (1977).*
27. Lumsden, J.B. and Szklarska-Smialowska, Z., *Corrosion 34, 169 (1978).*
28. Yu Yao, Y. -F., *J. Phys. Chem. 67, 2055 (1963).*
29. Piersma, B.J., "Electrosorption," E. Gileadi, Ed., Plenum Press, New York, 1967, p. 19.
30. Szklarska-Smialowska, Z. and Wieczorek, G., *Corrosion Sci. 11, 843 (1971).*
31. Kelly, E.J., *J. Electrochem. Soc. 112, 124 (1965).*
32. McCafferty, E. and Hackerman, N., *J. Electrochem. Soc. 119, 999 (1972).*
33. Darwish, N.A., Hilbert, F., Lorenz, W.J. and Rosswag, H., *Electrochim. Acta. 18, 421 (1973).*
34. Kelly, E.J., *J. Electrochem. Soc. 115, 1111 (1968).*
35. McCafferty, E. and Hackerman, N., *J. Electrochem. Soc. 120, 774 (1973).*
36. Hackerman, N., Snavely, E.S. Jr. and Payne, J.S. Jr., *J. Electrochem. Soc. 113, 677 (1966).*
37. Iofa, Z.A., Nikoforova, Y.A. and Batrakov, V.V. "Proceedings of the Third International Congress on Metallic Corrosion," Moscow, 1966, Vol. 2, p. 40, (1969).
38. Cavallaro, L. Felloni, L., Trabanelli, G. and Pulidori, E., *Electrochim. Acta. 9, 485 (1964).*
39. Hackerman, N. and McCafferty,"Proceedings of the Fifth International Congress on Metallic Corrosion,"NACE, 1974, p. 542.
40. Antropov, L.I., "First International Congress on Metallic Corrosion," Butterworths, London, 1962, p. 147.

41. Perkins, R.S. and Andersen, T.N., "Modern Aspects of Electrochemistry," No. 5, J. O'M. Bockris and B.E. Conway, Eds., Plenum Press, New York, 1969, p. 203.
42. Donahue, F.M. and Nobe, K., *J. Electrochem. Soc. 112, 886 (1965).*
43. Akiyama, A. and Nobe, K., *J. Electrochem. Soc. 117, 999 (1970).*
44. Eldakar, N. and Nobe, K., *Corrosion 33, 128 (1977).*
45. Annand, R.R., Hurd, R.M. and Hackerman, N., *J. Electrochem. Soc. 112, 144 (1965).*
46. Hackerman, N., Hurd, R.M. and Annand, R.R., *Corrosion 18, 37t (1962).*
47. Hackerman, N. and Hurd, R.M., "Proceedings of the First International Congress on Metallic Corrosion," Butterworths, London, 1962, p. 166.
48. Aramaki, K. and Hackerman, N., *J. Electrochem. Soc. 115, 1007 (1968); 116, 568 (1969).*
49. Salem, R.R., *Russ. J. Phys. Chem. 43, 1615 (1969).*
50. Shafrin, E.G. and Zisman, W.A. in "Monomolecular Layers," H. Sobotka, Ed., AAAS, Washington, D.C., 1954, p. 129.
51. Nikonov, V.Z. and Sokolov, L.B., *Russ. J. Phys. Chem. 43, 581 (1969).*
52. Kaganovich, R.I., Damaskin, B.B. and Andrusev, M.M., *Soviet Electrochem. 5, 695 (1969).*
53. Kaganovich, R.I., Damaskin, B.B. and Ganshina, I.M., *Soviet Electrochem. 4, 784 (1968).*
54. Grigorev, N.B. and Machavariani, D.N., *Soviet Electrochem. 5, 1270 (1969).*
55. McCafferty, E.,"Rept. of NRL Progress,"May 1974, p. 48.
56. Shafrin, E.G. and Zisman, W.A., "Upper Limits for the Contact Angles of Liquids on Solids, " NRL Report 5985, Naval Research Laboratory, Washington, D.C., 1963.
57. McCafferty, E., "Corrosion Inhibition by Hydrophobing Compounds," Paper No. 72, presented at Corrosion/74, Chicago, March 4-8, 1974.
58. McCafferty, E. and Harvey, M.K., "Rept. of NRL Progress," Nov. 1976, p. 23.
59. Riggs, O.L. in "Corrosion Inhibitors, " C.C. Nathan, Ed., NACE, Houston, Texas, 1973, p. 7.
60. McCafferty, E. and Zettlemoyer, A.C., *Discussions Faraday Soc. 52, 239 (1971).*
61. Zettlemoyer, A.C. and McCafferty, E., *Croatica Chem. Acta. 45, 173 (1973).*
62. Thomas, J.G.N., *Br. Corrosion J. 1, 156 (1966).*
63. Mayne, J.E.O. and Page, G.L., *Br. Corrsion J. 7, 115 (1972).*
64. Evans, U.R. "The Corrosion and Oxidation of Metals: First Supplementary Volume," St. Martin's Press, New York, 1968, p. 64.
65. Evans, U.R., "The Corrosion and Oxidation of Metals: Second Supplementary Volume, Edward Arnold, London, 1976, p. 94.
66. Thomas, J.G.N. in "Corrosion," L.L. Shreir, Ed., Newnes-Butterworths, London, 1976, Vol 2, pp. 18-34.
67. Brasher, D.M., Reichenberg, D. and Mercer, A.D., *Br. Corrosion J. 3, 144 (1968) (See references therein to earlier papers in this series).*
68. Gouda, V.K. and Sayed, S.M., *Corrosion Sci, 13, 841 (1973).*
69. Leckie, H.P. and Uhilg, H.H., *J. Electrochem. Soc. 113, 1262 (1966).*
70. Vijh, A.K., *Corrosion Sci, 12, 105 (1972).*
71. Vijh, A.K., *J. Electrochem. Soc. 119, 1498 (1972).*
72. Vijh, A.K., "Electrochemistry of Metals and Semiconductors," Marcel Dekker, New York, 1973, p. 167.
73. Weisstuch, A., Carter, D.A. and Nathan, C.C., *Mater. Prot. Per. 10(4), 11 (1971).*
74. Zecher, D.C., *Mat. Per. 15(4),33 (1976).*
75. LeRoy, R.L., *Corrosion 34, 98 (1978).*
76. Rajan, K.S. and Ase, P.K., "Development of Corrosion Inhibitors and Adhesion Promoters," National Technical Information Service Report AD-770 627 (1973).

77. Tirbonod, F. and Fiaud, C., *Corrosion Sci. 18, 139 (1978).*
78. Hayakawa, Y. and Ida, I., *J. Electrochem. Soc. Japan, (Overseas Ed.) 26, (10-12), E-197 (1958).*
79. Dwyer, F.P. and Mellor, D.P., "Chelating Agents and Metal Chelates," Academic Press, New York, 1964.
80. Klyuchnikov, N.G. and Uchenina, V.F., "Uch. Zap. Mosk. Gos. Pedegog. Inst.," Defense Research Institute Translation No. 3079, April 1973, p. 214 (1969).
81. Ushenina, V.F. and Klyuchnikov, N.G., *J. Appl. Chem. USSR 44, 185 (1971).*
82. McCafferty, E., *J. Electrochem. Soc. 121, 1008 (1974).*
83. McCafferty, "Corrosion of Iron in Crevices, " NRL Report 7781, Naval Research Laboratory, Washington, D.C., 1974.
84. Matsuda, S. and Uhlig, H.H., *J. Electrochem. Soc. 111, 156 (1964).*
85. Strehblow, H.H. and Titze, B., *Corrosion Sci. 17, 461 (1977).*
86. Gileadi, E., Kirowa-Eisner, E. and Penciner, J., "Interfacial Electrochemistry," Addison-Wesley, Reading, MA, 1975, p. 78ff.
87. Pourbaix, M., "Localized Corrosion," NACE, Houston, Texas, 1974, p. 12.

THE MECHANISMS OF CORROSION PREVENTION BY INHIBITORS IN PAINTS

J. David Scantlebury

Corrosion and Protection Centre
University of Manchester Institute
of Science and Technology,
England

J. D. SCANTLEBURY

ABSTRACT

This paper begins with a brief summary of the types of aqueous inhibitor which have been investigated and the mechanisms which have been put forward to explain their successful behavior. Inhibitive materials in coatings will then be discussed with a review of typical inhibitive compounds with the emphasis on the mechanism of inhibition. A preliminary investigation into the incorporation of novel inhibitive materials into paints will finally be described.

The theories proposed to explain the mechanisms of inhibitor action in paints rely heavily on the theories proposed for the mechanism of corrosion inhibition in aqueous solution. It is therefore instructive and essential to our understanding of the mechanism of inhibitor action in paints to look at the various sorts of inhibitor which are used in aqueous systems and the theories which have been proposed to explain them.

ACID INHIBITORS

Aqueous inhibitors may be conveniently divided into acid and neutral. Acid corrosion and its inhibition may be characterized by the following features, namely the presence of the corroding metal which is free from its oxide film and the major cathodic corrosion process being the discharge of the proton. Acid corrosion inhibitors are almost invariably organic and either possess unsaturated linkages or contain elements which have lone pair electrons, e.g., elements from Groups V and VI. The effectiveness of a particular metal/inhibitor combination is governed by a variety of factors, *inter alia,* the zero charge potential of the metal, the molecular structure of the inhibitor and the capacity of the adsorbed inhibitor to be reduced on the metal surface. The zero charge potential has been investigated by Antropov[1]. He suggested that since the initial approach of the inhibitor molecule to the metal surface involves an adsorption process then the electrical charge on the metal solution interface determines whether anionic or cationic groups may be adsorbed. As well as electrostatic adsorption electron transfer may well occur between the inhibitor molecule and the metal. Consequently, inhibitor structure is directly linked with the ability of the inhibitor to donate electrons. In general, substituent groups which increase the electron density at a molecule increase the probability of electron transfer and hence increase inhibitor efficiency.[2]

A variety of theories have been proposed to explain why the presence of an adsorbed layer of organic molecules on a bare metal surface should decrease its rate of corrosion compared with the metal surface without inhibitor. Some workers[3] have suggested that certain inhibitors may form or decompose into thin films which act as a very thin coating and restrict the access of ions to the metal surface, in which case a mechanism similar to that proposed by Mayne[4] may be thought to be applicable. Hoar[5] has discussed the process of acid inhibition, especially at low inhibitor levels in terms of the blocking of active anodic sites, Russian workers[6, 7] have suggested that the rate determining step in the acid corrosion of mild steel is the combination of hydrogen ad-atoms to form hydrogen molecules. Furthermore, the function of certain inhibitors is to retard the production of the hydrogen ad-atoms.

NEUTRAL INHIBITORS

The simplist inhibitors to visualize and understand are those known as cathodic inhibitors.[8] They are frequently divalent metal cations which are thought to migrate to cathodic areas on the metal surface and under the alkali conditions found there, precipitate as insoluble compounds, thereby screening the cathodic areas from oxygen and water. It is interesting to note that divalent cations have become more popular as possible modifications to anti corrosive additives in paints. Invariably if anionic compounds have been considered, the next investigating step seems to be to look at zinc, calcium or magnesium salts.[9]

The other types of neutral inhibitor operate by maintaining or even strengthening a film, usually oxide, on the metal surface. Furthermore, the process of inhibition is accompanied by an ennoblement[10] of potential and in the case of iron and steel into the region of stability of iron.[III] These inhibitors are usually weak acid anions which either need some quantity of oxygen in solution or themselves possess some oxidative capacity. Non oxidizing inhibitors appear to work by plugging weak points in oxide films and thereby encouraging film repair.[11] Oxidizing inhibitors, however, appear to operate by being reduced in an extra cathodic reaction and being precipitated on the metal surface and reinforcing the air-formed oxide film. The mechanism has been summarized[12] by Stern.

Many typical anti-corrosive pigments and additives in paints are based on oxidizing and non-oxidizing anionic inhibitors; chromates and phosphates have been used for many years and more recently metaborates and molybdates have come to the fore.

INHIBITIVE PIGMENTS AND ADDITIVES IN PAINTS

The classical mechanisms developed and expounded by Mayne and his coworkers[13] for red lead pigments in drying oil vehicles involve the presence in the drying oil of degradation acid products, especially azelaic acid which not only is inhibitive in its own right but as a lead soap provides outstanding inhibitive qualities in aqueous solution. Other soaps of azaleic acid, namely, zinc, magnesium, and calcium have also been investigated.[9]

Modern industrial paint finishes, however, are frequently formulated not to involve air-drying processes and consequently there is an absence of acid anionic degradation products; chlorinated rubber, chlorinated poly (proplylene) and epoxys are obvious examples. We cannot, therefore, rely on soap formation to provide successful inhibition, and must rely on pigments which, in their own right are inhibitive to corrosion. This model proposes a uniform distribution of pigment in the vehicle and on water permeation into the coating film, small quantities of inhibitive material dissolving in this water and permeating through to the metal surface. At the metal surface the inhibitive material is thought to modify the oxide film on the metal and make the film more protective. Evidence for this mechanism has recently been produced[14] where iron in a zinc chromate pigmented acrylate film, after exposure and film stripping has been shown to have present on the surface Cr^{III}. A rapid assessment of the anti-corrosive properties of a paint coating has been proposed by Dévay and his colleagues[15] whereby the paint coating after exposure had been removed and the reactivity, or otherwise, of the metal substrate was assessed by conventional polarization resistance techniques.

An interesting example of the combination of inhibitor studies and coating science leading to a successful organic coating may be seen with benzotriazole (BTA) and copper. The mechanisms whereby BTA inhibits the corrosion of copper in aqueous solutions has been extensively studied and a comprehensive summary has been given by Ashworth.[16] They concluded that BTA acted as a mixed inhibitor in neutral and acid solutions and furthermore in neutral solutions BTA was electroreducible. Not only is BTA well used as a research tool but also it is extensively used commercially as an additive to central heating systems. BTA has also been developed as an additive to a lacquer by the British Non-Ferrous Metals Research Association. They have formulated a lacquer for copper and its alloys based on an acrylate system. The BTA is dissolved in the solvent and on drying the BTA is precipitated together with the lacquer. This lacquer has been shown to be very successful with copper alloys and this development may well point the way future work may proceed with new protective coatings for other metals.

Many additives in paints have been proposed that do not necessarily possess an inhibitive capacity but merely are capable of reacting with aggressive ions and tying them down as insoluble and hence less aggressive species. Compounds which form insoluble chlorides and sulphates have frequently been suggested.[8] Indeed a further reason why red lead systems are so successful is probably due to the low solubilities of lead sulphate and chloride.

EXPERIMENTAL INVESTIGATION

It is appropriate at this stage to report some interim findings from an investigation which is being carried out at UMIST.[17] This investigation is concerned with the incorporation of novel inhibitive materials into a typical coating and assessing, using conventional accelerated tests and atmospheric exposure techniques the effects, if any, these additives have on coating performance. The purpose of such an investigation is to look for possible alternatives to the heavy metal pigment systems in common use, which would be environmentally more acceptable and which would possess adequate or superior anti-corrosive properties. Furthermore, the investigation would lead to an increase in understanding of the mechanisms whereby similar additives would prevent corrosion.

Six compounds were chosen for this study.

(1) substituted benzimidazole (SBA)
(2) benzotriazole (BTA)
(3) sodium benzoate
(4) sodium dihydrogen phosphate
(5) sodium azelate
(6) azelaic acid

The last two compounds were chosen because of their links with the previous work carried out on dr~~ying~~ oils.

The first heterocyclic compound was chosen since previous work had indicated that such compounds possessed a strong capacity to coordinate with dissolved iron. BTA was chosen since previous work[16] had shown this molecule to be mildly inhibitive to the corrosion of mild steel and also much was known about the chemistry and interaction between BTA and copper. Sodium benzoate and dihydrogen phosphate were chosen as being typical of those class of compounds referred to before as non-oxidizing neutral inhibitors. The precise formula of the substituted benzotriazole will not be revealed since because of its excellent behaviour in the tests to be described, it is the subject of possible commercial exploitation. An initial test in aqueous solution ($0.01M\ Na_2SO_4 + 100\ ppm\ Cl^-$) of the inhibitive qualities of these compounds on (*25mm x 25mm*) grit blasted mild steel was carried out. Inhibitor concentrations used were 50, 25, and 10m Mol dm^{-3} for the readily soluble inhibitors (2,3,4,5) and saturation, 1/3 and 2/3 saturation for the slightly soluble inhibitors (1,6). Weight losses in grams after thirty days immersion are given in the following table:

Concentration *Inhibitor*	*Low*	*Medium*	*High*
1	0.95	0.90	0.87
2	0.25	0.10	0.10
3	1.2	0.75	0.70
4	0.1	0.14	0.12
5	0.80	0.75	0.01
6	0.45	0.80	0.90
—	—	1.00	—

Several interesting points are worth noting. Firstly, the slightly soluble inhibitors have very little effect in aqueous solution. Secondly the highly inhibitive nature of sodium azelate at the highest concentration is surprising in view of the results with phosphate and benzoate.

The compounds were then incorporated into a vinyl copolymer lacquer, 86% vinyl chloride, 13% vinyl acetate and 1% maleic anhydride. An inhibitor concentration of 0.048M/100g of resin was chosen and was dissolved or dispersed in the solvent, methyl isobutyl ketone. The coatings were applied to grit blasted steel panels to give a dry film thickness of 50μm. The coatings were assessed by three separate exposure methods,

(a) exterior exposure in Manchester (three sites) for three months to one year.
(b) Salt spray testing.
(c) Salt water immersion.

An electrochemical investigation is also underway into the same coating systems but the results are not yet complete.

The salt spray testing was carried out for 350 hours at 50°C and the salt water testing was carried out for 1167 hours in *0.01M* Na_2SO_4 *+ 1000ppm* Cl^-. The solution was changed weekly to minimize interference from inhibitors leaking into the solution. The results of the exposure tests are summarized in the following table. The figures range from 10 — excellent to 1 — complete breakdown.

Exposure Inhibitor	*Exterior Exposure*			*Salt Spray*	*Immersion*
	Site 1	*Site 2*	*Site 3*		
1	10	10	10	10	10
2	4	9	3	8	10
3	8	8	6	4	6
4	2	6	4	6	5
5	8	7	7	4	3
6	6	4	5	2	3
blank	2	2	2	8	10

Clearly there are some interesting features about these tests which are worthy of note.

1. The blank experiment shows the importance of carrying out exposure testing as well as artificially accelerated tests. Whereas in the exposure tests the presence of any one of the inhibitors improves the coating corrosion resistance compared with the blank, in the accelerated test, the blank behaves significantly better than many of the other systems.

2. System 1 (SBA), even though in aqueous solution was hardly inhibitive to bare steel, when incorporated into the lacquer, exhibited outstanding behaviour. This seems to suggest that the model proposed previously whereby the inhibitor operates in a paint by leaching into the small amount of water present in the coating and conferring to that water an inhibitive character, may well be incorrect. It may be that direct adsorption of the inhibitor takes place at the coating metal interface thereby reducing the overall activity of the metal surface.

CONCLUSIONS

Present methods of anti-corrosion technology when applied to anti-corrosive coatings are both crude and inefficient. Invariably, the primer coat is loaded with large quantities of inhibitive pigment, most of which will either remain locked in the polymer and therefore unused or will be lost into the environment. This investigation has shown that to provide an effective inhibiting system, only small quantities of active compound need to be present in the coating system, the essential feature being to arrange that this compound be deposited at the right place, namely the metal surface, during the drying process. Clearly the possible different systems which could be tried are considerable and the rate determining step in the investigation no longer becomes one of producing a new coating but of assessment and the development of new methods for the rapid assessment of coating performance will continue to intrigue corrosion workers for a long time to come.

REFERENCES

1. Antropov, L.I., "Proc. First Int. Cong. on Metallic Corrosion," London, 1961.
2. Hackerman, N., *Corrosion 18, 332 (1962)*.
3. Machu, W., "Proc. First European Sym. on Corrosion Inhibitors," Ferrara, 1960.
4. Mayne, J.E.O., *Chem. Ind., 293 (1951)*.
5. Hoar, T.P. and Holliday, R.D., *J. Applied Chem. 3, 502 (1953)*.
6. Antropov, L.I. and Savgira, Y.A., *Prot. Met. 3, 597 (1967)*.
7. Grigoryev, V.P. and Osipov, O.A., *Prot. Met. 16, 473 (1971)*.
8. Evans, U.R., "The Corrosion and Oxidation of Metals,"Edward Arnold, London, 1960.
9. Heyes, P.J. and Mayne, J.E.O., "Proc. Sixth European Cong. on Metallic Corrosion," London, 1977.
10. Hancock, P. and Mayne, J.E.O., *J. Chem. Soc. 9, 345 (1959)*.
11. Mayne, J.E.O. and Page, C.L., *Br. Corros. J. 7, 115 (1972)*.
12. Stern, M., *J. Electrochem. Soc. 105, 635 (1958)*.
13. Mayne, J.E.O., "Corrosion," L.L. Shreir, Ed., Newnes-Butterworth, London and Boston, 1976.
14. Defossé, C. and Piens, M., "Proc. Fourteenth FATIPEC Cong.," Budapest, 1978.
15. Dévay, D., Mészáros, L. and Janászik, F., "Proc. Fourteenth FATIPEC Cong.," Budapest, 1978.
16. Ashworth, V., Kokoszka, C.L. and Tohala, N.H., "Proc. Sixth Int. Cong. on Metallic Corrosion," Sydney, 1975.
17. Johnson, W., Ashworth, V. and Scantlebury, J.D., unpublished work.

CORROSION-INHIBITIVE SEALANTS AND PRIMERS

Dr. Robert N. Miller

Lockheed-Georgia Company,
Marietta, GA

ROBERT N. MILLER

ABSTRACT

The issuance of Military Specification MIL-S-81733, "Corrosion Inhibitive Sealing and Coating Compounds," by the Navy and MIL-P-87112, "Corrosion Inhibiting Polysulfide Primer," by the Air Force has been a giant step toward the alleviation of major aircraft corrosion and maintenance problems. The chromate-inhibited polysulfide materials prevent exfoliation corrosion of fastener holes on wing and fuselage surfaces, faying surface corrosion between adjacent exterior panels, and extend the life of coating systems. Originally developed as a sealant by the Lockheed-Georgia Company, the corrosion inhibitive polysulfides were reduced in viscosity to make a very effective primer. The new primer, with a topcoat of MIL-C-83286 polyurethane, was applied to 50 B-52's of the Strategic Air Command in April of 1974. After 18 months the corrosion maintenance manhours required for the aircraft with the PR-1432GP primer was 80% less than was required for the aircraft with the MIL-P-23377 epoxy polyamide primer. The aircraft with the new coating system are still in good condition — more than four years after they were painted. Presentation includes a discussion of the role of the chromate inhibitors, results of laboratory tests, a procedure for the repair of corroded aircraft surfaces, and the results of field service tests on C-130, C-141, and B-52 aircraft.

INTRODUCTION

Within the last two years the Navy has issued a specification for a corrosion-inhibitive sealing and coating compound MIL-S-81733 (AS)[1] and the Air Force has a new specification for a corrosion-inhibitive primer, MIL-P-87112.[2] Both are for chromate-containing polysulfide elastomers. Why should this particular combination be so effective in minimizing corrosion? I would like to answer this question by sharing some of the highlights of the program which developed these unique corrosion-preventive compounds.

Chromates have been used as corrosion inhibitors in coatings and in cooling water for many years. One reason for their importance is their versatility in minimizing corrosion of both ferrous and nonferrous metals. They effectively protect steel and aluminum, singly or in combination. Figure 1 shows test specimens of 7075-76 aluminum, coupled with bare steel fasteners, which have been immersed in water for 18 months. The control specimen at the left is heavily rusted and a flocculant layer of ferric hydroxide has accumulated at the bottom of the jar. The specimen at the right was protected by a chromate inhibitor. Not only is the aluminum panel in perfect condition, but the steel fastener also is completely free from corrosion.

Fig. 1.—Steel fasteners in aluminum sheet after 18 months immersion. Specimen on right was protected by inhibitor.

Chromates are oxidizing agents but factors other than oxidizing power seem to be involved in their action. The physical and electrical properties of the oxide film formed on the surface of the metal is more important than the composition of the film. The results of a radioisotope study by Cohen and Beck[3] suggest that on iron, film formation proceeds in two steps; first, the formation of a relatively soluble ferrous hydroxide and, second, its conversion to ferric oxide by the action of the chromates. The passivity is attributed to a film at least 100 angstroms thick containing about 25 per cent CrO_3, the remainder being Fe_2O_3.

In the case of aluminum, the chromates function as anodic inhibitors. The negatively charged CrO_4 ions are attracted by the positively charged anode areas on the aluminum and coat them with a film of high electrical resistance. This film minimizes the flow of galvanic currents between the dissimilar phases in the microstructure of the aluminum alloy.

Work done by Lockheed-Georgia has demonstrated that the concentration of chromate ions is important.[4] At least 20 ppm are necessary for good protection against corrosion. Lower concentrations actually result in a greater corrosion rate than if no chromates were present.

I first became involved in the development of aircraft sealants and coatings in the early 1960's when we were having serious corrosion problems on the upper center wing of our Lockheed C-130's which were stationed at Okinawa and other bases in close proximity to salt water. Two types of corrosion were involved. The first type, exfoliation corrosion, is illustrated in Figures 2 and 3. It is a delaminating type of corrosion which originates at the fastener holes. Despite the fact that aluminum alloy surfaces are protected by anodizing or conversion coatings, the inside of the holes are susceptible to attack because the holes are drilled after the protective coating has been applied.

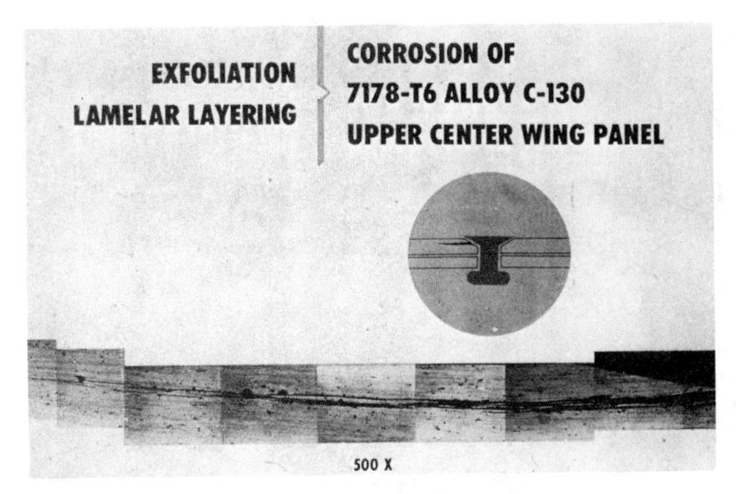

Fig. 2.—Cross section of exfoliated aircraft panel.

Fig. 3.—Cross section of edge of fastener hole showing wedgelike action of corrosion products.

A contributing factor to this corrosion problem is the presence of elements other than aluminum in high strength alloys such as Types 7178-T6 and 7075-T6. Because they are not homogeneous, these alloys develop anode and cathode areas when in contact with an electrolyte. Under service conditions electrolytes are present in the form of rain condensation, paint strippers, brighteners, and in coastal areas, sea water carried by the wind. Corrosion usually originates in the countersink areas because that is where initial penetration by the electrolyte occurs. The grain boundaries, being anodic to the grains, are converted to corrosion products. These products occupy more space than the original metal and exert tremendous pressures with subsequent expansion within the corroded area. The corrosion then follows a lamellar pattern along the grain boundaries and destroys the structural integrity of the metal. Figure 4 shows a section of an aircraft wing with exfoliation blisters adjacent to the fastener holes.

Fig. 4.—Exfoliation caused by expanding corrosion products beneath the metal surface.

Faying surface corrosion, illustrated in Figure 5, has a similar corrosion mechanism. Like the fastener hole corrosion, it occurs when electrolytes penetrate the space between adjacent wing and fuselage sections. This type of corrosion is especially serious because severe corrosion can cause a loss of structural strength.

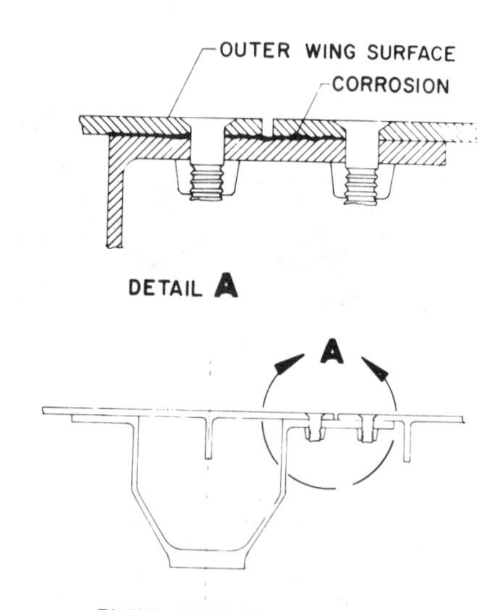

Fig. 5.—Faying surface corrosion between adjacent exterior panels.

DEVELOPMENT OF INHIBITIVE SEALANT

In 1961 the Lockheed-Georgia Company initiated a program to develop methods for preventing exfoliation and faying surface corrosion on aircraft. An obvious possible solution was to keep water and other liquids out of the fastener holes and wing joints by filling the space with sealants or other compounds. A series of tests was conducted in which rivets were wet-treated with polysulfide sealing compound or zinc chromate primer (MIL-P-8585) and mounted in 7075-T6 and 7178-T6 aluminum panels which were then subjected to simulated service conditions. The panels were exposed to a total of 1440 hours of salt spray, 896 hours of 100 per cent humidity, and 260 loading cycles to a surface stress of 30,000 psi. The results of the tests showed that the polysulfide sealant performed better than did the zinc chromate primer but neither material provided complete protection against exfoliation corrosion. Under the sustained high humidity conditions moisture eventually penetrated to the inside surface of the fastener holes and some corrosion occurred. Field service tests on a C-130 aircraft gave identical results. At this point it was realized that the moisture was not merely seeping between the sealants and the sides of the fastener holes, but was penetrating the test compounds. Laboratory tests on thin films of various types of polymers demonstrated that all sealants and polymers are permeable to moisture to some degree.

Figure 6 summarizes the results of water permeability tests on 3-mil thick films of Buna N, fluorocarbon rubber, silicone rubber, and polyurethane by ASTM Method E96-53T. The test films were placed over the top of

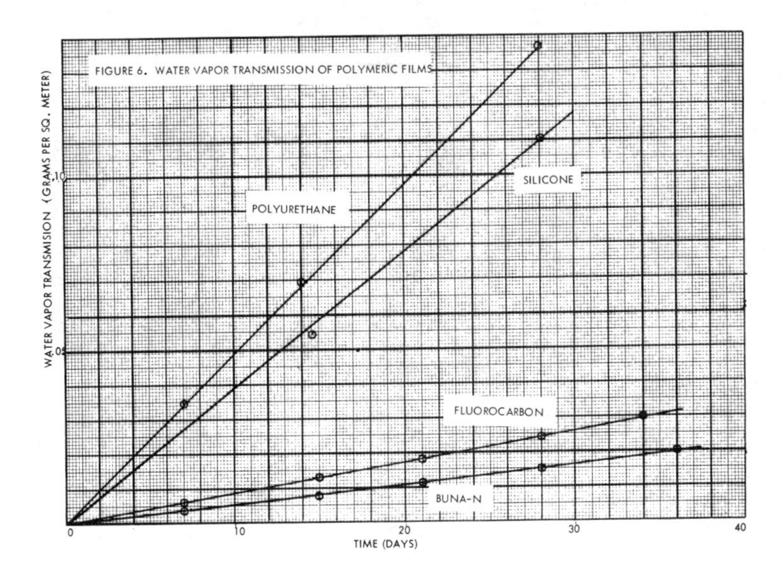

Fig. 6.—Water vapor transmission of polymeric films.

wide-mouth containers partially filled with water. The containers were placed in a humidity chamber at a temperature of 86°F and a relative humidity of 35 per cent and were weighed every seven days to determine the water vapor transmission through the films. Although there was considerable variation in the permeabilities of the various films, the most significant feature of the results is that all the films were permeable to moisture. This indicates that, under prolonged high humidity conditions, a sealing compound can slow down and minimize corrosion reactions but cannot completely prevent them.

The Lockheed-Georgia research team* then conceived the idea of turning this weakness of elastomers to an asset by incorporating a water-soluble inhibitor which dissolves in any moisture penetrating to the sealant-metal interface and converts it into a protective solution. Thirty-two potential inhibitors were investigated by means of potentiostatic polarization studies and by long-term immersion tests. Of the compounds tested, it was found that certain soluble chromates were the most effective. The specific inhibitors tested are listed in Table I.

TABLE I.—CORROSION INHIBITORS INVESTIGATED

Chromic Acid	Sodium Silicofluoride
Lithium Chromate	Magnesium Silicofluoride
Magnesium Chromate	Sodium Metasilicate
Potassium Chromate	Sodium Orthosilicate
Sodium Dichromate	Sodium Silicate
Strontium Chromate	Magnesium Silicate
Sodium Tetraborate	Sodium Tungstate
Sodium Molybdate	Sodium Metavanadate
Sodium Nitrite	Sodium Orthovanadate
Sodium Phosphate	Potassium Permanganate
Sodium Pyrophosphate	Ammonium Nitrate

It has been demonstrated, through radioisotope techniques, that the chromate ions are preferentially attracted by the anodic sites in the metal structure and so protect the areas which normally would be the first to corrode. Another good feature of chromate inhibitors is that the protective layer is self-healing. If a break occurs in the protective film it is instantly repaired as long as chromate ions are present in the environment.

The water-soluble chromates were added to polysulfide sealants. The resulting compound provided the reservoir of chromate ions which is necessary for complete protection of metal surfaces. The validity of the concept was then proven by simulated service tests with the inhibitive sealants. Panels of 7075-T6 aluminum, fastened together with sealed rivets and lockbolts, were exposed to a severe environmental test which included exposure to alkaline cleaners, paint strippers, and brighteners. They were then subjected to a fatigue test of 1800 cycles per minute with a 15,000 psi net area stress. Following this, they were exposed to simulated sunlight and rain in a weatherometer and to 5 per cent salt spray for 14 days. At the conclusion of three complete environmental cycles of exposure the control specimens were badly corroded but the insides of the fastener holes which had been protected by the inhibitive sealant were in perfect condition. Figure 7 shows the countersink area of one of these fastener holes.

Additional tests were made by placing buttons of inhibitive and regular sealants on the surface of 7075-T6 aluminum panels and then exposing the panels to 5 per cent salt spray for prolonged time periods. Figure 8

---

*See footnote at end of paper.

Fig. 7.—Countersink area of fastener hole sealed with chromate-inhibited sealant is in perfect condition at conclusion of test.

clearly demonstrates the protective action of the chromate inhibitor in preventing corrosion in the areas adjacent to the buttons.

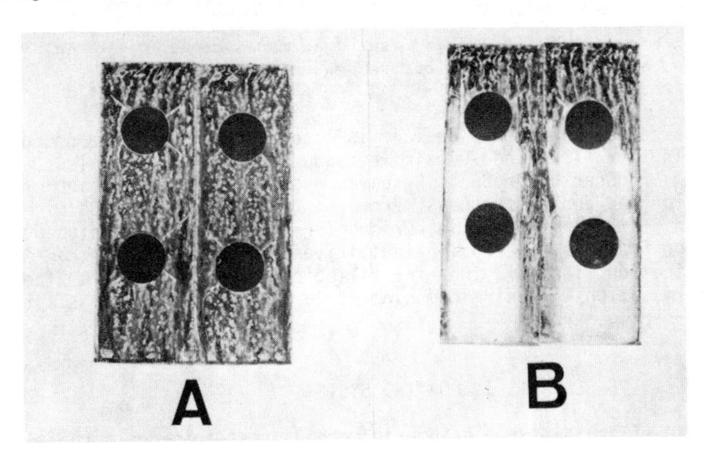

Fig. 8.—Aluminum panel 'A' with buttons of ordinary sealant corroded badly in salt spray test. Buttons of corrosion-inhibitive sealant protected areas both underneath, and adjacent to, sealant buttons on panel 'B.'

The inhibitive sealant was then subjected to field service testing as a part of an Air Force program to evaluate the performance of coated fasteners.[5] Fasteners were installed in an unpainted test section of a C-130 aircraft which has been stationed in the Azores. Fasteners wet-installed with the inhibitive sealant were still in perfect condition after two years of severe exposure to the sunlight and salt water spray of the South Atlantic.

As a result of the successful tests, both in the laboratory and in the field, the corrosion-inhibiting sealant was licensed for commercial production and is now being used on all aircraft produced by the Lockheed Corporation. These include the C-130, JetStar, L-1011, and SA-3 aircraft. The C-5 Galaxy was the first Lockheed aircraft in which the corrosion inhibitive sealant was extensively used. Figure 9 is a view of the C-5 on a test flight.

Fig. 9.—All exterior fasteners and faying surfaces on giant C-5 Galaxy are sealed with corrosion-inhibitive sealant.

The corrosion-inhibitive sealant is also being used in the production of the Boeing 747, 727, and 707 aircraft and the McDonnell-Douglas DC-10. It is being produced by Products Research and Chemical Corporation of Burbank, California, and by the Coast Pro-Seal Division of Essex Chemical Corporation. It is available in various formulations, each designed for specific applications. Corrosion-inhibitive compounds are produced for brush application, for spraying, for filleting, and for faying surfaces. A special formulation with aluminum pigment is available for use as a topcoat on corrosion-prone areas.

USE OF INHIBITIVE SEALANTS IN COATING SYSTEMS

The use of inhibitive elastomeric compounds as coatings paralleled their development and use as sealants. In 1959 some of the C-130 aircraft stationed in the highly corrosive environment of Air Force bases in the South Pacific were having a serious center wing corrosion problem. The zinc chromate primer and acrylic nitrocellulose topcoat then in use did not give adequate protection to the 7178-T6 upper center wing surfaces.

An initial attempt to solve the problem was the selection of a superior finish system from those commercially available but the results were extremely unsatisfactory because of the cracking which occurred around fastener holes as the result of flexing of the wing in flight.

An analysis of the requirments of an ideal coating system dictates the following requirements:

1. The system should have sufficient resiliency to prevent cracking in flexed portions of the aircraft and around fastener heads.

2. It should contain a reservoir of soluble corrosion-inhibitive ions which are present in sufficient quantity to protect the metal against diffused moisture and also to prevent corrosion of bare metal exposed by scratches or gouges in the paint system.

3. The soluble inhibitive ions must be retained on the surface of the metal and should not be depleted by rain and humid weather.

4. Walk-way coatings along the tops of the wings and fuselage must also be resilient to prevent premature failure of the walk-way by cracking. (A brittle coating, over a resilient substrate, will "mud crack" and lose most of its functional and aesthetic qualities.)

5. The coating, to be suitable for use on an aircraft which has experienced corrosion, must be capable of arresting the existing corrosion and preventing any additional degradation of the metal.

A coating system which meets all of these requirements became available when the inhibitive sealant was perfected. A slight reduction in viscosity was the only modification required for its successful use as a coating material. This corrosion-inhibited polysulfide primer is designated PR-1432GP.

Repair of Corroded Aircraft

Following are the details of the coating system and application procedure developed by W. A. Boggs of the Lockheed-Georgia Company for the repair and protection of corroded aircraft surfaces:

(a) Clean and passivate the metal surface with activated conversion coating such as Alodine 1200.

(b) Remove any remaining corrosion (e.g., exfoliation corrosion) by grinding.

(c) Passivate the metal in "grind-out" areas with conversion coating.

(d) Apply chromate-inhibited sealant to all depressions left by metal removal. Fill screw heads with same compound. Level the compound with a spatula or similar tool.

(e) Apply a 5 mil coating of chromate-inhibited sealant to the clean passive metal using special care to assure complete wetting and adhesion.

(f) Apply a 5 mil top coating of aluminized chromate-inhibited sealant to prevent leaching of chromates from the base coat.

(g) Apply elastomeric walk-way coating to areas where anti-slip or abrasion resistance is needed.

This procedure was used to repair and protect a severely corroded upper wing surface of a C-130 aircraft. At the conclusion of two full years of service with the Air Rescue and Retrieval System at Lajes Air Force Base in the Azores, the protective film was stripped from the aircraft wing to enable a thorough inspection of the metal surfaces. The surface of the aluminum and the fastener heads were in perfect condition. Figure 10 is a photograph of the area which was protected by the inhibitive elastomeric coating system. Figure 11 is a view of a control area which had been coated with zinc chromate primer.

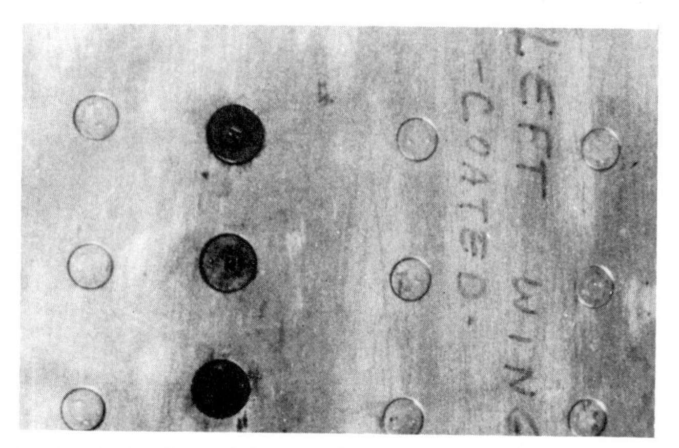

Fig. 10.—Heads of aluminum and steel fasteners on test area which had been coated with inhibitive sealant are in perfect condition.

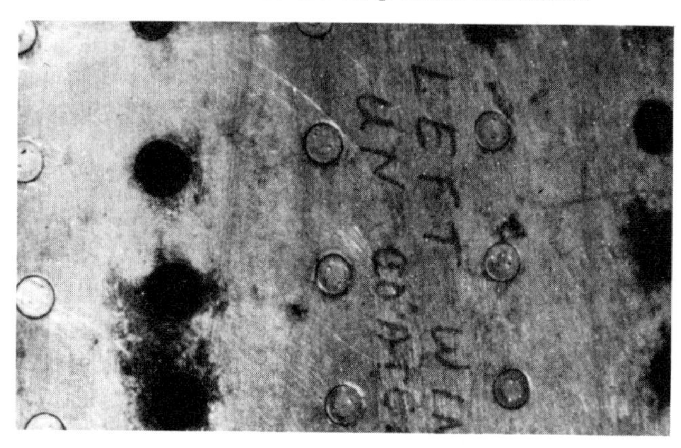

Fig. 11.—Test area which had been coated with zinc chromate primer aluminum panel is exfoliated and heads of steel fasteners are badly rusted.

Use of Corrosion-Inhibitive Polysulfides as Primer

Don Hazen at Warner Robins Air Force Base was also concerned with the cracking and corrosion around the fastener heads on C-141 and C-130 aircraft which came to the base for maintenance and repair operations. Knowing about the successful use of the chromated inhibited polysulfide material in protecting the center wing of the C-130's, he saw no reason why the protection should not be extended to the rest of the aircraft. He used the PR-1432GP corrosion-inhibitive primer coat on several C-130's and C-141's and applied the standard MIL-C-83286 polyurethane topcoat. These aircraft looked so good when they came back for their periodic maintenance checks and Don Hazen was so enthusiastic about the results that the Strategic Air Command painted 50 B-52's with the same coating system. The aircraft surfaces were first given an Alodine 1200 conversion coating treatment. The PR-1432GP chromate-inhibited polysulfide primer was then sprayed onto a wet film thickness of 3.0 to 3.5 mils. This results in a dry film thickness of 1.1 — 1.3 mils. The topcoat consisted of two coats of MIL-C-83286 aliphatic linear polyurethane.

I inspected two of these B-52's at Warner Robins in October of 1975 after they had eighteen months of service and about 500 flight hours. T. C. Merren and Lt. Letterman of Warner Robins had been accumulating data on the performance of the aircraft with the polysulfide primer. They told me that a SAC Commanding Officer wanted to know why some of the aircraft looked so much better than others. A check of the aircraft numbers revealed that the good looking aircraft had the PR-1432GP inhibited primer and had much less touchup after 18 months of service than the B-52's with the standard epoxy primer and polyurethane topcoat coating system. The polysulfide primer system was noticeably superior in resistance to erosion on the leading edges of wings and the air intake of the engines. The protection of fastener areas was excellent. Even in the flexed areas, there was no cracking adjacent to the fastener heads. The horizontal stabilizers of B-52's which had the epoxy-polyamide primer were badly checked and cracked.

A statistical analysis of the maintenance manhours spent on the two types of paint systems is summarized in Table II.

TABLE II.—ANNUAL FIELD MAINTENANCE PER B-52 AIRCRAFT
(Average for 10 Test and 10 Control Aircraft)

	Manhours Corrosion Control Related to Surface	Manhours Painting	Total Manhours
Epoxy System	13.6	39.1	52.7
Polysulfide System	2.5	7.0	9.5

These figures are based on the first 18 months of life of the paint system and show an 80% reduction in cost for the inhibited polysulfide. The B-52's which were painted in the original field service evaluation test are still in good condition after five years of service. Thus the actual cost saving to the Air Force is much greater than the figures in the table indicate because the service life of the paint system is almost doubled.

Based upon this type of demonstrated superiority of the chromated inhibited polysulfide primer over the epoxy-polyamide paint system, the Air Force issued Military Specification MIL-P-87112 and has made the new primer the standard material for use on all Materials Air Command and Strategic Air Command aircraft. The Naval Air Systems Command also recognized the dramatic decrease in corrosion achieved through the use of the corrosion-inhibitive sealants and, in June of 1976, issued a specification MIL-S-81733 to define their performance requirements.

The value of the MIL-S-81733 sealant and the MIL-P-87112 primer can be illustrated by two excerpts from Capt. Larry Griffin and Lt. Donald Latterman's prize winning paper on C-141 Service Experience.[6] They describe the use of the corrosion-inhibitive sealant and primers in solving corrosion problems on the C-141A aircraft. The C-141's were manufactured before the sealant and primer were perfected.

Upper Surface Wing Panels. "The integrally stiffened wing panels of the C-141A aircraft were manufactured from 7075-T6511 extrusion. The outer surface of these panels was shot peened and sulfuric acid anodized. The original finish system consisted of a MIL-C-8514 wash primer, followed by one coat of MIL-P-7962 primer and two coats of MIL-L-19537 acrylic nitrocellulose lacquer. In time, this system cracked around the heads of the countersunk steel Taper-Lok fasteners used to attach the panels to one another and to various internal structure. These fasteners were installed wet with MIL-S-8802 sealant, but eventually, the sealant-to-metal bond broke allowing moisture intrusion around the fastener heads. Water was then in contact with exposed end grain in the fastener hole wall setting up oxygen concentration cells and galvanic couples which initiated and supported an intergranular corrosion attack. This problem was recognized, and in 1970 the exterior surface finish was changed to one coat of MIL-P-23377 epoxy primer followed by two coats of MIL-C-83286 polyurethane. However, due to the brittleness of the epoxy primer, cracking of the coating around fastener heads continued. *This resulted in a change to the present system of two coats of MIL-S-81733B, Type III, corrosion-inhibitive coating followed by two coats of MIL-C-83286 polyurethane. Service tests of this system indicate that the elasticity of the polysulfide sealant and coating material has essentially resolved the upper wing panel corrosion problem.*

Fuselage Skins. "Most of the fuselage skins are clad 7079-T6 and, as delivered, the clad material was not painted. Consequently, acidic cleaning and brightening compounds were used during periodic washings to enhance the aircraft's appearance. These skins are attached with cadmium plated steel fasteners and aluminum rivets, and the periodic cleanings soon removed the cadmium plating from steel fastener heads, causing rusting of the fastener heads and pitting of the fuselage skin cladding. In 1970, the USAF started painting the entire aircraft with an epoxy primer, polyurethane topcoat system. As with the wing, the brittleness of the epoxy primer resulted in cracks in the protective system around the fastener heads. The problem was compounded by the breakdown of sealant around the countersunk fasteners, which in many cases were seated below the level of the cladding on the skins, thereby permitting the skin core metal to come into contact with the moisture and soils which accumulate on aircraft surfaces. This resulted in intergranular corrosion, and in many cases, stress corrosion cracking. *Service tests with the MIL-P-87112 corrosion inhibiting polysulfide primer followed by two coats of MIL-C-83286 polyurethane have shown a dramatic decrease in the incidence of fuselage skin corrosion. Starting in October 1977, the polysulfide primer will be used on all C-141A aircraft as they are repainted.*

In addition to its use on aircraft surfaces, the inhibitive elastomeric coatings are being used in the rework and refinishing of dry bays, the internal spaces of aircraft wings which are not filled with fuel. Corrosion of the dry bays is especially severe on those aircraft operating over the ocean or based near the sea coast. While salt spray is one of the prime offenders, corrosion is greatly accelerated by the presence of jet engine exhaust gases laden with sulfurous and sulfuric acid, water vapor and carbon particles.

The results of field service tests described in this paper indicate that the MIL-S-81733 chromate-inhibited sealants and the MIL-P-87112 elastomeric coatings are welcome additions to corrosion prevention technology.[7] These coatings, by providing an abundance of available completely soluble hexavalent chromium ions for the metal surfaces, effectively prevent corrosion.

The use of these two weapons in the anti-corrosion arsenal will save the Air Force and Navy millions of dollars in reduced maintenance costs. More important, the safety and reliability of our operational aircraft have been significantly increased.

CONCLUSIONS

1. All polymers are permeable to moisture to some degree and, therefore, cannot completely protect metal from corrosion.

2. Chromate-containing polysulfide sealants give complete protection by providing a reservoir of inhibitor ions.

3. Exfoliation corrosion of fastener holes may be prevented by wet-sealing all external fasteners with MIL-S-81733 corrosion-inhibitive sealant.

4. Faying surface corrosion may be prevented by sealing structural joints with the corrosion-inhibitive sealant.

5. The cracking of coatings on fastener areas of large aircraft is minimized and the life of the coating system is extended by using MIL-P-87112 corrosion-inhibitive polysulfide primer and a MIL-C-83286 polyurethane topcoat.

AREAS FOR FUTURE RESEARCH

Although the corrosion-inhibitive sealants and primers are doing an excellent job of preventing corrosion of aircraft, from an environmental point of view there are two potential areas for improvement. It would be desirable to substitute non-toxic inhibitors for part or all of the chromates. Also, the development of water dispersible polysulfide sealants and primers would reduce the quantity of solvent released to the atmosphere.

*SPECIAL ACKNOWLEDGEMENT

W. A. Boggs of Lockheed-Georgia was the other member of the research team which developed the corrosion-inhibitive sealant and primer system.

REFERENCES

1. Military Specification MIL-S-87133B (AS), "Sealing and Coating Compound, Corrosion Inhibitive," June 7, 1976.
2. Military Specification MIL-P-87112 (USAF), "Primer Coating, Elastomeric, Polysulfide, Corrosion Inhibiting," June 21, 1977.
3. Cohen, M. and Beck, A.F., *J. Electrochem. 62, 696 (1958)*.
4. Miller, R.N., "Aircraft Exfoliation Corrosion," *Mat. Prot. 6, 55 (1967)*.
5. Boggs, W.A. and Herron, W.C., "Evaluation of Fasteners and of Corrosion Prevention Techniques on HC-130H, S/N 64-14681," Lockheed-Georgia Engineering Report No. 9787-5 prepared under Contract LGD F09603-68-C-0771, Task 2, for Warner Robins Air Force Base, December 31, 1969.
6. Griffin, L. and Latterman, D., "C-141A Service Experience - Materials and Processes," *SAMPE Journal, March/April 1978, pp. 9-16*.
7. Ramsey, M.W., "Corrosion Inhibitors, Manufacture and Technology," Chemical Technology Review No. 60, Noyes Data Corp., Park Ridge, New Jersey, 1976.

CONTROL OF STRESS CORROSION CRACKING BY INHIBITORS

Charles S. O'Dell
and B. F. Brown

Department of Chemistry
The American University
Washington, D.C.

B. FLOYD BROWN

ABSTRACT

Studies of inhibitors of stress corrosion cracking (SCC) are reviewed for austenitic stainless steels, brass, titanium, aluminum alloys, and carbon and low alloy steels. There are at least five combinations of alloy and environment for which inhibitors have provided engineering solutions for the control of SCC, and others are not improbable at an early date. In view of the complexity of the problem and the comparatively few sustained research programs in this special field, it is not surprising that the fundamentals of inhibition of SCC are incompletely understood.

INTRODUCTION

Stress corrosion cracking (SCC) is a process involving the nature of an alloy (metallurgy), stress (mechanics), and environment (chemistry or electrochemistry). All possible countermeasures to SCC may be assigned to these three categories, metallurgy, mechanics and electrochemistry (including temperature), and in engineering practice measures from any or all three categories may be employed. One may for example select cupronickel condenser tubing to replace admiralty metal tubing to avoid ammoniacal SCC, representing a metallurgical approach to the problem. Or one may employ a thorough stress-relieving anneal of an assembled admiralty metal heat exchanger, representing a mechanics approach. And one may elect to apply cathodic protection, as by spraying stainless steel with lead to prevent chloride SCC, representing an electrochemical approach.

Changing the composition of the environment can also be regarded as a chemical or electrochemical approach, usually pursued by identifying a contaminant species responsible for SCC (such as chloride in austenitic stainless steel systems) and lowering the concentration of this species below what is hopefully a safe threshold value.

Another possible approach to controlling SCC by controlling the environment is by the use of additives. These additives may act in three distinct ways: (1) They may destroy or prevent the formation of the cracking species, as by the use of soda ash or ammonia to prevent polythionic acids cracking of sensitized stainless steel. (2) They may move the potential of the alloy into a safe range. Or, (3) they may enter directly into blocking of one or more of the chemical reactions responsible for SCC. This present survey of these additives, or inhibitors, will summarize the work to date regardless of which of the three categories of mechanisms above is operative, partly because in many instances the mechanism is unknown or only suggested. Although the present conference is concerned largely with coatings which are not appropriate to heat exchanger technology, the inhibitor work on heat exchanger materials is included on the chance that some of the principles involved may be useful in formulating thoughts about inhibitors in coatings.

The material is organized according to alloy family.

AUSTENITIC STAINLESS STEELS

These alloys undergo SCC in warm or hot chloride solutions, in hot caustic solutions, in strong acids containing chloride at room temperature, and (if sensitized) in polythionic acids at room temperature.

In hot chloride solutions Staehle, Beck and Fontana showed that reducing adventitious oxygen by reacting with hydrogen eliminated SCC,[1] perhaps by depressing the potential into a protective range. Other studies, by Williams and Eckel,[2] Rion,[3] Edeleanu and Snowden,[4] and Phillips and Singley[5] showed that deoxygenation by hydrazine was a countermeasure to SCC, again possibly by depressing the potential. Sulfite was found to be a somewhat less effective inhibitor.[2,4,5] Although this deoxygenation approach is not considered to belong to the strictly inhibitor literature, it is noted here as it suggests that inhibitors may be sought for some alloys by consideration of the effects of additives on potential. Actually, this approach (depressing the potential) was suggested by much earlier work by Mears, Brown and Dix who demonstrated the efficacy of cathodic protection against chloride SCC in stainless steel, with potential data.[6] H. van Droffelaar has shown that sprayed coatings of metallic lead will prevent chloride SCC of stainless steel;[7] O'Dell and Sarada have shown that this effect is due to cathodic protection and not to classic inhibition by lead species.[8]

Under various conditions, *chromate* may either be indicated to counter SCC,[9,10,11,12] or it may be counterindicated[4,13]. Perhaps this variability in response may reflect a competition between the beneficial effect of tending to stabilize a protective film versus the adverse effect of raising the potential into a hazardous range. Alkali *phosphates* have been reported to inhibit to various degrees chloride SCC of steel if unsensitized,[2,4,10,11,12,14] but perhaps the ratio of inhibitor to chloride is an important factor.[11,13]

Snowden found *nitrate* an effective inhibitor providing the molar ratio $NaCl/NaNO_2$ was 1/1, but not if the ratio was 2/1,[13] and Couper found it only slightly inhibiting at a much higher Cl^-/NO_2^- ratio, in 42% $MgCl_2$.[12]

Couper[12] and Uhlig and Cook[15] reported prevention of SCC in boiling 42% $MgCl_2$ with 2% $NaNO_3$. Fujii and Kumada reported work in 35% $MgCl_2$ showing that the inhibiting character of various polyoxyanions decreases as follows: $NO_3^- > H_2PO_4^- > CrO_4^{-2} \approx WO_4^{-2}$, from which they concluded that there is no relationship between oxidizing power and the inhibition.[16] Uhlig and Cook showed that the action of the nitrate is to shift the protection potential (or "critical cracking potential") with respect to the corrosion potential to put the corrosion potential in the protected range.[15] The efficacy of nitrate is judged to be sufficient for it to be called out as a preventive measure in NACE Recommended Practice RP-01-70.[17] If the nitrate is not present in sufficient concentration, SCC will occur, as shown by Uhlig and

Cook[15] and as may be inferred from the work of Edeleanu and Snowden.[4]

Uhlig and Cook also found iodide and benzoate to be inhibitive in *42%* $MgCl_2$.[15]

Holzworth and Symonds[18] found that dipping the steel in lithium *silicate* inhibited SCC in dilute acidified chloride solution at 90°C. Bergen found that sodium *metasilicate* somewhat inhibited SCC in $CaCl_2$ at 132°C, and burnishing the steel with silver inhibited SCC in short-time tests in that environment.[19]

Overman used radioactive tracers to study inhibition of SCC in Type 304 stainless steel in boiling solutions of 10 – 100 ppm chloride plus the selected substances. One to 50 ppm silicate, phosphate, nickelous, iodide, bromide or chromate ions inhibited crack formation in the standard solution. The tracer studies revealed that corrosion and crack formation were inhibited by cations (complexes?) or anions that compete with the chloride ions for the available corrosion sites on the surface of the steel.[20]

Podobaev and his associates studied the inhibition of SCC in $MgCl_2$ boiling at 153°C. KI acted as a partial inhibitor by itself; when combined with BA–12 (a condensation product of benzylamine and paraform-aldehyde) there is a synergistic action to effect complete protection against SCC. Another inhibitor designated PB–5) (not further described) also protected completely against SCC.[21] Katapin, a surface-active substance used as a pickling restrainer, retards the anodic process by forming a protective film, but it is incompletely effective in inhibiting SCC.[22]

Podobaev showed that $SnCl_2$ added to boiling $MgCl_2$ (153°C) produced a large increase in survival time of stainless steel; there was some inhibition from additions of $SnCl_4$, $PbCl_2$, $SbCl_3$, and $BiCl_3$. The addition of $CdCl_2$ at 45 millimolar concentration produced a 5-fold increase in survival time. Podobaev explained the results on the basis of changing cathodic polarization characteristics.[23] Beauchamp showed that if the chloride were present as 1705 ppm of $CdCl_2$ at 205°C, 18–8 stainless steel did not crack under a stress of 10 ksi in 8 hours, but if stressed at 26.5 ski, SCC occurred in 17 hours.[24] Leu and Helle found that the addition of $SnCl_2$, metallic magnesium or metallic zinc to boiling $MgCl_2$ prevented or spectacularly reduced SCC in type 304 stainless steel.[25] O'Dell's study of sensitized type 304 stainless steel in *4 N* $NaCl$ with pH 2.3 at 100°C found that the additives NaI, $SnCl_2$, $SnCl_4$, $FeCl_3$, $SnSO_4$, $CdSO_4$, and n-butylamine prevented SCC of U-bend specimens for 600 hours as compared with 24 hours required for SCC in the uninhibited system. All of these substances permitted pitting and/or general corrosion except for $CdSO_4$. Cd^{++} and SO_4^{--} were only moderately effective when added separately; however, $CdSO_4$ produced a synergistic effect preventing all forms of corrosion. Polarization data showed that the inhibitor has a pronounced effect on the anodic reaction. EDAX analysis of the material within cracks initiated in uninhibited solution, to which $CdSO_4$ was then added, showed high concentration of the inhibitor. The inhibition was attributed to the synergistic effects of the cadmium polymerizing with the chloride combined with a desorbing action of the sulfate ion upon the chloride.[26]

Although chloride is clearly a causative agent for SCC in stainless steel, the cation apparently has a significant role in the corrosion process. Varying the cation in chloride solutions can produce changes ranging from acceleration of SCC to blockage of the corrosion process.[27,28,29]

Several groups have studied the aggressive behavior of metal chloride solutions with varying results.[29,30,31,32,33] Perhaps the best order of ranking for four cations is as follows: Li^{+} (most aggressive), Mg^{+2}, Ca^{+2}, Na^{+}.[34]

Zucchi and associates have reported on a series of researches on inhibition of SCC of austenitic stainless steels in strong acid chloride solutions at room temperature, using various organic inhibitors.[35,36,37,38,39,40] Benzonitrile completely inhibited transgranular SCC in type 304 steel probably by interfering with the anodic reaction, and benzotriazole decreased SCC attack; the environment was $0.5\ N\ NaCl$ acidified with H_2SO_4.[35] In $2\ N\ HCl$, phenyl thiourea completely blocked attack, again by blocking the anodic reaction; di-n-butyl thiourea also blocked all attack, as did the combination of KI + n-decylamine.[36] In $2\ N\ HCl$, di-o-tolyl thiourea; benzoimidazole-2-thiol; N-N'-diphenyl thiourea; sulfadiazole-3-thiol; and benzothiazole-2-thiol completely blocked "fissuring" (cracking/).[37,38,39] Again in $2\ N\ HCl$ but this time with type 316 steel the inhibition of SCC by benzimidazole-2-thiol; benzothiazole-2-thiol; sulfathiazole-3-thiol; and N-N'-diphenyl thiourea is due to blocking the anodic reaction. All the organic substances examined altered the hydrogen reaction, but only those substances which inhibited the anodic reaction prevented SCC. The proposed mechanism is one of rapid adsorption onto the metal exposed by stress to block the anodic reaction. There does not seem to be any correlation between inhibiting general corrosion and inhibiting fissuring.[40]

Lafranconi et al. have studied SCC in austenitic stainless steel also in strongly acidic chloride media at room temperature, and they found inhibition by traces of Pb^{+2}. This inhibition they showed was due to effects on the cathodic reaction (hydrogen reduction), which they attributed to the effect of lead on hydrogen overvoltage.[41] Perhaps the apparent discrepancy between this finding and that of O'Dell and Sarada[8] can be assigned to a difference in cracking mechanisms, as the latter were working with (only slightly acidic) $MgCl_2$ at 153°C.

One inhibitor of SCC in stainless steel in a strong acid—white fuming nitric acid—which found its way into engineering practice but not into the technical literature is 1% HF.[42]

BRASS

Gisser found that continuous resin films such as phenol formaldehyde delayed attack on brass in ammonia.[43] Whether there is any intrinsic inhibiting effect of substances in the film or whether only barrier properties are involved does not seem to be established. Rowlands protected aluminum brass (C68700) against SCC in sulfur-polluted seawater with sodium dimethyldithiocarbamate which forms a protective copper-chelate film on the brass surface.[44] Poly-hexamethylphosphate and sodium chromate have also been found to prevent stress corrosion cracking of aluminum brass (C68700) in recirculating seawater but not in fresh water.[45] Uhlig and his students found that with 63-67 brass exposed to ammoniacal and to tartrate solutions, delay in cracking occurred with the addition of bromide ions and also chloride ions. The delay was attributed to the shift of the protection potential (or critical cracking potential) in the positive direction with respect to the corrosion potential.[46] Podobaev and his associates have worked on the problem of protecting cartridge brass against corrosion and SCC by protective coatings of greases and oils, with and without inhibitive additions. Though not perfectly protective, potassium ferrocyanide and dibutyl phthalate have inhibitive characteristics attributed to the formation of protective complex reaction products.[47] Walston attributed the SCC of an admiralty heat exchanger exposed to ammonium chloride to ammonia released by reaction of the chloride with the brass, combined with the absence of hydrogen sulfide which would have acted as an inhibitor.[48]

TITANIUM

Kiefer and Harple showed that in red fuming nitric acid 1% *NaBr* could inhibit SCC in commercially pure titanium[75A] at room temperature.[49] Rittenhouse found that the addition of water to red fuming nitric acid would control SCC in pure titanium and in the *Ti–8% Mn* alloy.[50] A Military Specification has been written calling for up to 14% NO_2 and 2.5% H_2O, balance HNO_3.

Haney and Wearmouth showed that additions of water inhibited SCC initiative in titanium in methanol.[51] This inhibition is not effective in the presence of a crack or of chloride.

Pressurized fuel tanks of *Ti-6%Al-4%V* experience SCC if they contain N_2O_4 of sufficiently high purity, but the addition of NO inhibits the SCC.[52] This inhibition is sufficiently practical as a solution to the SCC problem that a NASA Specification (PPD-2B) calling for 0.60 – 1 per cent NO was written and has been successfully applied in the U. S. space program.

The addition of oxygen gas has been found to inhibit cracking of titanium tanks containing gaseous hydrogen.[53]

The inhibition of SCC in titanium alloys is poorly understood, and little has been suggested concerning the mechanism(s).

ALUMINUM

Protection of high strength aluminum alloys from SCC in moist atmospheres is promoted by a primer containing the inhibitor strontium chromate (MIL-P-23377), widely used in aerospace weapons systems. Zinc chromate primer confers some considerable protection for atmospheric service, and even more so when further covered by an epoxy paint system.[54]

Galvanic coatings are practical solutions to the SCC problem for high strength aluminum alloys. Zinc-rich epoxy, zinc electroplate, and metallized 7072 coatings have been shown to greatly extend the life of 2014 and 7079 alloys.[54] Magnesium pigment added to inorganic binders[55] or epoxy primer[56] has given good results.

A corrosion prevention compound has been developed by personnel of the Navy Air Development Center which contains two unidentified inhibitors.[57] Although only corrosion tests, not stress corrosion tests, have been reported to date, the results of the corrosion tests are sufficiently impressive to suggest that tests of the compound as a countermeasure to SCC would be interesting. The compound, named AMLGUARD, is covered by military specification MIL-C-85054(AS).

CARBON STEELS AND LOW ALLOY STEELS

Caustic cracking of boiler plate was once a serious problem for which inhibitor technology contributed many countermeasures. Straub lists phosphate, tannate, chromate, nitrate, acetate, and arsenic as inhibitors of this failure mode. He also showed that a ratio of sulfate to alkalinity of 2.1 would control cracking.[58] Schroeder and Berk found that in high pressure boilers the addition of nitrates to 40% of the alkalinity inhibited cracking, and sodium sulfate was also effective. Quebracho inhibited cracking by removing oxygen and by keeping the calcium phosphate and calcium carbonate dispersed. The addition of waste sulfite liquor to the water in locomotive boilers inhibited cracking.[59] Several inhibitors of cracking in boiler plate were discussed in a symposium on caustic embrittlement. These additives included R_2O_3 (where R=Fe and Al),[20] organic extracts,[60] nitrate,[61,62,63] phosphate,[61] waste sulfite liquor[62] (for oxygen removal), quebracho,[63] myrabalans,[63] and combinations of sodium nitrate and lignin sulfonate[63] and of sodium nitrate and tannin.[64] No mechanism of inhibition was proposed at this symposium. Extensive use has been made of coordinated phosphate and pH treatment to protect boilers against cracking.[63]

Ammonia storage tanks of mild steel can be protected against SCC by the addition of 0.2% water,[65] and this procedure is in widespread use. It is required by the Department of Transportation for transport of ammonia in quenched and tempered steen tankages;[66] the technical background is given in reference 67. But condensation of vapor on cold walls of the vapor space above water-inhibited ammonia can produce a film of insufficiently inhibited ammonia, leading to SCC; the practical solution to this sort of problem is to maintain the temperature of the liquid ammonia a few degrees cooler than the walls on which condensation would otherwise occur.[68] The addition of ammonium chloride or oil will also retard SCC in liquid ammonia storage tanks.[69] The addition of 0.05% refrigerator oil which coats the steel surface and prevents SCC is a patented procedure.[70]

Ammonium chloride will also inhibit SCC of low carbon steel exposed to nitrate solutions.[71]

Dunlop was able to show that SCC of ASTM A203 Grade E steel (a low strength nickel-bearing ferritic steel) in sulfide environments was an anodic process and not one of hydrogen embrittlement; morpholine was found to be an effective inhibitor.[72]

Carbonate-bicarbonate solutions have been shown capable of producing SCC in line-pipe steel, and there is considerable incentive to develop an inhibitor effective against this attack and compatible with primer materials and with cathodic protection conditions. Berry and his associates have shown that in a standard carbonate-bicarbonate solution the following substances are effective inhibitors of SCC: Sodium chromate, potassium dichromate, and zinc chromate (0.1 wt. %, based on constant strain rate tests on smooth specimens); sodium monobasic phosphate, calcium monobasic phosphate, sodium tripolyphosphate, and potassium silicate (1.0 wt. %, based on constant load tests using edge-notched flat tensile constant load tests). Phosphate inhibitors incorporated in asphalt-based primer and coal-tar based primer were shown to be effective countermeasures to SCC in laboratory tests.[73]

Anthraquinone derivatives adsorbed on AISI 4340 steel (smooth specimens) retarded SCC presumably by acting as hydrogen acceptors[74] in a system in which SCC is probably a hydrogen-assisted cracking process. $NaOH$, Na_2CO_3, Na_3PO_4, Na_2SiO_3, or $Na_2B_4O_7$ when introduced with $NaNO_2$ produced a synergistic inhibition of SCC in AISI 4340 steel in 10% KNO_3 at 100°C.[75] Several authors have found that organic amines, including aniline and hexamethylenetetramine can inhibit SCC in high strength steels by adsorbing on the metal surface and interfering with the transport processes.[76,77,78,79,80] Using pre-cracked specimens, Parrish et al. showed that oxidizing inhibitors could increase KISCC for a high strength steel (D6ac), but that in the presence of chloride, dichromate is not a perfect inhibitor.[81] C. T. Lynch et al., working with a variety of high strength steels, were able to reduce cracking rates in most of the steels with a multifunctional inhibitor such as borax-nitrate.[82] Brown, using precracked specimens, showed that the addition of enough sodium acetate to prevent the crack tip from going acid, could inhibit SCC in high strength steel.[83] Johnson found that the addition of a trace of oxygen was sufficient to inhibit the growth of cracks in high strength steel under stress in gaseous hydrogen at one atmosphere pressure.[84]

Humphries and Parkins [85] showed that $NaNO_3$ and $KMNO_4$ inhibit the cracking of annealed mild steel in *NaOH* by moving the potential out of the cracking range; they also found that valonea and quebracho tannin with NaH_2PO_4 prevent SCC even in the SCC potential region. These substances may interfere with the formation of an Fe_3O_4 film which may play a vital role in crack propagation in caustic.[86] Berry has found eleven inhibitors of SCC in line pipe steels in (1) $Na_2CO_3/NaHCO_3$, (2) *33% NaOH*, and *20%* NH_4NO_3, as follows: CAH_2PO_4, $CaSO_4$, potassium chromate, NaH_2PO_4, sodium orthosilicate (not effective in *NaOH*), sodium tetraphosphate, and tripolyphosphate.[87]

DISCUSSION

Considering the small number of studies there have been on inhibiting SCC compared with the vast literature on SCC in general, it is somewhat surprising that there have been so many systems (combinations of alloy and environment) in which SCC is controlled in engineering applications by the use of inhibitors: water in ammonia in steel tankages, NO in N_2O_4 in titanium tankages, HF in white fuming nitric acid in stainless steel tankages, and chromate against atmospheric moisture for high strength aluminum aerospace structures. The striking thing about the inhibitor approach is that when a good inhibitor is found, it is often the only measure needed for technological success. By contrast, when other control measures are applied, they are often partial measures and are often combined with partial measures, such as various combinations of altered alloy composition, reduced strength, changed microstructure, reduced stress, and modified environment and temperature.

Stress corrosion cracking is scientifically a complex process, and it is a nucleation-and-growth process. In very few of the studies reviewed here have nucleation and growth been differentiated.

Despite long years of work by brilliant and dedicated investigators, there is hardly a single SCC system for which a mechanism of SCC has been proposed which has met with overwhelming acceptance by most of the workers in the field. It is therefore not surprising that our knowledge about the mechanisms of inhibition of SCC is so incomplete.

ACKNOWLEDGEMENT

The support of the Office of Naval Research, under Contract N 00014-75-C-0799, NR 036-106, in the writing of this review is gratefully acknowledged. The final typescript was prepared by M. H. Wason. It is a pleasure to acknowledge the contributions of Dr. R. M. Hurd, who drew to the author's attention several obscure and important references. Professor R. T. Foley kindly read the draft and made many helpful suggestions.

REFERENCES

1. Staehle, R.W., Beck, F.H. and Fontana, M.G., *Corrosion 15, 373t (1959)*.
2. Williams, W.L. and Eckel, J.F., *J. Am. Soc. Naval Engrs. 68, 93 (1956)*.
3. Rion, W.C. Jr., "Stainless Steel Information Manual for the Savannah River Plant," AEC Research and Development Rept. DP-860 (1964).
4. Edeleanu, C. and Snowden, P.P., *JIS[I] 186, 406 (1957)*.
5. Phillips, J.H. and Singley, W.J., *Corrosion 15, 450t (1959)*.
6. Mears, R.B., Brown, R.H. and Dix, E.H., "Symp. on Stress Corrosion Cracking of Metals," ASTM, Philadelphia, 1945, p. 337.
7. van Droffelaar, H., *CEBELCOR Rapports techniques 112, 172 (1970)*.
8. O'Dell, C.S. and Sarada, T., unpublished work, The American University, 1978.
9. Heger, J.J., *Metal Progress 67(3), 109(1955)*.
10. Scharfstein, L.R. and Brindley, W.G., *Corrosion 14, 588 (1958)*.
11. Neumann, P.D. and Griess, J.C., *Corrosion 19, 345 (1963)*.
12. Couper, A.S., *Mater. Prot. 8, 17 (1969)*.
13. Snowden, P.P., *Nucl. Eng. 6, 409 (1961)*.
14. Williams, W. L., *Corrosion 13, 539t (1957)*.
15. Uhlig, H.H. and Cook, Jr., E.W., *J. Electrochem. Soc. 116, 173 (1969)*.
16. Fujii, N. and Kumada, M., *J. Japan Inst. Metals 35, 560 (1971)*.
17. Recommended practice RP-01-70, National Association of Corrosion Engineers, Houston, 1970.
18. Holzworth, M.L. and Symonds, A.E., *Corrosion 25, 287 (1969)*.
19. Bergen, C.R., *Nuclear Application 1, 484 (1965)*.
20. Overman, R.F., *Corrosion 22, 48 (1966)*.
21. Balezin, S.A. and Podobaev, N.I., *J. Appl. Chem. USSR 33, 1287 (1960)*.
22. Podobaev, N.I. and Balezin, S.A., *J. Appl. Chem. USSR 33, 2258 (1960)*.
23. Podobaev, N.I., *J. Appl. Chem. USSR 36, 341 (1963)*.
24. Beauchamp, R.L., M.S. thesis, Ohio State Univ., 1963, cited by R.M. Latanision and R.W. Staehle, "Proc. of a Conf. on Fundamental Aspects of Stress Corrosion Cracking," R.W. Staehle, A.J. Forty and D. van Rooyen, Eds., NACE, Houston, 1969, p. 229.
25. Leu, K.W. and Helle, J., *Corrosion 14, 249t (1958)*.
26. O'Dell, C.S., Ph.D. Dissertation, The American University, 1978.
27. Scheil, M.A., Zmeskal, O., Waber, J. and Stockhausen, F., *Welding J. 22, 499 (1943)*.
28. Franks, R., Binder, W.O. and Brown, C.M., "Symposium on Stress Corrosion Cracking of Metals," ASTM, Philadelphia, 1945, pp. 411-420.
29. Logan, H.L. and Sherman, Jr., R.J., *Welding J. 35, 389 (1956)*.
30. Thomas, K.C., Ferrari, H.M. and Allio, R.J., *Corrosion 20, 89t (1964)*.
31. Brunet, S., Coriou, H., Grall, L., Mahieu, C. and Pelras, M., "Eur. Symp. Fresh Water Sea Preprints," cited in *Chem. Abstr. 68 : 52431q*.
32. Rhodes, P.R., *Corrosion 25, 462 (1969)*.
33. Truman, J.E., *Corr. Sci. 17, 737 (1977)*.
34. Berg, S. and Henrikson, S., *Jernkont. Ann. 144, 392 (1960)*.
35. Zucchi, F., Trabanelli, G. and Frignani, A., *Atti della Accademia delle Sceince di Ferrara 49 (1971-72)*.
36. Zucchi, F., Trabanelli, G. and Sezione, V., *Annali dell Universita di Ferrara, Chimica Pura ed Applicata 3(5), 62 (1973)*.
37. Zucchi, F., Trabanelli, G., Frignani, A. and Zucchini, M., *Ann. Univ. Ferrara Sez. 5, 4(1), 12 (1976)*.
38. Trabanelli, G., *CEBELCOR Rapports techniques 129, 94 (1976)*.
39. Zucchi, F., Trabanelli, G. and Frignani, A., "Extended Abstr. No. 3-14," 6th International Congress on Metallic Corrosion, Sydney, 1975.
40. Zucchi, F., Trabanelli, G., Frignani, A. and Zucchini, M., *Corr. Sci. 18, 87 (1978)*.

41. Lafranconi, G., Mazza, F., Sivieri, E. and Torchio, S., *Corr. Sci. 18, 617 (1978).*
42. Cohen, B., personal communication to B.F. Brown, Dec. 21, 1977.
43. Gisser, H., "Symposium on Stress-Corrosion Cracking of Metals," ASTM, Philadelphia, 1945, 199-209.
44. Rowlands, J.C., *J. Appl. Chem. 15, 57 (1965).*
45. Sato, S. and Nasetani, T., *Sumitomo Keidinzoku Giho 10, 175 (1969). (Chem. Abstr. 72:15156R).*
46. Uhlig, H., Gupta, K. and Liang, W., *J. Electrochem. Soc. 122, 343 (1975).*
47. Podobaev, N.I., Balezin, S.A. and Romanov, V.V., *J. Appl. Chem. USSR 33, 1297 (1960).*
48. Walston, K.R., *Corrosion 17, 492t (1961).*
49. Kiefer, G.C. and Harple, W.W., *Metal Progress 63(2), 74 (1953).*
50. Rittenhouse, J.B., *Trans. ASM 51, 871 (1959).*
51. Haney, E.G. and Wearmouth, W.R., *Corrosion 25, 87 (1969).*
52. Vance, R.M., "Proc. 1st Joint Aerospace and Marine Corrosion Techn. Seminar, 1968," NACE, Houston, 1969, pp. 34-35. See also, "Corrosion," Vol. 2, 18:15, L.L. Shreir, ED., London, Newnes-Butterworths, 1976.
53. Williams, D.P. and Nelson, H.G., cited by T.R. Beck, "The Theory of Stress Corrosion Cracking in Alloys," J.C. Scully, Ed., NATO, Brussels, 1971, p. 18.
54. Lifka, B.W. and Sprowls, D.O., "Shot Peening—A Stress Corrosion Cracking Preventative for High Strength Aluminum Alloys," paper presented at the 26th Annual Conference of the NACE, March, 1970.
55. Brown, S.R. and Ketcham, S.J., "Evaluation of Sermatel 385 Coating for Protection of Aluminum Alloys Against Stress Corrosion Cracking," Rept. No. NADC-MA-6930, Naval Air Development Center, Oct. 24, 1969.
56. Ketcham, S.J., private Communication to B.F. Brown, 1971.
57. Knight, W.E., "AMLGUARD, A Corrosion Prevention Compound for Military Applications," paper presented at the 1977 SAMPE Meeting.
58. Straub, F.G., Univ. of Ill. Eng. Expt. Sta. Bull. No. 216, 1930.
59. Schroeder, W.C. and Berk, A.A., "Intercrystalline Cracking of Boiler Steel and Its Prevention," Bur. of Mines Bull. 443, 1944.
60. Straub, F.G., *Trans. ASME 64, 393 (1942).*
61. Purcell, T.E. and Whirl, S.F., *Trans. ASME 64, 397 (1942).*
62. Bardwell, R.C. and Laudermann, H.M., *Trans. ASME 64, 403 (1942).*
63. Bird, P.G. and Johnson, E.G., *Trans. ASME 64, 409 (1942).*
64. Partridge, E., Kaufman, C.E. and Hall, R.E., *Trans. ASME 64, 417 (1942).*
65. Loginow, A.W. and Phelps, E.H., *Corrosion 18, 299t (1962).*
66. Department of Transportation, 49 CFR, part 173 (as revised, Federal Register 40 No. 113, June 11, 1975).
67. Lyle, Jr., Fred F., "A Study of Stress Corrosion Phenomena Resulting from Transportation of Anhydrous Ammonia in Quenched-and-Tempered Steel Cargo Tanks," Report No. DOT-FH-11-8568, Feb. 1976.
68. Ludwigsen, P.B. and Arup, H., *Corrosion 32, 430 (1976).*
69. Imagawa, H. and Nakamura, K., *Nippon Kinzoku Gakkaishi, 40,1250 (1976).* Cited in *Chem. Abstr. 86:58823n.*
70. Imakawa, H. and Kano, T., Japan Patent 77 42,140 (CL C23Fa5100), 22 Oct. 1977. Cited in *Chem. Abstr. 88:138341k.*
71. Noninski, K. Raichev, R. and Vodenicharov, P., *God. Vissh. Khim-Teknol. Inst., Sofia 22, 127 (1977). Cited in Chem. Abst. 88:67140f.*
72. Dunlop, A.K., *Corrosion 34, 88 (1978).*
73. Berry, W.E., Barlo, T.J., Payer, J.H., Fessler, R.R. and McKinney, B.L., "Evaluation of Inhibitor-Containing Primers for the Control of SCC in Buried Pipelines," CORROSION/78 Paper No. 64.
74. Tirman, A., Fugassi, P. and Haney, E.G., *Corrosion 25, 434 (1969).*
75. Baker, H.R. and Singleterry, C.R., *Corrosion 28, 385 (1972).*

76. Azhogin, F.F. and Pavlov, Yu.K., *Zashch. Metal. 2, 533 (1966).*
77. Gutman, E.M., Petrov, L.N. and Karpenko, G.V., *Fiz.-Khim. Mekh., Mater. 4, 149 (1968).*
78. Grigor'ev, V.P., *Fiz.-Khim. Mekh. Mater. 6, 54 (1970).*
79. Kanlarova, A.G., et al., "Korroz. Zashch. Neftegazov Prom-sti, 1977," cited in *Chem. Abstr. 88:108893.*
80. Eidel'man, N.G. et al., *Fiz.-Khim. Mech. Mater. 13, 93 (1977).*
81. Parrish, P.A., Chen, C.M. and Verink, E.D. Jr., "Stress Corrosion—New Approaches," H.L. Craig, Jr., Ed., STP 610, ASTM, Philadelphia, 1976, pp. 189-198.
82. Lynch, C.T., Vahldik, F.W., Bhansali, K.J. and Summitt, R., Abstracts 1977 TMS-AIME Fall Meeting, to be published by AIME.
83. Brown, B.F., "The Theory of Stress Corrosion in Alloys," J.C. Scully, Ed., NATO, Brussels, 1971, p. 186.
84. Hancock, G.G. and Johnson, H.H., *Trans. AIME 236, 207 (1966).*
85. Humphries, M.J. and Parkins, R.N., *Corr. Sci, 7, 747 (1967).*
86. Venczel, J. and Wranglen, G., *Corr. Sci. 4, 137 (1964).*
87. Berry, W.E., "5th Symposium on Line Pipe Research." American Gas Assoc., Cat. No. L30174, V-1 (1974).

A NEW METHOD TO SOLUBILIZE INHIBITORS IN COATINGS

Kenneth G. Clark

Naval Air Development Center,
Warminister, PA

KENNETH CLARK

ABSTRACT

The use of phase transfer catalysts to effect the solution of inorganic ions in organic solvents of low solvating ability has recently received considerable attention. Preparation of organo-soluble inhibitors using these techniques with macrocyclic crown polyethers and quaternary ammonium salts has been successful. Recent applications in coatings and crack arrestment compounds for high strength steel are noted.

INTRODUCTION

The development of organo-soluble corrosion inhibitors has been stimulated in recent years by the need for innovative protective materials as well as advances in chemistry. Applications for solubilized inhibitors now exist in transparent coatings, adhesive primers, lubricating preservatives, laser hardening coating systems, and metallic crack arrestment compounds. While the solubilized inhibitors have not shown significant promise in high performance aluminum and steel primers, they may be used to impart an intermediate level of corrosion inhibition to a material by the simple mixing addition of an appropriate compound, as opposed to the energy-consuming milling of solid pigment. Indeed, pigmented inhibitors are highly detrimental in certain applications. This paper discusses the use of phase transfer catalysis to achieve non-polar organic solubility with chromate salts and describes a recent application to other anions.

Chromate salts are among the best corrosion inhibitors for aircraft metals. However, they are insoluble in non-polar solvents. In order for chromates to be effective in organic coatings, for example, they must be dispersed throughout the film by milling at high concentrations. By minimizing the distance between each vehicle-encapsulated salt clump in the dry film, efficient leaching of the inhibitor is assured in the event of a film discontinuity such as scratching or cracking. A logical extension of this concept would be a coating containing a solid solution of *Cr(VI)* inhibitor.

Organic chromates and dichromates were known over 100 years ago but only within the last 40 years have they been used as corrosion inhibitors. Aniline and guanidine salts were being made at the turn of the century. Ammonium chromates were patented in 1940.[1] During the 1960's the Associated

Chemical Companies investigated chromate and dichromate salts of biguanides and triazines.[2] In 1973, Clark and Ohr studied chromate-macrocyclic polyether complexes[3] and in 1977 demonstrated the effectiveness of quaternary ammonium phase transfer catalysts for solubilizing chromates and dichromates.

Phase Transfer Catalysis

The concept of phase transfer catalysis originated in 1967 with Pedersen's discovery of the unusual properties of macrocyclic polyethers, not the least of which was their unusual ability to form strong stoichiometric complexes with alkali metal, alkaline earth, and some transition element cations. Such phenomena until then had been observed only in biological materials. The first polyethers (called crowns) were neutral compounds possessing a heterocyclic ring containing four to twenty oxygen atoms separated by ethylene bridges and often aromatic or cycloaliphatic moeities external to the ring. The ion-binding properties of the electro-negative ring depended upon the "fit" of the cation; the type, number and location of the heteroatoms; the solvent, the charge of the cation; and the anion.

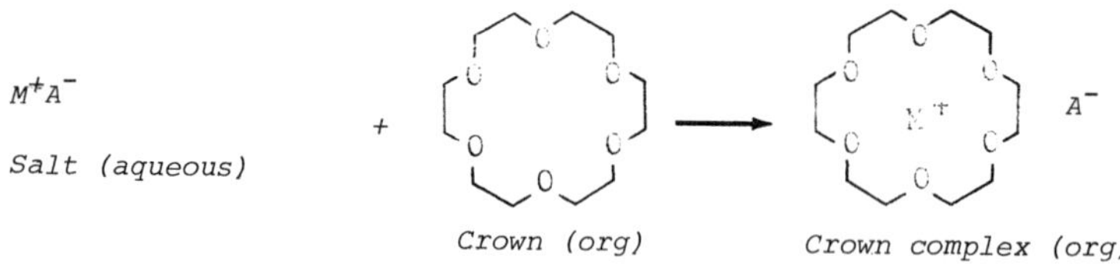

M^+A^- *Salt (aqueous)* + *Crown (org)* → M^+ A^- *Crown complex (org)*

The most interesting manifestation of crown complex formation was the solubilization of phase transfer of ionic compounds from their aqueous solution into organic solvent.

This concept of phase transfer was extended when it was realized that quaternary ammonium compounds could achieve similar results at a remarkably lower cost. A variety of alkyl and alkyl-aryl quaternary ammonium hydroxide, chloride, bromide, and iodide salts are currently available from Aldrich Chemical Co.[4] Phase transfer ability depends on the nature of the organic substituents, the original quaternary salt anion, the anion being transferred, the aqueous concentration, and the solvent.

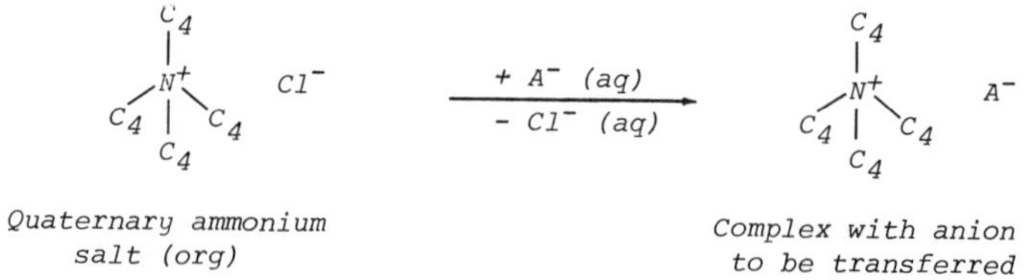

$$(C_4)_4N^+\ Cl^- \xrightarrow[-\ Cl^-\ (aq)]{+\ A^-\ (aq)} (C_4)_4N^+\ A^-$$

Quaternary ammonium salt (org) — *Complex with anion to be transferred*

In either case, crown polyether complex or quaternary ammonium complex, the preparation is usually carried out in a two-phase aqueous/organic solvent system — the water phase containing the anion to be transferred and the organic phase (usually methylene chloride) containing the phase transfer catalyst (PTC). Due to its slight solubility in water, a small amount of PTC diffuses into the aqueous phase where it is able to form a complex with the anion to be solubilized in the organic phase. The new PTC complex is then able to diffuse back into the organic phase where it is available for chemical reaction with an organic solute or for storage as a solubilized anion.

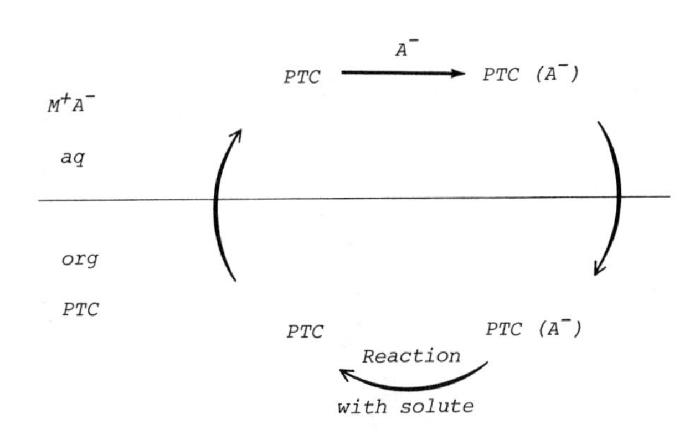

Phase Transfer Catalysis

The extent to which any one step of this sequence occurs is determined by equilibrium, but because the solubility of PTC or of the PTC complex is small in the aqueous phase, nearly all of the original PTC may be finally associated with A^- in the organic phase.

In some instances, phase transfer can be effected by adding a finely ground inorganic solid to an organic solution containing the catalyst. If ionic forces in the crystalline lattice are not too great, a weak solution is formed.

Phase transfer catalyst applications in organic synthesis have been reviewed by Dehmlow[5] and will not be treated here. Suffice it to say that increased yields and rates of reaction are generally due to the lack of a solvation sphere about the transferred reactive anion, i.e., the "naked" anion effect.

CHROMATE AND DICHROMATE SOLUBILIZATION

1. Crown Complex. Three crown polyethers were used.

Dibenzo - 18 - crown - 6

Dicyclohexano - 18 - crown - 6

18 - crown - 6

The 18-atom crown ring had been found an optimal size for the potassium cation.[6] Potassium chromate, therefore, was used as the aqueous solution from which a *Cr(VI)* anion was to be extracted. However, extraction did not occur at alkaline pH. Since the *Cr(VI)* species found in aqueous solution depends upon the following equilibria:

$$CrO_4^{=} + H^+ \rightleftharpoons HCrO_4^-$$

$$2HCrO_4^- \rightleftharpoons Cr_2O_7 + H_2O$$

acidification produced increasing *Cr(VI)* extraction by forming the more polarizable, "softer" (lower charge to size ratio) anions,[7] $Cr_2O_7^{=}$ and $HCrO_4^-$. The complex was isolated by evaporation and subsequent crystallization, but was found to be insoluble in toluene and quite photosensitive — the orange crystals darkened to a light brown in several minutes under a fluorescent light tube. This problem of stability arises again and again in the field of organic chromates: chromates and dichromates are potent oxidizers and as such tend to oxidize any available cooperative species whether it be the solvent or the complexing agent itself.

It should be pointed out that phase transfer can also occur from the solid ($K_2Cr_2O_7$, finely ground) to an organic solution of crown polyether, although only dilute solutions have been prepared in this way.

2. *Quaternary ammonium complex.* These phase transfer catalysts appear to be more suitable to the problem of *Cr(VI)* solubilization because of their relative stability in oxidizing media. Eleven quaternary salts, four organic solvents, and two aqueous solutions were investigated. Two important effects were noted. First, the catalyst must have good organic solubility. While tetrabutylammonium chloride was successful in transferring chromate and dichromate, tetraethylammonium chloride could do neither. In fact, only one quaternary salt was effective in all solvents (methylene chloride, chloroform, toluene and hexane) — Adogen 464[8] known as methyltrialkyl (C_8-C_{10}) ammonium chloride. Second, the original quaternary salt anion must be "harder" (higher charge to size ratio) than the anion to be transferred. For example, while tetrabutylammonium *chloride* transferred chromate and dichromate, tetrabutylammonium *iodide* transferred only dichromate. Thus, the order of preference for extraction is $Cr_2O_7 => I^- > CrO_4= > Cl^-$.

As with the crown polyethers, phase transfer can also occur from solid $K_2Cr_2O_7$ to an organic solution of quaternary salt but only to produce weak *Cr(VI)* solutions.

The Adogen 464 — dichromate complex was prepared in better than 90 per cent yield by extracting a 60% aqueous solution of sodium dichromate with a 17% methylene chloride solution of Adogen 464. Rotary evaporation of the methylene chloride phase yielded a clear, dark brown syrup which proved to have excellent storage stability at room temperature. Differential scanning calorimetry showed the initial decomposition temperature to be 213°C, slightly better than any of the stable organic chromates tested by Rice et al.[9] Although solutions in oxygenated solvents are stable for at most several days, stability in hydrocarbon solvents is excellent. The complex structure below has been confirmed with ultraviolet spectroscopy.

APPLICATIONS

1. *Modification of MIL-C-85054* (metal preservative). Originally formulated to be water white for use on aircraft exterior rivet patterns when corrosion was detected, this compound contained only a single organic inhibitor, barium petroleum sulfonate. Modifications of the preservative with Adogen 464 — dichromate complex (464-D) were tested in the ASTM 5% salt fog cabinet on AISI 1010 steel with the following results:

	Time to Failure (More than (5% Rusted Area)
MIL-C-85054	3 days
MIL-C-85054 with 0.1% 464D	7 days
MIL-C-85054 with 1.0% 464D	9 days

2. *Modification of an Epoxy Primer.* The strontium chromate primer MIL-P-23377 was reformulated by replacing the only inhibitor strontium chromate with titanium dioxide pigment to achieve an equivalent pigment volume concentration. Modification of this formulation with 464D produced the following salt fog results on AISI 1010 steel (coatings scribed with an "X"):

	3 hours	10 days
MIL-P-2337	Severe rusting in scribe	Severe generalized rusting, small blisters
Reformulation	Little rusting in scribe	Severe rusting in scribe, many small blisters
Reformulation with 1.0% 464D	Little rusting in scribe	Severe rusting in scribe, no blisters away from scribe

3. *Oxidations in Organic Solvents.* Hutchins et al have used benzene solutions of 464D (orange benzene) as a mild oxidizing agent to synthesize aldehydes and ketones from conjugated alcohols.[10] They point out that Adogen 464 can also be used to solubilize $Ce(NO_3)_6^{=}$, $[ON(SO_3)_2]_2^{-4}$, and $ON(SO_3)_2^{-2}$ to form yellow, orange and purple benzene solutions.

4. *Crack Arrestment in High Strength Steel.* Agarwala and De Luccia have used the phase transfer technique to create solutions of inorganic inhibitors (dichromate, nitrite, borate and molybdate) in water-displacing organic solvents for injection into the notch of AISI 4340 precracked corrosion fatigue specimens.[11] A four-fold reduction in crack growth rate was achieved using a solution of dichromate (anodic inhibitor), nitrite (oxygen scavenger), and borate (buffer). Cycles to failure data are as follows:

Solution prepared by contacting 0.1M Adogen 464 in xylene with solid sodium salt	Cycles to failure in 90% RH environment (Kilo cycles).
Control (no compound added)	1.8
Nitrite and Borate	3.3
Molybdate	4.5
Dichromate	6.8
Dichromate and Nitrite and Borate	9.0

CONCLUSION

To date there are nearly 100 phase transfer catalysts which are commercially available, although future impact of the technique on industrial organic synthesis promises many more, such as the phosphonium catalysts. The Adogen 464 — dichromate complex was prepared in high yield and was shown to be a corrosion inhibitor. Further materials engineering applications are anticipated in the oxidative curing of coatings and sealants and in the solution of special corrosion problems.

REFERENCES

1. Sloan, C.K., U.S. Patent No. 2,270,386 dated January 20, 1940.
2. Vessey, C.A. and Harris, A.S., "Development of New Organic Compounds Containing Chromium as Corrosion Inhibitors," Harrowgate, Yorkshire.
3. Ohr, J. and Clark, K., Naval Air Development Center Report No. 74017-30 dated March 5, 1974.
4. Aldrich Chemical Company publication, *Aldrichimica Acta 9(3), 35 (1976)*.
5. Dehmlow, E.V., *Chem. Tech., 210, April 1975.*
6. Petersen, C.J., *J. Am. Chem. Soc. 89, 7017 (1967)*.
7. Petersen, C.J. and Frensdorff, H.K., *Angew:Chem. 2(1), 16 (1972)*.
8. Adogen 464 is a product of Ashland Chemical Co.
9. Rice, K.K.et al., Air Force Materials Laboratory Report No. TR-77-121 dated July 1977.
10. Hutchins, R.O. et al., "Tetrahedron Letters,"No. 48, 1977, p. 4167.
11. Agarwala, V.S. and DeLuccia, J.J., to be presented at NACE Corrosion/79 Conference, Atlanta, Ga.

VI. DESIGN OF COATINGS FOR OTHER THAN CORROSION CONTROL

OPTICAL PROPERTIES OF COATINGS

Eugene M. Allen

Center for Surface and Coatings Research,
Lehigh University,
Bethlehem, PA

EUGENE ALLEN

ABSTRACT

Industry is concerned with the color of a coating in two respects — first, as something to be matched when the product is first formulated, and second as something to be controlled when the product is manufactured. Color is expressed through the use of the CIE system of color specification. Three values are used, called the tristimulus values; these quantities define color as the eye sees it. In order to match a color, one matches the tristimulus values. Computer programs have been written that match the color of any standard by a suitable mixture of specified colorants. Other computer programs show how to correct the color of a batch or other item during production.

Quality control schemes ensure that the color of the manufactured article stays within specified limits. These procedures make use of so-called color difference formulas, which tell how much visual difference there is between a standard and a production sample.

INTRODUCTION

In discussing the reflective properties of coatings, I will emphasize the appearance properties, and will not specifically mention engineering considerations related to reflectivity. Thus I will not be concerned directly with reflective or antireflective properties of coatings as collectors of solar energy, for example. Among appearance properties, I will restrict my remarks to the coloration of opaque film only, and will not consider thin film technology. We should bear in mind, however, that the reflectance relationships that we will mention are equally suited to applications other than color and appearance.

Industry is concerned with the color of a coating in two respects — first, as something to be matched when the product is first formulated, and second as something to be controlled when the product is manufactured. To understand either of these applications we will have to look at the CIE system of color specification, used the world over as the fundamental way of expressing the color of an object.

An object that reflects light appears colored because it does not reflect all of the wavelengths of light equally. If it reflects more red than it does blue, it may appear red. If it reflects mostly in the middle of the spectrum, it will appear green. The complete physical specification of the color of an object resides in its spectrophotometric reflectance curve, which is a record of the fraction of light that strikes it that is reflected at each wavelength in the visible spectrum. This record could be plotted in the form of a curve that shows percent reflectance versus wavelength in nanometers. It could also be presented in the form of a table that shows the reflectance at every ten or twenty nanometers.

In addition to the reflectance curve of an object, we are also vitally concerned with the distribution of energy in the light striking the object. It never happens that a light source has the same energy at all wavelengths. In daylight, for example, there is usually more energy in the blue end of the spectrum than there is in the red end. In the light from a tungsten lamp, on the other hand, the energy distribution climbs up steeply toward the red end of the spectrum. In any color calculation, it is essential to take the energy distribution of the light source into consideration. We do this by multiplying, wavelength by wavelength, the light source energy by the reflectance of the colored object; the result of such a calculation gives the distribution of energy in the radiation leaving the colored object and entering the eye.

Now once this energy gets into the eye, it falls upon three different kinds of cone in the retina. These three kinds of cone have three different spectral sensitivity curves, and flash three different messages to the brain. It is these three messages taken in conjunction that determine the color of the object. If we knew these spectral sensitivity curves exactly, we would have a direct method of calculating the three numbers that determine the color in the form of the three specific messages sent by the cones to the brain.

We do not, unfortunately, know the spectral sensitivity curves of the three kinds of cone. We do, however, have three other functions that we use in place of the spectral sensitivities. These functions were determined by an indirect method, but are nonetheless exactly representative of human vision as it exists. These three functions, together, constitute the so-called standard observer, established in 1931. We use them in the same way that we would use the spectral sensitivity curves of the cones if we had them. We multiply, wavelength by wavelength, the distribution of energy leaving the colored object and entering the eye calculated as described above by the value of each of the three functions at that wavelength. We finally sum up the products for each of the three functions. This calculation scheme gives three numbers representative of the color of the object, designated X, Y and Z, and collectively called the tristimulus values of the color. If $\beta(\lambda)$ represents the reflectance values of a sample as a function of wavelength λ, $S(\lambda)$ represents the energy distribution of the light source, and $\bar{x}(\lambda)$, $\bar{y}(\lambda)$ and $\bar{z}(\lambda)$ represent the three functions that together constitute the standard observer, then the calculation of the tristimulus values can be represented by the following equations:

$$\begin{aligned} X &= k \sum_{\lambda} \beta(\lambda)\, S(\lambda)\, \bar{x}(\lambda), \\ Y &= k \sum_{\lambda} \beta(\lambda)\, S(\lambda)\, \bar{y}(\lambda), \\ Z &= k \sum_{\lambda} \beta(\lambda)\, S(\lambda)\, \bar{z}(\lambda), \end{aligned} \tag{1}$$

where k is a normalizing factor chosen in such a way as to make Y for a perfectly reflecting sample at all wavelengths equal to 100.

If two colors have the same tristimulus values, they match to the standard observer and under the light source for which the calculation was made. This last point is all important, and must be emphasized; I refer to the fact that the color calculation refers to only one specific light source, given by the factor S (λ) in the equations, and to one specific observer, the standard observer. Different tristimulus values would be obtained under some other light source, or by a different observer whose functions differ appreciably from $\bar{x}$, $\bar{y}$ and $\bar{z}$. Two colors that match under one light source as indicated by equal tristimulus values may possibly no longer match under a different light source, as indicated by unequal tristimulus values. It should, however, be pointed out that if two colors have identical spectrophotometric reflectance curves it follows from the way the calculation is done that they will always match under any light source. This is called an invariant match, and is the best possible kind of match you can get. On the other hand, it is quite common to find cases where two samples with entirely different spectrophotometric reflectance curves have identical tristimulus values under, let us say, a daylight source but different tristimulus values under, let us say, tungsten light. This would mean that the two samples would match under daylight but would not match under tungsten light. Such a match is called a metameric match.

To summarize this section very briefly, we can say that if we are trying to match a color we should always strive for a complete spectral match if possible. If we cannot achieve a complete spectral match we must settle for a tristimulus match, but we should at least try to make the spectrophotometric reflectance curves as much alike as possible when we match the tristimulus values. If we don't do this, we will be in trouble either when we try to change the light source or when we try to change the observer.

COLORANT FORMATION AND BATCH SHADING

The Relationship between Reflectance Value and Pigment Composition

In the last section we have considered the calculation of the spectrophotometric reflectance curve into the tristimulus values that represent color as the eye sees it. We must now consider how the pigmentary composition affects the color. The first thing we must emphasize is that the various relationships and calculations that we are going to describe apply to a single wavelength at a time. This means, for example, that if we wish to calculate the tristimulus values for a certain pigment mixture we must first calculate the reflectance values for the mixture at each wavelength and then use equations (1) to calculate the tristimulus values from these reflectance values.

By far the largest proportion of computer color matches today are based on a theory known as the Kubelka-Munk theory. A clear exposition of this approach can be found in Judd and Wyszecki's book[1]. We will not discuss who the actual originator of this approach was; suffice it to say that in their original paper Kubelka and Munk set up and solved the basic differential equations and showed how the algebraic solution to these equations can be used to determine the reflectance of a colored film[2]. What they did was to assume that a turbid film, not necessarily opaque, is irradiated by diffuse light at one surface of the film. Some of the light comes back out the top, some penetrates the film and comes out the bottom, and some is absorbed within the film. We make the simplifying assumption that the light goes either up or down, so that we have only two channels to consider; an upward channel and a downward channel. We now concentrate on a vanishingly thin element of this film somewhere inside the film, so that whereas if the entire film has a thickness X, this element has a thickness dx. Light striking dx from the top suffers weakening in passing through the element dx for two reasons: some of it is absorbed in dx and some of it is scattered backward into the upward channel by means of dx. On the other hand, the light coming out the bottom of dx is reinforced by the light which is scattered back in the downward direction from the upward beam. Exactly the same thing happens in the reverse direction, so that we have a pair of simultaneous differential equations. Solution of these equations gives the following expression:

$$R = \frac{1 - R_g\,(a-b\,\coth bSX)}{a - R_g + b\,\coth bSX}, \tag{2a}$$

where

$$a = 1 + (K/S), \qquad b = \sqrt{a^2 - 1}. \tag{2b}$$

R is the reflectance of the film of thickness X, K and S are, respectively, the absorption coefficient and the scattering coefficient of the film, and R_g is the reflectance of the background over which the film lies. We will see later how K and S of pigments can be determined.

In most practical colorant-formulation and shading work, we deal with opaque films; that is, we deal with films of such thickness that increasing the thickness would have no effect on the reflectance values. If we let X approach infinity in equation (2a), we obtain the following famous algebraic expression:

$$R_\infty = 1 + \frac{K}{S} - \left[\left(\frac{K}{S}\right)^2 + 2\left(\frac{K}{S}\right)\right]^{1/2} \tag{3}$$

The inverse of this equation is even better known:

$$\frac{K}{S} = \frac{(1-R_\infty)^2}{2R_\infty} \tag{4}$$

R_∞ in these equations refers to the reflectance of the film at infinite thickness. We can see, therefore, that if we know K and S we can calculate the reflectance and vice versa. Remember that equations (3) and (4) apply to opaque films only; if the film is translucent we need equations (2a, 2b). Also, all these equations apply only to a single wavelength of light; when we work with the complete spectrum we have to make similar calculations wavelength by wavelength and then use equations (1) to get the tristimulus values.

Now what makes all this work out for colorant formulation is the fact that both the absorption and the scattering coefficients can be built up from the individual absorption and scattering coefficients of pigments or dyes in the film. We have the following useful relationship:

$$\frac{K}{S} = \frac{c_1k_1 + c_2k_2 + \ldots + K_{substrate}}{c_1s_1 + c_2s_2 + \ldots + S_{substrate}}, \tag{5}$$

where c represents concentrations of colorants, k represents unit absorption coefficients of the colorants, s represents unit scattering coefficients of the colorants, and the subscripts 1, 2, ..., are the identifiers for the individual colorants in the film. We can therefore see that if we assume that we have a mixture of four pigments, and if we know the values of k and s of the pigments at a certain wavelength as well as those of the substrate, and also if we know the concentrations of the pigments in the film, we can then calculate the reflectance of this film at the given wavelength by means of the equations just given. If we know the values of K and S at each wavelength throughout the spectrum, we can calculate the whole reflectance curve. Knowing the whole reflectance curve, we can, of course, calculate the tristimulus values of the match and compare them with those of the standard sample.

There is an additional complication that must be taken into account in making these calculations: this is the refractive index discontinuity between air and the film. If we are working with a paint film, the refractive index of the film would be somewhere in the neighborhood of 1.5; that of air is 1.0. A parallel beam of light from a spectrophotometer striking such a film first suffers partial reflection at the boundary. The amount reflected is somewhere around 4%. The rest of the light enters the film and is diffused by the pigments. The light emerging from the film again suffers partial reflection at the boundary. This time, however, the light traveling upward and striking the boundary falls at all possible angles to the boundary, and it is a consequence of the laws of optics that the reflectivity increases with angle until, at angles greater than the critical angle, all the light is reflected and none emerges. The average reflectance of this diffuse light is therefore much greater than 4%, and in fact if the light is completely diffuse, the reflectivity would be 60%. The light reflected back into the film is again reflected by the pigments, again attempts to emerge, and again suffers partial reflection back into the film. Mathematical treatment of this situation results in an equation, commonly known as the Saunderson correction[3] that can be used to eliminate the error that would otherwise be introduced by the existence of the refractive-index boundary. This correction is only approximate but serves for most formulation and shading work. The equation is

$$R_m = K_1 + \frac{(1 - K_1)(1 - K_2)R_t}{1 - K_2R_t}, \qquad (6)$$

where R_m is the measured reflectance, R_t is the so-called theoretical reflectance that calculated by Kubelka-Munk theory, K_1 is the front-surface reflectance of the film, and K_2 is the internal reflectance of the film. The inverse of this equation is

$$R_t = \frac{R_m - K_1}{1 - K_1 - K_2 + K_2R_m} \qquad (7)$$

Let us summarize what we have said so far about Kubelka-Munk theory. We have a relatively simple theory, which nevertheless seems to work quite well. A turbid film is irradiated by diffuse light traveling downward in channel 1. Inside the film we have light traveling in two directions: downward in channel 1 and upward in channel 2. Some of the light that is traveling upward comes out the top of the film and is counted as reflected light. From a pair of simple differential equations we can derive an expression which tells us how much reflected light we have at each wavelength as a function of two quantities: the absorption coefficient and the scattering coefficient of the film. These two coefficients can be built up from the corresponding coefficients of the individual pigments that are in the film. We characterize the pigments by determining their absorption and scattering coefficients at each wavelength.

What we have been describing so far is the two-constant Kubelka-Munk theory, the two constants being the absorption coefficient and the scattering coefficient. In certain applications, a simplification of the theory known as single-constant theory can be used. For pastel shades of paints, for example, the largest single constituent of the pigment mixture is titanium dioxide or some other high-scattering white pigment. Adding very small amounts of colored pigment hardly changes the scattering power of the film, which is largely due to the white pigment. The overall scattering coefficient can therefore be considered to be a constant, and accordingly the denominator of equation (5) is a constant. This means that the K/S ratio can be considered to be built up in the following manner:

$$\frac{K}{S} = c_1\left(\frac{k}{s}\right)_1 + c_2\left(\frac{k}{s}\right)_2 + \ldots + \frac{K}{S}_{substrate} \qquad (8)$$

The ratio k/s is now a parameter on its own even though it is a ratio, and really amounts to only a single constant instead of the two we had before. This simplifies the mathematics considerably, but even more important, it makes it possible to characterize a colorant with only one parameter rather than two. We must be sure, of course, that if we decide to use single-constant theory we will be dealing only with pastel shades, or else that some other situation holds that ensures no change in scattering power of the mixture. An example of the latter that is removed from the paint field is the coloration of textiles, where we have a collection of textile fibers that are dyed by appropriate textile dyestuffs. These dyes can be imagined to be dissolved in the fiber, and therefore have no scattering power of their own. All the scattering is accomplished by the fibers themselves.

Characterization of Colorants

It is clear from what we have said so far that before we can use Kubelka-Munk theory to do color formulation or shading we must characterize our pigments. That is, we must determine the absorption and scattering coefficients of our set of pigments at each wavelength for use with the two-constant theory, or else we must determine the k/s ratio of our set of pigments for use with single-constant theory. How do we calibrate our colorants and determine their Kubelka-Munk constants? The only workable method for making this determination is to work backward from reflectance values.

Let us consider the single-constant case first. If we look at equation (8) and consider the case where we have only one colorant, we have

$$\frac{K}{S} = c\left(\frac{k}{s}\right) + \left(\frac{K}{S}\right)_{substrate} \qquad (9)$$

where c is now the concentration of the single colorant. This equation shows that if we plot the K/S value against concentration of pigment we should get a straight line. The slope of this line is the unit k/s value required to characterize the pigment.

In determining the absorption and scattering coefficients separately for use with two-constant theory, the investigator can use several different methods; I will present the most popular of these here. In this presentation I will assume that we are dealing with opaque films and can therefore use the simplified Kubelka-Munk equations presented as equations (3) and (4). It is quite common for work with opaque films to make the assumption that the scattering coefficient of titanium dioxide white pigment at all wavelengths is unity, and then determine all the other absorption and scattering coeffients relative to this assumption. In the description that follows I will be talking about a single wavelength, but we should understand that the procedure must be repeated at each wavelength of interest.

The first step is to determine the absorption coefficient of the white pigment. This is easy; we make a drawdown of the white pigment, measure its reflectance, convert to K/S by equation (4), and then since we assume that S is unity we say that $K/S = K$, the absorption coefficient of the white.

We will now describe the tint-ladder method, used by Cairns and Spooner[4] of the du Pont Pigments Department. Mixtures of the colored pigment to be characterized and titanium dioxide are made at different ratios of colored pigment to white. About five to eight such mixtures are made, including a masstone (no titanium dioxide). The reflectance values for each of these mixture drawdowns is measured, and converted to K/S by equations (7) and (4) in tandem. We therefore have, for each pigment at each wavelength, a series of $(K/S)_i$ values and a corresponding series of c_i values, the latter being the fraction of colored pigment in the mixture of colored pigment and white. The subscript i refers to the ith mixture of colored pigment and titanium dioxide.

Let us now rewrite equation (5) for the present situation. We will let the subscript 1 refer to the colored pigment and subscript 2 to the white pigment, these being the only two pigments present. We assume that there is no substrate to speak of; any absorptive or scattering properties of the vehicle can be neglected. We thus have

$$\left(\frac{K}{S}\right)_i = \frac{c_i k_1 + (1 - c_i)k_2}{c_i s_1 + (1 - c_i)s_2} .$$

If we multiply throughout by the denominator of the right-hand side and rearrange, we get

$$c_i k_1 - c_i (K/S)_i s_1 = (1 - c_i)[(K/S)_i s_2 - k_2]. \tag{10}$$

This equation has two unknowns, k_1 and s_1; everything else is known. If we had only two such equations, we could solve them simultaneously for these two unknowns. But since we have five to eight mixtures, we have as many equations, and we therefore have an overdetermined system that we can solve by least squares. The solution gives us k_1 and s_1, the unit absorption and scattering coefficients of the colored pigment.

The tint-ladder method gives good results except in regions of the spectrum where the colored pigment has a high reflectivity approaching that of the white. For example, chrome yellow absorbs as little light as titanium dioxide does in the red region of the spectrum, and is also a good scatterer, as is titanium dioxide. Therefore the spectrophotometer cannot easily distinguish between chrome yellow and titanium dioxide at these wavelengths, and the simultaneous equations become indeterminate for that reason. Despite this difficulty, however, workable results are obtained by the tint-ladder method and pigments standardized in this way give good results in computerized color formulation.

Other investigators, however, prefer to eliminate the kind of indeterminacy just described by using a black reduction as well as a white reduction. The thinking is that if you mix the colored pigment with a black instead of a white, using enough black to reduce the reflectance to about

50%, you have much more accuracy where the colored pigment does not absorb. You can still use the white pigment to work on the regions of the spectrum where the reflectivity is low; the general idea is to mix unlike pigments so as to obtain maximum accuracy. The procedure, then, would be to prepare three drawdowns: a masstone, a mixture with white, and a mixture with black. The masstone and the white reduction would be used to determine the absorption and scattering coefficients of the colored pigment in the strongly absorbing regions of the spectrum; the masstone and the black reduction would be used in regions where the colored pigment hardly absorbs any light. In both cases, equation (10) would be used and solved simultaneously, two equations at a time. For the black reduction, the subscript 2 now refers to the black pigment instead of the white pigment.

In order to make the black-reduction method work, a carefully controlled and standardized black pigment is necessary. Many black pigments suffer from agglomeration difficulties, and some investigators find that a preformulated black paint works best for this purpose. The absorption and scattering coefficients of the black must be determined from a masstone and a reduction with white in the usual manner.

An important part of the standardization problem is the setting up of workable Saunderson constants k_1 and k_2. The k_2 constant has received special attention. We saw previously that $k_2 = 0.60$ would be correct if the light inside the film were completely diffuse. It is not, however, and many workers recommend lower values for k_2. Saunderson himself recommended 0.4 for pigmented plastics[3] and Mudgett and Richards[5] have worked out a theoretical basis for preferring a k_2 value of about 0.4 to higher values.

Many workers use k_2 as another constant to be fitted to the experimental results. A good example is the work of Andrade[6] who found that he was able to write equations that expressed k_2 as a function of pigment volume concentration and Kubelka-Munk scattering coefficient.

Visual Methods for Formulation and Shading

Having established the basic mathematics and technology used for industrial color control, we will now look at the main production problems for which these methods are used. First is the difficult activity of colorant formulation (setting up dye or pigment recipes) and the allied problem of batch control.

What does a color formulator in the paint industry, for example, do when he receives a sample to match? He knows from long experience that it takes four pigments in general to produce a color match under a particular light source, say, daylight. Let us imagine that he has fifteen pigments of various colors at his disposal with the proper physical and chemical characteristics that would make them suitable for the particular paint under consideration. With this number of pigments there are many formulas that he could use to match the color, even though each of the formulas contains four pigments. In fact, if the color happens to be not too bright, and if the fifteen pigments provide a fairly good color gamut it can be shown by computer methods that the number of possible formulas would extend into the hundreds. Which one of the tremendous number shall he choose and on what basis shall he make his choice? Remember, we are assuming that each of the possible formulas will provide a perfect match to the formulator's eyes and in daylight, so that this is not a criterion.

The two main considerations that will guide him are metamerism and cost. Metamerism is a phenomenon arising from the fact that different colorants are used in the match than were used in the standard. If the matching was skillfully done, this difference of colorants does not create too much of a problem. But if the match was poorly done, you might have a good match under one light source but a poor match under a different light source. Another manifestation of this effect is that one observer in daylight may see the match as a good one whereas a different observer may not see the two colors as matching at all. In either case, the match is termed a metameric match and is to be avoided by all means. We have already considered metamerism in our discussion of the CIE system, where we found that a metameric match results when standard and match have the same tristimulus values under a given set of conditions but different spectrophotometric curves. It usually works out that if we change either the light source or the observer the tristimulus values are no longer equal and the match falls apart.

The other major consideration is the cost of the match. The different matching formulas have different price tags associated with them. There is no way of telling in advance what the cost of a match will be other than determining the matching formula and then making a cost calculation. The most important determinants of the cost of any particular formula are first of all the cost of the individual pigments that go into the formula and the amounts of each required, and second the covering power of the pigments in combination which will dictate how much pigment must be put into the paint in order to obtain adequate coverage of the painted surface.

The formulator cannot possibly make up each of the hundreds of formulas that would match the color. He must, instead, rely on his years of experience plus his intuition to decide which combination of pigments he will choose. Once he has made this choice, he must determine, by trial and error, the amounts of each to use in order to obtain a match. He may require as many as seven or eight tries before he has achieved a match to his and his supervisor's satisfaction. Each trial involves making a mixture of the pigments in the proper proportion and then incorporating these into a paint with the desired vehicle. The resulting trial paint is then compared for color against the standard by putting a dab of the paint on paper or sheet aluminum, passing a broad bladed knife down the dab of paint to remove the excess, and finally allowing the paint to dry. The color cannot be judged before the paint is dry, and this drying step may take quite some time. Finally, a color judgment is made against the standard, and if the match is not exact another trial is performed. The colorist must make sure that the final sample matches the standard not only in daylight but under any other possible light source that might be used. You can easily see that with the number of trials required and the time required for mixing and drying each trial the process of obtaining a good matching formula may take many hours if not several days. This may be an important consideration if competitive bidding against other paint companies is involved.

The activity just described is an example of colorant formulation. This activity means obtaining a match for a new sample submitted by a customer. Always involved is the problem of getting the best formula (i.e., best pigment combination) out of a large number of possibilities. Always involved also is the problem of determining the proper ratio of pigments in the chosen combination in order to achieve a match. Before the advent of computer methods, there was no certainty that the best possible formula had been chosen; it was just a question of having faith in the excellence of the colorant formulator's ability.

Batch shading is the operation that ensures that a paint batch manufactured in the plant winds up by having the proper color. It is similar to colorant formulation in that a proper ratio of pigments must be chosen. In this case, however, the correct combination of pigments is already known and it is only a matter of determining the amount of each to use. Even though the formula used for making the previous batch of paint is known exactly, automatic use of the same formula will not necessarily produce the same color; in fact, it is quite seldom that this would happen. Different batches of pigments are available now compared with the time of the last production. Also, differences in dispersion of the paint in the vehicle may occur and these have a profound effect on the color. There are also mechanical errors and inaccuracies in weighing and mixing that would affect the color. Accordingly, it is very seldom that a paint manufacturer merely weighs up the amounts of pigment called for by the paint formula and grinds these amounts directly into the paint batch. If the color winds up too red, indicating that too much red pigment has been used, it is impossible to remove the red from the batch once it has been put in. The only kind of correction possible is to increase the batch size by going up on the amounts of the other pigments, which is not a good procedure.

It is therefore customary to begin the production of a batch by using a smaller amount of all of the pigments than is called for by the formula, say, 80%. A color comparison is made against the standard, and then an add is made which would bring the amount of pigment up to about 90%. This add is color adjusted so as to compensate for any color inequity in the first try. In this way, the correct color is achieved by several stepwise additions of pigment. For batch shading, too, each addition of pigment and color examination requires the preparation of a drawdown and examination of this drawdown after it is dry. Here again, the drying procedure entails considerable hold-up of production because of the times involved. It is not uncommon to have residence times of a paint batch in the plant of one or more days. With reputable paint manufacturers exerting stringent control over the color, this color adjustment process may be the critical factor that limits profits for the company.

Computer Programs for Formulation and Shading

In view of the complexity of the color formulator's task, it is not surprising that computer methods of doing this work were eagerly sought. Today the larger paint companies all use various kinds of computer hardware and software to help in this task.

The most widely used kind of program for colorant formulation is the combinatorial program, so called because it tries combination after combination of colorants. It works as follows: the color to be matched has three tristimulus values under a standard light source, say, daylight. The matching combination of colorants must have the same tristimulus values or else there is no match. In order to equal three tristimulus values, we must be able to vary three things independently -- three equations, three unknowns. For opaque paints, we need four pigments to give us the required three degrees of freedom. The colorist supplies the computer with a list of all the pigments that he feels can be used for the particular application at hand (usually 15 or less). The combinatorial program goes through all the possible combinations of the pigments on the list taken four at a time. By procedures that will be mentioned later, a tristimulus match to the standard under daylight is calculated for each of these combinations, where possible. When we say that a tristimulus match is calculated, we mean specifically that we determine the relative proportions of the four pigments needed to give a color that has the same tristimulus values as the standard. Also, we say 'where possible' because some of these combinations cannot possibly provide a match; for example, you cannot match a blue sample with a combination consisting of four reds. But enough of the combinations are usually left to provide a choice of match formulas.

The formulation program must then decide which of these matches is the best. The choice is made on the basis of the criteria mentioned previously. The criterion that the match should be non-metameric requires that the computer program should calculate some sort of a metamerism index for each match and then choose one with a low index. A commonly used index of metamerism is given by taking each formula that has given a match under daylight and calculating the color difference between the formula and the standard under another quite different light source, usually tungsten light, some fluorescent-lamp source, or both. The greater this color difference, the more metameric the match by this convention.

Another type of metamerism index which might be used was suggested by Nimeroff and Yurow[7]. It is based on the fact that the reflectance curves of match and standard may differ from each other even though the tristimulus values are the same. A general index of metamerism, which would not be tied down to specific light sources or in fact to any light sources at all, would then be given by calculating the mean square deviation between the two reflectance curves. If desired, these deviations might be weighted by the color-mixture functions in accordance with Nimeroff and Yurow's suggestions.

The cost of each matching formula is very easy to calculate if we know the price of each colorant. The formulation program would then give a metamerism index and a cost for each of the matches, and the programming can be done in such a way as to print out the matches in any desired order based on these two criteria.

Computer shading programs differ from the formulation programs in several important respects. There are two samples to consider: a trial batch sample that is to be shaded over to the standard; and the standard itself The problem is how much of each of the colorants to add to the batch in order to match the standard. We are not concerned here with alternative formulations from which we must select the best, which means that we are also not concerned with metamerism. The reason for this last statement is that in formula shading the standard is not the original sample supplied by the customer but rather a production sample that has been set aside as standard — perhaps the first batch made. For this reason the matching is entirely non-metameric.

Algorithms for Colorant Formulation

We have now standardized all our pigments that are going to be used for colorant formulation for a certain type of paint, and will soon receive our first shade to match. How can we write a computer program that will calculate the concentration of four pigments needed to give the same tristimulus values as those of the standard being matched?

The basic mathematics for making this calculation have already been presented. Consider a specific combination of four pigments. Let us suppose that we will guess at the proportions of the four pigments which we think will give a close match to the tristimulus values. Since we know the unit K and S values of each of the pigments at each of the wavelengths, we can use equation (5) at each wavelength to calculate the K/S of the mixture. We then use equation (3) followed by equation (6) to convert K/S to reflectance at each wavelength. This gives us the reflectance curve of our postulated mixture. We can then calculate the tristimulus values of the mixture and see how close they come to those of the standard. If our guess has been fortunate, they will match. If not, we must make a correction to the formula and start over again.

Now there are algorithms available that make it unnecessary to guess at the three starting concentrations for a tristimulus match. Unfortunately, these algorithms do not give a direct answer, and the three starting concentrations usually do not produce an exact match to the tristimulus values. But the match is much closer than would be obtained by merely guessing. Furthermore, the same algorithm tells how to correct the three concentrations to get a much closer match. The program works by iteration and stops when the tristimulus values of the match equal those of the standard to within a preassigned tolerance. I have published such an algorithm for use with single-constant Kubelka-Munk theory[8] and, more recently, a similar algorithm for use with two-constant theory[9].

COLOR DIFFERENCE AND QUALITY CONTROL

In the control of the color of a manufactured article, the measurement of the color difference between the production sample and the standard is of great importance. If this difference is greater than a specified value, the production sample will be rejected. The problem can be stated very simply in the following way: given the tristimulus value of a standard and the tristimulus values of a sample, what is the difference in color between them?

We might consider two separate ways of solving this problem. The first way would be to obtain a single number representing the color difference between the standard and the sample. If this number is bigger than a previously agreed upon tolerance, the sample is to be rejected. The second way, which is more sophisticated and probably more generally useful, is to admit that a single number is not sufficient for determining the acceptability of a sample. The nature of the color difference has a lot to do with it. For example, a sample of butter might be very slightly redder than standard and yet be acceptable. If the sample were greener than standard by the same amount, however, it might be unacceptable even though the numerical value of the color difference would be the same in the two cases. It therefore becomes expedient to visualize a "color space" in three dimensions (since there are three tristimulus values) and to create a geometric figure in this color space which surrounds the position of the standard. All batches whose color falls within this figure would be acceptable. The figure would not necessarily have to be symmetrical around the standard; for example, in the case of the butter standard just mentioned it would extend much further on the red than it would on the green side.

It is not within the scope of this paper to present any detail on the constitution of either color difference formulas or uniform color spaces. The standard texts on color science should be consulted for more detail on these topics.

REFERENCES

1. Judd, D.B. and Wyszecki, G., "Color in Business Science and Industry," 3rd edition, John Wiley & Sons, New York, 1975.
2. Kubelka, P. and Munk, F., *Z. Tech. Phys. 12, 593 (1931).*
3. Saunderson, J.L., *J. Opt. Soc. Am. 32, 727 (1942).*
4. Cairns, E.L. and Spooner, D.L., private communication.
5. Mudgett, P.S. and Richards, L.W., *J. Paint Technol. 45, 43 (1973).*
6. Andrade, D., private communication.
7. Nimeroff, I. and Yurow, J.A., *J. Opt. Soc. Am. 55, 185 (1965).*
8. Allen, E., *J. Opt. Soc. Am. 56, 1256 (1966).*
9. Allen, E., *J. Opt. Soc. Am. 64, 991 (1974).*

ULTRAVIOLET CURING COATINGS

Robert W. Bassemir

Sun Chemical Corp., Graphic Arts Lab,
631 Central Ave.,
Carlstadt, N J

ROBERT W. BASSEMIR

ABSTRACT

An introduction to the use of ultraviolet coatings in industry is presented. The four most widely used types of curing chemistry are discussed along with their advantages and disadvantages.

The commonly used photoinitiatory systems are described along with the most probable mechanisms by which they initiate polymerization. An introduction to the various types of actinic light sources is given and the effect of coating parameters on cure rate and film properties is discussed in detail. The paper concludes with a presentation of the future needs of the UV coating industry, if the use of this technique is to penetrate the metal coatings market.

INTRODUCTION

There has been concentrated discussion during this meeting about the use of coatings for corrosion protection. It may also be worthwhile to look at the various purposes for which surface coatings are utilized.

Coatings for Decorative Purposes:

1. To convey information — as in printing inks, stencil paints or color codes;
2. To impart color or contrast;
3. To make the surfaces coated attractive and pleasant in appearance;
4. To enhance or subdue the gloss of the surface.

Coatings for Protection or Function:

1. To impart abrasion resistance;
2. To impart corrosion protection for substrates;
3. To impart weather resistance for substrates;

4. To impart a controlled surface friction (slip);

5. To provide anti-fungal action.

The design of U.V. cured coatings to suit these purposes has been successful in some cases and has yet to be accomplished in others. In this paper, a review of the use of U.V. coatings in industry will be given along with their composition and characteristics.

The use of radiant energy for curing coatings has been known for many centuries. However, it was only recently, in the early 1940's, that the first reductions to commercial practice occurred and these were in the printing ink field. These were abandoned after industrial tests failed to reach their technical and economic goals.

Renewed efforts in the 1960's in both the U.S.A. and Europe produced commercially successful U.V. curing surface coatings. In Europe, these were utilized mainly as filler coats for wood paneling while in the U.S. printing inks for paper and paperboard were successfully commercialized. These developments were stimulated by different needs. In the U.S. federal legislation was introduced to limit air pollution, while in Europe there was an economic need for a low cost, rapid and effective method for sealing and decorating open grain surfaces.

Both of these applications as well as others have achieved significant commercial sales and are growing markets for U.V. curable materials.

ADVANTAGES

The advantages of using U.V. activated coatings are usually specifically related to the industry and substrate for which they are used. However, there are several advantages which apply irrespective of the specific area of utilization. These are:

1. Elimination of effluents;

2. The rapid speed of cure;

3. The generally good crosslinked properties obtained.

There are, of course, disadvantages, the most notable of which are:

1. Generally higher costs of raw materials for the coatings, due to elimination of solvent;

2. High levels of opaque pigmentation tends to slow down cure rates;

3. Thick films also tend to slow down cure rates.

CHEMICAL MECHANISMS OF U.V. CURING[1,6]

A number of different chemical approaches to obtaining cured films quickly through the use of U.V. irradiation have been developed during the past decade. These may be categorized as follows:

1. Free Radical Addition Polymerization and/or Crosslinking

The curing chemistry of this type most utilized is the acrylate class because of very rapid reaction rates, ease of polymerization, moderate prices and good availability in commercial quantities.

A second type of chemistry is that used mainly in wood coatings. This system is composed of unsaturated polyesters mixed with low boiling reactive monomers, mainly styrene. These coatings are very slow curing and the monomer is very volatile before curing.

A third type of curing chemistry using free radical initiation and propagation is the unique, patented thiol-ene reaction developed by W. R. Grace Company. This system uses various mercaptans which can react with unsaturated resins or oligomers to rapidly form a solid. The odor and high cost of these systems has limited their widespread use.

2. Ionic Polymerization

A number of patents and publications have appeared describing the cationic polymerization of epoxy-bearing compounds by the use of U.V. light-generated Lewis or Broensted acids. The backbone vehicles have been various types of epoxy resins which can be very quickly cured in the presence of these powerful catalysts. Some of these systems have very short pot lives and are two-part systems. These systems are being commercially supplied by American Can Company, 3M Corporation, and General Electric.

U.V. INITIATORY PROCESSES

The major classes of photoinitiators used for U.V. free radical polymerizations are the aryl ketones and the acyloins. These are typified by the classic examples of benzophenone and benzoin ethers, respectively. The aryl ketones function by undergoing intermolecular hydrogen abstraction, while the benzoins undergo homolytic fission when irradiated (1A)[17], (Figure 1). Some widely used initiators are listed in Table I.

The ionic polymerization previously described uses diazonium salts, diaryl iodonium salts and most recently trialkyl sulfonium salts,to generate the catalysts for curing of the epoxy resins used in these coatings.[1,6] (Figure 2).

Some unusual effects are notable when photoinitiators are used in combination. These have been described as synergistic and a number of patents mention combinations of two or three different photoinitiators as being superior in performance. (1A)(1F)[8,12].

Amines are well known accelerators for the free radical type of chemistry and typical among the amines used are dimethylethanolamine and triethanolamine. (1A)[16,28,37].

Halogenated photoinitiators have also been mentioned in a number of patents but they must be used with great care, particularly on metal substrates, since they may tend to cause corrosive reactions after irradiation with U.V. due to the halogenated products released during the reaction. (Figures 3 & 4) Incorporation of acid scavengers, such as glycidyl ethers can mitigate this effect, or eliminate it, if properly chosen.

TYPICAL COATINGS COMPONENTS

The typical U.V. coating will usually consist of the following:

1. Oligomers or prepolymers;

Using Intermolecular Hydrogen Abstraction:

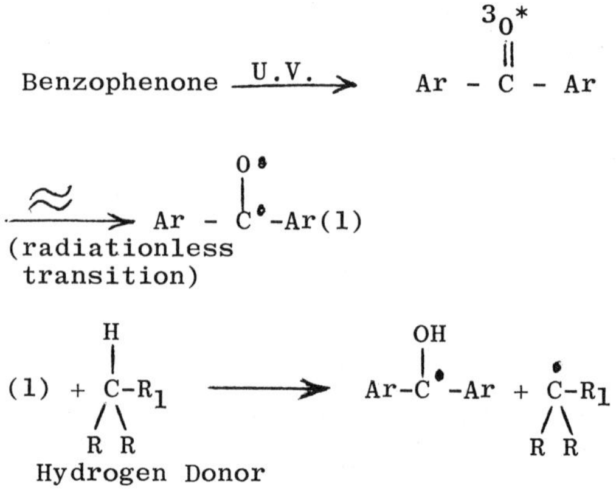

$$\text{Benzophenone} \xrightarrow{U.V.} Ar - \overset{^{3}O^{*}}{\overset{\|}{C}} - Ar$$

$$\xrightarrow[\text{(radiationless transition)}]{\approx} Ar - \overset{O\bullet}{\overset{|}{C^{\bullet}}} - Ar \ (1)$$

$$(1) + \underset{\underset{\text{Hydrogen Donor}}{R\ R}}{\overset{H}{\overset{|}{C}} - R_1} \longrightarrow Ar - \overset{OH}{\overset{|}{C^{\bullet}}} - Ar + \underset{R\ R}{\overset{\bullet}{C} - R_1}$$

Using Intramolecular Cleavage:

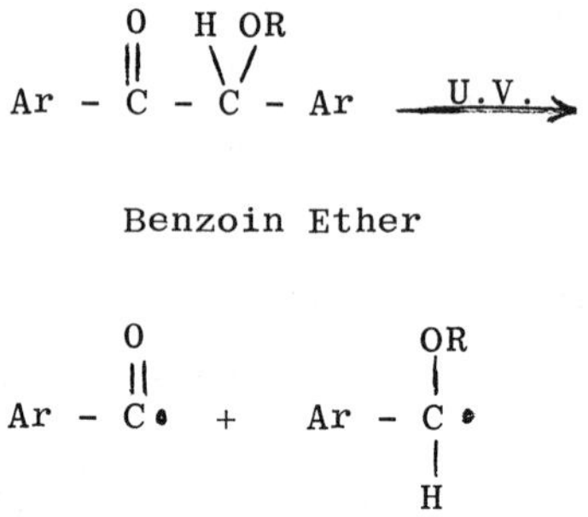

$$Ar - \overset{O}{\overset{\|}{C}} - \overset{H\ OR}{\overset{\backslash /}{C}} - Ar \xrightarrow{U.V.}$$

Benzoin Ether

$$Ar - \overset{O}{\overset{\|}{C}}\bullet + Ar - \underset{\underset{H}{|}}{\overset{OR}{\overset{|}{C}}}\bullet$$

Fig. 1.—Photoinitiation mechanisms.

2. Monomers and/or reactive reducers;

3. Photoinitiators;

4. Pigments and/or colorants;

5. Special additives.

(1) Oligomers:

These materials are the backbone of the coating and are analogous to the resins or binders used in conventional coatings. They provide film toughness, pigment wetting and the proper rheology for good transfer to the substrate. Some of the common types which have been used in acrylic U.V. coatings are:

(a) Epoxy Acrylates:

TABLE I.—COMPOUNDS WHICH HAVE BEEN USED AS PHOTOINITIATORS IN ACRYLIC SURFACE COATINGS

4-Acetyl biphenyl
Anthraquinone
Benzoin ethers
Acyl chlorides
Benzil
Benzophenone
4,4'-Bisdimethylamino benzophenone (Michler's Ketone)
4,4'-Bis-Trichloromethyl benzophenone
2-Chlorothioxanthenone
Dibutyl phenyl phosphine
6,6,-Diethoxy acetophenone
p-Dimethylamino benzaldehyde
2-Ethyl anthraquinone
Michael amine adducts
Thioxanthenone
2-Benzoyl benzoic acid esters
Benzoyl biphenyl
1-Benzoyl napthalene
4-Benzoyl pyridine
2-Chloranthraquinone
Chlorobenzophenone
p-tert-butyl 6,6,6,-trichloroacetophenone
1,3-Di(N-Beta-hydroxyethyl-4-piperidyl) Propane
6,6,-Dimethoxy-2-phenylacetophenone
p-Dimethylamino benzoic acid esters
Hexachloro-para-xylene
Sufonyl chlorides
6,6,6,-Trichloroacetophenone
Amine accelerators (hydrogen donors):
Diethanol amine
Dimethyl ethanol amine
Triethanol amine

1. Epoxy drying oil acrylates;

2. Bisphenol A diglycidyl ether acrylates;

3. Modified Bis A epoxy acrylates;

4. Aliphatic epoxy acrylates.

(b) Polyol Acrylates:

1. Pentaerythritol triacrylate/tetraacrylate;

2. Trimethylol propane triacrylate;

3. Dipentaerythritol hexaacrylate.

1. Aryldiazonium Salts:

$$Ar - \overset{+}{N}{\equiv}N \quad BF_4^- \xrightarrow{U.V.}$$

$$Ar - F + N_2\uparrow + BF_3$$
(Lewis Acid)

2. Diaryliodonium Salts:

$$Ar_2 \; I^+ \; BF_4^- + R\text{-}H \xrightarrow{U.V.}$$

$$ArI + Ar\bullet + R\bullet + HBF_4$$
(Brønsted Acid)

3. Triarylsulfonium Salts:

$$Ar_3S^+ \; BF_4^- + R\text{-}H \xrightarrow{U.V.}$$

$$Ar_2S + Ar\bullet + R\bullet + HBF_4$$
(Brønsted Acid)

Fig. 2.—Photoinitiators for cationic polymerization.

(c) Urethane Acrylates:

These are acrylated carbamates. They are ordinarily reaction products of a hydroxyl containing acrylate, a di or poly isocyanate and in some cases another hydroxyl containing moiety such as a polyester resin, an alkyd, a fatty alcohol, a melamine resin or a polyether having terminal hydroxyl.

(2A) Monomers

Monomers in coatings allow the formulation to be brought to the proper viscosity and flow for smooth application. In addition, the proper selection of monomers is required to provide high conversion percentages. A choice of mono, di and occasionally trifunctional monomers may also be made to regulate the degree of crosslinking obtained in the finished coating film.

(a) Monoacrylates:

hydroxy butyl acrylate, phenoxy ethyl acrylate, cyclohexyl acrylate, dicyclopentadiene acrylate, Cellosolve acrylate.

(b) Diacrylates:

1,6 hexanediol diacrylate, triethylene glycol diacrylate, tripropylene glycol diacrylate, neopentyl glycol diacrylate.

(c) Triacrylates:

trimethylol propane triacrylates, pentaerythritol triacrylates.

Fig. 3.—SEM photo @150X of U.V. coating on steel, 1 hour after irradiation. Coating was formulated with halogenated photoinitiator and no acid scavengers.

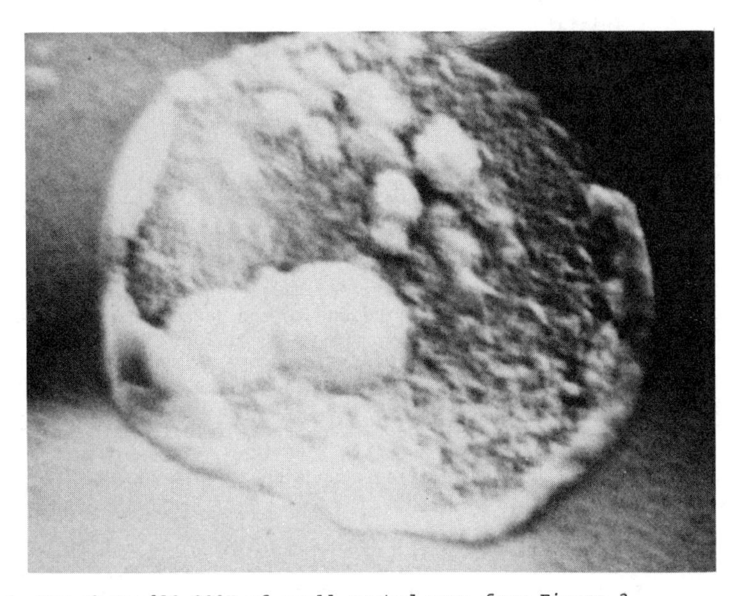

Fig. 4.—SEM photo @10,000X of small rusted area from Figure 3.

(2B) Reactive Reducers:

These are primarily used for viscosity adjustment when very low consistencies are required. Common types are vinyl acetate, n-vinyl pyrolidone and 2 ethyl hexyl acrylate.

(3) Photoinitiators:

Figure 5 lists a number of photoinitiators which have been used in U.V. coatings. In addition to these chemically identified compounds, there are a number of commercially available photoinitiators, some of which have not been chemically identified by the vendor, but which do provide efficient curing for some types of U.V. coatings.

Benzophenone
Michler's Ketone
Thioxanthone
Chlorothioxanthone
Methylthioxanthone
Esters of p-Dimethylamino Benzoic Acid
Benzoin Methyl Ether
Di ethoxy acetophenone
Trichlor acetophenone Derivatives
Phenyl propanedione benzoyloxime
Irgacure 651 (Ciba-Geigy)
Sandoray 1000 (Sandoz)
Initiator FI-4 (Eastman)
Trigonal initiatiors (Akzo)
Quantacure initiators (Aceto)

Fig. 5.—Widely used photoinitiators for acrylic U.V. curable coatings.

(4) Pigments and Colorants:

The choice of pigments and colorants for U.V. coatings must be made with several additional considerations in mind due to the potential reactivity of these materials with the U.V. vehicles. For example, some pigments may catalyze dark polymerization and provide poor shelf life. Impurities that may be present in some organic pigments such as residual reactants or unwashed salts, may also cause instability and it is, therefore, advisable to pretest the pigments which are chosen with the particular vehicle system in order to ascertain whether a stable condition exists. Further effects of pigmentation on cure rate are discussed in a later section of this report.

(5) Additives:

Many of the same types of materials used in conventional coatings for controlling gloss, wetting, dispersion, scuff resistance, slip and viscosity are also useful in U.V. coatings. In addition, most coatings include an inhibitor to prevent premature thermal dark polymerization. In these cases, substituted phenol types of inhibitors have been found most useful. It is also possible to include small amounts of non-reactive modifying resins and plasticizers in order to obtain a desirable balance of film properties or adhesion to specific substrates.

Some typical white coatings formulae are given in Figure 6.[22]

ACTINIC LIGHT SOURCES (1C)[16]

1. Arc Lamps

(a) Electrode type:

This type of arc lamp has proven the most popular in commercial usage and both medium and low pressure gas filled Mercury lamps have been used. The medium pressure quartz tubular lamp, at a 200Wts/inch rating, is probably the most widely used in industrial curing equipment for coatings and inks. It is typically housed in an elliptical reflector and the surface to be cured should be located at the focus. (Figures 7 & 8)

(b) Radio frequency excited lamps:

The development of an electrodeless lamp, fired by RF excitation allows more flexibility in the choice of material with which the tube is filled, thus giving a choice of spectral outputs. It is also cooler in operation, longer-lived, but somewhat more expensive in capital cost.

(c) Pulsed lamps:

The use of Xenon flash lamps with high repetitive pulsing rates has also been recommended for U.V. curing, particularly with thick films.

2. Lasers

During the past few years, U.V. lasers of reasonable power output and efficiency have become available. Lab tests are now being conducted in the use of these tools for curing coatings and some of the early work done in our laboratory looks extremely promising from a technical viewpoint. The cost of this equipment in commercial quantities is presently unknown and will likely be a major factor in determining whether they are ultimately used for curing coatings.

EFFECTS OF COATING PARAMETERS

1. Film Thickness

Since U.V. coating utilizes electromagnetic radiation for curing it is obvious that the thickness of the coating film will attenuate the intensity of the radiation as it penetrates the film. Because of this, there has been much attention paid to the curing of thick films at the surface and at the bottom. Several papers[8,23] have estimated the ratio of surface to substrate layer curing based upon the absorption and scattering of radiation as it penetrates the film. Some estimates have shown a ratio of cure of more than one thousand to one. (Figure 9)

In pigmented thin films (12 μm and less) it was found[1,8] that the log of cure speed was inversely proportional to film thickness (Figure 10).

Epoxy based white UV coatings

Materials	*A* Wt %	*B* Wt %	*C* Wt %
Acrylated epoxy (DRH-303.1, Shell Chemical Co)	50.00	50.00	47.00
Rutile titanium dioxide (RTC-2, Tioxide of America)	25.47	25.47	23.13
Hexanediol diacrylate	10.19	10.19	10.29
Trimethylolpropane triacrylate	—	10.19	10.29
Pentaerythritol triacrylate	10.19	—	—
Polyethylene powder (Polymist B-6, Allied Chemical Co)	—	—	5.14
Igepal CO-430 (GAF Corp)	0.13	0.13	0.13
2-Chlorothioxanthone (Sherwin-Williams)	0.02	0.02	0.02
Hexachloroethane (American Firstoline, Div of Instel Corp)	2.00	2.00	2.00
Dimethyalminoethanol (Pennwalt Corp)	2.00	2.00	2.00
Physical properties			
Tukon hardness	12.6	7.0	6.7
Reverse impact (pass/fail, in-lbs)	0-4	4-8	4-8
Conical bend (failure, inches)	1-1/8in	7/8in	1in
Gloss 60°	94.8	95.2	82.8
Gloss 20°	81.0	85.8	44.2

Note: 1.0mil coating cured in air by exposure to a 200watt/linear inch UV source for 3 seconds

White pigmented UV coating system

Coating composition	Weight %
DRH-303 (Shell) epoxy diacrylate	40.0
2-Ethylhexyl acrylate	2.5
N-Vinyl pyrrolidone	7.5
Trimethylolpropane triacrylate	25.0
Rutile titanium dioxide (0.2–0.25μ)	25.0

Photoiniator packages		phw
2-Chlorothioxanthone	A	3.0
Dimethylaminoethanol	A	2.0
2-Chlorothioxanthone	B	1.5
Carbon tetrachloride	B	1.5
Dimethylaminoethanol	B	2.0

Note: 1.5mil coatings cured in air by exposure to a 200watt/linear inch UV source for 3 seconds (3 passes at 20 feet per minute).

Photo-initiator package	Pigment grade*	Tukon hardness	Conical Mandrel (inches)	Reverse impact (inch-pounds)
A	R-CR 3	9.5	Fail	0-4
B	R-CR 3	13.9	Fail	0-4

Fig. 6.—Typical white coating formulae.

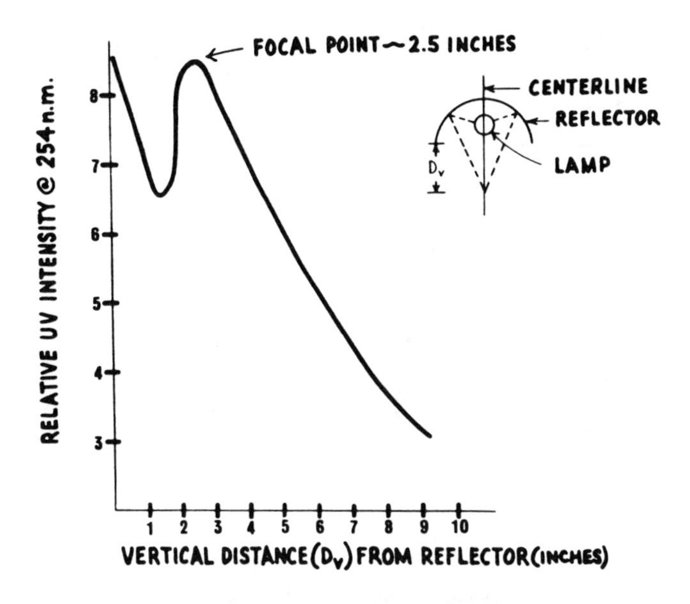

Fig. 7.—Lamp U.V. intensity vs. vertical distance.

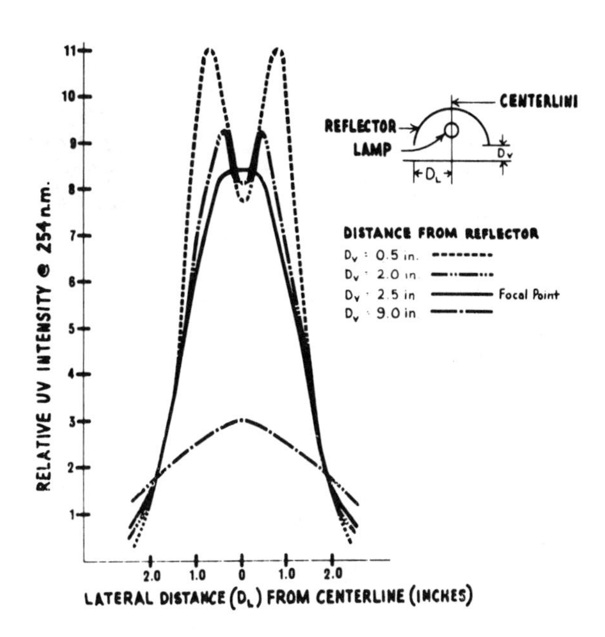

Fig. 8.—Lamp U.V. intensity vs. lateral distance.

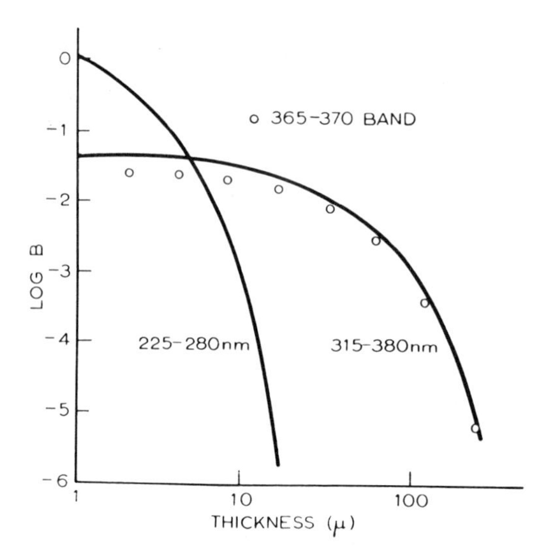

Fig. 9.—Relative cure in bottom layer of white coating.

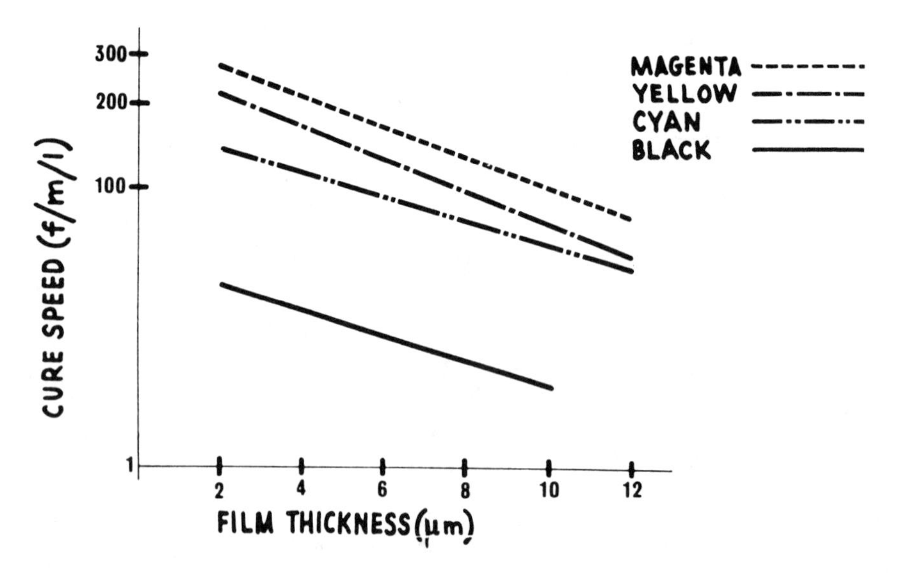

Fig. 10.—Cure speed vs. coating film thickness.

These films, however, did not contain highly scattering pigments such as TiO_2.

2. Pigment Concentration

The pigment loading of a coating will cause increased scattering and absorption of the incident U.V. light. The scattering will be regulated largely by the particle size and by the refractive index difference between the pigment and vehicle surrounding it. Since many pigments absorb strongly in the U.V. region they will compete with the photoinitiator for the U.V. radiation, thus further slowing the rate of generation of free radicals in the lower layers of the film.

3. Color and Type of Pigmentation

The U.V. absorptivity of pigments varies sharply depending on their chemical constitution. Many organic pigments are quite transparent to a major portion of the U.V. spectrum, while others (carbon black) can be nearly opaque. A study of the U.V. transmission of some common pigments was made for thin films (1, Chapter 8) and showed that yellow and red pigments were reasonably transparent while phthalo cyanine blue and carbon black transmitted rather poorly (Figure 11).

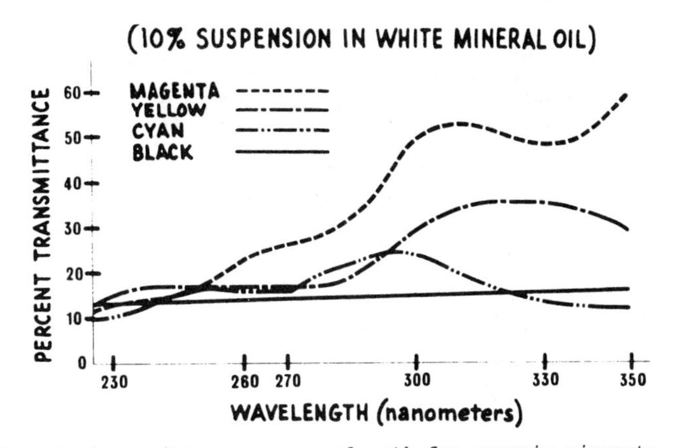

Fig. 11.—U.V. transmittance vs. wavelength for organic pigments.

In the area of white coatings a number of studies have been made of alternatives to the use of Rutile TiO_2 since it is the most widely used opaque white pigment in coatings. The use of Anatase TiO_2, Zinc Oxide and a variety of other white inorganic pigments have been examined. (1, Chapter 3)[2,7,19]. Despite the fact that many of them transmit U.V. better than rutile, in most cases, rutile is still the best compromise pigment due to its other superior characteristics.[22]

4. Wetting, Adhesion and Shrinkage

A number of studies of the wetting of metal substrates by U.V. coatings have been made in our laboratory. We have been able to correlate contact angles of a variety of metal substrates with the smoothness of lay of surface coatings and inks (Table II). It was also found that in many cases the adhesion of the U.V. cured coatings followed the ease of wetting as determined by contact angle measurements, although other factors can also affect

TABLE II.—SURFACE PARAMETERS VS. ADHESION OF U.V. COATINGS

Sample	Critical Surface Tension Dynes/CM	Slope of Zisman Plot ($X10^3$)	Adhesion of U.V. Ctg.
Metal A	35	-14	Good
Metal K	29	-14	Fair
Metal S	31	-17	Fair
Coating M	34	-20	Poor
Coating D	35	-13	Good
Coating T	34	-8	V. Good
Coating P	35	-5	Excell.

adhesion of U.V. coatings.

In coating of thin metal substrates, with U.V. coatings, it is common practice in some industries to use a thermal treatment after the U.V. irradiation. This serves a number of purposes, but the main effect on the film is to dramatically increase its toughness and adhesion. Many authors have attributed this to completion of reaction and to a relaxation effect in the U.V. cured film which relieves the stresses induced by the shrinkage, which occurs when the crosslinking and polymerization take place.

It has been observed in many cases that very thick films of some formulations may cure in such an uneven manner from surface to substrate, that severe wrinkling of the surface will occur. In these cases, it is sometimes effective to slow down the overall cure rate via formulation changes so that the shrinkage is lessened or at least occurs over a long time span thus minimizing internal film stresses.

5. Flexibility and Rate of Cure

It is desirable, in many cases, to have a reasonable degree of flexibility of the surface coating and it is particularly true where the substrate must undergo stresses or forming operations after curing. Most of the rapid curing systems achieve this speed by a high crosslinking density with consequent film brittleness, however. The use of urethane chemistry allows the synthesis of oligomers which have much improved flexibility and toughness. (1D)[11,20,22] Cationically cured epoxy based coatings[6] may also offer good adhesion and improved flexibility when properly formulated. It has also been reported that the use of thiol-ene chemistry for U.V. cured coatings on bottle caps and other deep formed metals has been successful[20]. The incorporation of long chain moieties, such as vegetable oils, in U.V. curing oligomers is another way to get improved flexibility and toughness. (1-Chapter 6)[20,36].

NEEDS FOR THE FUTURE

If U.V. curing of surface coatings for metal substrates is to become a widely used commercial process, there are a number of needs that must be fulfilled. While much has been accomplished, wider use can only occur by a concerted effort on the part of the research and development groups of coatings suppliers and also of the vendors who supply the U.V. coatings industry. The most important of these needs are:

1. New types of reactive monomers and oligomers to provide more versatility in formulation for desired film properties;

2. Better and more efficient, non-yellowing, photoinitiators;

3. Lower cost raw materials;

4. Monomers and oligomers having low toxocity and skin irritation potential;

5. A variety of reactive reducers which can be used to adjust viscosity of coatings;

6. Vehicle systems which will have less permeability to water vapor and good resistance to weathering.

Fulfilling these needs poses some formidable problems for the coatings chemist, but much progress is being made in vehicle research. This research is certain to increase the market penetration of U.V. coatings and their application in protection and decoration of metal substrates.

REFERENCES

1. Pappas, S.P., Ed., "U.V. Curing-Science & Technology," Chapters 1A, 3B, 6C, 7D, 8E, Publisher-Technology Marketing Corp., Stanford, Conn., 1978.
2. Hird, M.J., *J. Coat Technol. 48(620), 75 (1976).*
3. Koch, S.D., "Metal Decorating Advances (Non-Container)," SME Tech. Paper No. FC75-329, p. 6PP.
4. McGinniss, V.D., "Ultraviolet Curing of Pigmented Coatings," ACS Symp., Ser. 76, 25.
5. Wismer, M. and Parker, E.E., "Forming a Non-Glossy Film," PPG Industries, U.S. Patent No. 3953622.
6. Schlesinger, S.I., "Epoxy Resin Coatings Cured With Phototropic Aromatic Nitro Compounds," American Can Co., U.S. Patent No. 3949143.
7. Pappas, S.P. and Kuhhirt, W., *J. Paint Technol. 47(610), 42 (1975).*
8. Wicks, Jr., Z.W. and Kuhhirt, W., *J. Paint Technol. 47(610), 49 (1975).*
9. Koch, S.D., "Metal Decorating Advances (Non-Container)," Soc. Manuf. Eng., FC, 75-329, 1975.
10. Nakabayshi, S., "The U.V. Curing of Opaque Film," Soc. Manuf. Eng., FC, 75-322, 1975.
11. Ketley, A.D., "Ultraviolet-curable Resins for Wire & Cable Coating," Soc. Manuf. Eng., FC, 75-331, 1975.
12. McGinniss, V.D., "Photopolymerizing Ethylenically Unsaturated Vehicles Using a Thioxamthone-Halogenated Naphthalene Sensitizer System," SCM Corp., U.S. Patent No. 3876519.
13. Hulme, B.E., "Curing of Coatings with Ultraviolet Radiation. II. Film Properties," *Paint Manuf. 45(4), 12 (1975).*
14. Hulme, B.E., *Paint Manuf. 45(2), 9-13, 16 (1975).*
15. McGinniss, V.D. and Ting, V.W., *J. Radiat. Curing 2(1), 14 (1975).*
16. McGinniss, V.D., *J. Radiat. Curing 2(1), 3-4, 6-13 (1975).*
17. McGinniss, V.D. and Dusek, *J. Paint Technol. 46(589), 23 (1974).*
18. Hencken, G., *Farbe und Lack 81, 916 (1975).*
19. Hird, M.J., *J. Coat. Technol. 48(620), 75 (1976).*
20. Pelgrims, J., *J. Oil Colour Chem. Assoc. 61, 114 (1978).*
21. Davis, M., et al., *J. Oil Colour Chem. Assoc. 61, 256 (1978).*
22. Rybny, C., *J. Oil Colour Chem. Assoc. 61, 179 (1978).*
23. Rubin, H., "U.V. Curing: Role of Absorption by Photo-Initiator," TAGA Proceedings, 1976, p. 279.
24. Berner, G., *J. Oil Colour Chem. Assoc. 61, 105 (1978).*
25. Hulme, B. and Hird, M., "Factors Influencing the Cure Response of Pigmented U.V. Coatings," SME Technical Paper, FC, 76-492 (1976).
26. Kinstle, J. and Watson, S., *J. Rad. Curing 2, 7 (1975).*
27. McGinniss, V.D., "Ultraviolet and Laser Curing of Pigmented Polymerizable Binders," SCM Corp., U.S. Patent No. 3847771.
28. Pappas, S.P., "Photochemical Aspects of U.V. Curing," Soc. Manuf. Eng., FC, 74-501 (1974).
29. Wicks, Jr., Z.W., "Defining Radiation Curing," Soc. Manuf. Eng., FC, 74-501 (1974).
30. Morishita, H., "Light-curing Pigmented Coating Compositions," Kansai Paint Co., Ltd., Japanese Patent No. 7354135.
31. Schroeter, S.H., "Ultraviolet Curing of Coatings," Proc. ACS Symp. 1973, Plenum Press, New York, pp. 109-133.
32. Sandner, M.R. and Osborn, C.L., "Acetophenonetype Photosensitizers for Radiation-Curable Coatings," Union Carbide Co., U.S. Patent No. 3715292.
33. Due, G.B. and Guarino, J.P., "Radiation Curable Coating," Mobil Oil Corp., U.S. Patent No. 4072592.
34. Parrish, M.A., *J. Oil Colour Chem. Assoc. 12, 474-78 (1977).*

35. Wismer, M. and Parker, E.E., "Non-glossy Film," PPG Industries, Inc., U.S. Patent No. 4020193.
36. Wicks, Jr., Z.W. and Grunewalder, J.F., *Rad. Cur. 1, 8-11, 14-17 (1977).*
37. McGinniss, V.D., "Photoinitiation of Acrylate Systems for U.V. Curing," SME Tech. Paper FC76-486, 1976.
38. Krajewski, J., Packer, E. and Yamazaki, T., *Met. Finish. 74(11), 58 (1976).*
39. Krajewski, J.J., Packer, E.S. and Yamazaki, T., *J. Coat. Technol. 48 (623), 43 (1976).*
40. Shahidi, I.K. and Zeliznak, K.J., "Low Gloss Ultraviolet Curable Coatings Utilizing alpha, alpha, alpha, trichlorotoluene as a Photo-initiator," Celanese Corp., U.S. Patent No. 3992275.
41. McGinniss, V.D., "Ultraviolet Curing of Pigmented Coatings," *Am. Chem. Soc. Div. Organic Coatings Papers 35(1), 118 (1975).*
42. McCarty, W.H., "Radiation Curable Coatings," Tech. Paper, Res. Tech. Conf., Nov. 13-15 1976, pp. 205-214.

ANTIFOULING COATINGS—PART I. PREPARATION AND TESTING OF ANTIFOULING COATINGS BASED ON TRIBUTYLTIN FLUORIDE

Nadim A. Ghanem and
Mounir M. Abd El-Malek

Laboratory of Polymers and Pigments,
National Research Center,
Dokki, Cairo, Egypt

NADIM GHANEM

ABSTRACT

In view of the limited period of protection of cuprous oxide-based antifouling coatings in fouling-rich regions and the danger of copper accumulation in waters and sediments in semi-enclosed areas, need for longer protection periods with less polluting effects has emerged. In this paper tributyltin fluoride is used as an alternative to cuprous oxide in preparation of a variety of paints meant for fouling resistance for long periods. The results of testing in natural environment of high fouling intensity are very promising with respect to animal organisms but not very much so with respect to paint fouling.

INTRODUCTION

In a previous publication[1] dealing with antifouling coatings depending on vinyls as binders and cuprous oxide as toxic pigment, it was determined that the maximum period of protection in fouling-rich regions was thirteen months. It was claimed that research should be directed to areas where other binder systems are used or higher efficiency toxins may supplement or replace cuprous oxide. This trend is prompted by recent investigations[2] in which copper at two orders of magnitude higher than the natural limit was determined in the waters and sediments of the Suez Canal. This threatening level was attributed - solely - to copper leached out from ships traversing the international pathway at a rate estimated at 3000 vessels per month. The phenomenon of increased copper levels was also observed in other places such as the harbors of Antibes, La Pallice and La Rochelle in South France[3]. The high copper concentrations were attributed to the great number of pleasure boats which use copper and cuprous oxide antifouling coatings. High copper concentrations were also determined in the waters of the docking areas of San Francisco Bay[4]. The same author found copper contents in abundant marine algae, *Ulva sp.* ranging from 10.6 to 53.4 μg/g tissue (dry weight). Further alarm was provided by Disalvo[5], where the effect of copper on the survival and functioning of the lower — but most important organisms —, the bacteria, was examined. The results indicated variable susceptibility of bacteria to the presence of copper and suggested the possibility of cumulative synergistic effect of heavy metals on bacterial motility. It was estimated that 5000 tons of copper are released yearly from copper-based antifouling coatings in the oceans and seas of the world[6].

In the present investigation the binder mixture — based essentially on a vinyl copolymer — was kept the same, while cuprous oxide was replaced entirely by one of the recently developed organotin compounds. Trialkyl- and triaryltin compounds have been gaining ground since their fungicidal effects were discovered in 1954. They were recommended for antifouling applications because of their slow but effective release, because of the long periods of protection of as much as two years and because they do not accelerate corrosion. An instance of protection against fouling for 92 months has been documented[7] where tributyltin compounds were included in a thick rubber sheet.

The type of plasticizer for the vinyl binders was shown to play an important role in antifouling coatings depending on cuprous oxide as toxic ingredient[1]; its chemical composition controls the leaching rate of copper and hence the prevention of fouling in long-term environmental immersion. The present investigation includes the use of three different commercial plasticizers — hydrolyzable, nonhydrolyzable and nonhydrolyzable but swellable plasticizer — to examine their role in conjunction with tributyltin fluoride as toxic ingredient.

While it was hoped to conduct the environmental testing for much longer period, the test was terminated after 211 days due to a sudden storm which sank the raft. However, the photographic recording of the last inspection could provide evidence of further protection periods.

MATERIALS

The vinyl copolymer was Laroflex MP 35 from BASF AG, Ludwigshafen, Germany. The plasticizers were: chlorinated paraffin (nonhydrolyzable) n 50 from Farbwerke Hoechst AG, Germany, tritolyl phosphate (hydrolyzable) and polyvinyl methyl ether (nonhydrolyzable but swellable) Lutonal M 50 from BASF AG. The modifying resins were ester gum; a product of Carless Capal, England, and a phenolic resin which was Wresnyl 265, a product of Resinous Chemicals, England. Thixtrol ST, a hydrogenated castor oil used as anti-settling agent was also obtained from BASF, AG. Tributyltin fluoride was obtained from Berk Co., Ltd., England. Titanium dioxide was of the rutile type, a product of Titangesellschaft, AG, West Germany. Iron oxide was of the type 130 F from Farbwerke Bayer, AG, Germany. China clay was obtained from English China Clays, Sales Division, England. Suitable mixtures of xylene, white spirit (145 - 185°C), butyl acetate and butanol were used as solvents.

EXPERIMENTAL

Paint Preparation

Formulations in the amount of 300 g each were prepared by mixing the selected amounts of ingredients. The mass was matured and then dispersed in a centrifugal ball mill. The paints were diluted to the application viscosity using a suitable blend of solvents, filtered twice, and stored in tins with tight lids in a cool place. The storage time before application ranged between 2 and 4 weeks; the paints were free from sediments or any other defects after several months of storage.

Test Panels

Sheets, 3 mm thick, of pigmented impact-resistant polystyrene were cut to panel dimensions of 20 x 15 cm, the edges were tapered and the surfaces were coated with a thin layer of an epoxy paint, left to dry and then roughened with sandpaper. The use of polymer panels avoided corrosion interference and complications of incomplete compatibility with the anticorrosive layer.

The panels were coated in duplicate, front and back, with two successive coats allowing an interval of 6 hours for drying between the two coats; the average film thickness was 90μ m ± 10μ m. Particular care was paid to the edges to eliminate pores and weak points.

The coated panels were connected to boat-like cages with nylon threads through holes bored in the panels. Each composition was represented by two panels; one placed in a vertical position and the other at 60° to the horizontal.

Environmental Exposure

The cages were submerged in the test area of the raft[8] in Alexandria, Egypt (eastern harbor) on 24 May at a depth of about 150 cm from the surface. Periodic visual and biological examination and photographic recording of the condition of the panels were performed. The region is known for its abundant fouling organisms almost all the year round; there are reasons to believe that the rate of growth of most foulers has doubled since the extensive survey concluded by Ghobashy[9] in 1975. Periodic testing of water salinity, temperature, and pH was performed.

RESULTS AND DISCUSSION

Paint Compositions

Ten paint compositions were prepared with a common binder mixture containing mainly a neutral vinyl copolymer. A slight acidity was obtained by a small amount of ester gum resin (acid number 3-5 mg KOH/g) and some improvement of hardness was imparted by a small content of a phenolic resin. A traditional constituent in antifouling coating, rosin, was absent from all formulations. It was established long ago that rosin plays an important role in the performance of both soluble-matrix and contact-leaching copper antifouling coatings. In trialkyl- or triaryltin compounds, the mechanism is believed to be one of slow-release which does not depend on the dissolution of rosin but rather on migration of a particle of the compound from the interior of the coating to its surface and then to the water[10].

Tributyltin fluoride was introduced in the formulations in four different amounts, namely, 15, 20, 25 and 30% by weight of the total dry film. Values in some of the literature available[11,12] on tributyltin fluoride antifouling coatings range between 16 - 20%; values of 30 - 40% are given in other literature[13] but these amounts were considered to be impractical. The wide range used in the present work was designed for the severe fouling conditions at the test site. Talc and China clay were included in substantial amounts not only for improving paint properties but also for their favorable control of the rate of release of the toxic ingredient. The coating compositions are given in Table I which also includes a calculation of the total pigment volume concentration and the toxic (pigment) volume concentration.

Test Results

All coatings were completely intact and free from fouling at the end of the first two months of immersion. A general feature was the formulation of slime and attachment of slight algal growth. Only a few barnacles were observed at the edges and around the holes of the panels due to bare points and some detachment of the coating during manipulation.

TABLE I.—DRY FILM COMPOSITION OF ANTIFOULING COATINGS

AF. No. Constituent	1	2	3	4	5	6	7	8	9	10
Vinyl Copolymer	15	15	15	15	17	17	17	16	16	16
Polyvinyl Methyl Ether	4.5	4.5	4.5	4.5	--	--	--	--	--	--
Chlorinated Paraffin	--	--	--	--	1.7	1.7	1.7	--	--	--
Tritolyl Phosphate	--	--	--	--	--	--	--	2.5	2.5	2.5
Phenolic Resin	3	3	3	3	3.3	3.3	3.3	3.2	3.2	3.2
Ester Gum	2.5	2.5	2.5	2.5	3	3	3	3.4	3.4	3.3
TBTF	15	20	25	30	20	25	30	20	25	30
Iron Oxide Red	14	14	14	14	14	14	14	14	14	14
Titanium Dioxide	20	20	20	20	20	20	20	20	20	20
Talc	15	10	5	--	10	5	--	10	5	--
China Clay	11	11	11	11	11	11	11	11	11	11
PVC	58	60	61.6	63.0	60.5	61.9	63.4	60.2	61.9	63.3
TOXIN VOLUME CONCN	24	31	37.3	43	31.3	37.4	43.2	31.3	37.4	43.1

From the beginning of the third month until the test was terminated, the intensity of the slime film on all plates was moderate but the film tended to be thicker on coatings with lower contents of the tin compound. No serious attack of the test panels by barnacles, tube worms or other foulers was observed throughout the whole period of immersion of 211 days. Little scattered settlements took place at the edges only.

Photographic recording of the test is shown in Figure 1. The great contrast between the cleanliness of the coated panels and heavy fouling on the supporting cage is clearly demonstrated. The latter carried such heavy loads of barnacles and tube worms, reaching 12 to 15 cm in thickness at the last inspection, that parts of the test panels area were hidden behind the intensive growth. Continuation of the test would have left very little room for comparison between the individual formulations for their relative long-term efficiency because of encroachment of fouling. A new design of supporting frames which provided larger space between the test panel and other objects was developed and used[1].

The test period (211 days) was either too short to provide differentiation between the plasticizer type or the protective function in the case of organotin compounds was independent of the chemical nature of the plasticizer. These results are in contrast to cuprous oxide formulations where antifouling function was lost much earlier in the aforementioned period when a nonhydrolyzable plasticizer was used[14]. The second poltulation is more plausible and supports the view that the mechanism of protection by organotin formulations depends on the ability of the toxic particle to diffuse from the interior of the film to its surface. This diffusion is controlled by the texture and configuration of other components of the coating[15] rather than on the swellability of the plasticizer[1] or the solubility of other components like rosin.

One of the disadvantages of slow-release organotin formulations is their very short range of protection; the repellancy to fouling larvae takes place in the vicinity of the surface only in contrast to copper formulations which leach out a soluble toxic layer which has its effect over a much longer range. Full protection of the edges and corners was therefore much more difficult in organotin formulations than in their copper counterparts.

Another drawback is the ineffectiveness of organotin formulations on plant foulers relative to copper formulations. Although brown algae was detected in the present series, serious plant fouling did not appear because the experiment was terminated before the season of plant fouling which begins early in the spring; plant fouling was detected and characterized in a later series[16]. One of the solutions of this drawback is to use combined cuprous oxide-organotin formulations.

CONCLUSION

Vinyl binder coatings containing 15 - 30% by weight tributyltin fluoride are effective antifoulants and satisfy standard specification[18] of six months protection in natural environment. They function by a slow-release mechanism which is independent of the presence of a soluble ingredient, and thus their effective lifetime is prolonged in comparison with compositions based on cuprous oxide as toxin and rosin as soluble ingredient. A further advantage of the former lies in their less hazardous effects on marine ecology, particularly in harbors and narrow water pathways. The economy seems favorable with increasing production of organotin compounds.

Fig. 1.—Environmental testing of tributyltin fluoride antifouling coatings. Samples are identified in Table I. From left to right they are: blank, 1, 2, 3, 4, 5, 6, 7, 8, 9, 10. The rows are identified as follows: 1st row, appearance at start on May 24; 2nd row, after 61 days, 3rd row, after 88 days; 4th row, after 139 days; 5th row, after 201 days.

ACKNOWLEDGEMENT

This work represents part of the activities of Contract N00014-75-C-1112 between the National Research Center of Egypt and Office of Naval Research of the Department of the Navy, U.S.A.

The authors are also indebted to BASF, AG, Ludwigshafen for provision of starting materials of their own and other Companys' products, and for useful discussions with Dr. Neubert, Dr. Brussmann and Dr. Morcos of AWETA (Applied Research Department) of BASF, AG.

REFERENCES

1. Ghanem, N.A., Abd El-Malek, M.M., Abou-Khalil, M.A., and El-Awady, M.M., *Ind. and Eng. Chem. Prod. Res. and Dev. 17, 44 (1978)*.
2. Ghanem, N.A., El-Awady, M.M. and Abd El-Malek, M.M., to be published.
3. Callame, P., Report 75.066 presented to the Inter. Comm. for Research on the Preservation of Materials in Marine Environment (COIPM), May 20, 1975.
4. Hilderbrand, R.L., A report to COIPM, June 10, 1976.
5. Disalvo, L.H., Report to COIPM, June 10, 1976.
6. Phillip, A.T., OCCA Proc. and News, July 17, 1973.
7. Cardarelli, N.F. and Caprette, S.J. Jr., U.S. Pat. No. 3,426,473, Feb. 11, 1969.
8. Abd El-Malek, M.M. and Ghanem, N.A., *J. Paint Technol. 47(608) 75 (1975)*.
9. Ghobashy, A.G.A., "Proceedings of the 4th International Congress on Marine Fouling and Corrosion." Juan-les-Pins, France, June 14-18, 1976.
10. Bollinger, E.H., "Proceedings of the Symposium on Controlled Release Pesticides," The University of Akron, Akron, Ohio, Sept. 16-18, 1974. Paper 19.
11. Lorenz, J., Ciba-Geigy Marienberg CMBH, Switzerland, private communication, 1972.
12. Beiter, C.B., "Proceedings of the Symposium on Controlled Release Pesticides," The University of Akron, Akron, Ohio, Sept. 13-15, 1976, p. 2.
13. Mearns, R.D., *J. Oil Colour Chem. Assoc. 56, 353 (1973)*.
14. Abd El-Malek, M.M. and Ghanem, N.A., "Proceedings of the 4th International Congress on Marine Fouling and Corrosion," Juan-les-Pins, France, June 14-18, 1976.
15. Bennett, R.F. and Zedler, R.J., *J. Oil Colour Chem. Assoc. 49, 928 (1966)*.
16. Ghanem, N.A. and Abd El-Malek, M.M., "Part II. Proceedings of the 4th International Congress on Marine Fouling and Corrosion," Juan-les-Pins, France, June 14-18, 1976.
17. Ghanem, N.A. and Abd-Malek, M.M., to be published.
18. Egyptian Standard Specifications No. 197, 1962 and No. 765, 1966. The Egyptian General Authority for Standard Specifications. Ministry of Industry, Cairo, Egypt.

ANTIFOULING COATINGS—PART II. PREPARATION AND TESTING OF ANTIFOULING COATINGS BASED ON TRIPHENYLTIN FLOURIDE

Nadim A. Ghanem and
Mounir A. Adb El-Malek

Laboratory of Polymers and Pigments,
National Research Center
Dokki, Cairo, Egypt

ABSTRACT

Antifouling paints were prepared using a vinyl copolymer as main binder and triphenyltin fluoride alone as the toxic ingredient. The paints were tested in a heavily infested marine locality with the aim of throwing light on the following questions pertaining to long-term protection: 1. The function of the nature of the organic residue in the organotin compound; 2. The optimum concentration of the organotin compound in the formulation; 3. The contribution of the classical soluble ingredient, rosin, in slow-release formulations employing an organotin compound for fouling prevention; 4. The role of the pigment extender in the proper functioning of the antifouling coating.

Photographic follow-up of the condition of test panels during 238 days of continuous immersion allowed decisive conclusions with respect to some of the aforementioned questions which should help in producing modern, economic, durable and less hazardous organotin antifouling paints for use in fouling-rich regions.

INTRODUCTION

Notwithstanding current attempts to create nontoxic antifouling coatings[1], all commercial antifouling coatings until the present day depend on the development of a toxic laminar layer capable of killing or repelling the larval attachment. Among the thousands of toxic compounds in different classes of organic, heterocyclic and inorganic compounds, copper compounds (mainly cuprous oxide) meet most of the requirements and have dominated the scene in commercial formulations[2] for many years for reasons stated in a previous publication[3]. However, cuprous oxide-based paints began to lose ground in recent years for compositions based on certain organotin compounds, alone or mixed with cuprous oxide. The reasons were a combination of factors which, although invisible in the commercial literature, are possible to trace in independent research efforts. One of the reasons is inherent, depending on the necessity of consumption of copper from the film at a rate not less than 10μg/cm^2/day irrespective of the mechanism, and thus connecting the protection period to the full depletion of the poison store or the clogging the pores through which the poison may pass. In a recent publication, the period of protection of a highly recommended class of cuprous oxide-based antifouling compositions in a region of intensive fouling was set at a maximum of

13 months[3]. Another reason may be extracted from hydrological studies where alarming copper concentrations were found in the waters of some harbors in the south of France[4], in organisms in the waters of San Francisco Bay[5] and in the waters and organisms in the Suez Canal[6].

The above problems have been solved, at least in part, by the advent of organotin compounds and polymers. One of these compounds, bis-tributyltin oxide, a liquid which received considerable success at one time, was however banned from use in some countries. Among the rest of the organotin compounds tributyltin fluoride (TBTF) and triphenyltin fluoride (TPTF) are outstanding until the present time for being solids which can be incorporated at desired ratios, for their slight but effective solubility in sea water, and for their reasonable dissolution in common paint solvents. In contrast to matrix dissolution or contact leaching mechanism of cuprous oxide coatings, the mechanism in organotin coatings is believed to be one of slow-release depending on the migration of toxic particles from the interior of the film to its surface and then to the water[7], thus allowing a much longer period of protection[8]. Much higher toxic efficiency of organotin compounds compared with cuprous oxide, estimated at five times, is not only an economic advantage but also a non-pollution factor especially as most organotins are believed to be eventually converted in sea water to inorganic tin compounds such as the carbonate which, contrary to other metal inorganic compounds like those of copper, arsenic, and mercury, are non-toxic[9].

In a previous publication[10], formulations based entirely on tributyltin fluoride for toxicity were prepared and tested in heavily infested natural environment[11,12]. Their efficiency against plant fouling was doubtful in view of the absence of flourishing algal growth, i.e., spring and early summer, from the total test period. Otherwise, it was concluded that coatings containing 15 - 50% by weight of TBTF are effective and satisfy standard specifications[13] of keeping the surface completely clean for 6 months in a natural environment.

In the present work, TBTF was replaced by TPTF and the immersion period was extended to include the season of extensive plant fouling. The aim was to throw more light on the nature of the mechanism of fouling prevention; TPTF being claimed to be more stable to heat, U.V. light and hydrolysis than its tributyl counterpart and less soluble in sea water[14]. The studies are supplemented by a series employing graded amounts of the sea-water-soluble ingredient, rosin, which helps release in copper formulations. A series is also devoted to examining the role of 6 different pigment extenders, the particle size, shape and texture of which are believed to contribute in controlling the mobility of the toxic particle while traversing the film bulk to its surface.

MATERIALS

A full description of all materials except TPTF and the pigment extenders was given in Part I[10]. TPTF was obtained as a fine powder from M & T Chemicals, Inc. (Subsidiary of American Can Company), Rahway, New Jersey, U.S.A. China clay was obtained from English China Clays Sales, England. The other pigment extenders were obtained as trade samples through BASF, AG, West Germany; their sources were not given.

EXPERIMENTAL

The Paints

Three groups of a total of 12 antifouling compositions based mainly on a neutral vinyl copolymer were prepared. Group I comprised 3 compositions containing graded amounts of TPTF which were 10, 15 and 20% by weight of the dry film, while the binder, pigment and extenders were fixed except the talc from which content the toxin was substituted. The formulations are given in Table I.

Group II consisted of 3 compositions differing only in the rosin content which was present in graded amounts of 15, 30 and 45% by weight of the binder mixture, while the toxin and other ingredients were kept constant. The compositions of Group II paints are shown in Table II.

Group III comprises 6 compositions in which the contents of the nontoxic pigments were kept the same while a different pigment extender was used in each formulation, i.e., six extenders were incorporated in this group, namely, talc, china clay, barytes, asbestine, lithopone and magnesium trisilicate. The paint compositions (dry film) are shown in Table III.

Paint Preparation

Laboratory specimens of 300 g. each were prepared. The pigments were first mixed and then the binder mixture in part of the solvents mixture was added and the whole was homogenized in a centrifugal ball mill. The toxic agent TPTF was included in the pigment mixture from the beginning. When a fineness of the grind of less than 10μ m was reached using a Hegemann gauge, milling was stopped. The paints were diluted to the application viscosity using the same blend of solvents, filtered twice and stored in tins with tight lids. The storage time before application ranged from 2 to 4 weeks.

Panel Preparation

The test panels were cut to dimensions, 30 x 17cm, from pigmented, impact-resistant polystyrene sheets, the edges were tapered and the surfaces were coated with an epoxy paint, left to dry and then roughened with sandpaper. The epoxy layer served as protection against attack of the substrate by the aromatic solvents of the antifouling paint thus avoiding defects (cracks) in the dry antifouling film. The use of a polymer substrate avoided corrosion and rust problems and probably complications of incomplete compatibility with the anticorrosive layer.

Each composition is represented by two surfaces, front and back, of one panel hanging in a vertical position. Two successive coats were applied to each face at intervals of 8 - 14 hours to allow drying, the total average film thickness was 90μ m ± 10μ m. Particular care was paid to avoid pores and weak point at edges.

The coated panels were supported in the test stand by nylon threads through holes bored in the panels. The present testing stand shown in the figures is a modification of a previous one[10] that avoids encroachment of fouling masses from the ribs of the stand to the testing panels.

Raft Testing

The stands were submerged in the test area of the floating station[11] in eastern-harbor of Alexandria, Egypt, on September 30, 1977 at a depth of

TABLE I.—COMPOSITIONS OF GROUP I PAINTS. THE EFFECT OF TPTF CONTENT.

AF No. Constituent	1	2	3
Vinyl Copolymer	15	15	15
Polyvinyl Methyl Ether	4.5	4.5	4.5
Phenolic Resin	3	3	3
Ester Gum	2.5	2.5	2.5
Triphenyltin Fluoride	10	15	20
Iron Red	14	14	14
Titanium Dioxide	20	20	20
Talc	20	15	10
China Clay	10.9	10.9	10.9
Additives	0.1	0.1	0.1

TABLE II.—COMPOSITIONS OF GROUP II PAINTS. THE EFFECT OF THE ROSIN CONTENT.

AF No. Constituent	4	5	6
Vinyl Copolymer	13.75	11.50	9.00
Polyvinyl Methyl Ether	4.50	3.50	2.75
Phenolic Resin	3.00	2.50	2.00
Rosin	3.75	7.50	11.25
Triphenyltin Fluoride	15	15	15
Iron Red	14	14	14
Titanium Dioxide	20	20	20
Talc	15	15	15
China Clay & Additives	11	11	11

TABLE III.—COMPOSITIONS OF GROUP III PAINTS. THE EFFECT OF THE PIGMENT EXTENDER.

AF. No. Constituent	7	8	9	10	11	12
Vinyl Copolymer	15	15	15	15	15	15
Polyvinyl Methyl Ether	4.5	4.5	4.5	4.5	4.5	4.5
Phenolic Resin	3	3	3	3	3	3
Ester Gum	2.5	2.5	2.5	2.5	2.5	2.5
Triphenyltin Fluoride	15	15	15	15	15	15
Iron Oxides red	14	14	14	14	14	14
Titanium Dioxide	20	20	20	20	20	20
Talc	--	--	25.9	--	--	--
China Clay	25.9	--	--	--	--	--
Barytes	--	25.9	--	--	--	--
Asbestine	--	--	--	25.9	--	--
Lithopone	--	--	--	--	25.9	--
Magnesium Trisilicate	--	--	--	--	--	25.9
Additives	0.1	0.1	0.1	0.1	0.1	0.1

150 cm from the surface. Periodic visual and biological examination and photographic recording of the condition of the panels were performed. The region is known for its abundant fouling organisms almost all the year round; there are reasons to believe that the rate of growth of most foulers, especially the tube worms, has doubled since the extensive survey concluded by Megally[15] in 1970, and by Ghobashy in 1973 (published in 1976)[12].

RESULTS

The following results were derived from examination of the test panels at intervals during the immersion period.

Group I Paints:

1. The antifouling efficiency was directly proportional to the triphenyltin fluoride content in the composition. There were no remarkable differences in the condition of the surfaces of the three compositions during the initial period of 120 days immersion except for a few individual attachments of barnacles and tube worms at the edges of compositions 1 and 2 (see collective Figure 1). Aggressive growth of heavy colonizers from the ribs of the stand to the interior of these two panels covering about 50% of the surface areas was observed in the last inspection. Different foulers were detected, namely, barnacles, tube worms, ascidians and bryozoans, and these formed a thick layer of about 10 to 12 cm.

2. Thick slime films consisting of bacteria, diatoms and detritus materials formed over all the panels. Algal growth constitutes a great part of the slime film. Brown algae were detected in the early period of immersion; three species have been recorded: *Histochylum pastulatum, Jania subenus* and *Corallins* sp. It was reported[12,15] that the above individuals are present at the testing site all year round. During the winter months and in the beginning of spring, heavy green algal growth was abundant; the most predominant species were: *Ulva Lactuea* and *Enteromorpha clathrata*. It is observed that the attachment of the organisms begins at the upper parts of the panels and spreads downwards to the lower edges. The thickness of the slime film is dependent upon the toxin content in the composition. Three levels in the slime intensity can be differentiated ranging among heavy, moderate, and moderate to slight. The rate of slime formation on the testing panels can be arranged as follows: $AF_1 > AF_2 > AF_3$.

Group II Paints:

The three compositions[4,5,6] with increasing rosin contents are represented by three panels. The following results were derived from the condition of the panels during the period of test (see Figure 1).

1. The higher the rosin content, the thicker the slime film that formed and the heavier the attack by plant fouling.

2. The antifouling efficiency for animal foulers was generally diminished mainly due to the presence of rosin in the compositions. Composition 2 of Group I which contains the same toxin content, 15% TPTF, in the dry film but no rosin has better efficiency than any member of Group II. The presence of rosin accelerated the formation of heavy slime films and reduced the slow release of toxin particles and thus encouraged the settlement of foulers, namely, barnacles on the heavily slimed surfaces[9]. The condition of surfaces of composition 4, which contains the lowest amount of rosin, was nearly similar to that of composition 2 with no rosin but smaller TPTF content. Figure 3 shows the heavy attack by plant fouling after 177 days of

Fig. 1.—Collective view of panels after immersion times of 34 days (row 1), 120 days (row 2), 177 days (row 3), and 238 days (row 4). Samples are arranged in column form and from left to right are in the order: 1, 2, 3, 4, 5, 6, 7, 8, 9, 10, 11, and 12.

immersion.

Group III Paints:

This group includes 6 compositions each containing only one extender. The extenders used are included in the compositions in Table III. Figure 1 shows clear differences between the 6 antifouling systems which are unified in all constituents except the type of the pigment extender. Good antifouling efficiency was obtained with compositions 7 and 11 which contain China Clay and lithopone, respectively. The rest of the formulations have efficiencies ranging from moderate to low. The antifouling efficiency can be arranged in the following order:

AF_7 *(China clay)* > AF_{11} *(lithopone)* > AF_{12} *(Mg trisilicate)* >

AF_{10} *(Asbestine)* > AF_8 *(barytes)* > AF_9 *(talc)*.

DISCUSSION

Triphenyltin fluoride is one of the most promising organotin compounds to be used in antifouling coatings. It provides good resistance to attachment of many fouling organisms specially the serious kinds, namely, barnacles and tube worms, but it fails to prevent plant attachment, mainly, algae.

Vinyl-based antifouling compositions should employ triphenyltin fluoride in amounts near 20% by weight in the dry film; lower contents make the formulation parameters more critical to proper content of each of the other constituents employed. In contrast, systems based on Cu_2O with as high contents as 80% are easier to study. In a recent system employing 97% Cu_2O and polyisobutylene rubber as a sole binder[16] the role of the other constituents disappears completely but the poor adhesion on substrate of anti-corrosive layer provides difficulties.

Compositions 1, 2 and 3 conserve on the use of TPTF since these contain 10 to 20% by weight. It might have been preferable to increase the content to 25% or more in order to obtain maximum performance in environments of severe and serious fouling conditions. Composition 3, containing 20% TPTF, was successful in controlling serious fouling attack for a period of 8 months (see Figure 2), but there is no evidence of protection for longer periods, similar to the compositions which resisted attachment in panel exposure in Biscayne Bay, Miami, Florida, for two years. Such efficient compositions may be applied to idle ships of limited activity which remain still in warm, fouling-rich waters for long periods[17], like naval ships in time of peace and certain kinds of freighters which spend many weeks in ports; while compositions of less toxin content, i.e., 15 ± 3% TPTF may be used on vessels staying in ports for short periods only, such as tankers or ships which should go into dry-dock every 6 or 9 months for periodic maintenance.

Introduction of rosin in antifouling coatings is of great importance for its effectiveness in maintaining adequate release of toxins. It was established long ago that rosin plays an important role in the performance of both soluble matrix and contact-leaching copper antifouling coatings which contain mainly Cu_2O. In the case of systems based on organotin compounds the stand is different due to the wide difference in the physical and chemical properties of organotin antifoulants. For example, the most common organotins, namely, tributyltin oxide (TBTO), TBTF and TPTF have different solubilities in sea water of pH 8.2, 51.4, 6.0 and 0.7, parts per million, respectively. It is expected that each organotin compound has a certain behavior either in the wet paint or as painted films in contact with sea water. According to

Fig. 2.—The appearance of Group I coatings after underwater exposure for 238 days. From left to right samples are 1, 2, and 3.

the little literature available no sharp conclusions on their mode of action can be derived. According to Bennett and Zedler[18] good antifouling efficiencies of paints containing TBTO may be obtained with binder/rosin ratios equal to or greater than 1, but also with rosin-free paints. In another investigation[19] by De la Court and De Vries on the leaching of several organotin compounds from paints based on chlorinated rubber and vinyl binders, the toxin release appeared to be almost independent of rosin content but they could not establish a clear mechanism of leaching due to difficulties in the analysis of the organotin compounds, especially in sea water. Recent studies of effect of rosin on the release of TPTF from vinyl systems[20] showed that the more effective compositions are those containing rosin in an amount equal to 66% by weight of the binder mixture or 24% of the total dry paint film.

The results of Group II paints contradict the above findings; as the rosin content in the composition increased, the antifouling efficiency is reduced. Figure 3 illustrates that the largest unfouled area is on panel No. 4 of the least rosin content. This result may be due to deficiency of toxin concentration in the heavy slime layer adhering to the paint film. In view of the much greater dissolution of rosin than TPTF from these compositions in basic (pH 8.2) sea water, the former would preferentially react with calcium and magnesium salts forming insoluble soaps which would be deposited in the slime film along with other organic and inorganic detritus material and would interfere with the release of TPTF from the coating surface to the water. Any reaction between rosin and the hydrolysis product of TPTF, such as the hydroxide or the carbonate, would give organotin soaps of unknown toxicity. This is not the case in formulations containing near stoichiometric amounts of rosin and cuprous oxide, where readily formed copper rosinate contributes to the toxicity with ionic copper liberated through the catalytic action of chloride ion in sea water; both copper sources help in reducing the intensity of the slime film or keep it enriched with toxins at a level which prohibits fouling.

The results of exposure of compositions of Group III (see Figure 4) employing six different pigment extenders draws the attention to the importance

Fig. 3.—The appearance of Group II coatings after underwater exposure for 177 days. From left to right samples are 4, 5, and 6.

Fig. 4.—The appearance of Group III coatings after underwater exposure for 238 days. From left to right samples are 7, 8, 9, 10, 11, and 12.

of the role played by the morphology, particle size, shape and texture of a major component in antifouling coatings depending on a slow-release mechanism. There is evidence that China clay contributes to efficient release and helps in maintaining a clean surface for long periods. On the other hand, talc and barytes should be avoided in TPTF slow-release formulations. Other pigment extenders were ranked on a scale between 10 for China clay and 4 and 5 for talc and barytes, respectively. Selection of the best pigment extender needs further study.

CONCLUSION

Long-term raft testing in marine environment of intensive fouling provides practical data of great importance to the producer and the user of antifouling paints. Moreover, light can be thrown on the mechanism of function by proper design of the series investigated. In the above work, the level of incorporation of the active component, TPTF, has been set at 20% by weight and the change of the organic moiety in organotin fluorides has no role in controlling plant foulants. Contrary to previous belief, incorporation of rosin was found to be of disadvantage to organotin formulations depending on slow-release. Among the pigment extenders used, China clay and lithopone were the best for efficient long-term release; talc and barytes were the poorest.

ACKNOWLEDGEMENT

This work represents part of the activities of Contract N00014-75-C-112 between the National Research Center of Egypt and the Office of Naval Research of the Department of the Navy of the U.S.A.

The authors are also grateful to BASF, AG., Ludwigshafen for provision of starting materials of their own and other Companys' products, and for useful discussions with Dr. Neubert, Dr. Brussmann and Dr. Morcos of Aweta (Applied Research Department) of BASF, AG.

REFERENCES

1. Ghanem, N.A., and Gerhart, H.L., work in progress.
2. Abd El-Malek, M.M., Abou-Khalil, M.A. and Ghanem, N.A., *Paint Manufacture 40(10), 32 (1970).*
3. Ghanem, N.A. and Abd El-Malek, M.M., *J. Coatings Technol. (In the course of publication.)*
4. Callame,P., Report 75.006 presented to International Committee for Research on the Preservation of Materials in Marine Environment (COIPM), May 20, 1975.
5. Hilderbrand, R.L., A Report to COIPM, June 10, 1976.
6. Ghanem, N.A., El-Awady, M.M. and Abd El-Malek, M.M., to be published.
7. Bollinger, E.H., "Proceedings of the Symposium on Controlled Release Pesticides,"The University of Akron, Akron, Ohio, Sept. 16-18, 1974. Paper 19.
8. Cardarelli, N.F. and Caprette, S.J. Jr., U.S. Pat. No. 3,426,473, Feb. 11, 1969.
9. Woods Hole Oceanographic Institution, Marine Fouling and its Prevention, U.S. Naval Institute, Annapolis, Md., 1952.
10. Ghanem, N.A. and Abd El-Malek, M.M., Part I, these proceedings.
11. Abd El-Malek, M.M. and Ghanem, N.A., *J. Paint Technol. 47(608), 75 (1975).*
12. Ghobashy, A.F.A., "Proceedings of the 4th International Congress on Marine Fouling and Corrosion."Juan-les-Pins, France, June 14-18, 1976.
13. Egyptian Standard Specifications No. 197, 1962 and No. 765, 1966. The Egyptian General Authority for Standard Specifications. Ministry of Industry, Cairo, Egypt.
14. Beiter, C.B., Englehart, J.E., Freiman, A. and Sheldon, A.W., "Proceedings of the Symposium on Marine and Fresh Water Pesticides." Presented at the Meeting of the American Chemical Society, Atlantic City, New Jersey, August 8, 1974.
15. Megally, A.H., M. Sc. Thesis 1970, Faculty of Science, Alexandria University, Supervised by A. El-Maghraky and N.A. Ghanem.
16. Dear, H., Report on Design and Application of Antifouling Paint, David Taylor Naval Ship R&D Center, Annapolis, Md. Read at the 3rd International Conference on the Science and Technology of Surface Coatings, Athens, July, 1977.
17. Berendsen, A.M., "Ship Painting Manual," DeBoer Maritiem Verfinstitut TNO, Delft, 1975.
18. Bennett, R.F. and Zedler, R.J., *J. Oil Colour Chemists' Assoc. 49, 928 (1966).*
19. De la Court, F.H. and De Vries, H.J., *Progress in Organic Coatings 1, 375 (1973).*
20. Beiter, C.B., "Proceedings of the Symposium on Controlled Release," The University of Akron, Akron, Ohio, Sept. 13-15, 1976.

INTUMESCENT COATINGS

David F. Pulley

Aero Materials Laboratory
Naval Air Development Center
Warminister, PA

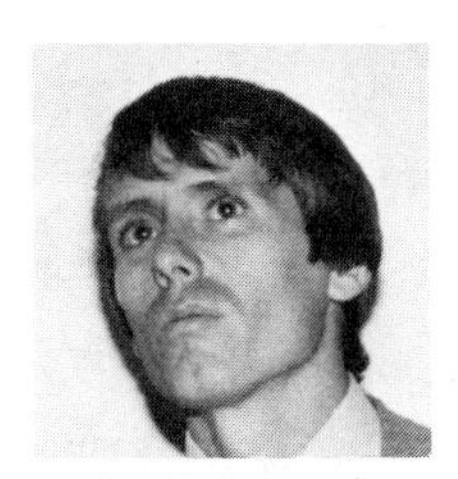

DAVID F. PULLEY

ABSTRACT

Intumescent coatings to protect weapons from detonating during a fire are being developed and tested. The test chamber being used at Naval Air Development Center is described. Thermal efficiency ratings are given for formulations based on the ammonium salt of p-nitroaniline-o-sulfonic acid, sodium tetraborate decahydrate, and triphenyl phosphite.

INTRODUCTION

"Intumescent" is an adjective that describes anything that swells or expands, particularly when heated. The means by which intumescent coatings provide fire-retardant, thermal insulation to a vulnerable structure has been known since 1938; but these materials have just begun to achieve wide acceptance in the last ten years. When exposed to intense heat, the coating functions by three separate reaction mechanisms.

1. The endothermic decomposition of the intumescent filler consumes heat energy.

2. Inert gases are evolved that expand the resulting carbonaceous char and drive back the convective air currents.

3. The thick, expanded char layer which is permeated with entrapped, multi-cellular, gas pockets serves as an effective heat insulator.

The figure on page 412 is a visual representation of the intumescent reaction in progress.

The Navy first became interested in such materials as a result of two disasterous fires aboard the aircraft carriers USS FORRESTAL (July, 1967) and USS ENTERPRISE (January, 1969). Both were caused by accidents on the flight deck, leading to fires spread rapidly by spilled aviation fuel. Weapons lying on the deck and hung under the wings of nearby aircraft soon became enveloped in these fires. Within one to three minutes after the flames first became evident, the weapons began to detonate or launch propulsively. Hundreds of people attempting to fight the spreading fires were killed or injured as a result; and damages estimated in the millions

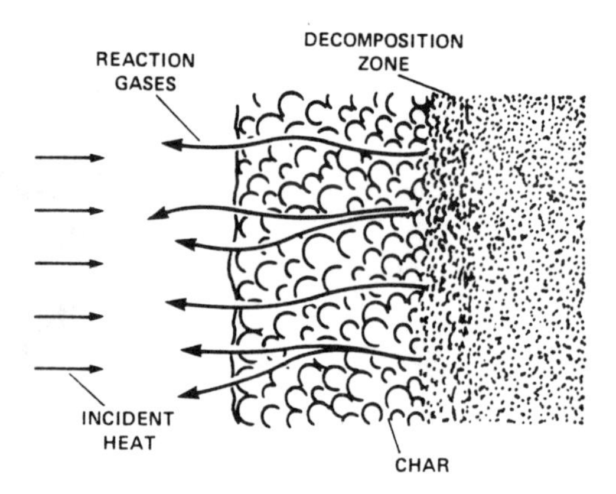

Fig. 1.

were incurred. A subsequent investigation by a congressional committee headed by ADM. J. S. Russell led to the recommendation that cook-off protection be provided for all naval weapons. In 1969, the Weapons Cook-off Improvement Program was established to accomplish two specific objectives: delay the ordnance reaction for at least five minutes and limit the extent of that reaction to a deflagration (case rupture and burning of explosive). Since then, the five-minute delay has been successfully achieved primarily through the use of ablative and intumescent coatings on the exterior skin of various bombs, rocket launchers, mines, missiles, etc. Intumescent coatings are the material of choice whenever high insulation efficiency from a limited film thickness is desired.

HISTORY

The first observation of carbonaceous intumescence was probably the novelty store "snakes" that, when heated to their activation temperature, begin to swell and foam into the shape of a long, snake-like body. While this phenomenon has been known for years, the first commercial reference to an expanding and foaming fire-retardant material did not occur until 1938. Tramm[1] described a coating containing diammonium phosphate, dicyandiamide, and formaldehyde that swelled to form a layer of carbon when heated. Ammonium phosphate, sulfate, chloride, and bromide were also suggested as possible substitues for the diammonium phosphate. Ten years later, Olsen and Bechle[2] first used the term "intumescent" to describe their phosphoric or sulfuric acid treated esters and alcohols in an asphaltic base.

From 1948 to 1950, Jones and coworkers[3] developed formulations similar to those in use today. Their intumescent systems consisted of "carbonifics" and "spumifics." A carbonific was defined as a carbon-yielding source and a spumific as a foam-producing substance. Their carbonific was a resinous mixture of an aldehyde and urea. They also discussed the use of non-resinous materials such as starches, carbohydrates, and proteins. Their spumifics were chosen from known fire-retardant compounds (acids, phosphates, sulfates, and borates). The resinous materials generally reacted at about 140°C; while the non-resinous ones reacted above 300°C.

Unfortunately, these coatings possessed little or no water-resistance. In an effort to replace the water-soluble spumifics, Nielson[4] developed phosphorylamide from a reaction between phosphoryl chloride and ammonia. After polymerization, this material is relatively insoluble in water. However, it was never commercialized due to processing and economic difficulties.

Sakurai and Izumi[5] studied urea formaldehyde, starch, and diammonium phosphate films in fire environments. They measured char height and volume as well as the length of time for the backside of a coated wooden panel to reach 260°C when the coating was subjected to an impinging 500°C flame. It was concluded that the optimum protection does not always coincide with the most voluminous char.

In 1953, Jones and coworkers[6] discussed the use of melamine phosphate, a compound that resists burning at temperatures below 750°C and degrades to a hard, carbon residue. It is still used in commercial solvent-based formulations, but not in aqueous mixtures because of its tendency to gel in water.

Christianson,[7] in 1954, reported the desirability of controlling the size of the cells in an intumescent char. He found that the char structure of a typical coating could be varied by the use of different blends of china clay and zinc oxide fillers, due to their foam-nucleating properties. Wilson and Marotta[8] were the first investigators to suggest the use of fibrous materials like asbestos and fiberglass to hold the char together.

Cummings[9] described an intumescent formulation as a ternary system where the preferred composition was within the shaded boundaries shown in the figure below.

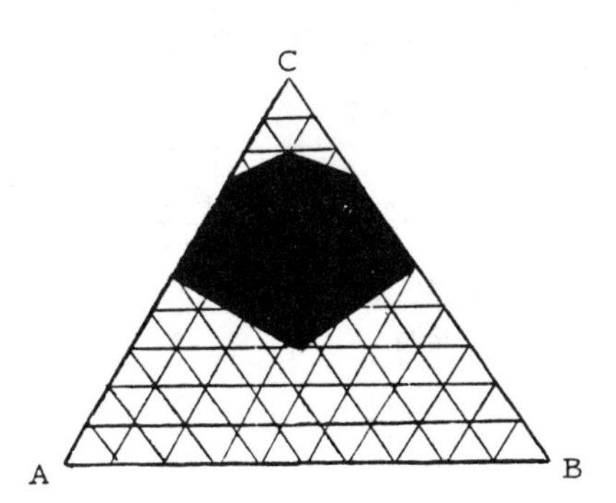

Fig. 2.

"A" is a non-resinous carbonific (dipentaerythritol), "B" is an organic nitrogen compound (dicyandiamide), and "C" is a spumific (ammonium phosphate). Optimum performance was obtained when A = 0-50%, B = 0-50%, and C = 30-80%. However, these ranges were further restricted with the introduction of a thermoset resin binder, due to its detrimental effect on intumescent expansion. Thermoplastic resins are generally favored in such coatings.

In 1961, Quelle[10] patented the first clear intumescent coating using a condensate resin formed from urea, formaldehyde, and phenol combined with butyl and dibutyl phosphate. Such formulations are uncommon because of the high loadings of resins and fillers having far different refractive indices.

In summary, traditional intumescent coatings are generally composed of the following:

Acid Source

Salts or esters of acids capable of dehydrating the carbonific are utilized. The relative effectiveness of these materials depends on the amount of acid character present (percentage of active element). A number of the more common acid sources are listed in the following table.

TABLE 1

Material	Active Element	Decomposition Temperature
melamine phosphate	14% phosphorus	300°C
urea phosphate	20% "	130°C
diammonium phosphate	24% "	87°C
ammonium phosphate	27% "	147°C
ammonium polyphosphate	32% "	215°C
ammonium sulfate	24% sulfur	
ammonium borate	16% boron	

Carbon Source (Carbonific)

The effectiveness of a carbon source is dependent on its carbon content and the number of reactive hydroxyl sites. Carbonifics are usually chosen from among the available sugars, starches, and polyhydric compounds listed below.

TABLE 2

Sugars	Formula	% Carbon	Reactivity (OH sites/100 g)
glucose	$C_6H_{12}O_6$	40	2.8
maltose	$C_{12}H_{22}O_{11}$	42	2.3
arabinose	$C_5H_6O_4$	45	3.0
Starches	$(C_6H_{10}O_5)_n$	44	2.1
Polyhydric Compounds			
sorbitol	$C_6H_8(OH)_6$	40	3.0
pentaerythritol	$C_5H_8(OH)_4$	44	2.9
dipentaerythritol	$C_{10}H_{16}(OH)_6$	50	2.5
tripentaerythritol	$C_{15}H_{24}(OH)_8$	53	2.4
resorcinol	$C_6H_8(OH)_2$	63	1.8

Nitrogen Source

Nitrogenous compounds, such as amides, are utilized as catalysts to promote the esterification of the carbonific by a dehydration process.

They are even more effective when used with orthophosphoric or sulfuric acid.

Blowing Agent (Spumific)

Blowing agents are needed to expand the melt formed during dehydration of the carbonific to a thick, multi-cellular char. They must decompose to release large quantities of non-flammable gases. Some of the more common spumifics are included in the following table.

TABLE 3

Material	Gaseous By-products	Decomposition Temperature
urea	NH_3, CO_2, H_2O	130°C
guanidine	"	160°C
glycine	"	230°C
melamine	"	300°C
chlorinated paraffin	HCl, CO_2, H_2O	160-350°C

In order for intumescence to proceed, several distinct reactions must occur almost simultaneously, but in the proper sequence. First, the acid salt must decompose to yield the dehydrating acid which must esterify the carbonific. (The nitrogenous compound promotes this reaction.) Next, the ester must decompose by dehydration resulting in the formation of a carbonaceous mass. Water vapor released from these reactions and gases evolving from the spumific cause the carbonaceous mass to foam and expand. As the reaction nears completion, solidification occurs and a thick, multicellular char is formed. If any of these reactions does not proceed at the required time, intumescence will not take place. Thus, the temperature at which the specified reactions occur is extremely important.

CURRENT STATE-OF-THE-ART

Nitro-aromatic Amines

In 1956, Alyea[11] described the formation of voluminous, black foam from the action of heat on a mixture of sulfuric acid and p-nitroacetanilide. Subsequently, Parker and coworkers[12] at NASA's Ames Research Center studied a number of substituted nitro-aromatic amines for their intumescent properties. Ortho and para-nitroanilines were found to yield chars of 70 to 240 times the original volume when heated within the temperature range of 390 to 500°F. These chars exhibited hard, crusty skins that resisted the erosive action of an impinging gas flame. Additional research on dry compounds containing the necessary acid character led to the development of the bisulfate salt of p-nitroaniline. When heated to temperatures above 430°F, it yielded a black char that was stable at 1000°F. Unfortunately, being the salt of a strong acid and a weak base, it hydrolyzed to an acidic solution in the presence of moisture. This adversely affected coating vehicles in which it was combined and corroded metallic substrates.

One of the primary reactions between sulfuric acid and nitroaniline is sulfonation of the ring to form p-nitroaniline-o-sulfonic acid. NASA found that, in its pure state, this compound intumesces at 450°F to give a 50% char yield by weight. It showed good hydrolytic stability, but the problem of acidity still remained. The problem was solved with the development of the ammonium salt of p-nitroaniline-o-sulfonic acid. The reaction temperature of this compound (572°F) was determined by thermogravimetric analysis, as shown below.

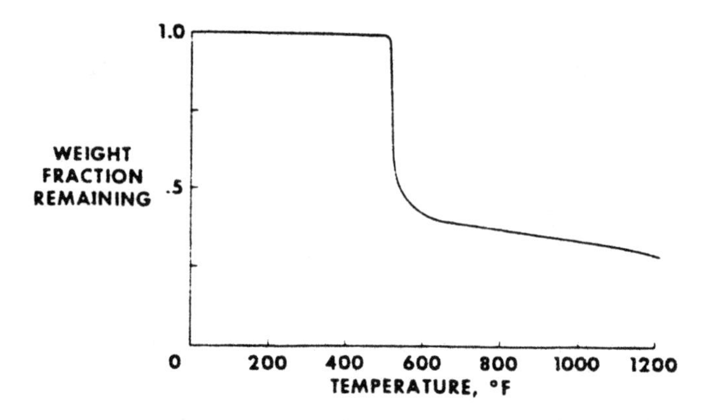

Fig. 3.

A coating formulation containing this intumescent salt in an epoxy/polysulfide binder system was exposed to a 100°F/90% RH environment for thirty days with no significant effect.

Sodium Tetraborate Decahydrate (BORAX)

Investigators at the Minerals, Pigments, and Metals Division of Pfizer, Inc. have patented[13] a process for the formulation of a coating filled with sodium tetraborate decahydrate. It is considered intumescent by virtue of its expansion in a fire to form a dense char two or three times the original coating thickness. However, the thermal protection is largely the result of transpirational cooling due to the sublimation of water vapor from the compound. Approximately 47% of its theoretical weight is water of crystallization. Although it is slightly water-soluble, sodium tetraborate decahydrate can be used in coatings subjected to exterior weathering if topcoated with a film of low moisture-permeability.

Triphenyl Phosphite

Triphenyl phosphite is a clear, water-immiscible liquid that has been used to reduce the viscosity and extend the shelf life of epoxy resins. In 1972, Blair and coworkers[14] at the Hooker Research Center reported the intumescent properties of basic coating formulations diluted with this material. Subsequent work at the Naval Air Development Center demonstrated a number of inherent advantages in such coatings. Because it is a liquid, triphenyl phosphite can be used to reduce a formulation to spray viscosity without the addition of organic solvents that can become entrapped in thick films, thus affecting the resulting thermal and physical properties. It provides the intumescent capability at low concentrations, so that the optimum strength and flexibility of the binder system can be obtained. In addition, since this filler is both clear and completely immiscible in water, effective coatings can be pigmented in any color desired and topcoating is not required.

LABORATORY COOK-OFF SIMULATION

Definitive fast cook-off tests are accomplished by suspending a coated weapon three feet over a 2000-gallon "pool" of JP-4 aviation fuel and measuring the time between ignition and the first reaction of the explosive or propellant. Naturally, such testing is expensive to run and is not a practical means of screening a large number of coating candidates. For this purpose, a small-scale facility capable of exposing flat panels to a realistic cook-off environment was constructed at the Naval Air Development Center during the early phases of the program. As shown in Fig. 4, this facility consists of a six cubic foot furnace lined with fire-brick and enclosing a horizontal rotary oil-burner of the mechanical atomization type which is capable of using JP-4 or JP-5 aviation fuel. During operation, air is circulated from two vents on the lower side up through the furnace by an overhead exhaust fan. The port located at area 1 is used for all coating evaluations. The heat flux at this point is 90% radiative and is readily adjusted to 10 BTU/ft^2-sec with the use of an asymptotic calorimeter.

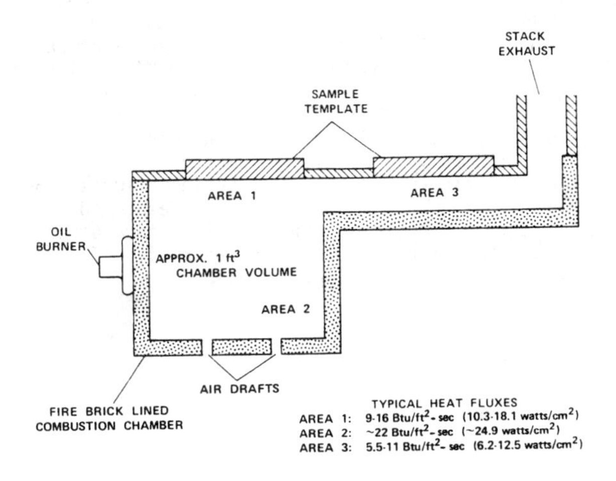

Fig. 4.

Test panels (ranging from three to eight-inch squares) are constructed by cutting out a round disc from the center of each panel, shaving 1/16 of an inch from the circumference, and cementing it back in place with a high-temperature epoxy adhesive. This procedure serves to insulate the disc from heat conducted away from the mounted edges of the panel.

The coating to be tested can be applied to either side of the test panel. When fully cured, the film thickness over the center disc is measured and recorded. A thermocouple is attached to the backside of the disc, in order to monitor its temperature. At time (t = 0), the panel is placed, coated side down, over the preheated furnace. The back of the panel is covered with 6 lb/ft^3 refractory fiber insulation. A strip-chart recorder measures the temperature rise as a function of time. From this data, a thermal efficiency rating is determined as the time in seconds/mil thickness required for the center disc to reach a given temperature.

Approximately 600 tests have been conducted during the course of the program on more than 175 different coatings. Typical thermal efficiency ratings obtained for similar formulations containing the ammonium salt of p-nitroaniline-o-sulfonic acid (NASA 169), sodium tetraborate decahydrate (Pfizer RX-2370A), and triphenyl phosphite (NADC TP-62) are given below.

TABLE 4

Formulation	Film Thickness	Thermal Efficiency to: 500°F	1000°F
NASA 169	85 mils	2.7 sec/mil	5.4 sec/mil
Pfizer RX-2370A	90	2.5	3.7
NADC TP-62	85	2.2	3.4

CONCLUSION

Intumescent coatings have been applied to weapons in production and proposed for such military applications as aircraft and ship fire-containment, missile heat-shields, laser hardening, etc. Specifications TT-P-26, TT-C-1883, MIL-C-46081, MIL-C-81945, MIL-C-81946, and MIL-C-81947 have been established by government agencies for various compositions. These materials have demonstrated the potential of intumescent coatings to provide short-term thermal protection for vehicles and structures with minimum additional weight. In the private sector, intumescent coatings have been applied to steel superstructures and aircraft loading bridges. Many other possible applications exist that have not been fully explored. It is hoped that, in the future, stricter building codes and insurance underwriting practices will lead to greater usage of these materials. The resultant savings in lives and property would be inestimable.

REFERENCES

1. Tramm, H. et al., U.S. Patent No. 2,106,938, Feb. 1938.
2. Olsen, J. and Bechle, C., U.S. Patent No. 2,442,706, June 1948.
3. Jones, G. et al., U.S. Patent No. 2,452,054/5, Oct. 1948.
4. Nielson, W., U.S. Patent No. 2,596,937, May 1952.
5. Sakurai, T. and Izumi, T., *J. Chem. Soc. (Japan) Ind. Section 56, 156 (1953)*.
6. Jones, G. et al., U.S. Patent No. 2,628,946, Feb. 1953.
7. Christianson, C., U.S. Patent No. 2,681,326, June 1954.
8. Wilson, I. and Marotta, R., U.S. Patent No. 2,702,283, Feb. 1955.
9. Cummings, I., *Ind. Eng. Chem. 46, 1985 (1954)*.
10. Quelle, G. et al., British Patent No. 862,569, March 1961 and U.S. Patent No. 3,077,458, Feb. 1963.
11. Alyea, H., *J. Chem. Educ. 3, 3 (1956)*.
12. Parker, J. et al., *SAMPE Journal, August/September, 1968*.
13. U.S. Patent No. 4,001,126, Jan. 1977.
14. Blair, N. et al., *J. Paint Technol. 44(573), 75 (1972)*.

COATING SURFACES UNDERWATER

Richard W. Drisko

Civil Engineering Laboratory,
Port Hueneme, CA

RICHARD W. DRISKO

ABSTRACT

There has long been a need for systems to control corrosion and biological fouling on underwater surfaces. Of the many concepts studied, underwater coatings are the ones that best achieve both needs.

The first underwater-applicable coatings developed were the viscous epoxy-polyamide materials commonly called "splash-zone compounds." These materials, which are based on a Shell Chemical Company formulation, were slowly applied by the palm of the hand at a thickness of 1/8 to 1/4 inch. Curing occurred slowly at temperatures below 60°F(16°C), and, until cured, the coatings were susceptible to wave damage.

More recently, thinner, brushable coatings have been investigated. Studies at the Civil Engineering Laboratory (CEL) have been directed at solvent-free epoxy formulations cured with an amine adduct catalyst. Blown fish oil and an organotin additive have been found to be effective wetting agents that permit underwater application. Such formulations cure at temperatures as low as 40°F(4°C), although curing proceeds much slower at this temperature.

Several biocide combinations were incorporated into CEL underwater-brushable coating formulations to prevent the attachment and growth of marine fouling organisms. Only organotin compounds produced long-lasting fouling control. Virtually complete elimination of fouling for 3 years occurred with coatings having a tin metal content of 6%. These coatings compared quite favorably with the standard Navy copper-based antifouling paint used for comparative purposes.

BACKGROUND

A system for applying protective coatings to underwater surfaces has long been sought. Such a system would permit damaged or weathered coatings to be repaired on fixed or floating structures without removing them from the water. It would also permit touch-up or renewal of hull coatings without having to drydock the vessels, or it would allow one to coat the keel block areas that were inaccessible during drydocking.

An old concept for underwater coating that still finds occasional use is that of a cofferdam (Figure 1) (1, 2). A metal, wooden, or plastic cofferdam can be placed against or around a structure, such as a piling, that is normally underwater and pumped dry so that the exposed metal can be cleaned and coated using conventional methods. CEL found (1) that high solids coatings (such as Steel Structures Painting Council Specification No. 16, coal-tar epoxy-polyamide) applied in this manner will cure under seawater after the cofferdam has been removed. Such a cofferdam approach is limited to depths close to the surface. However, it has been suggested that for greater depths, the cofferdam could be secured upside down and air pumped into it to displace the water.

Fig. 1.—Coating steel piling using CEL-developed cofferdam.

In 1962, the Shell Chemical Company (3, 4) offered coating and adhesive manufacturers a two-component epoxy-polyamide formulation designed for application on damp and underwater surfaces. Shortly thereafter several coating manufacturers began marketing products generally called "splash-zone compounds" based on this formulation. Equal volumes of two differently colored components are usually mixed before being applied at a thickness of 1/8 to 1/4 inch by pressing the palm of the hand against the substrate or using a fiberglass, canvas, or burlap backing. Although abrasive blasting is the preferred method of surface preparation (5), cleaning with a needle gun with needles of 2 mm diameter can provide a suitable surface (6). Because the mixed resins are slightly soluble in water, splash-zone compounds slowly erode when worked to give a smooth finish. This erosion also permits one to clean the equipment with water before curing occurs. A military specification (MIL-P-28579) has been prepared for such a product.

Because of their high viscosities, the two components of splash-zone compounds are difficult to mix to a uniformly colored blend. Prewetting of the two components with water permits easier mixing, but results in a mix that bonds much more poorly underwater to steel. The high cost of materials, the large film thickness, and the slow rate of application by divers make the use of splash-zone compounds very expensive. Also, curing occurs to a significant extent only at temperatures above 60°F (16°C) and generally requires 24 hours of curing before strong bonding is achieved. During this period, the soft epoxy is susceptible to wave damage. Thus, it is a common practice with splash-zone compounds to coat steel piling above water as far as possible at low tide and then to coat underwater at high tide. Despite the limitations of splash-zone compounds, several are marketed today for special field use that is difficult to duplicate with other materials.

More recently, several experimental underwater-brushable coatings have been developed (6, 7). Most of them are solvent-free formulations, but some contain small quantities of solvent, such as n-butyl alcohol or butyl cellosolve, that are sufficiently water soluble to permit curing underwater. They are usually based on epoxy resins, but a few polyester and coal tar epoxy formulations have been developed for underwater application. The polyester materials generally are easier to apply than other generic types, but tend to be softer and more easily damaged. Because the products developed to date have neither the desired ease of application nor the desired durability on curing, the search for a better underwater-brushable material continues.

FORMULATION EXPERIMENTS

Resin Systems

Many resin systems were considered for underwater-brushable coating formulations. The epoxy-polyamide system used in splash-zone compounds seemed to impart neither the brushing nor the curing properties desired. A Celanese Chemical Company epoxy resin system cured with an amine-adduct catalyst seemed to give the best combination of properties. Many different epoxy/catalyst variations were examined, but none seemed to be a significant improvement over the system recommended by Celanese.

Because of the good underwater application properties of polyester coatings, several polyester variations were investigated. While it was not difficult to formulate such products that could be readily applied underwater, it was not possible to get them to cure to a hard, tough finish in a reasonable amount of time. Attempts at combining epoxy and polyester resins to give a product with the good properties of each resin were unsuccessful.

Wetting Agents

In order for underwater-applicable epoxy formulations to wet immersed surfaces and displace surface water, special wetting agents must be added. Of a wide variety of natural and synthetic wetting agents tested, blown fish oil gave the best results. It was subsequently found that the addition of about 4% by weight of the reaction product of bis (tri-n-butyltin) oxide and the free fatty acids of linseed oil (hereafter referred to as "organotin wetting agent") significantly increased the ease of application.

A few wetting agents were tested in polyester formulations, but they did not seem to impart any beneficial properties.

Pigments

Lead silica chromate was generally used in experimental epoxy formulations to impart corrosion inhibition. It performed well in this manner, but tended to settle to the bottom of the can during prolonged storage. On cathodically protected or otherwise electrically charged underwater steel surfaces, it was quite difficult to wet the surfaces. In order to resolve this problem and investigate new concepts, formulations were developed using conductive graphite and magnetic iron oxide. Neither yielded products with improved properties. Aluminous cement was also investigated as an additive because it cured underwater by reaction with water. It did not, however, produce any beneficial results.

Basic Formulation Tested

After extensive laboratory testing of different coating formulations, the one listed in Table 1 was used by itself or along with an organotin wetting agent in field testing of surface preparation and coating application requirements.

SURFACE PREPARATION EXPERIMENTS

Field experiments were conducted to determine optimum methods and relative speeds of surface preparation of steel for underwater coating. The quality of surface preparation was measured by the adhesion of the underwater-applied coating of Table 1. Different methods of surface preparation included abrasive blasting, waterblasting (Figure 2), wirebrushing (Figure 3), and needlegun (Figure 4) cleaning. In all cases, the violent agitation of the water greatly limited visibility and, thus, increased cleaning time. As a result, all cleaning rates were about the same. Abrasive blasting gave the best surface for bonding; waterblasting at 10,000 psi was almost as good; needlegun cleaning was almost as good if small diameter (e.g., 2 mm) needles were used so as to leave a textured surface; and hydraulically powered wirebrush equipment was significantly poorer because it polished the surface rather than leaving it textured.

APPLICATION EXPERIMENTS

Laboratory and field experiments were conducted by applying underwater the coating of Table 1 by brush (Figure 5), roller (Figure 6), or special plastic applicator (Figure 7). The brush for best application was relatively stiff and of medium bristle length; the best roller cover had a medium nap; and a special applicator was fabricated by the Naval Coastal Systems Center (NCSC) by bonding a handle to the interior surface of a half section of plastic pipe. NCSC personnel also developed a system using compressed air to feed the coating to the surface of a roller (Figure 6) or brush. It performed effectively but was very difficult to clean after use. All three application systems were effective in applying the coating, with the preferred choice of applicator varying with the diver. Application rates of 10 to 25 sq ft per hour were achieved by different divers.

TABLE 1.—BASIC EPOXY FORMULATION[a]

Epoxy Portion	
Component	Parts by Weight
Epoxy Resin (Epon 828)[b]	42
Lead Silica Chromate (Permox 143)[c]	38
Blown Fish Oil (Pacific Vegetable Oil Z-7-1/2)	17
Adduct Convertor Portion	
Component	Parts by Weight
Amine Curing Agent (Epicure 8701)[d]	11.6
Amine Curing Agent (Epicure 874)[d]	1.9
Epoxy Resin (Epon 828)[b]	2.5

[a]While sources of raw materials are listed for complete identification, equivalent material from other sources may be used.

[b]Trade name of Shell Chemical Company.

[c]Trade name of Eagle-Picher Industries, Inc.

[d]Trade name of Celanese Chemical Company

FOULING RESISTANCE

The formulation of Table 1 had significant resistance to the attachment and growth of marine fouling organisms when 4% by weight of organotin wetting agent(corresponding to 1% of tin metal) was added. A number of formulation variations with organotin additives were tested for long-lasting fouling resistance. While all formulations with at least 1% tin had significant resistance, only those with as much as 6% tin gave resistance comparable to that of a conventional cuprous oxide antifouling paint (MIL-P-15931) — 3 years of essentially complete absence of fouling. The addition of rosin to accelerate leaching had no significant effect.

Formulations similar to that of Table 1 with cuprous oxide rather than lead silica chromate pigmentation did not have significant fouling resistance unless loaded with leaching agent. Addition of up to 12% of 2, 4, 5, 6 tetrachloroisophthalonitrile was not effective in reducing fouling even when loaded with leaching agent (8).

Fig. 2.—Waterblasting of steel underwater.

Fig. 3.—Hydraulically powered wirebrush equipment used for cleaning steel underwater.

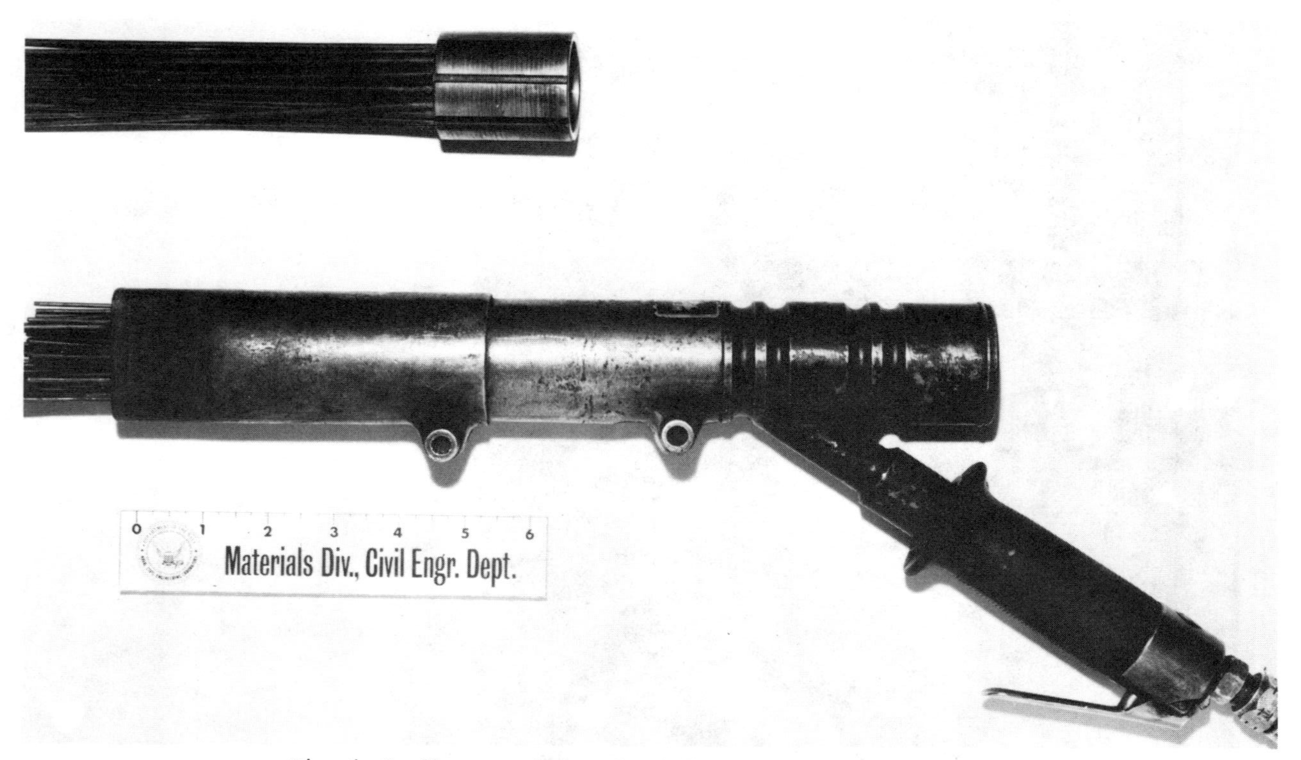

Fig. 4.—Needlegun used for cleaning steel underwater.

Fig. 5.—Brush application of coating underwater.

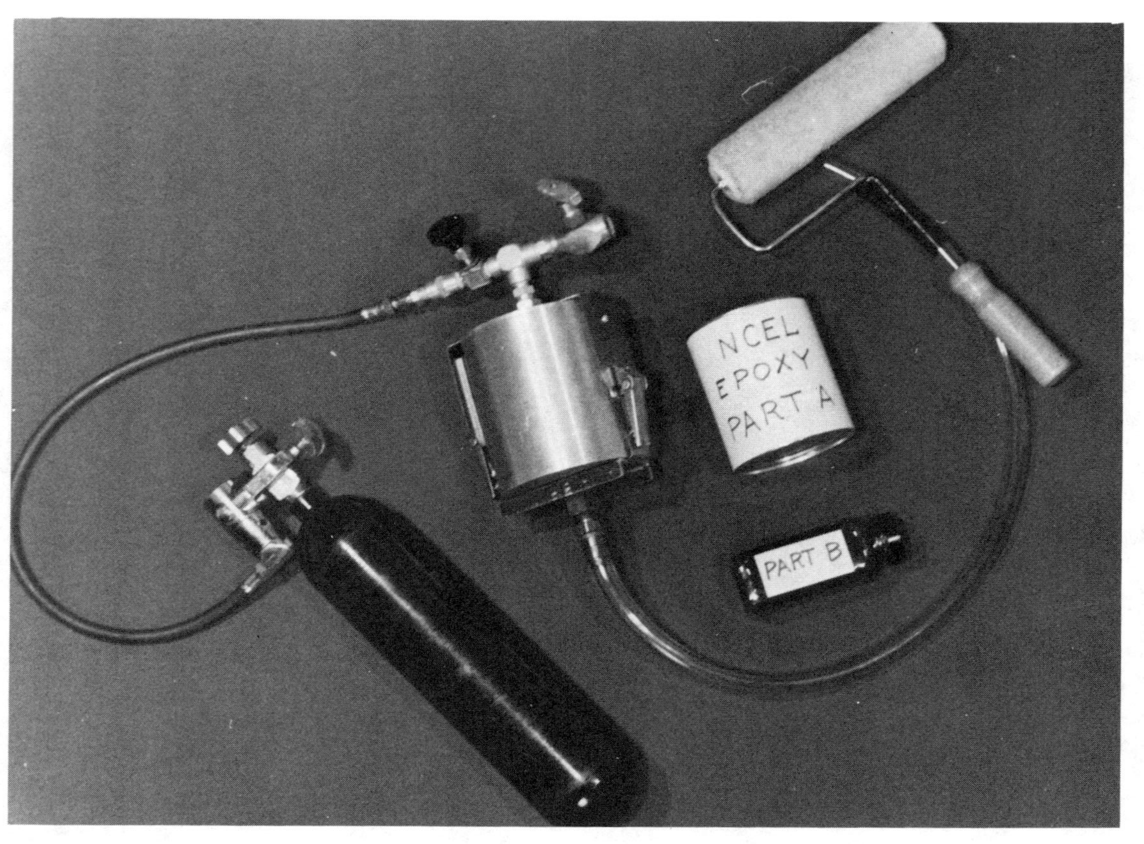

Fig. 6.—Roller equipment for applying coating underwater.

Fig. 7.—Curved plastic applicator for applying coating underwater.

REFERENCES

1. Brouillette, C.V., *Mat. Per. 10(4), 32 (1971).*
2. Schultz, E., "Steel Sheet Piling Corrosion at Pier C Berth 23, Long Beach Harbor," Paper 186, Corrosion/76, National Association of Corrosion Engineers, Houston, Texas.
3. Jorda, R.W., *Mat. Per. 2(3), 56 (1963).*
4. Staff Feature. "Splash zone coating protects corroded steel piling," *Mat. Per. 2(10), 81 (1963).*
5. Drisko, R.W., *Paint and Varnish Production 58(7), 31 (1968).*
6. Drisko, R.W., *J. Coatings Technol. 47(600), 40 (1975).*
7. Drisko, R.W. and Brouillette, C.V., *Mat. Per. 10(4), 32 (1971).*
8. Drisko, R.W., "Protective coatings and antifouling paints that can be applied underwater," Offshore Technology Conference, 9th, 1977 Proceedings, Dallas, Texas, pp. 419-421.

VII. SURFACE AND COATINGS SCIENCE

THE COMPOSITION OF METAL SURFACES AFTER ATMOSPHERIC EXPOSURE

James E. Castle

University of Surrey, Guildford,
Surrey, England

JAMES CASTLE

SUMMARY

This review considers the evidence produced by electron spectroscopy for the nature of the layers formed on metals and alloys of industrial importance by atmospheric exposure. It concludes that the 'oxide' layer is more properly described as a hydroxy oxide. However, this layer is itself normally coated with a layer of similar thickness, formed from adsorbed water molecules and organic contamination. This layer is not removed by simple preparative techniques and probably remains after application of organic coatings.

INTRODUCTION

A decade ago in the symposium sponsored by General Motors, on interface conversion for polymer coatings Eirich[1] described the hierachy of layers (Figure 1) by which the surface of a metal becomes terminated after a period of atmospheric exposure. At about the same time the discovery of the principles of surface analysis by Auger electron spectroscopy (AES)[2] and X-ray photoelectron spectroscopy (ESCA or XPS)[3] provided the means for obtaining direct analytical confirmation of the structures shown in Figure 1. The possibilities were reviewed in 1972 by Sparnaay[4] in his introduction to a Nato Study Group on surface coatings. He dwelt more on the possibilities for understanding the basal plane of the hierarchy, i.e., the clean metal surface, rather than the layers formed on it. The sensitivity of AES to elements such as sulphur and chlorine which have poor photoemission cross-sections in XPS[6,7] has given this technique an important role in this context. Unfortunately, for the purposes of this review, it was soon realized that the 'hydrocarbon contamination' has to be removed in order to obtain AES spectra of the metal surface. Thus, while the ability to detect contamination by AES, and to remove it by successive cycles of ion-etching and annealing, has perfected those measurements made of necessity on clean surfaces, e.g., surface diffusion and segregation, for the coating technologist the baby has been thrown out with the bath water. The possibilities inherent in XPS for the study of the outermost layers of Eirich's hierarchy were not appreciated until more recently. As will be shown, its ability to recognize differing chemical forms of oxygen has proven of particular value.

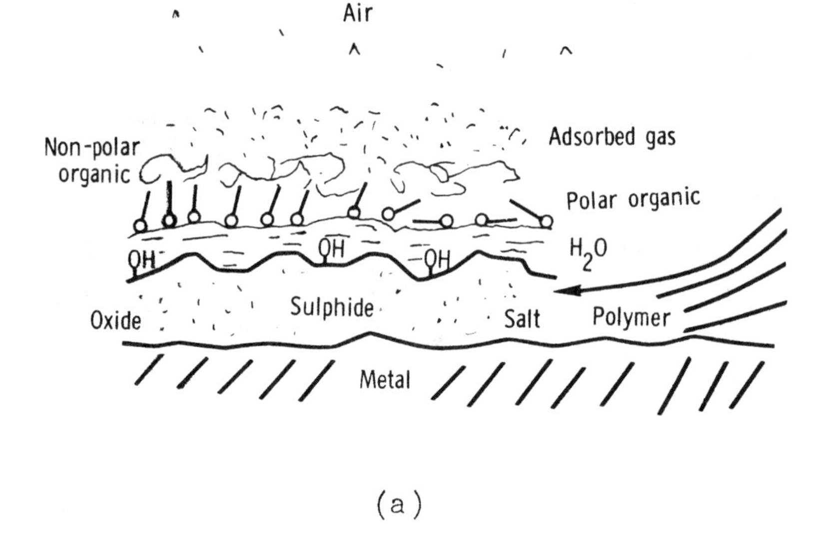

(a)

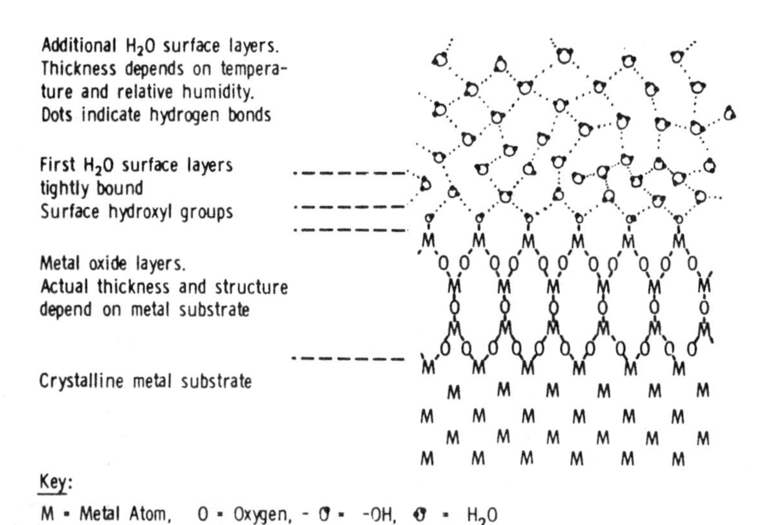

(b)

Fig. 1.—Surface layers as understood prior to the general use of surface analysis.
(a) after Eirich[1]
(b) after Bolger and Michaels.[5]

This review will consider the evidence, which the use of AES and XPS has produced, of the nature of the layers formed on metals and alloys of industrial importance by atmospheric exposure, i.e., the layers which might be expected to be present after routine preparation of the surface for coating. It does not take account of the evidence produced by secondary ion techniques nor is it concerned with the breakdown of the passivating layer. This latter step enters the realm of corrosion which has already been the subject of a review[8] to which this is complementary.

THE METAL SURFACE

The metal surface is the basal plane of the surface layers and to some extent the baseplate which guides their structure by virtue of the phenomenon of epitaxy. Interest in its structure, studied by LEED predates the ability to analyze it chemically. A useful review of surface structural defects has been given by Rhead[9] and monolayer segregation to the surface has been reviewed by Seah.[10] Combined use of an ion gun and AES on, for example, atmospherically oxidized low carbon steel, soon reveals the presence of impurities such as sulphur and chlorine concentrated close to the metal/oxide interface and beneath the outer hydrocarbon and water contamination layers, Figure 2. The LVV Auger electrons from these elements have inelastic mean free paths of the order of only 0.5nm but this is compensated by their large cross-section for Auger emission. XPS, which routinely provides analyses of both the layer structure and the underlying metal interface[11,12,13,14] is very insensitive to these elements and their presence is probably overlooked in many cases. The investigation of sulphur and chlorine on unoxidized metal surfaces has a history as long as AES itself.[15,16,17] It is now beyond doubt that they arise from the metal matrix and their concentration depends on its thermal history. Their investigation has been complicated by effects arising from local heating of the surface by the electron beam[18] and by the tendency for the same impurities to be generated from the filaments of gauges and guns within the vacuum system. The segregation of non-metallic and metallic elements to free metal surfaces[19] and to grain boundaries[20] in high vacua is of immense importance in our understanding of metallurgical phenomena such as temper brittleness[21] and of some types of stress corrosion cracking. Nevertheless their effect on the formation of an oxide layer may be minimal and the reviewer knows of only one instance in which sulphur, for example, has been proven to influence the growth kinetics of a relatively thick oxide, in this case on nickel.[23] Nor does interface segregation appear to have been indicted in any case of coating disbondment. Their small effect may be because of the much larger forces available for alloy restructuring in the presence of oxidizing gases or because they are trapped at the alloy/oxide interface in much the way that they are trapped at a grain boundary,[10] Figure 3. The segregants may influence adsorption, e.g., calcium in gold influences water adsorption strongly,[24] but this is no proof that they influence the outer layers of the passive surface film to which coatings are applied.

ADSORPTION FROM THE GAS PHASE

The uptake of monolayers of oxygen by clean metal surfaces at low or ambient temperature has been a favored topic of study.[25] Much of the work has had the aim of elucidating the reactivity of metal surfaces and thus of the mechanism of adsorption. A useful review has recently been given by Joyner.[26] Considerable success has been had in distinguishing dissociative from non-dissociative adsorption of molecules. However, from the viewpoint

TABLE I.

Type of Oxygen / Region of Growth	Binding Energy, 1s, (eV) 530 532 534	Identification	Ref
Frozen water	×	$H_2O_{(s)}$	31
Contamination	o	O_2....MO	34
	o	H_2O...MO	34
	+	H_2O...MO	40
	×	H_2O...MO	48
	+	H_2O...MO	39
	× × ×	H_2O...MO	*
Atmospheric oxidation	o	$Ni(OH)_2NiCo_3$	49
	□ □	CoOOH	49
	o o	NiO $Ni(OH)_2$	30
outer scale	× ×	FeOOH	48
	× ×	FeOOH	43
inner scale	× ×	Fe_2O_3	*
	o	NiO	49
	×	Fe_3O_4	49
	‡	$Cr_2O_3Cr(OH)_3$	39
Thick oxide or reference material	◁	Cu_2O	34
	+ +	CrOOH	40
	× ×	FeOOH	71
	×	Fe_2O_3	71
	×	Fe_3O_4	71
	×	FeO	71
	□	Co_3O_4	44
	o	NiO	44
	× ×	FeOOH	43
	×	$Fe_3O_4Fe_2O_3$	43
Very thin film (<5nm)	+	-	53
	×	-	53
	o	NiO	30
	+	Cr_2O_3	45
	▷	Mn_3O_4	47
	×	Fe_2O_3	31
	×	$Fe_3O_4Fe_2O_3$	43
Adsorbed gas or mono layer (θ <2)	o o o o	$.15<\theta<1.6$	30
	×	$\theta<0.2$	38
	×	O=	
	× × ✱	O- O_2Ads.	31

Substrate key: + Cr, ▷ Mn, × Fe, □ Co, o Ni, ◁ Cu.

*This paper.

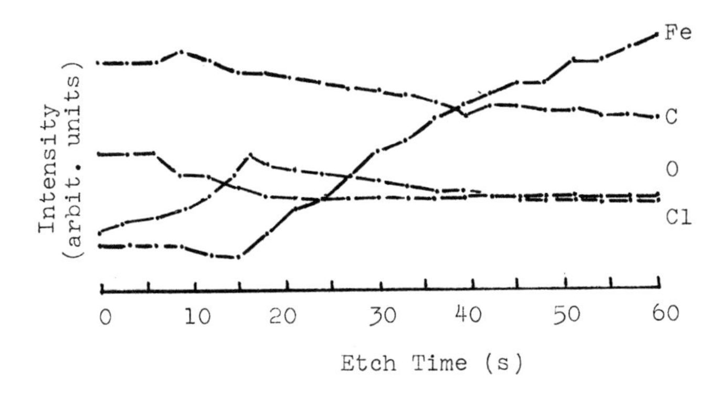

Fig. 2.—Typical Auger etch profile showing chloride at metal/oxide interface (sample c of Figure 8A).

of coating promotion the extent to which 'back-bonding' stabilized adsorption by donation of electrons from the metal to the bonding orbitals in the adsorbed molecule is of interest. Figure 4, taken from the work of Joyner and Roberts[26] shows how back bonding gives rise to a relationship between the heat of adsorption of CO on various metals and the binding energy of the O_{1s} electron. This class of adsorption, which is likely to include hydrocarbons, is likely to be important in determining the most suitable molecules for use as adhesion promotors or corrosion inhibitors. The phenomenon was invoked by Castle and Bailey[27] to explain why the N_{1s} binding energy in butylamine, apparently chemisorbed on clean iron, did not differ from that reported for the free molecule. Some doubt has been thrown on this interpretation, however, by the finding of Kishi and Roberts that back-bonding was unimportant in the adsorption of a number of nitrogen containing molecules.[28] This latter study[28] brought to light one further fact of importance in the establishment of surface layers: that the strength of the N-Metal bond is lowered by subsequent adsorption of the more electronegative oxygen molecules.

THE OXIDE LAYER

The further growth of adsorbed monolayers into three dimensional oxide has now received considerable study by way of in situ oxidation of clean metals. Table 1 contains a collection of some of the data obtained for the principle steel forming and non-ferrous alloying elements. Generally, the results show growths reaching a thickness of ≃2nm with kinetics not markedly different from those described by Cabrera and Mott's equations for very thin films, i.e., a rate of growth showing an inverse exponential dependence on time. Fuggle[29] however, has summed up a decade of progress by pointing out that the number of states of oxygen postulated in describing transitions from adsorbed oxygen to oxide film is still chaotic. Norton et al[30] have made a similar point. Nevertheless, there are some general features, e.g., whilst adsorption of oxygen at low temperatures gives an oxygen 1s peak at ≃531.5eV, conversion to oxide by warming to ambient temperature leads to a shift to ≃530.0eV. A weak peak at ≃533eV often seen at low temperature disappears on heating and has been ascribed to adsorbed molecular oxygen.[31]

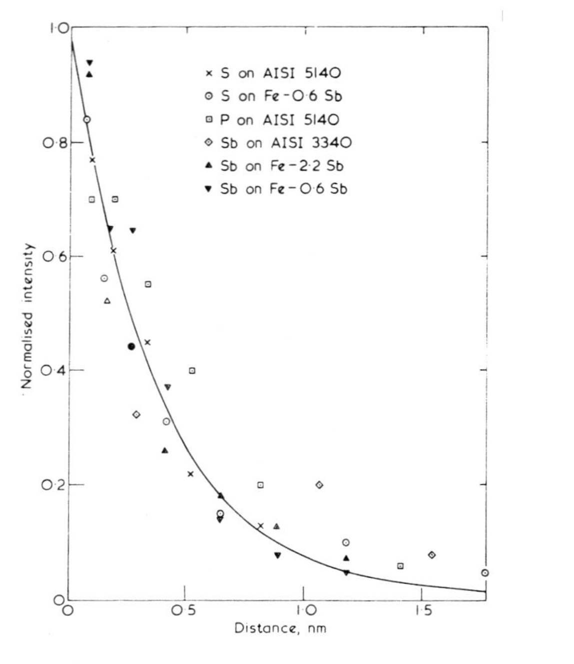

Fig. 3.—Segregation at a metal surface after Seah.[10]

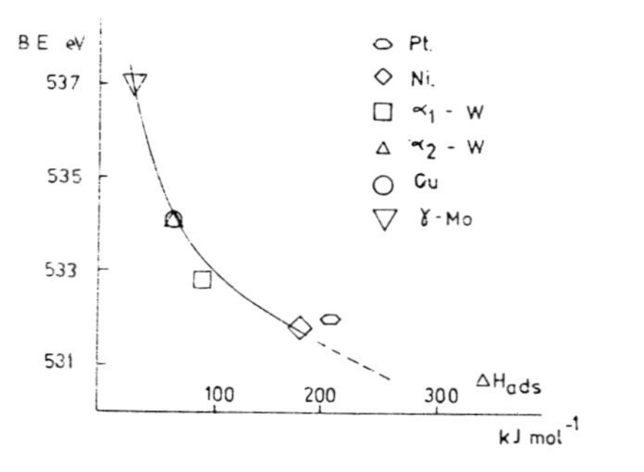

Fig. 4.—The relationship between heat of adsorption of carbon monoxide and the binding energy of the O_{1s} electron. After Joyner.[26]

Fromm and Mayer[32] have attempted to classify the oxidation of thirteen transition metals into 5 groups. They dissent strongly from the use of Cabrera and Mott's equations in which the limiting thickness is determined by the quantum mechanical tunnelling limit for electrons and point out that the same limit is reached in the nitridation of metals even though many of the nitrides are metallic conductors.

All told, there is probably little of surprise to the coating technologist from the work on pure metals. The films reach a limiting thickness when they are exceptionally thin but that was well established prior to the use of surface analysis. Surface and interface energies appear not to have greatly disturbed thermodynamic equilibria, for example: FeO is not found as a surface phase as had once been considered possible and the nickel oxide (NiO) layer is probably oxidized to Ni_2O_3 at the oxygen interface.[30,34] However, there is relatively little change in the spectra on heating to moderate temperatures, i.e., 400-500°K.

There has been very much less work on the early stages of oxidation of alloys but here the particular advantage of XPS of recognizing the oxidation state of each alloying component in addition to monitoring the uptake of oxygen has given useful results. For example, the surface oxide formed on 70/30 Cu/Ni alloy is exclusively nickel oxide[36] whilst the surface layer on 80/20 Cu/Zn alloy is cuprous oxide.[37] High temperature (473°K) oxidation of the cupro-nickel promotes diffusion of copper and the formation of a thick outer layer of cuprous oxide: similar treatment of brass causes zinc oxide to grow as an inner layer. With many steels the oxide has a composition closely similar to that of the steel itself.[11,12,13] Further interest has been aroused by the low temperature at which solid state reactions occur in the absence of atmospheric oxygen. For example, annealing cupronickel leads to growth of nickel oxide at the expense of copper oxide[36] at only 478°K. Similarly chromium and especially manganese will diffuse to the surface of a steel under the oxygen potential available from magnetite, Fe_3O_4.[41,42] Such reactions might well be important when stoving a coating onto a metal surface.

In distinguishing oxides from each other and from their parent elements, standard spectra of well defined surface compounds are obviously necessary. These are now available for many metals. Figure 5 shows data produced by in situ crushing of single crystal material (Brundle[43]) and Figure 6 gives similar data for nickel oxides (McIntyre[44]). Notice how detailed attention to fine-structure enables Ni^{++} to be distinguished in the oxide and hydroxide forms.

EXPOSURE TO WATER VAPOR

There have now been several well documented investigations into the behavior of metals in water vapor.[31,45] These, of course, hold especial interest because of the adverse effect of water vapor on coating adherence. Both Fabian and co-workers,[31] Roberts[46] and Dwyer et al[47] have shown that the oxidation of iron is arrested at a much smaller layer thickness (0.2nm) in water vapor. In all cases oxygen and water vapor exposures were carried out with the same materials and in the same equipment and the contrasting results in the gases cannot be explained by differences in technique. Roberts concludes that the unusual passivation arises from the formation of FeOOH because of the stoichiometric growth of two oxygen peaks. Fabian suggests the formation of an oxide, Fe_3O_4 or Fe_2O_3, with approximately 90% release of hydrogen. This group found no difficulty in observing the growth of AlOOH on aluminum which makes the different result on iron the more significant.

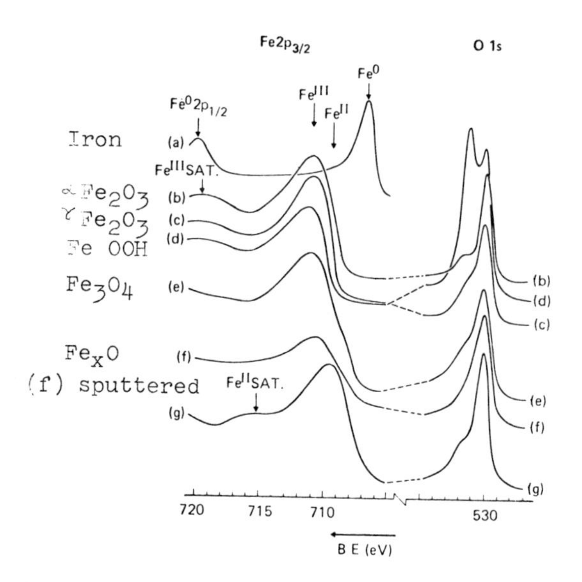

Fig. 5.—O_{1s} and Fe_{2P} photoelectron peaks on a series of iron oxides (after Brundle et al[43]).

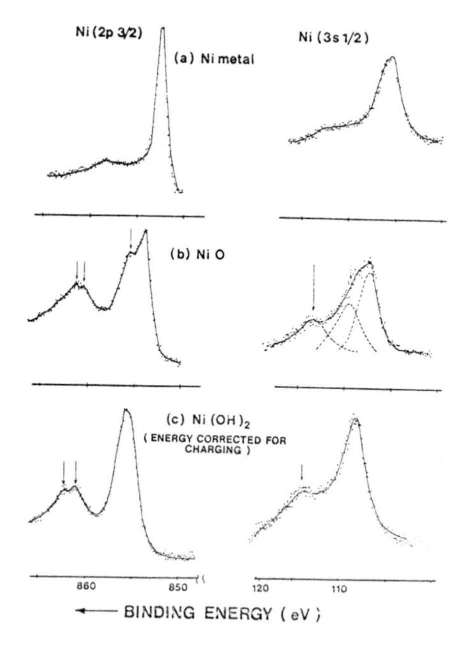

Fig. 6.—Ni_{2P} and Ni_{3S} photoelectron peaks (after McIntyre et al[44]).

Neither of these groups found oxygen peaks corresponding to adsorbed water. Fabian et al found the O_{1s} peak in frozen water to be 535eV but even at temperatures as low as 123°K this peak converted to 530.3eV ($O^=$) plus 532.5eV (OH^-). A low temperature peak at 533eV was ascribed to physisorbed molecular oxygen but, N. B. Asami et al[48] ascribe a peak at 533 to adsorbed water molecules. Norton et al[30] made the interesting and significant observation that films of NiO formed in oxygen convert on exposure to water vapor for periods of several days to a hydroxide form (O_{1s} = 531.3eV) completely although the thickness does not increase, Figure 7.

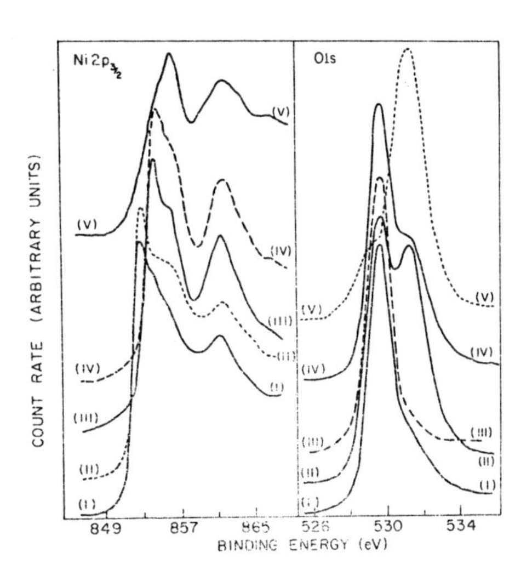

Fig. 7.—Conversion of Ni^o to $Ni(OH)_2$ by atmosphere exposure (after Norton et al[30]).

ATMOSPHERIC EXPOSURE

Apart from the initial, almost inadvertent, exposure to the atmosphere there have been few reported studies of the effect of atmospheric oxidation. Norton et al[30] showed that exposure of preformed oxides to the atmosphere led to the formation of hydroxides with little change in total layer thickness. Barr[49] has, however, recently reported an investigation of the surface layers formed on a large number of metals after atmospheric exposure. The metals were each etched back to the zero valent state and then exposed to air (humidity 35%) for as long as was necessary to produce a passivating or 'terminil' layer. The film thickness varied from Ca.2nm (Cu) to 10nm (Zr) and were thus not dissimilar to those found by other workers in pure oxygen. However, the outermost layers were, as might be expected from work in water vapor, rich in hydroxides. The most interesting observation was that the behavior of the metals could be described in two groups which had a close parallel with the groupings used by Zachariasen and by Mott[50] to describe network modifiers and network (glass) forming elements. Table II reproduced from Barr's study shows his results. Group A metals, which include iron, cobalt, nickel and copper, formed an oxide in its lower normal valence state with a terminal layer of a higher valent state. Group B metals, including aluminium, tin and molybdenum, formed surface hydroxides or hydrated oxides as a terminal layer. The division is based not only on the final

TABLE II.—SURFACES OXIDES AFTER BARR[49]

Group	Metal	Bulk Oxide (Hydroxide)	Saturation (Skin) Oxide (Hydroxide)
A Network Modifiers	Ce	Ce_2O_3 ($Ce(OH)_3$)	CeO_2, $CeO(OH)_2$ ($Ce(OH)_4$)
	Pt	PtO	PtO_2 ($Pt(OH)_4$)
	Pd	PdO	(PdO_2) $Pd(OH)_4$
	Fe	FeO	Fe_3O_4, Fe_2O_3 ($Fe(OH)_3$ or (FeOOH)
	Co	CoO ($Co(OH)_2$)	Co_3O_4, Co_2O_3 CoOOH
	Ni	NiO	Ni_2O_3, (NiOOH or $Ni(OH)_2$)
	Cu	Cu_2O	$Cu(OH)_2$, (CuO)
B Network Formers	Si	SiO_2	SiO_2 $Si(OH)_4$
	Al	Al_2O_3, AlOOH	Al_2O_3, $Al(OH)_3$
	Zr	ZrO_2	ZrO_2, $Zr(OH)_4$
	Sn	SnO_2	SnO_2, $Sn(OH)_4$
	Mo	MoO_3 Mo_2O_5	MoO_3, mixed oxide-hydroxide
	W	WO_3	WO_3, (mixed oxide-hydroxide)
	Y	Y_2O_3	Y_2O_3 YOOH
	La	La_2O_3 (LaO)	La_2O_3, LaOOH
	Rh	Rh_2O_3	Rh_2O_3, RhOOH

state but also on the fact that the Group A oxides grew in the lower valent condition and only oxidized to the higher state when the growth had virtually ceased. Many of the metals in Group B, however, do not naturally form two valence states and the oxides of SnO and SnO_2 are very difficult to distinguish.[51] Iron in Group A apparently had a rather thick Fe outer layer. Thus the division into groups is not exact but the concept is useful and in line with current thinking concerning the role of high valence ions, chromate, and molybdate for example on the formation of passive oxides.[52] Barr noted the presence of hydrocarbon contamination on the surface and also the occasional formation of carbonate (<5%) but paid no especial attention to these. Chromium was apparently not included in the study. This important steel forming element could perhaps be found in either group but curiously it alone in Fabian's study[31] did not form a hydroxide. The atmospheric oxidation of iron chromium alloys and stainless steels has been commented on by Asami,[48] Olefjord,[53] Holm & Storp,[13] Castle and Clayton[14] and McIntyre et al.[54,55] At ambient temperature none of the authors found selective oxidation of chromium relative to iron in either dry or moist air. Nickel, however, is not normally oxidized and may become enriched to a detectable extent at the alloy/oxide interface.[56] However, as was the case with cupro nickel alloys above redistribution of the alloying elements within the oxide can be observed at surprisingly low temperatures. McIntyre[55] observed the selective oxidation of chromium at only 373°K.

None of the above authors ascribes any of the oxygen peaks to water. This may be a correct interpretation when the samples have been prepared by a 'clean' method, such as ion etching, before atmospheric exposure but may not be so when the surface has a hydrocarbon contaminent layer. Castle and Clayton[14] showed a dependence of OH^- like oxygen on the degree of surface contamination after exposure to water. Castle and Epler[57] studying brass electrodes showed that removal of ≈0.25nm of contamination reduced the carbon signal to 0.66 and the oxygen signal to 0.92 of the original peaks respectively whilst other ionic peaks of similar kinetic energy increased in value (see Castle[8]). The fact that there is an association between hydrocarbon and water retention on the surface is shown by the results of an unpublished study of surface preparation. Coupons of plain carbon steel were prepared by five methods; a) abrasion under water to 600 grade silicon carbide, b) dry abrasion with 400 grade energy, c) wire brushing, d) ultrasonic cleaning in water containing 'Decon' detergent, e) ultrasonic cleaning in acetone. These were left in the atmosphere for 70 hours. A further five samples were then prepared by the same methods and all ten loaded together into the spectrometer. The iron peaks of both sets show features from metallic iron and the ferric ion: clearly surface preparation has far more effect on total film thickness than environmental exposure. The oxygen peaks are reproducible from set to set (Fig. 8) showing that the different methods are consistent in their effect on the surface, and have two components in the oxygen spectrum which bear a close resemblance to those found from oxy-hydroxides. On the emery prepared surface which had the most intense integrated iron peak the two components are of equal intensity. All other samples have an excess of the higher binding energy peak and when this excess is included in a plot of normalized peak intensities, Figure 9, there is a correlation with the carbon intensity. There is also a steady shift in the peak position from 531.2eV (C_{1S} = 285) to 532.4eV at high carbon values. At low carbon coverages (dry emery) the peak is principally due to that of the hydroxyl ion (531.5eV)[48] whereas the high coverage peak (BE >532.4eV) represents adsorbed water associated with the organic molecules on the surface. We have found this characteristic association of a high binding energy oxygen with organic molecules on many samples. It is removed by the lightest ion etch as can be seen from the layer structure on an 18/8 stainless steel exposed by ion etching at 5s

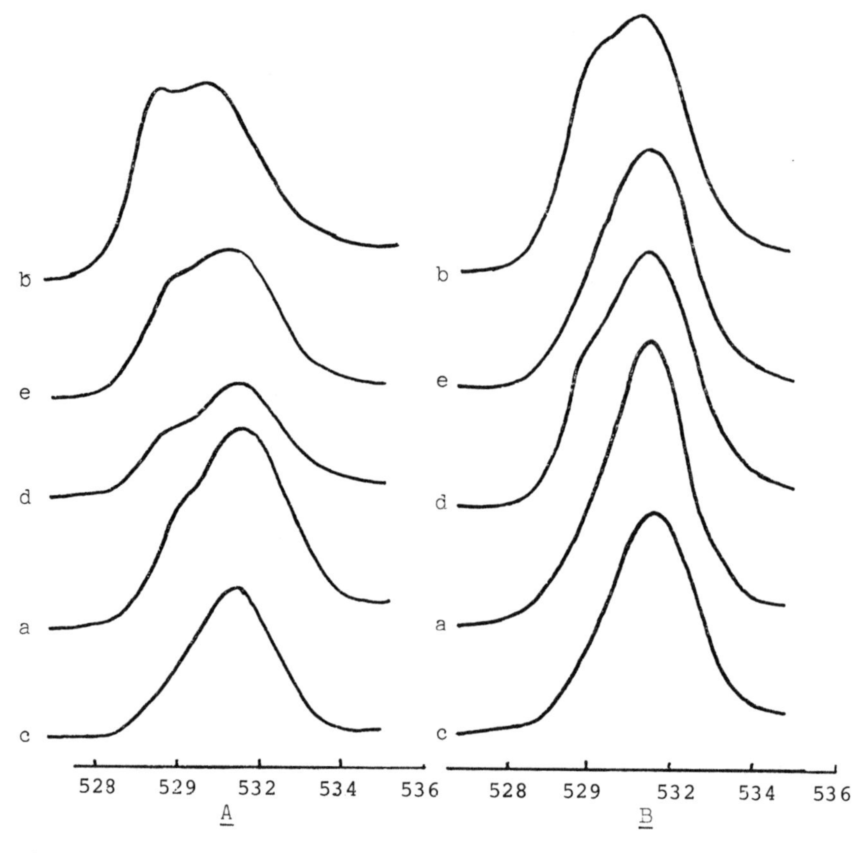

Fig. 8.—Binding Energy (eV)—The O_{1S} peaks on steel after differing surface treatments; A) immediately after preparation, B) after 70 hours. For details of treatment see text.

intervals, Figure 10. Because of its lightly bonded state the ion etch results underestimate its thickness and analysis of XPS spectra,[14] suggest that its thickness is similar to that of the underlying oxide.

INTERACTION WITH LIQUID ENVIRONMENTS

a) Water

Although this paper deals with the atmospheric formed surface layers some mention of their response to water and its solutions should be made. All authors[13,14,58] have found steels to lose the iron constituent in their oxide by selective solution of Fe^{++} ions. Similarly zinc is leached from brass.[59,60] However, the oxide formed on aluminium brass (4at% Al in 70/30 brass) in sea water retains zinc and magnesium as a mixed hydroxide with aluminium.[61] This mixed hydroxide has ion exchange properties which act to protect the brass from sea water by a chemical buffering action.[62] XPS

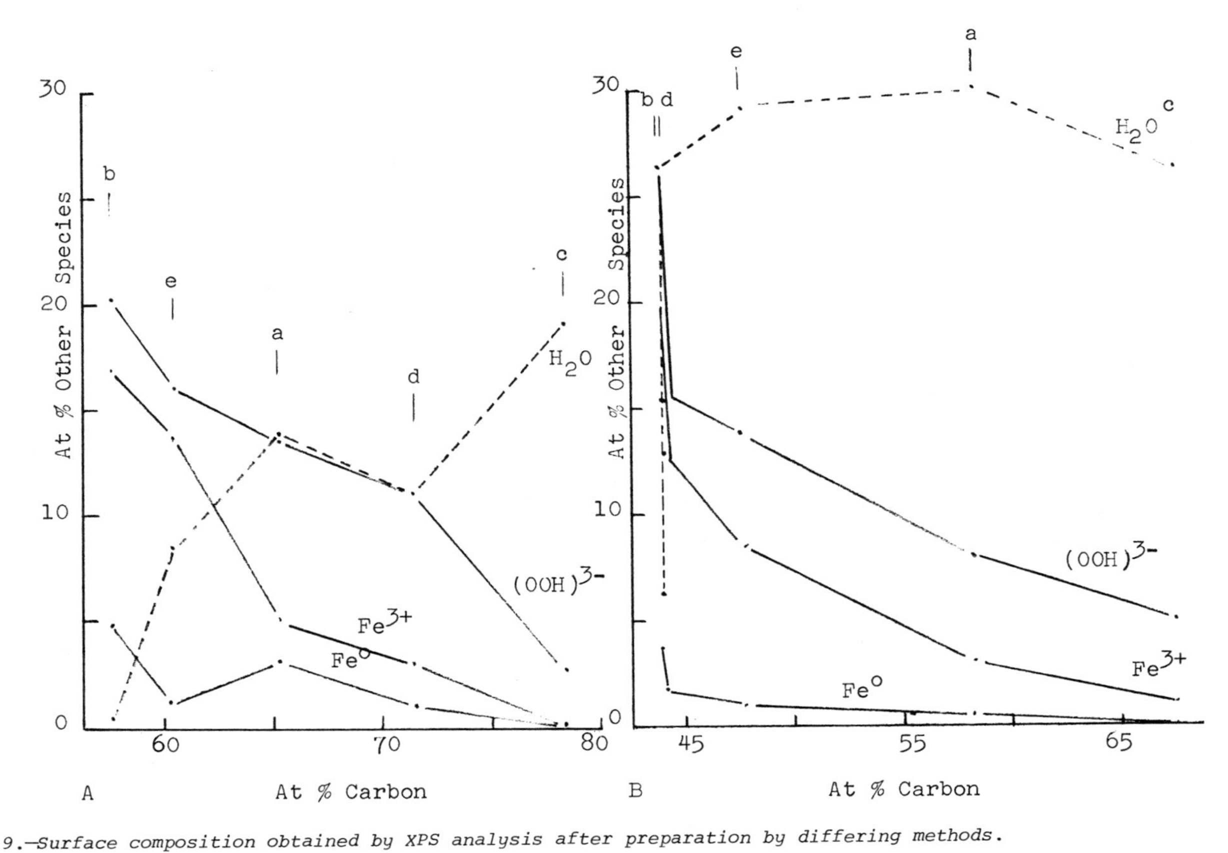

Fig. 9.—Surface composition obtained by XPS analysis after preparation by differing methods.
(A) immediately after treatment,
(B) after 70 h water is adsorbed strongly on all surfaces excepting the clean surface prepared by dry emery.

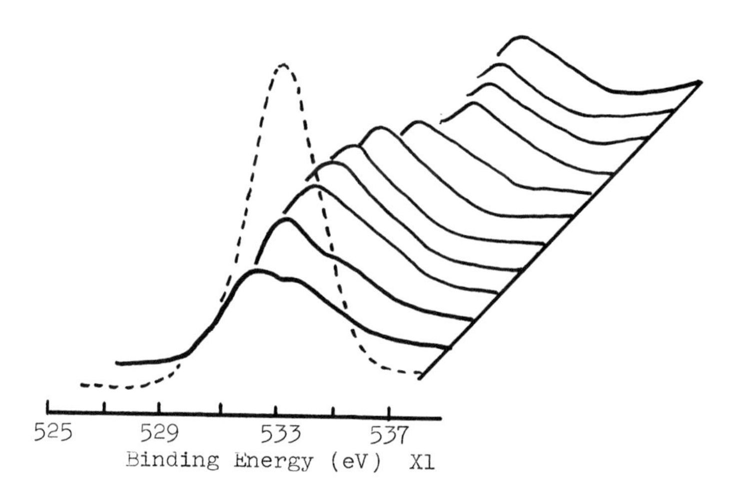

Fig. 10.—Montage of oxygen 1s peaks obtained at 5s intervals (= 0.1nm) during removal of 'oxide' layers by ion etching. Each peak is off-set for clarity. The initial contamination peak is shown in broken line but to the same scale. (Castle and Shakeshaft unpublished.)

results showing the change in surface composition as a function of pH are given in Figure 11. Uhlig[63] suggested on the basis of electrochemical data that the oxy-hydroxide layer on steel may behave in the same way. It has also been shown that[65] the film on stainless steel is able to absorb chloride ions rapidly and throughout its structure without increasing in thickness.

b) Non-Aqueous Solvents

The interaction between the fully passivated metal surface and solutes in organic solvents is typified by the adsorption of adhesion promoters such as the ethoxysilanes. Studies of their adsorption on iron[27] showed that a Tempkin isotherm could be obtained using pure solvents, indicative of a non-uniform heat of adsorption, Figure 12a. However, when solvents were used which contained impurities, notably lead and tin from the joints in metal drums, these competed with the silane for the same range of sites. The resulting surface concentration depended on the solution composition as shown in Fig. 12b. The presence of electronegative substituent groups in the organic radical reduced adsorption.

c) Adhesive Systems

There has again been little study of the influence of organic adhesives on the nature of the surface layers. Anderson and Swalen[66] have shown that acid-base reactions do occur when cadmium eicosanoate is applied to a variety of clean surfaces. This work is still remote from commercial adhesive systems. However, Solomon and McDevitt[67] have introduced a promising technique in which thin evaporated films of aluminum are given conventional pretreatments, coated and cured and only then stripped from a titanium supporting substrate. The advantage of this technique is that

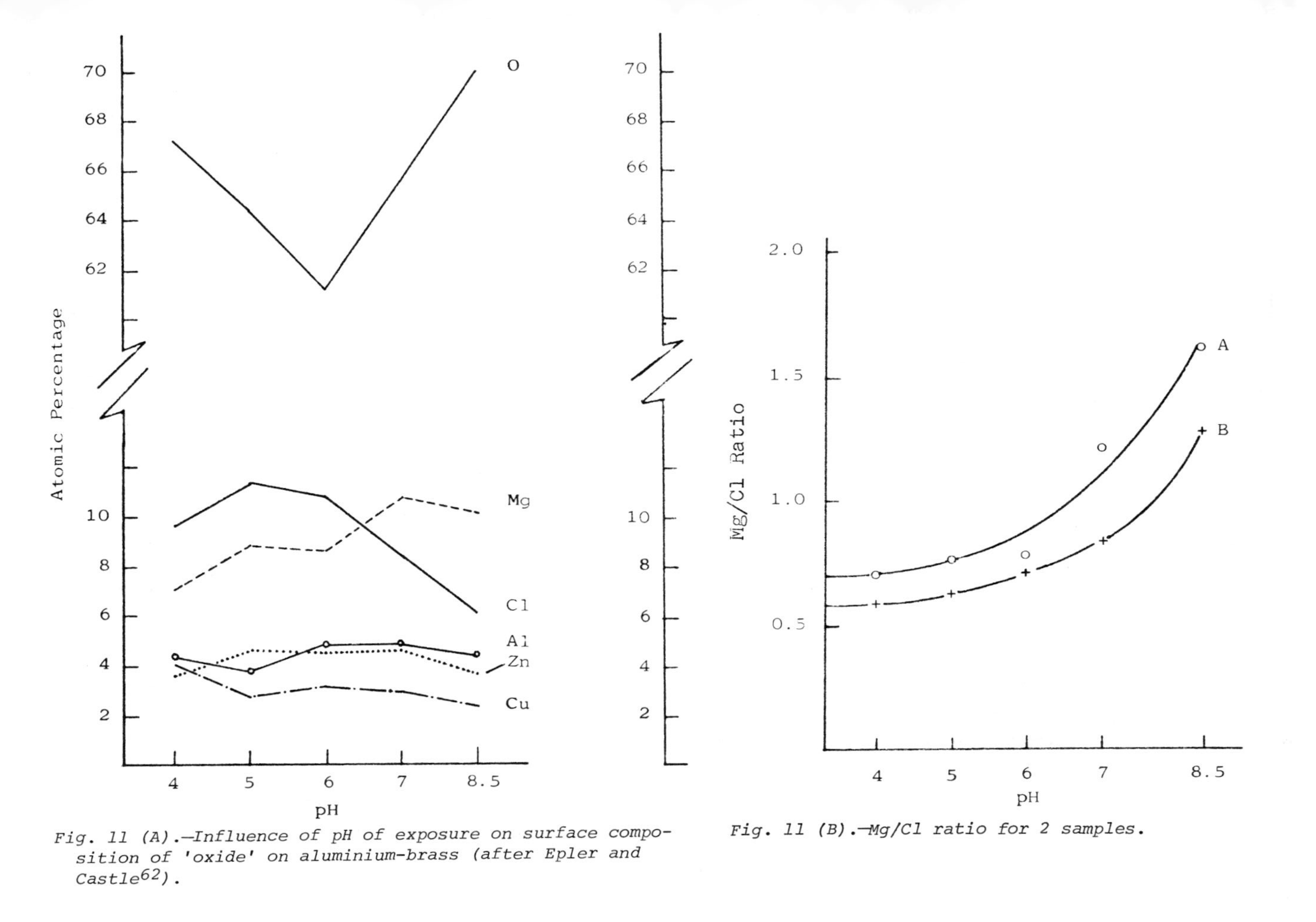

Fig. 11 (A).—Influence of pH of exposure on surface composition of 'oxide' on aluminium-brass (after Epler and Castle[62]).

Fig. 11 (B).—Mg/Cl ratio for 2 samples.

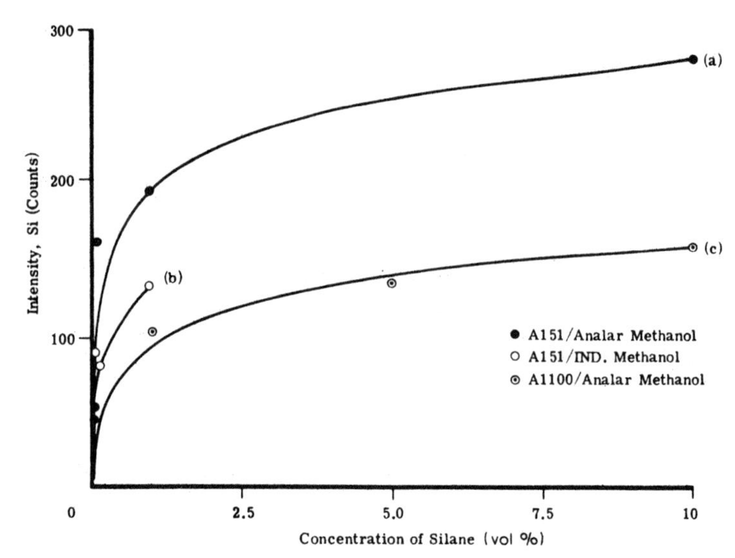

Fig. 12 (A).—Adsorption of ethoxysilanes on iron, isotherms at 25°.

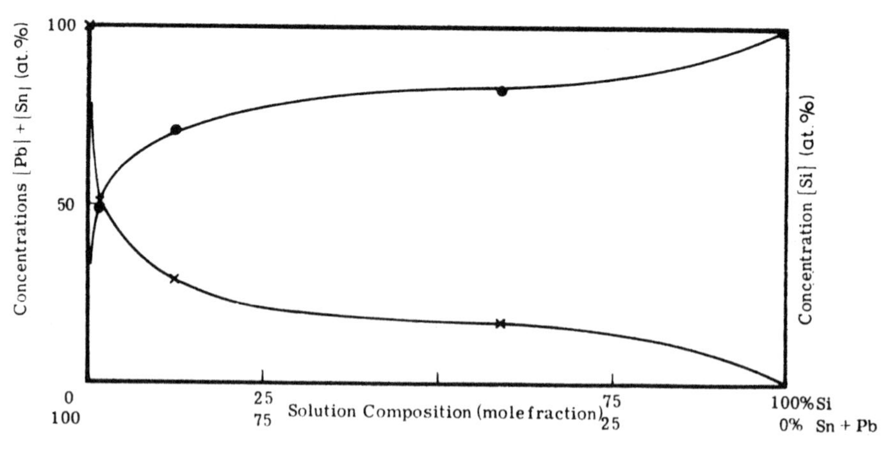

Fig. 12 (B).—Adsorption of ethoxysilanes on iron, competition with inorganic impurities in solvent (after Castle and Bailey[27]).

Auger depth profiles can be made from the metallic side of the interface, with less consequential damage. By this means they illustrated the good penetration of the polymer into the pretreated (porous) zone. Interesting work is now coming from combined XPS and Sims studies of the bonding region which tends to give direct evidence of chemical bonding in the presence of adhesion promotors.[68] Baun et al[70] have studied a number of metals of commercial importance.

CONCLUSIONS

Can we now paint a characteristic picture of the passivated metal surface which advances on that of a decade ago and typified by Figure 1? We can indeed. The composite diagram in Figure 13 summarizes some of the processes which from the results cited above we now know to be important. Centrally we isolate a submonolayer of impurity atoms on the nude metal surface and grain boundaries. In the presence of oxygen the oxidisable components of the alloy form an oxide in which little structural differentiation can be made but which may be terminated by a higher oxide. The interface between oxide and metal probably breaks up into a zone across which the oxygen potential of the oxide drops to that of the alloy and within which segregated impurities would be lost by dispersion over the many boundaries. In the presence of water vapor the oxide converts to hydroxides and in the presence of water it will exchange with ions, yet not necessarily increase the thickness; major oxide components may be lost by selective dissolution. Organic molecules, especially those derived from the aqueous phase stabilize an extended structure of water molecules with a thickness similar to that of the oxide. This may be pierced by strongly adsorbing organic molecules but it is not readily removed by preparative techniques such as abrasion or wire brushing.

This structure has fewer formal layers than that of 1968. Water in particular extends in some form continuously throughout the structure and probably accounts for its reactivity in terms of bulk absorbtion. Because of its lack of sharp divisions it seems unlikely that the extended bound water structure can be displaced by polymer application and in the absence of direct interfacial evidence we must assume it to be present as a discontinuity in most organic coating applications. This is clearly the most important area for further work using the power of XPS although ingenuity will be required to expose the region for analysis.

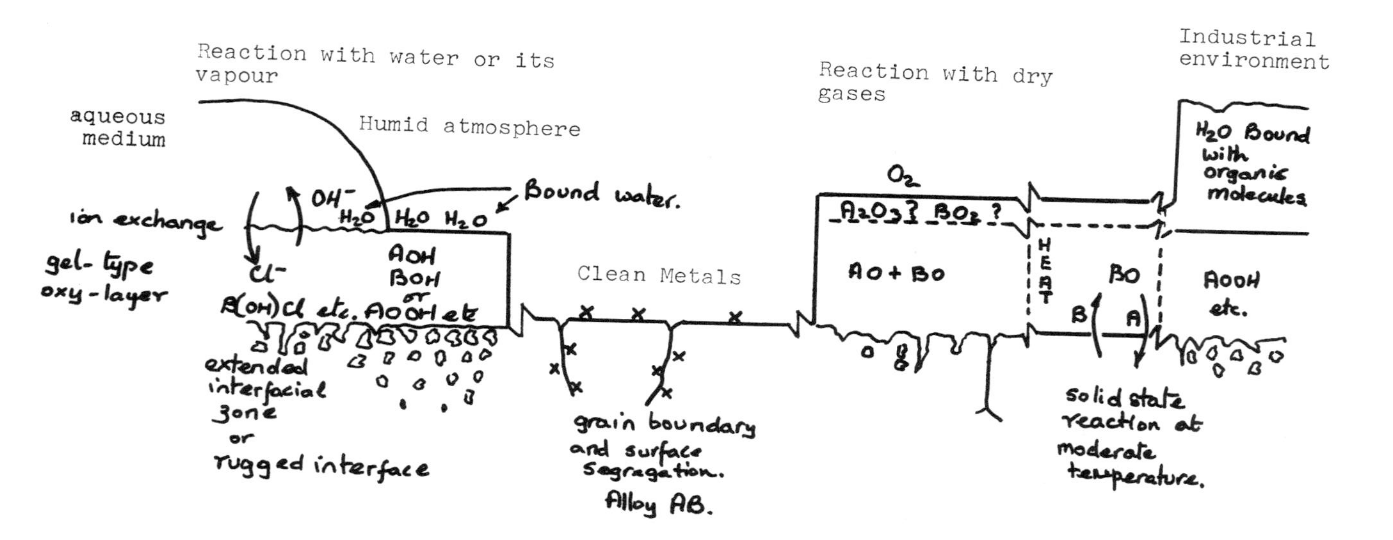

Fig. 13.—Surface structures including the evidence from surface analysis.

REFERENCES

1. Eirich, F.R., "Interface Conversion for Polymer Coating," P. Weiss and G.C. Cleever, Eds. Elsevier, New York, 1968.
2. Harris, L.A., *J. Appl. Phys. 39,1419, 1428 (1968)*; see also J.C. Riviere, *Contemp. Phys. 14, 513 (1973)*.
3. Siegbahn, K., Nordling, C., Fahlman, A., Nordberg, R., Hamrin, K., Hedman, J., Johansson, G., Berghark, T., Karlson, S., Lindgren, I., Lindberg, B., "ESCA, Atomic, Molecular, and Solid State Structure Studies by Means of Electron Spectroscopy," Almquist and Wiksells, Uppsala, 1967; see also T.A. Carlson, Photoelectron and Auger Spectroscopy, Plenum Press, New York, 1975.
4. Sparnaay, M.J., "Science and Technology of Surface Coating," B.N. Chapman and J.C. Anderson, Eds., A.P. London and New York, 1974.
5. Bolger, J.C. and Michaels, M.S., "Interface Conversion for Polymer Coating," P. Weiss and G.D. Cleever, Eds., Elsevier, New York, 1968, p. 12.
6. Jorgensen, C.K. and Berthou, H., *Anal. Chem. 47, 482 (1975)*.
7. Schofield, R., *J. Electron Spectros. and Related Phenomena 8, 129 (1976)*.
8. Castle, J.E., *Surface Science 68, 583 (1977)*.
9. Rhead, G.E., *Surface Science 68, 20 (1977)*.
10. Seah, M.P., *Surface Science 53, 168 (1975)*.
11. Asami, K., Hashimoto, K. and Shimodaira, S., *Corros. Sci. 17, 713 (1977)*.
12. Olefjord, I., *Corros. Sci. 15, 687 (1975)*.
13. Storp, S. and Holm, R., *Surf. Sci. 68, 10 (1977)*.
14. Castle, J.E. and Clayton, C.R., *Corros. Sci. 17, 9 (1977)*.
15. Palmberg, P.W. and Marcus, H.L., *Trans. A.S.M. 62, 1016 (1969)*.
16. Sikafus, E.N., *Surf. Sci. 19, 181 (1970)*.
17. Bishop, H.E., Riviere, J.C. and Coad, J.P., *Surf. Sci. 24 (1971)*.
18. Gettings, M., Coad, J.P. and Riviere, J.C., private communication.
19. Wynblatt, P. and Ku, R.C., *Surf. Sci. 65, 511 (1977)*.
20. McMahon, C.J. Jr. and Feng, H.C., *Proc. Ann. Conf. Microbeam, Anal. Soc. 11, 39 (1976)*.
21. Coad, J.P., Riviere, J.C., Guttman, J.C. and Krahe, P.R., *Acta. Met. 25, 161 (1977)*.
23. Hales, R., Hill, A.C. and Wild, R.K., *Corros. Sci. 13, 325 (1973)*.
24. Schroder, M.E., *Colloid Interface Sci. 50, 105 (1976)*. M. Kerker, Ed., Academic, New York.
25. Simmons, G.W. and Dwyer, D.J., *Surf. Sci. 48, 373 (1975)*.
26. Joyner, R.W., *Surf. Sci. 63, 291 (1977)*; see also Joyner, R.W. and Roberts, M.W., *Chem. Phys. Lett. 29, 447 (1974)*.
27. Bailey, R. and Castle, J.E., *J. Mat. Sci. 12, 2049 (1977)*.
28. Kishi, K. and Roberts, M.W., *Surface Sci. 62, 252 (1977)*.
29. Fuggle, J.C., *Surface Sci. 69, 581 (1977)*.
30. Norton, P.R., Topping, R.L. and Goodale, J.W., *Surface Sci. 65, 13 (1977)*.
31. Gimzearski, J.K., Padalia, B.D., Affrossman, S., Watson, R.M. and Fabian, D.J., *Surface Sci. 62, 386 (1977)*.
32. Fromm, S. and Mayer, O., *Surface Sci. 74, 259 (1978)*.
34. Brundle, C.R. and Carley, A.F., *J. Chem. Soc. Farad. Disc. 60, 51 (1976)*; See also Dickinson, T., Povey, A.F. and Sherwood, P.M.A., *J. Chem. Farad. Trans. I.73, 327 (1977)*.
36. Castle, J.E. and Nasserian-Riabi, M., *Corros. Sci. 105, 537 (1975)*.
37. Van OOij, W.J., *Surface Sci. 68, 1 (1977)*.
38. Brundle, C.R., *Surface Sci. 66, 581 (1977)*.
39. Asami, K., Hashimoto, K. and Shimodaira, S., *Corros. Sci. 17, 713 (1977)*.
40. Allen, G.C., Curtis, M.T., Hooper, A.J. and Tucker, P.M., *J. Chem. Soc. Dalton Trans., 1675 (1973)*.

41. Asami, K., Hashimoto, K. and Shimodaira, S., *Corros. Sci. 18, 125 (1978).*
42. Castle, J.E. and Durbin, M.J., *Carbon 13, 23 (1975).*
43. Brundle, C.R., Chuang, T.J., Wandelt, K., *Surf. Sci. 68, 459 (1977).*
44. McIntyre, N.S., Rummery, T.E., Cook, M.G. and Owen, D.G., *J. Electrochem. Soc. 123, 116 (1976).*
45. Fuggle, J.C., Watson, L.M., Fabian, D.J. and Affrossman, S., *Surf. Sci. 49, 61 (1975).*
46. Roberts, M.W. and Wood, P.R., *J. Elec. Spec. 11, 431 (1977).*
47. Dwyer, D.J., Simmons, G.W. and Wei, R.P., *Surf. Sci. 64, 617 (1977).*
48. Asami, K., Hashimoto, K. and Shimodaira, S., *Corros. Sci. 16, 35 (1976).*
49. Barr, T.L., *J. Phys. Chem. 82, 1801 (1978).*
50. Fehlner, F.P. and Mott, N.F., *Oxidation of Met. 2, 59 (1970).*
51. Lan, C.L. and Wartheim, G.K., *J. Vac. Sci. Tech. 15, 622 (1978).*
52. Hoar, T.P., Pallad. Medal. Addr., *J. Electrochem. Soc. 117, 17C (1970).*
53. Olefjord, I. and Elfstrom, B.O., *React. Solids. 8, 791 (1977).* J. Wood, O. Lindquist, C. Helgesson, Eds., Plenum Press, New York.
54. McIntyre, N.S., Zetaruh, D.G. and Owen, D., *J. Electrochem. Soc.* (to be published.)
55. McIntyre, N. S., Zetaruh, D.G. and Owen, D.G., *Proc. Int. Vac. Congr. (Vienna) 7 (1977).*
56. Lee, L.Y. and Eldridge, J., *J. Electrochem. Soc. 124, 1747 (1977).*
57. Castle, J.E. and Epler, D.C., *Surface Sci. 53 (1975).*
58. Olefjord, I. and Fischmeister, H., *Corros. Sci. 15, 697 (1975).*
59. Storp, S. and Holm, R., *Proc. Int. Vac. Congr. 7, 2255 (1977).*
60. Fort, W.C. III and Verink, E.D. Jr., *C.R. Congr. Int. Corros. Mar. Salissures 4, 179 (1977).* Pub. Contr. Rech. Etudes Oceanogr. Boulogne.
61. Castle, J.E., Epler, D.C. and Peplow, D.B., *Corros. Sci. 16, 137 (1976).*
62. Epler, D.C. and Castle, J.E., *Corrosion*, (to be published.)
63. Uhlig, H.H. "Corrosion and Corrosion Control," 2nd ed., J. Wiley & Sons, New York, 1971, p. 98.
64. See also, Sahashita, S. and Sato, N., *Corros. Sci. 17, 473 (1977).*
65. Castle, J.E. and Elias, S., unpublished work.
66. Anderson, Jr., H.R. and Swalen, J.D., *J. Adhes. 9, 197 (1978).*
67. Solomon, J.S. and McDevitt, N.T., ASTM STP (1978) Pittsburgh Meeting.
68. Gettings, M., Baker, F.S. and Kinlock, A.J., *J. Appl. Polymer Sci. 21, 2375 (1977).*
70. Bann, W.L., McDevitt, N.M. and Solomon, J.S., ASTM STP 596, 86 (1976).
71. McIntyre, N.S. and Zetaruk, D.G., *Anal. Chem. 49, 1521 (1977).*

A MODEL FOR MASS TRANSPORT IN PAINT FILMS

R. T. Ruggeri
and T. R. Beck

Electrochemical Technology Corp.
3935 Leary Way N.W.
Seattle, WA

ROBERT RUGGERI

ABSTRACT

A review of paint literature indicates that there are many parameters which may be important in understanding how paint protects metal surfaces from corrosion. Most research in this area has been focused on a single aspect such as osmotic pressure or electrolyte concentration, and there is a clear need to establish a broader framework for analysis of paint performance. To this end a one-dimensional model of the mass transport through paint films was developed. All mobile species, ionic, solute, and solvent, have been included. The paint phase is considered to be a concentrated, multicomponent solution for which appropriate flux equations have been written. By combining these flux equations with the usual species continuity equations, and with the inclusion of an equation of state and the electroneutrality condition, a complete set of coupled nonlinear differential equations was obtained. The solution of this equation set was accomplished by an iterative technique which allowed for metal-surface kinetic effects through the boundary conditions. This model has identified a complete set of relevant parameters (diffusion coefficients, transference numbers, etc.). The complete description of an unpigmented polyurethane coating is currently underway.

INTRODUCTION

The initiation of metallic corrosion under paint is known to involve many factors. Various authors have investigated the effects of such important parameters as osmotic pressure (1-5), electrolyte concentration (6-8), diffusion resistance (9-15), and adhesion (16-19). A thorough investigation of the paint literature leads to the conclusion that all these factors and more are involved in the corrosion of painted metals.

At the present time there is no comprehensive theoretical framework for understanding and correlating corrosion protection by paint. There are many reasons for this, but paramount among them is the complexity of the three-phase systems involved. Even the simplest model of the corrosion mechanism must include the interactions of such factors as mass transfer resistance, phase equilibrium, and electrode kinetics. Only in recent years have the mathematical tools been available to treat such complex systems. A secondary reason for the lack of understanding is that paint formulations are rapidly changing. This fact, coupled with long duration experiments and many required parameters, has led to incomplete data for each paint-metal-electrolyte system. Consequently it has been impossible to correlate accurately service performance with either physical parameters or laboratory experiments such as salt spray tests. The purpose of the present work is to develop a comprehensive theoretical framework to describe the transport processes that determine the protective properties of paint.

MODEL FORMULATION

It has long been recognized that one way paint reduces corrosion is by acting as a mass transport barrier (20). The present model primarily quantifies diffusion through the membrane (paint); the effect of metal-surface reaction kinetics enters through the boundary conditions, but it remains to be determined how the paint film affects surface kinetics. In the interest of simplicity a one-dimensional model has been developed. The extension of the principles presented here to more complex two- or three-dimensional cases is also possible, but it must be recognized that numerical solution of even the one-dimensional model is not trivial.

The system of interest consists of three phases: metal, paint, and aqueous electrolyte. The model quantitatively describes the transport of all mobile species, gases, ions, salts, and solvents, through the paint film. The properties of the electrolyte and metal phases enter only through the boundary conditions. This model has the advantage of being based on general principles. It can, therefore, be applied to a wide variety of paint systems and conditions. The primary deficiency of the model is its one-dimensional nature. The accurate description of a blister, for example, will surely involve more complex mathematics. There is reason to believe, however,that the corrosion mechanism changes after blister formation, and that the one-dimensional model is satisfactory for the initiation of corrosion.

Several authors have proposed that paint may be porous and fail by mass transport through pores. Although the one-dimensional model cannot describe diffusion in the vicinity of a pore, it can describe the macroscopic behavior of such a system containing many pores. At some later time when pore structure is defined quantitatively, a three-dimensional model can be formulated to treat the microscopic pores. In the meantime there is no justification for other than a macroscopic one-dimensional treatment.

The principles which form the theoretical basis of this model have been succinctly described by Newman (21). The application of these principles to this particular circumstance is straightforward, and only the major points will be outlined here. Although the aforementioned principles are quite general, a particular system has been chosen in order to facilitate presentation. The example system is composed of six mobile species: Na^+, H^+, H_2, $ZnCl_2$, Cl^-, and H_2O, as well as the fixed paint phase. This system represents a painted zinc sample immersed in deaerated, acidified (HCl), salt (NaCl) solution. Under these conditions the anodic reaction produces $ZnCl_2$, and the cathodic reaction produces H_2.

The theory of concentrated solutions states that:

$$N_i = - \sum_{k=1}^{n} L^{\circ}_{ik} \, c_k \nabla \mu_k \qquad (1)$$

where N_i = the flux of the i$\underline{th}$ species, relative to the paint,
c_k = the concentration of the k$\underline{th}$ species,
$\nabla\mu$ = the gradient of chemical potential, and
L°_{ik} = the coefficient of the k$\underline{th}$ driving force for the i$\underline{th}$ flux.
The L°_{ik} coefficients are similar to diffusion coefficients, but equation 1 represents a much more general case (a concentrated multi-component solution) than can be successfully represented by Fick's law type equations. There are of course as many equations 1 as there are mobile components; in this case there are six.

Before equations 1 can be used, some means must be found to describe the chemical potentials in terms of measurable quantities, i.e., concentrations, temperature, and pressure. This is exactly the field of equilibrium thermodynamics, and a complete knowledge of the thermodynamics of the paint phase will be necessary to obtain useful results. Usually the chemical potentials will be expressed in the following form:

$$\mu_i = \mu_i^{\circ} + RT\ell n f_i c_i \qquad (2)$$

where μ_i° = reference state chemical potential,
f_i = activity coefficient, and
c_i = concentration of the i$\underline{th}$ component.
For ionic species equations 2 must be modified slightly to include the potential, which for this model is defined as the potential of an ideal chloride reversible reference electrode.

After establishing the paint-phase thermodynamics, equations 1 and 2 can be combined with the usual species continuity equation:

$$\frac{\partial c_i}{\partial t} = -\nabla\cdot N_i + R_i \qquad (3)$$

in which R_i = the rate of generation of species i, and the other terms have been defined previously. Once again there are as many equations 3 as mobile species (six). Thus far the system (equations 3 after combination with equations 1 and 2) consists of six equations in eight unknowns. The seventh unknown is the potential (Φ) which entered because some of the mobile species are ions. The eighth unknown, pressure, enters because the activity coefficients are functions of all the thermodynamic variables including temperature, pressure, and the six concentrations. Temperature has been eliminated in this model because no large temperature gradients are expected across the thin paint film.

A seventh equation is readily available in the form of the electroneutrality condition:

$$\sum_{i=1}^{n} z_i c_i = 0 \qquad (4)$$

The solution of the problem now requires a single additional equation, which manifests itself in the form of an equation of state for the paint phase. This is not surprising because a knowledge of all six equations 2 is equivalent to knowledge of n-1 = 5 equations of state. What we require at this point is simply the last independent equation of state representing the relationship between pressure and component concentrations in the paint phase. In general the pressure can be expressed as follows:

$$P = P(V, c_1, \ldots, c_6) \qquad (5)$$

where P = pressure, and V = volume.

Equation 5 constitutes the eighth equation necessary to form a complete set. If the rate of generation of all species inside the paint phase is zero, this set of equations has the following form:

$$\frac{\partial c_1}{\partial t} = RT \sum_{k=1}^{n} \nabla (L^{\circ}_{1k} c_k \nabla \ln f_k c_k)$$

$$\vdots$$

$$\frac{\partial c_6}{\partial t} = RT \sum_{k=1}^{n} \nabla (L^{\circ}_{6k} c_k \nabla \ln f_k c_k) \qquad (6)$$

$$\sum_{i=1}^{n} z_i c_i = 0$$

$$P = P(V, c_1, \ldots, c_6)$$

Because the volume change is small (1.5% for polyurethane), the volume has been assumed constant. Under these conditions this set of equations has been solved to yield the six independent concentrations, potential and pressure, at each point within the paint.

Although there is no direct correspondence between the L°_{ij} of equations 6 and the more conventional diffusion coefficients, the parameters L°_{ij} are similar to diffusion coefficients, and must be known in order to obtain numerical solutions to the set of eight equations. By using Onsager reciprocal relations and introducing the current density into the set of equations, some simplification is possible. After proceeding in this manner, the total number of unknown parameters required for solution of equations 6 has been reduced to fifteen. These parameters represent specific combinations of the L°_{ik}'s and can be summarized in more conventional terms as follows:

D_1-D_6 = diffusion coefficients for the six mobile species,
B_1-B_5 = constants relating the moles of the ith component moving with each mole of water through the paint,
t_3, t_4, t_5 = transference numbers indicating the moles of H_2, $ZnCl_2$, and H_2O transported as a result of the charge passed, and
K = electrical conductivity of the paint.

Once these fifteen parameters have been evaluated, equations 6 can be solved for the concentration profiles. Then the fluxes can be calculated using equations 1 and 2. Because the fluxes at the paint-metal interface are precisely related to the rate of corrosion, our objective is at hand.

This model assumes that the paint phase represents the major resistance to mass transfer. Consequently all concentrations in the aqueous electrolyte phase have been assumed known; they affect the solutions through the boundary conditions. At the electrolyte-paint interface, thermodynamic equilibrium is assumed and can be formalized as follows:

$$\mu_i^e = \mu_i^p$$

$$\Phi^e = \Phi^p \qquad (7)$$

$$P^e = P^p$$

where superscripts e and p represent electrolyte phase and paint respectively. These are the usual equilibrium conditions and lead to Donnan equilibrium equations for dilute aqueous solutions. These equations then constitute the boundary conditions for the aqueous electrolyte side of the membrane. At the paint-metal interface the simplest case is for fast electrode kinetics. Under these conditions the fluxes are related to each other by stoichiometry, and these relations constitute the boundary conditions. Generally, then, the model is designed to handle boundary conditions of two types: known concentration, or known flux. Any independent combination of these two types of boundary conditions can also be used at either boundary of the paint film.

Equations 6, in general, constitute a set of coupled nonlinear differential equations. In order to solve these equations a method has been chosen by which they are first linearized and then solved as a set of coupled linear equations. The solution of the linear set is used as a basis for calculations involved in the linearization process and the new set of equations is solved again. This method has proved quite successful, usually providing a converged solution, at each time step, in fewer than four iterations. The basic method for effecting a solution is given by Newman (22) and employs the subroutine band (J).

CORRELATION WITH EXPERIMENT

The evaluation of the fifteen parameters (diffusion coefficients, etc.) requires a minimum of fifteen experiments. The fifteen data points must be linearly independent, i.e., representing different experimental conditions. At the same time there are only three basically different types of experiments which can be performed. All three types involve free films of paint separating aqueous solutions of known composition. This configuration requires slightly different boundary conditions than the example of the painted metal given earlier; however, they are clearly within the stated guidelines for the most general case. The first of the three types of experiments is similar to the conventional Hittorf method in which changes in the salt concentration of the aqueous phases are determined as functions of current passed. In the second type of experiment, dialysis, the membrane (paint) separates solutions of different salt concentration. In this case a potential is set up due to the unequal rate of diffusion of different ions through the paint. The third type of experiment involves the application of a substantial pressure gradient across the membrane. This type is similar to a reverse osmosis experiment and may be thought of as such. There are of course other experiments which can be imagined; however, they can invariably be simplified to some combination of the three basic experiments which are sufficient to determine all required parameters.

The model presented above has been successfully tested on a single set of data representing two dialysis and two Hittorf experiments. In principle these experiments contain enough information to evaluate four of the required parameters, provided the other eleven are known. As a preliminary test, eleven of the variables were fixed at estimated values. No data were available for the particular polyurethane paint used here, although diffusion coefficients for some similar formulations were found. The five B parameters and transference numbers were based on available data for cellulose acetate reverse osmosis membranes, and the conductivity was made consistent with the values of the diffusion coefficients used. The four variable parameters were adjusted until the computer results (fluxes, potentials, etc.) matched the corresponding experimental data.

The results are important in that the model predicts three fluxes and the potential between two silver-silver chloride electrodes; all are simultaneously near the observed values. No attempt was made to achieve precise correspondence, but the model and experimental results agree within a factor of two. Based on previous experience this simultaneous agreement with four important observables is very encouraging; however, a larger data set is required before a critical evaluation of the model can be made.

The six mobile-component example described above represents a relatively simple system. Nevertheless, fifteen parameters must be evaluated in the laboratory prior to any quantitative description of the corrosion process. This is exactly the situation alluded to at the beginning of the discussion regarding the apparently large number of important variables. What is now clear is exactly which variables may be considered to be independent, which are dependent, and the minimum number of laboratory experiments required to totally describe these systems. Although this knowledge is of considerable importance, the primary question remains whether or not the model of a one-dimensional mass transfer resistance can accurately describe metallic corrosion under paint films. The answer to this question is currently under investigation, and preliminary indications are that this model is appropriate. This is an encouraging fact because only after the model has been proven can we begin addressing the larger problem of quantitatively describing the service performance of painted metals. Our current efforts are thus aimed at establishing a foundation for the achievement of this ultimate goal.

ACKNOWLEDGEMENT

The authors are indebted to Professor Douglas N. Bennion for his guidance and advice during formulation of the model. This work was supported by Air Force Office of Scientific Research Contract No. F49620-76-C-0029.

REFERENCES

1. Perera, D.Y. and Heertjes, P.M., *J. Oil Colour Chem. Assoc. 54, 546 (1971).*
2. Perera, D.Y. and Heertjes, P.M., ibid. *54, 589 (1971).*
3. Mayne, J.E.O., ibid. *33, 538 (1950).*
4. Kittelberger, W.W. and Elm, A.C., *Ind. Eng. Chem. 38, 695 (1946).*
5. Kittelberger, W.W. and Elm, A.C., ibid. *39, 876 (1947).*
6. Cherry, B.W. and Mayne, J.E.O., "First International Congress on Metallic Corrosion," Butterworth, London, 1962.
7. Cherry, B.W. and Mayne, J.E.O., *Official Digest 33, 469 (1961).*
8. Cherry, B.W. and Mayne, J.E.O., ibid. *37, 13 (1965).*
9. Bacon, R.C., Smith, J.J. and Rugg, F.M., *Ind. Eng. Chem. 40(1), 161 (1948).*
10. Perera, D.Y. and Heertjes, P.M., *J. Oil Colour Chem. Assoc. 54, 313 (1971).*
11. Perera, D.Y. and Heertjes, P.M., ibid. *54, 395 (1971).*
12. Perera, D.Y. and Heertjes, P.M., ibid. *54, 774 (1971).*
13. Maitland, C.C. and Mayne, J.E.O., *Official Digest 34, 972 (1962).*
14. Wormwell and Brasher, D.M., *J. Iron Steel Inst. 164, 141 (1950).*
15. Rothwell, G.W., *J. Oil Colour Chem. Assoc. 52, 219 (1969).*
16. de Bruyne, N.A., *J. Appl. Chem. 6, 303 (1956).*
17. Walker, P., *Official Digest 37, 1561 (1965).*
18. Phillips, G., *J. Oil Colour Chem. Assoc. 44, 575 (1961).*
19. James, D.M., ibid. *39,39 (1956).*
20. Mayne, J.E.O., ibid. *40, 183 (1957).*
21. Newman, J.S., "Electrochemical Systems," Prentice-Hall, Inc., New Jersey, 1973, Chapt. 12.
22. Newman, J.S., ibid., Appendix C.

WEATHERING OF ORGANIC COATINGS

Richard M. Holsworth
and Theodore Provder

Glidden Coatings and Resins
Division of SCM Corporation
16651 Sprague Road
Strongsville, OH

RICHARD M. HOLSWORTH

ABSTRACT

An introduction to the effects of the environment on organic coatings is presented. Physical properties of several exterior coatings were investigated as a function of accelerated aging in the Atlas Weather-Ometer. Thermal properties, as measured by differential scanning calorimetry and thermomechanical analysis, as well as water vapor permeability are compared to mechanical property data from the torsion pendulum and Instron Tensile Tester. In general, the increasing embrittlement of the coatings, caused by the "aging" process, is followed by these various methods. Surface morphology as a function of aging has also been examined by scanning electron microscopy. Scanning electron microscopy is used to study problems associated with organic coatings on wood, aluminum siding and can bodies.

Coatings like many other materials are attacked by the elements of weather. Evaluating and solving the coatings weathering problem is definitely not a simple matter. There are many reasons for this. Three main reasons are:

1. There are a large number of coatings materials and formulations all of which are affected in different ways by the elements of weather.

2. There is a lack of standardized methods for determining how weather affects the physical, mechanical and chemical properties of coatings.

3. The effects that the elements of weather have on coatings and resins are not understood.

Despite these obstacles, significant advances are being made. In recent years numerous companies and industry groups such as the Federation of Societies for Coatings Technology, the National Paint, Varnish and Lacquer Association, and American Society for Testing and Materials have intensified their efforts to develop more meaningful weathering and aging tests for coatings. There have been many publications which discuss polymer, coating and "paint" durability in the literature in the past several decades. A state of the art survey in 1968 on predicting the service life of organic coatings cited one article in 1921, one in 1943, one in 1954,

56 in 1956, 68 in 1958, 59 in 1960, 86 in 1963, 113 in 1964, and 124 in 1966. The increase in literature has undoubtedly continued to grow at this fantastic rate. In addition to the published literature there are no doubt lab book upon lab book and file upon file of data relating to coatings aging and service life capabilities.

In addition to studies already underway, new polymers and coatings formulations are continually being developed in anticipation of more effectively resisting the elements of weather.

To better understand the problems associated with weathering of coatings one must look at the nature and elements of weathering and the methods used in evaluating the weatherability of coatings. Weathering can be defined as the effect of outdoor exposure on a material's properties and service performance. In the case of coatings weather can cause changes through the following chemical and physical reactions:

(1) Volatilization of leaching of plasticizers and solvents;

(2) Chemical decomposition of plasticizers, pigments, etc.;

(3) Breakage of main polymer chains in the coating vehicle;

(4) Splitting off of side groups along the main polymer chain;

(5) Reactions among the new groups or residues formed;

(6) Reaction of reactive groups such as residual double bonds in the vehicle;

(7) Oxidation of susceptible groups catalyzed by metal driers and catalyst residues;

(8) Stressing and deformation of the coating due to dimensional changes in the substrate, swelling by water, or temperature changes.

Weather includes many elements, among which are sunlight, oxygen and ozone, humidity, precipitation, wind, and atmospheric contaminants. All these elements contribute individually and in combination to change the properties of the coating through the reactions listed above.

Further, none of the reactions necessarily increase linearly with variations in the intensity of the element or elements that cause the reaction. Although individual elements of weather can be quite destructive, their combined effect can be even more harmful to coatings. In the final analysis, the sum total of all the effects on the coatings under consideration will determine whether the coating is suited for the intended use.

Water or moisture can have a great effect on the degradation of coatings. Chemically, hydrolysis of the polymer can occur which can result in degradation and physically, the pigment-polymer bond can be attacked which can result in premature chalking and short coatings life. The absolute humidity determines the total quantity of water available to cause shrinking or swelling of the coating. The relative humidity will influence the amount of condensation on the coating surface. *Condensation* on the coating surface, or *dew*, can result in certain materials in the coating being extracted and deposited on the coating surface. This can result in not only a poor appearance of the coating but may also have a detrimental effect on the polymer life. Water droplets can also act as magnifying lenses to increase the effect of solar radiation.

Precipitation may occur as rain, snow, or hail. All forms of precipitation are measured as a vertical depth of water but they vary in their effects on coatings. Rain and hail make a relatively short contact with the coating but snow can remain on the coating surface for longer periods. Hail can physically damage the coating and substrate. The particle size and velocity of hail or rain drops can be significant in determining the type of damage produced. *Wind* in itself is not particularly destructive of coatings, but the atmospheric impurities such as sand, dust, fungi carried by the wind can cause damage. The wind also influences the rate of deposition and removal of impurities. The drying rate and effective temperature of the coating surface are also influenced by the wind. Microcracking or poor film formation can result if proper drying rate and surface temperature conditions are not maintained. This can cause premature physical failure of the coating.

Impurities carried by the wind can either degrade or protect the coating surface. Dust on the coating surface can protect the coating by screening out solar radiation. Wind-blown sand and dirt can erode the coating. Chemical impurities in the wind can have an adverse effect on coatings. Salt air is a well known corrosive atmosphere. Industrial gasses and pollutants such as SO_2 can cause chemical degradation of the polymer. Wind borne spores can produce mildew and mold growth which give the coating an aesthetically poor appearance. Mold growth can also physically disrupt the coatings film permitting moisture to penetrate the film and possibly having a detrimental effect on the coating appearance and performance.

Under certain exposure conditions, which would include coating color and heat, solar and/or thermal, a coating film may reach temperatures of 140°F to 170°F. This temperature alone is not usually sufficient to cause bond cleavage in the polymer. However, it is sufficient to cause migration of low molecular weight polymer and plasticizers leading to micro-cracking failure, and surface tack and dirt pick-up due to the exuded material. Heat also degrades polymers by accelerating processes such as hydrolysis, chemical reactions or oxidation of trace contaminants.

The ultimate test of a coating designed for exterior use is its successful performance on the object coated. Since it is generally not feasible to test a coating on all objects to which it will be applied, certain test procedures have been devised to test weathering characteristics. These general practices are stated in ASTM Standards.

Exposure time is the most commonly considered variable in evaluation of weathering results. The time of year chosen for initial exposure is also important, especially for short exposure periods. Coatings placed outside in the fall frequently show better exterior durability than those exposed initially in the spring. Exposure data for 3 month or 6 month periods can be very misleading unless substantiated by longer exposure times because of seasonal variations in the climate.

Test sites with different climates are frequently chosen to better determine the effects of individual elements of weather on coatings. Florida and Arizona exposures are commonly made because of the total year-round sunlight of high intensity. Particular environments are often chosen because they accelerate a particular type of failure, chalking, mildew, etc.

In general, multiple site testing is required to evaluate the durability of coatings, especially for products that will be sold nationally.

Test site terrain can also affect exterior exposure results. Vegetation, plowed land, concrete, blacktop, and buildings all vary in their ability to absorb and reflect sunlight. The amount of wind blown dust and moisture can also be influenced by the prescence of trees and vegetation.

Accelerated aging tests are intended to provide quick answers on weathering affects by establishing general trends. Recommended practices for accelerated testing of coatings are described by the ASTM. The speeding up of results can be accomplished in two ways. One method is by continuous exposure to the weathering condition rather than waiting for the intermittent effects of direct exposure. For example, rain is an occasional occurrence and sunlight varies in intensity during the day. Each of these occurrences can be approximately duplicated by the use of specialized equipment and applied to the coating in an orderly cyclic arrangement. The other method of accelerated testing is to increase the severity of the exposure condition above the level normally found in nature. Both these methods have limitations but the increased intensity method is the least recommended. The increased intensity method may lead to degradative mechanisms which do not occur at the low intensities of outdoor exposure. Generally the rate of degradation is not proportional to the intensity of the exposure condition and in many instances the breakdown of the coating will occur at a vastly increased rate.

In spite of the difficulties in correlating accelerated tests and actual exterior exposure the use of accelerated testing will continue. This is because normal aging is too slow. It is not possible to wait 5, 10, or 20 years to determine if a coating has satisfactory exterior durability. Also reasonable correlations between accelerated testing and performance has been found for certain coatings and types of failure. The results must, however, be interpreted from background knowledge of how specific coatings degrade and the factors that are most influential in causing degradation.

To determine the effects of weather on coatings, the change in several properties is measured using long term outdoor or accelerated weathering conditions. Usually, visual properties such as chalking, checking and cracking, erosion, blistering, color change, etc. are used to measure the effect of weather. A trend in the coating industry is to augment the visual tests with measurements of bulk film properties.

A variety of visual properties are normally measured to determine whether the weathering performance of a coating is satisfactory. Sometimes in the case of metal panels scribe marks are placed on the panel and the corrosion characteristics of the panel are evaluated. In addition, a variety of tests such as impact tests, hardness tests, etc. are also required to determine the suitability of the coating for a specific application. In the past these latter tests have been performed on the unweathered samples but rarely on weathered panels. A trend in predicting durability involves the measurement of bulk properties on a regular time schedule. In some cases it is found that relatively short range measurements can rank coatings from a durability standpoint. This type of testing is most reliable in cases in which small changes in a given formula are to be screened. In cases where chemically different coating types are being rated the results have a more qualitative significance.

For some time, mechanical tests have been used to measure the influence of weather on coatings (1, 2, 3, 4, 5). Workers at Sherwin-Williams have proposed that stress strain measurements offer a means for predicting exterior durability. A similar method was investigated in the thirties by Jacobsen. The work of Schurr, Jay, and VanLoo indicates that this neglected method has considerable value. Their accelerated weatherometer results indicate that 100 hours in the Twin Arc Weatherometer are equivalent to one month outside. The pattern of weathering behavior for poor and good conditions deserves careful attention although the underlying reasons for this behavior are not understood at the present time.

Dynamic mechanical behavior measured via the torsion pendulum show very definitely that the type of degradation of alkyd coatings in the weatherometer is different than that observed outside. The alkyd primers investigated embrittle much more rapidly in the weatherometer than they do on Cleveland test fences. The exterior results indicate that the coatings with glass transitions higher than 45°C are too hard for exterior exposure. The results also indicate that alkyd coatings become very highly crosslinked and their mechanical properties become quite insensitive to temperature after they are exposed to weathering for a sufficient period of time. Often the onset of cracking and checking failure becomes quite pronounced beyond this point. It is probable that this is a result of the films inability to expand and contract with the dimensional changes of the substrate, induced by moisture and temperature changes.

It should be remembered that changes in the bulk properties of the polymer do not necessarily indicate change in gloss retention, chalk resistance, color retention or other visual film surface properties. A most valuable tool for the study of changes in the coatings surface is the scanning electron microscope (SEM). Procedures, techniques, and interpretation are most important in the use of SEM. Once established, the SEM gives an unparalleled look at the surface of the polymer film. The use of energy dispersive x-ray analysis in conjunction with the SEM allows the coating surface to be qualitatively and quantitatively analyzed for changes in the pigment distribution of systems with age.

In the work presented in this paper, several methods have been used to monitor the physical changes in coatings upon aging. Thermal Mechanical Analysis (TMA) is used to follow changes in the bulk properties of the coating, and Differential Scanning Calorimetry (DSC) is used to follow changes in transitions, such as the glass transition temperature (T_g), melts and cure exotherm. These thermal methods of analysis are compared to the mechanical methods of analysis, such as tensile strength and elongation at break, torsional shear modulus, and logarithmic decrement. SEM is used to follow changes in the surface morphology of the coating, such as the onset of chalking, micro-checking, mildew growth, etc. Data are also presented for the Moisture Vapor Transmission (MVT) of several coatings upon accelerated aging.

Several years ago tensile strength and per cent elongation data were obtained on a series of primers (6). One of these primers, primer #2, is an exterior primer with a long oil alkyd-tung oil varnish vehicle and has proven exterior durability and an exposure history. This coating will be used as a reference material. Data from another primer from this series, primer #4, will also be included here. Primer #4 is an interior sanding primer with short oil alkyd-phenolic varnish which has shown poor exterior durability under severe climatic conditions.

Figure 1 shows the change in tensile strength vs. hours of exposure in the Weather-O-Meter for alkyd primer #2. A steady increase in tensile strength from an initial value of 900 psi to over 3300 psi at 806 hrs. of Weather-O-Meter exposure is observed. A small change in the rate of increase of tensile strength at 400 hrs. of exposure is observed but, otherwise, no obvious change is apparent. Figure 2 shows per cent elongation at break vs. Weather-O-Meter exposure for primer #2. The per cent elongation decreases from an initial value of 14% to about 10% at 250 hrs. of exposure. At 400 hrs. of exposure, the elongation has decreased to approximately 2%. This primer was observed to have failed by cracking after 600 hrs. of exposure when applied over yellow pine. A more stable cedar substrate showed no failure at this point. What should be noted from these results is the dramatic decrease in per cent elongation which takes place at 400 hrs. of exposure.

Torsion shear modulus and log decrement data for primer #2 is shown in Figure 3. The shift in the log decrement peak from an initial T_g of about 15°C to over 60°C at 394 hrs. of exposure agrees well with the changes observed in the tensile and elongation data. The material embrittles with aging, showing the effects of crosslinking. At 806 hrs. of exposure, the height of the log decrement peak has decreased significantly. This effect is common among very high molecular weight crosslinked materials. The results of aging can also be seen in the shear modulus data. A decrease in the change of the modulus over the temperature range tested is observed. The change in the slope of the modulus curve at T_g becomes less pronounced. At 806 hrs. of exposure, the values of the modulus curve above T_g have increased significantly, indicative of the increased degree of crosslinking upon aging.

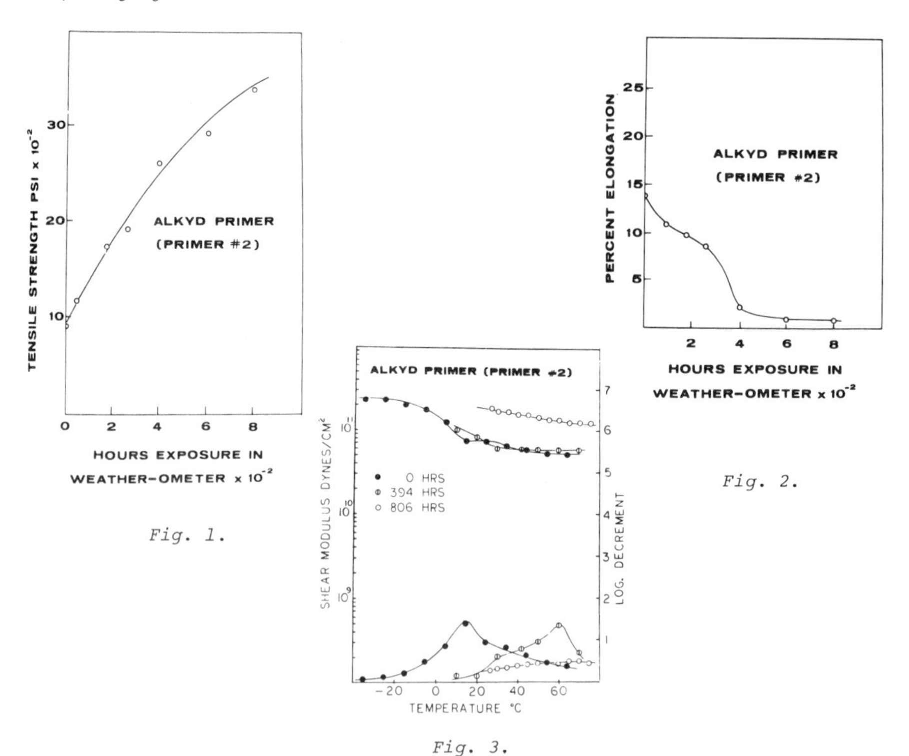

Fig. 1.

Fig. 3.

Fig. 2.

Figure 4 shows the effects of exterior exposure on the shear modulus and log decrement for primer #2. An overall view of the changes which have taken place in this film on exterior exposure would seem to be quite similar to the accelerated aging test results. The rapid shift of the high damping peak to higher temperatures that was seen on accelerated aging is not apparent. For example, at six months' exposure, a high, broad damping peak is observed; at nine months, the modulus and damping peak look suspiciously like those at 806 hrs. of Weather-O-Meter exposure. This might suggest a correlation at the nine-month or one-year exterior exposure to 806 hrs. of Weather-O-Meter for primer #2, although it would appear that the two systems arrived at this point via different mechanisms.

Figure 5 shows the shear modulus and log decrement data for the interior primer, primer #4, on exterior exposure. The initial tensile strength of this primer was about 2700 psi and the elongation was less than 1%. This sample embrittled so much on accelerated aging that handling of the free film was impossible. As can be seen in Figure 5, the initial T_g of the sample is above 40°C. Upon aging for one year, the T_g shows a dramatic increase to over 80°C. This coating is a good example of the effect of crosslinking on the glass transition temperature of a coating. This again is seen by the steady increase in T_g, the shift of the damping peak to higher temperatures and the resulting decrease in overall area under the damping peak at high molecular weight, and the flattening out of the modulus-temperature curve.

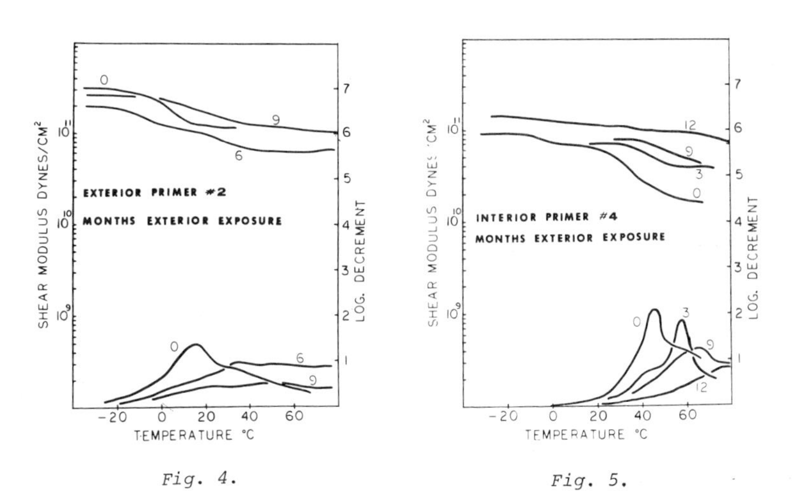

Fig. 4. *Fig. 5.*

Figure 6 shows tensile strength data for a series of coatings: linseed, alkyd, acrylic and vinyl. As was noted with Figure 1, tensile strength can be important, but does not sufficiently define bulk changes within a coatings film. The scatter in the data of Figure 6 agrees with this idea. An increase and decrease in tensile strength of a coating upon aging can be significant, depending on the initial T_g of the system and on the rate of increase of the T_g upon aging. However, scattered data of a cyclic nature can lead only to further confusion.

Figure 7 shows the per cent elongation data for this series of coatings.[9] The general trend to decreased elongation on exposure is observed in all cases. The small per cent elongation for #1, the linseed oil primer, can be expected after short exposure because of the rapid oxidation and crosslinking of the vehicle. The low value for #12, vinyl acetate, is somewhat unexpected, but at 50 PVC, this coating may be over the CPVC which could result in a low per cent elongation.

Thermal Analysis

The application of thermal analysis to coatings has been discussed previously (9). The ability to determine T_g, reaction exotherms and other reaction kinetics makes this a very useful tool to follow the properties of coatings upon aging.

Thermal properties of the previously discussed coating were investigated by DSC and TMA in the penetration mode. Figure 8 shows the DSC results for coating #5, the exterior alkyd white (medium oil length). Zero

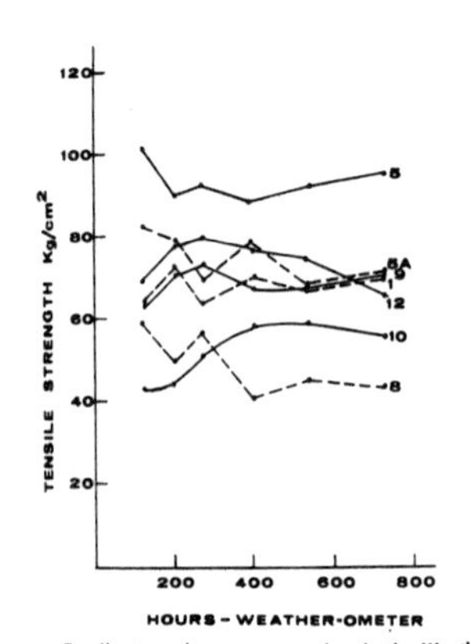

-Tensile strength vs. exposure time in the Weather-Ometer for a series of coatings based on various vehicles: (1) linseed oil; (5) alkyd; (5A) alkyd; (8) acrylic & alkyd; (9) acrylic; (10) vinyl & alkyd; (12) vinyl

Fig. 6.

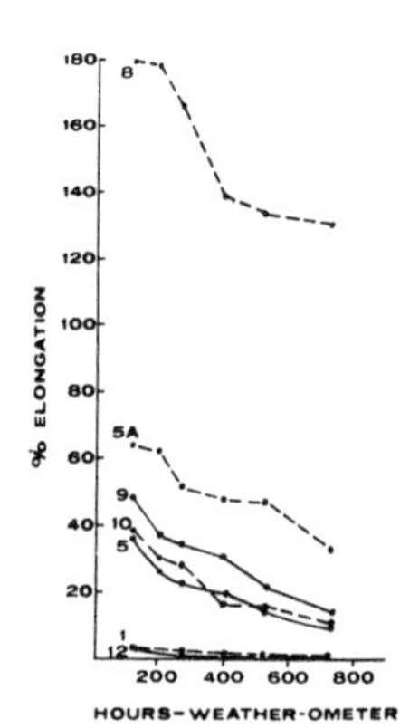

-Percent elongation at break vs. exposure time in the Weather-Ometer for a series of coatings based on various vehicles: (1 linseed oil; (5) alkyd; (5A) alkyd; (8) acrylic & alkyd; (9) acrylic; (10) vinyl & alkyd; (12) vinyl

Fig. 7.

hours exposure shows a T_g of 0°C and an exotherm that starts at about 100°C and peaks at 160°C. At 200 hrs. of Weather-O-Meter exposure, multiple transitions are observed at -10°C and 30°C. The exotherm peak has decreased in intensity from the initial test. After 500 hrs. of exposure, the multiple transitions are more clearly defined at 10°C and 50°C, while the exotherm has decreased even further. It would appear that the exotherm is caused by vehicle which has not undergone polymerization to this point in its exposure history, and that heating through this temperature causes thermal polymerization of the unpolymerized portion of the vehicle.

Figure 9 shows the DSC curve of the initial unexposed sample and an immediate re-run of the same sample. As can be seen in the lower curve, the T_g has shifted to 8°C and the exotherm has disappeared. The change in the ΔT curve at 175°C is probably due to the onset of degradation. The change in slope at -50°C may be caused by degradation products resulting from the high temperature cure, >200°C, of the initial sample run.

In any case, the multiple transitions seen after Weather-O-Meter exposure could be due to different mechanisms of polymerization, i.e., thermally induced and radiation-induced polymerization.

Figure 10 shows the DSC curves for coating #5A, exterior alkyd white (long oil length). The initial T_g of -15°C increases to multiple transitions of 2°C and 38°C at 500 hrs. of exposure, and 2°C and 60°C after 800 hrs. of exposure. The exotherm, similar to that discussed in Figures 7 and 8, is seen also to diminish with exposure time. Figure 11 shows results like those in Figure 9. After heading through the exotherm, the T_g has increased to -5°C and the exotherm has disappeared.

Thermal Mechanical Analysis (TMA) also was used to investigate the change in properties of coatings upon aging. The displacement profile of a penetration probe on a film was investigated with a programmed increase in temperature. TMA permits investigation of the coating while it is attached to the substrate; it is not necessary to prepare free films. Small samples can be cut from a metal panel, either steel or aluminum, and the panel returned for further exposure. Figure 12 shows the results of TMA

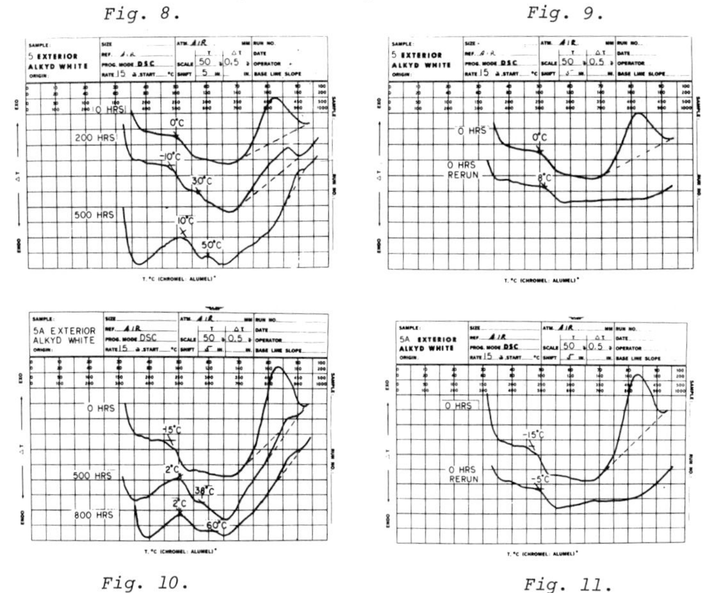

Fig. 8. *Fig. 9.*

Fig. 10. *Fig. 11.*

on the hard interior alkyd primer #4. "Base" is the expansion profile of the substrate metal panel. TMA was run on panels exposed for 200, 400, and 600 hrs. At zero hours of exposure, the T_g was 31°C; at 200 hrs., 45°C; at 400 hrs., 51°C; and at 600 hrs., 60°C. These results show the same increase of T_g with aging as was observed by torsion pendulum. The degree of crosslinking can be related to the amount of penetration above T_g. The area under the log decrement curve also is related to the degree of crosslinking in the system.

Results of TMA on coating #5, the medium oil length exterior alkyd, are shown in Figure 13. Testing of this coating was done at 200, 600, 1000, and 1900 hours of exposure. At 200 hrs. of Weather-O-Meter exposure, this coating has a T_g of 19°C; at 600 hrs., 16°C; at 1000 hrs., 20°C; and at 1900 hrs., 33°C.

Figure 14 shows the results of TMA on #5A, the long oil length exterior alkyd. At 200 hrs. of exposure, the value of T_g is 8°C; at 600 hrs., 20°C; at 1000 hrs., 15°C; and at 1900 hrs., also 15°C. The longer oil length alkyd retains a low T_g over the exposure time, and would probably enjoy a longer service life than the other alkyd, #5, if conditions were such that the coating would fail.

Fig. 12.

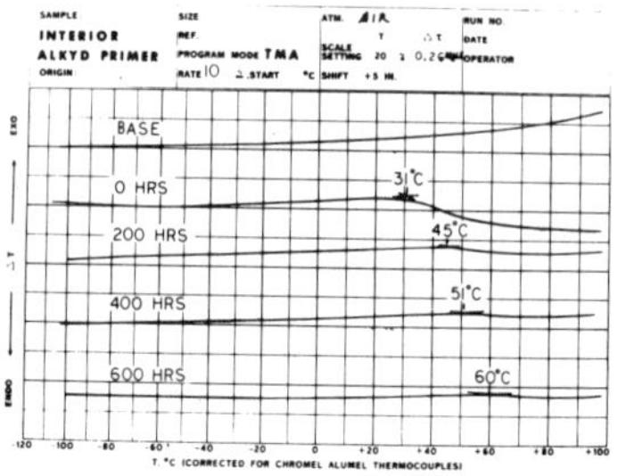

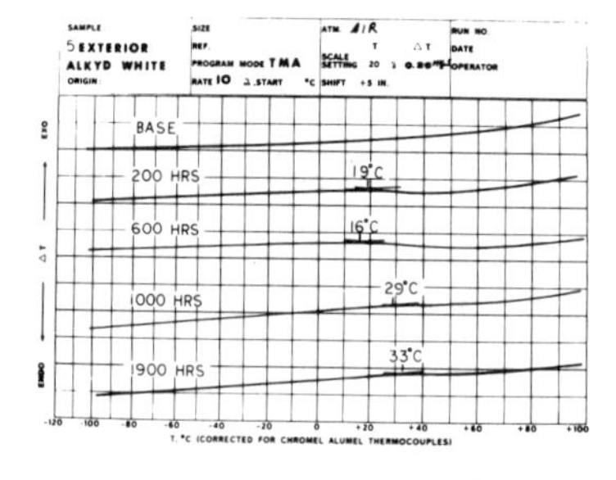

Fig. 13.

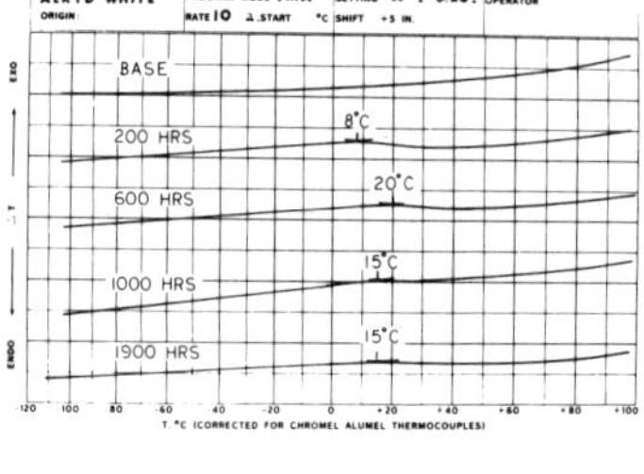

Fig. 14.

Figure 15 shows the results of TMA on coating #8, an alkyd modified acrylic latex. This coating retains an effective T_g of less than 13°C throughout the 1900 hrs. of accelerated aging. At 200 hrs., the T_g is 5°C; at 600 hrs., 11°C; at 1000 hrs., 11°C; and at 1900 hrs., 13°C.

Figure 16 shows TMA data for coating #9, the straight acrylic latex. This system has higher T_g's than the alkyd modified system over the exposure time covered in the test. At 200 hrs., T_g is 16°C; at 600 hrs., 17°C; at 1000 hrs., 18°C; and at 1800 hrs., 23°C. The T_g is lower in coating #8 probably because of the plasticizing effect of the alkyd modification on the acrylic latex. The lesser amount of penetration in #9 may be due to the harder polymer, but it is probably due to the higher level of pigmentation - #8 is 30 PVC, and #9 is 40 PVC.

Moisture Vapor Transmission

Resistance to weathering by paint films requires that the films be able to withstand several environmental factors, one of which is water. Water is generally considered to be an agent which destroys the integrity of coatings systems. It may act by causing blisters, loss of adhesion, and corrosion, or it may accelerate the action of other factors such as UV radiation, oxygen, micro-organisms, etc. These effects of water require that it be present in the paint film and at interfaces. The quantity of water in paint films and transported through paint films is determined by the permeability, diffusion and solubility properties of the films. Normally, water vapor transport measurements are made on unweathered films. Data resulting from such measurements (although useful) has not correlated effectively with paint film performance.

This study was undertaken to determine the water vapor permeability of paint films after they have been exposed to accelerated weathering devices (5). Permeability measurements were taken initially (no weathering) and after various exposure periods up to 2000 hrs. The resulting data show that water vapor permeability of paint films changes with weathering. The permeability constant (expressed in units of Std·cc·cm/cm^2·sec·cmHg) vs. the hours of Weather-O-Meter exposure for the samples in the study is shown in Table 1.

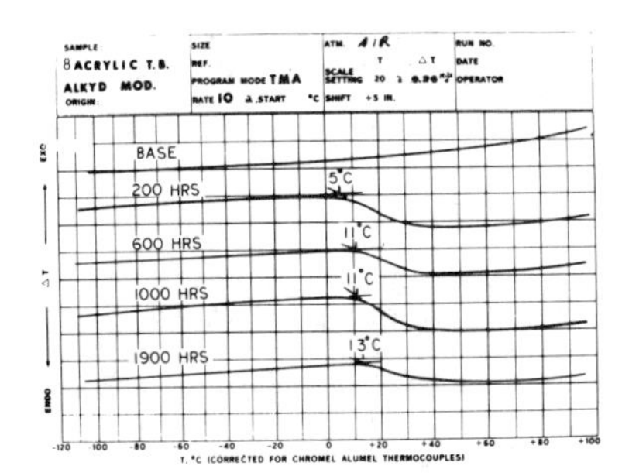

Fig. 15.

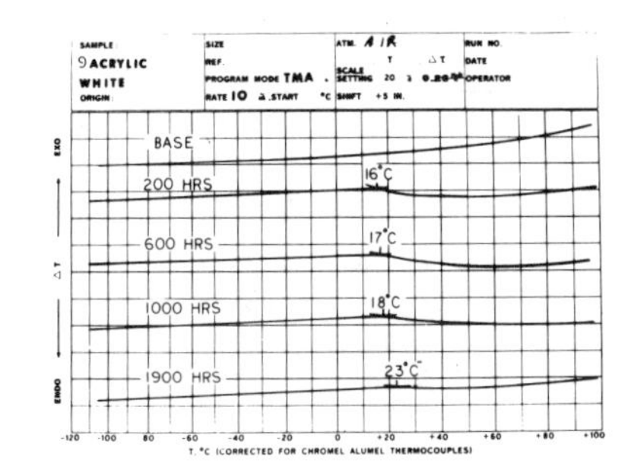

Fig. 16.

The permeability constant of three solvent-based paints vs. exposure hours is shown in Figure 17. Permeability of the three samples decreases with time (an observation made for all paints in this study). The linseed house paint, sample #1, decreases at a high rate initially. After 500 hrs., the linseed paint film checked and cracked, which made additional measurements impossible. Sample #5A is more permeable after 200 hrs. than it is before exposure. After 200 hrs., it decreases to approximately the same value as for sample #5.

The same general trends observed for the solvent-based paints are also observed for latex house paints as shown in Figure 18. All these paints become less permeable with exposure time and do so either by a gradual decrease from an initial measurement, or by increasing during early exposure hours and then gradually decreasing. It is interesting to note that the high PVC film decreased to a value comparable with the solvent-based paints. However, after 800 hrs., this film was so brittle that additional testing was impossible. At 2000 hrs., samples #8 and #10 remain intact and, interestingly enough, the modified acrylic and PVC show approximately identical water vapor permeabilities. Initially, the modified acrylic was approximately three times as permeable as the vinyl acetate house paint.

A commercial alkyde primer and topcoat were also evaluated. The permeability profiles for these two systems are shown in Figure 19, and are remarkably similar. Early in the weathering profile, both coatings are more permeable than the unweathered films. The higher PVC primer is slightly less permeable than the low PVC topcoat.

-Permeability Constant [Std-cc-cm/cm²-sec-cmHg) x 10⁷]

Sample	Initial	Hours of Exposure 200	300	500	800	1000	2000
#1	2.09	1.18		0.872			
#5	1.26	1.08		0.913		0.141	
#5A	1.78	2.26		1.64		0.146	
#8	9.29	6.47		3.52		2.84	2.79
#9	5.90		6.10		0.852		
#10	319		5.03		3.94		2.56
Commercial Primer	1.51	2.93		0.55		0.079	
Commercial Topcoat	1.65	2.97		1.87		0.115	

TABLE I.

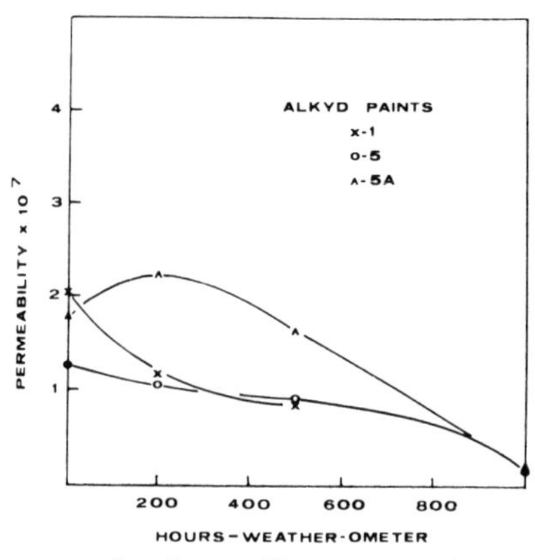

—Plot of permeability constant vs. hours of Weather-Ometer exposure for the solvent-based coatings #1, #5, and #5A

Fig. 17.

Water vapor transport in paint films is a complex phenomena which is affected by the nature of both pigment and binder. For example, it is known that transport varies with pigment concentration and pigment dispersion. Different binders are also known to affect transport in polymer films. Adding the effect of accelerated weathering complicates the process even further. In the present study, all weathered paint films were observed to be ultimately less permeable than unweathered films. In some cases, the weathered films became more permeable after 200-300 hrs. of weathering and then became less permeable. Again, as is the case with most tests on coatings, a single-point reading may give erroneous results on the ultimate properties of the system. Because of the changes in permeability that are observed with aging, it is reasonable to expect that permeability measurements on unweathered films will not necessarily correlate effectively with ultimate paint film performance.

The accelerated weathering process subjects the films to exposure of UV radiation, light and temperature, as well as water. For those films which are observed to have larger permeabilities after initial exposure, it is suggested that the H_2O exposure causes an apparent plasticizing effect on the binder portion of the paint films. Eventually, the UV and heat exposures cause the films to become less permeable to water vapor, probably because of crosslinking of the binder. During Weather-O-Meter exposure, the paint films are subjected to temperatures of up to 65°C. All films observed in this study are, therefore, subjected to a temperature above the T_g of the paint films. Polymer flow may be partially responsible for the observed lower permeabilities after weathering.

Scanning Electron Microscopy

Over the past several years, SEM has been demonstrated to be an excellent tool for use in the investigation of coatings surface morphology. The previously investigated systems were studied by SEM in an attempt to see whether or not interesting and pertinent morphological data could be obtained on these systems as a function of accelerated aging.

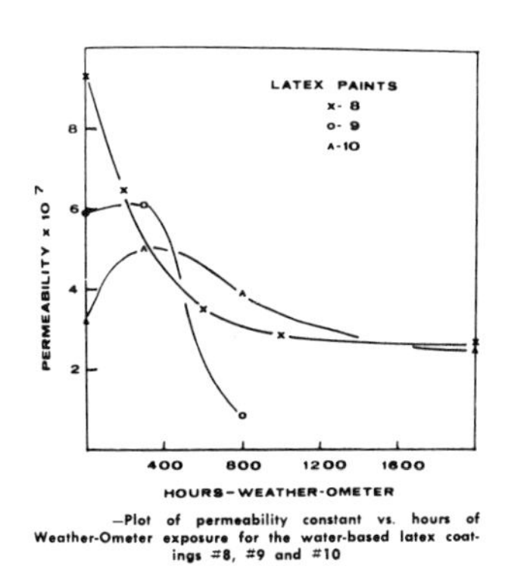

—Plot of permeability constant vs. hours of Weather-Ometer exposure for the water-based latex coatings #8, #9 and #10

Fig. 18.

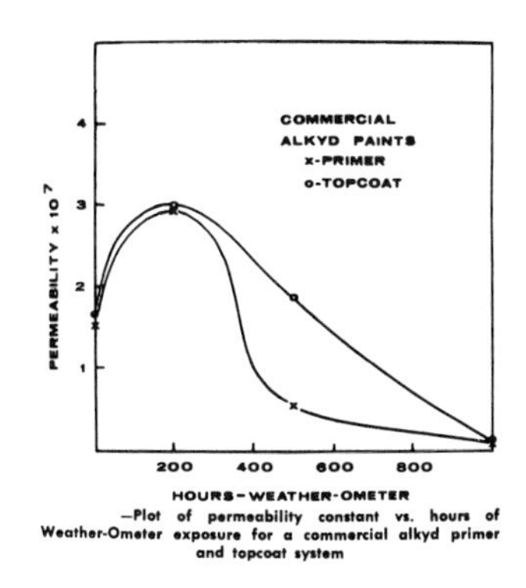

—Plot of permeability constant vs. hours of Weather-Ometer exposure for a commercial alkyd primer and topcoat system

Fig. 19.

Figure 20 shows the interior primer at 2000x magnification at different stages of Weather-O-Meter aging. At 200 hrs. of exposure, the surface appears to be very porous with many voids and some micro-cracks and micro-checks. The appearance of a continuous phase or binder is evident from the 2000x photo. At 800 hrs. of exposure the surface appears even more porous and loosely bound. At 2000x, this surface porosity is more evident and is shown by a dramatic decrease in the amount of continuous phase or surface binder. At 1900 hrs., even more degradation of the surface is apparent. The 2000x photo of this exposure shows essentially a complete lack of surface binder. The coating is chalking quite heavily at this point.

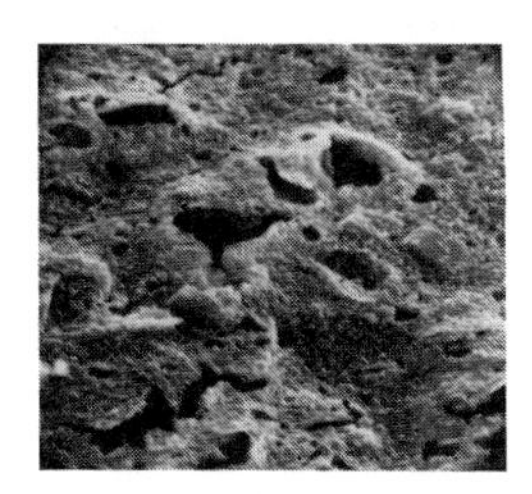

200 hrs.

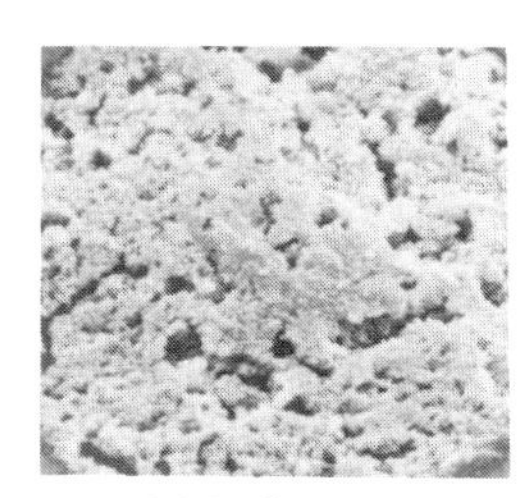

800 hrs.

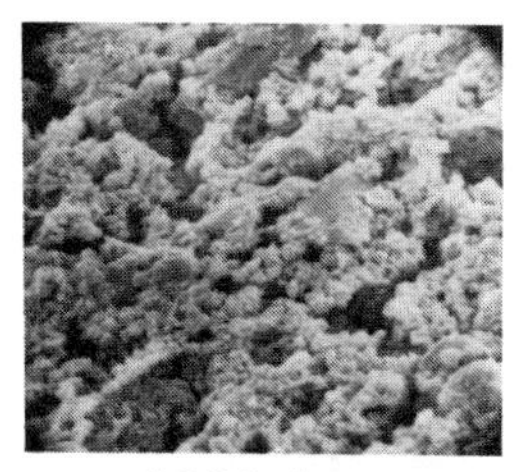

1900 hrs.

Fig. 20.

Figure 21 shows paint #1, the linseed oil-based paint, at different stages of aging. At 200 hrs. of exposure, the continuous binder phase is very evident. At 800 hrs. of exposure the coating has started to chalk, and the surface is characterized by fine, loosely packed particles. At

1900 hrs. of exposure, the surface has chalked heavily; there is no evidence of binder, and the pigment and filler particles can be seen.

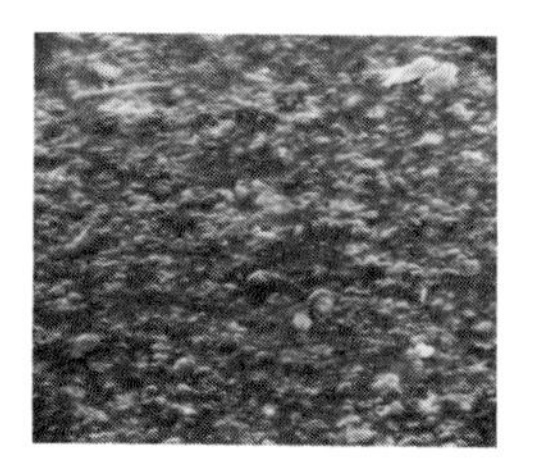

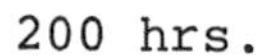

800 hrs. 1900 hrs.

Fig. 21.

Figure 22 is a composite of photographs at 200, 800 and 1900 hrs. of exposure at 2000x for sample #5, the exterior alkyd white. Photographs at 200 hrs. show a very continuous surface. At 800 hrs., the binder has eroded to a point where fine, uniform-sized pigment particles can be seen. This coating is pigmented with TiO_2 only. No significant different can be seen at 1900 hrs. of exposure.

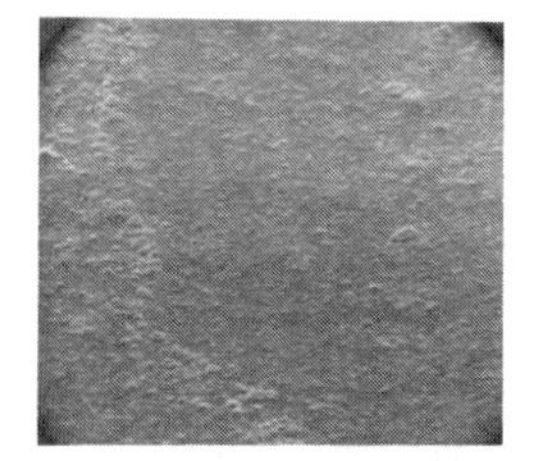

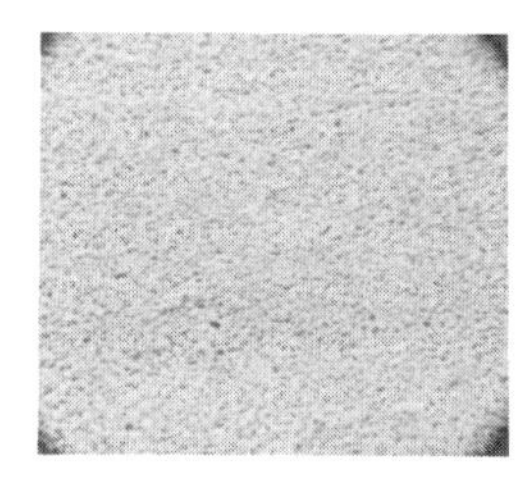

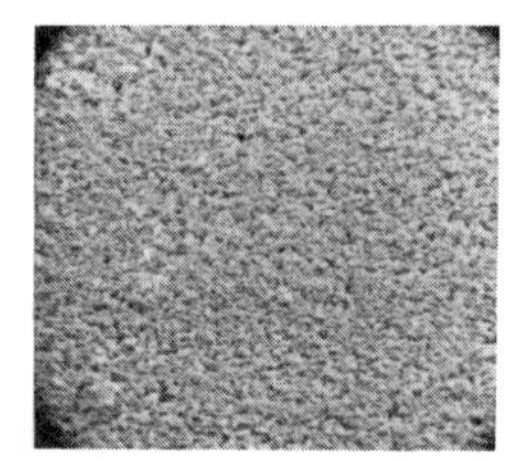

200 hrs. 800 hrs. 1900 hrs.

Fig. 22.

Figure 23 shows a similar series of exposures for sample #5A, the exterior long oil alkyd white. Again, at 200 hrs. of exposure, the continuous binder phase can be seen with some pigmentation through the surface. At 800 hrs. of exposure, the surface is chalking and microchecks have developed. At 1900 hrs. of exposure, the checks appear to be less deep, which would indicate surface erosion. There are more small checks with some larger ones in evidence; the surface is very heavy with chalk.

Figure 24 shows the surface morphology of sample #8, the alkyd modified acrylic latex tint base. In addition to the obvious void that is present, probably the result of air entrapment, the surface appears to be binder-rich and continuous. At 800 hrs. of exposure, the surface is still continuous, with the pigmentation protruding in some areas. At 1900 hrs. of exposure, the surface has taken on the rough appearance characteristic of the onset of chalking, but the surface does not appear to be heavily eroded or chalky. This acrylic system, then, chalks at a rate of about 2-3 times slower than the alkyd or linseed oil systems studied.

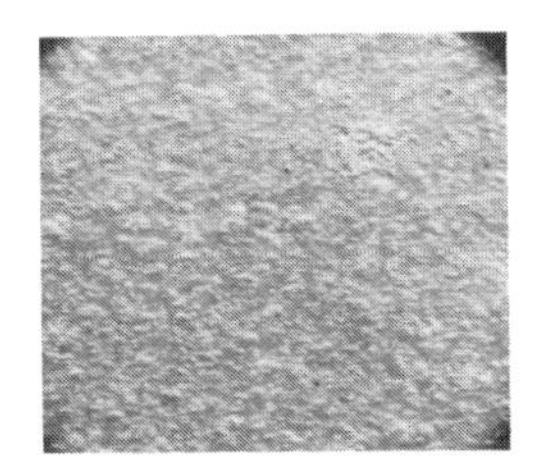

200 hrs.

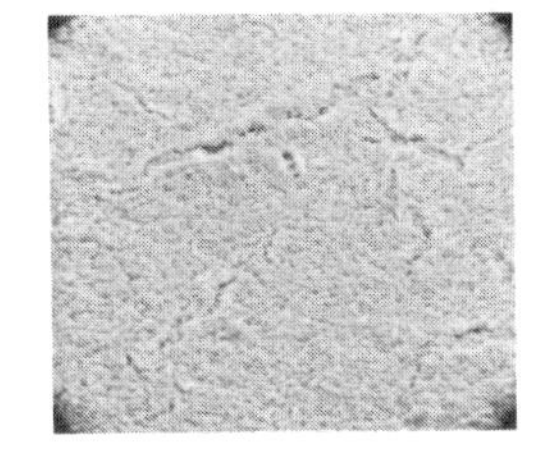

800 hrs.

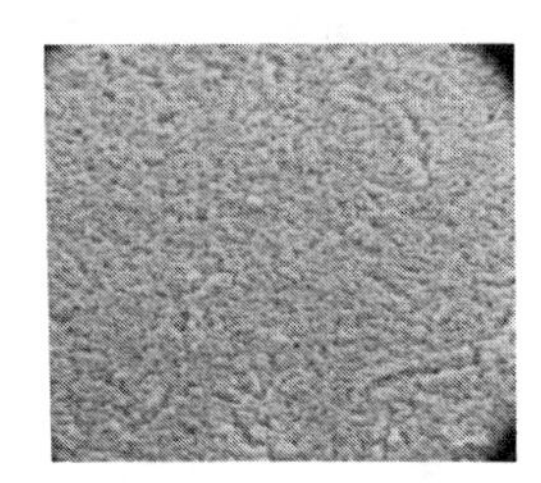

1900 hrs.

Fig. 23.

It has been shown that the effects of aging on coatings and polymer systems can be monitored by the use of physical property measurements, thermal property measurements, bulk transport measurements, and the study of the surface morphology. Because of the similarity in nature of the results of the various tests applied to the coatings, it is possible to correlate results between tests. It should be cautioned, however, that correlating results from coatings which are based on different types of vehicles can result in gross errors in the evaluation of a new coating. For example, if a new coating is to be evaluated, its test results should be compared with an established coating based on a similar vehicle system. Using the results of the different test methods described in this paper should aid the coatings chemist in formulating property optimized coatings for specific applications.

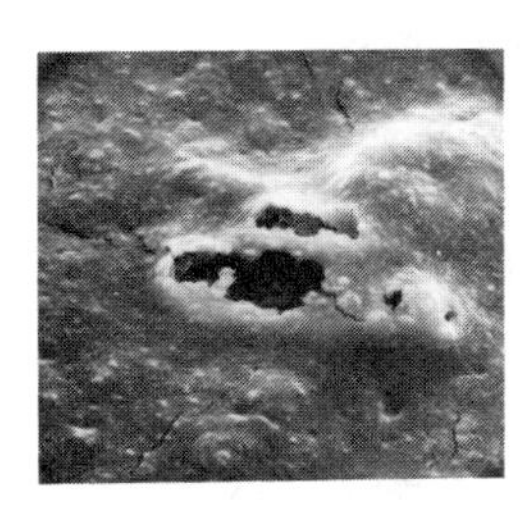

200 hrs.

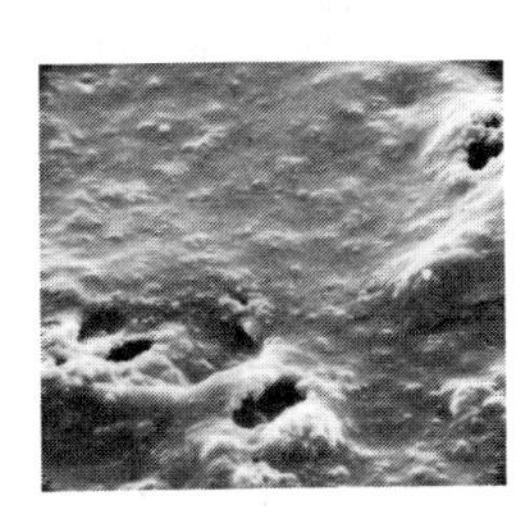

800 hrs.

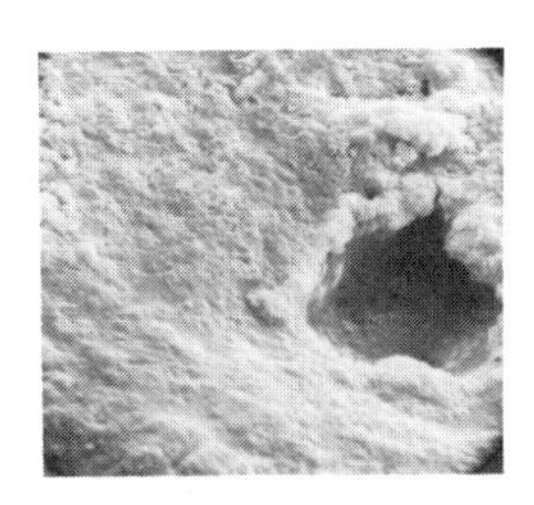

1900 hrs.

Fig. 24.

The topcoat alone is not the complete coatings system and therefore cannot always be blamed for the coating failure. The primer, substrate and even the surface treatment can be the cause of "paint failure." Figure 25 is a SEM photomicrograph of a panel from the exposure fence which appeared to have a very heavy dirt pickup. There were tiny black spots on the surface which did not wash off. The black spots did but did not behave like dirt. The black spots, upon visual observation, did not look like mildew. Under the SEM at 400x magnification, it can be seen that the surface of the film appears to be broken and that mildew is growing out onto the surface of the coating. In Figure 26, at 700x magnification, this can be seen even better. The coating is being pushed off or broken by the eruption and growth of the mildew from beneath the topcoat. Figure 27 is 2000x magnification of the same panel. The coating can be seen to be fractured by the eruption of the mildew colony. Whether the spore got into the primer by way of small foam or air voids in the topcoat through to the substrate, or whether it was on the plywood substrate itself before it was painted, of course, is not known. The mildew does not seem to be starting its growth on the surface, but instead from under the topcoat of latex paint.

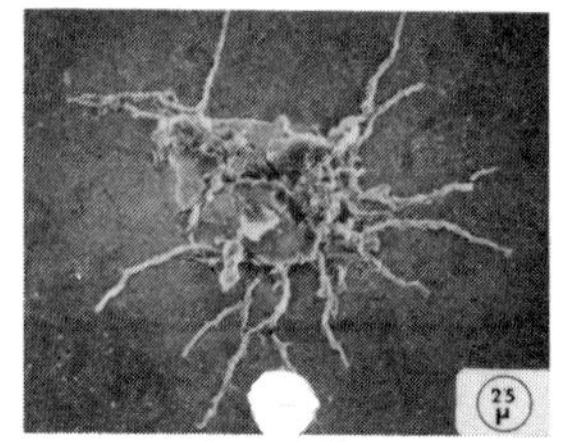

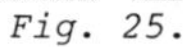
Fig. 25.

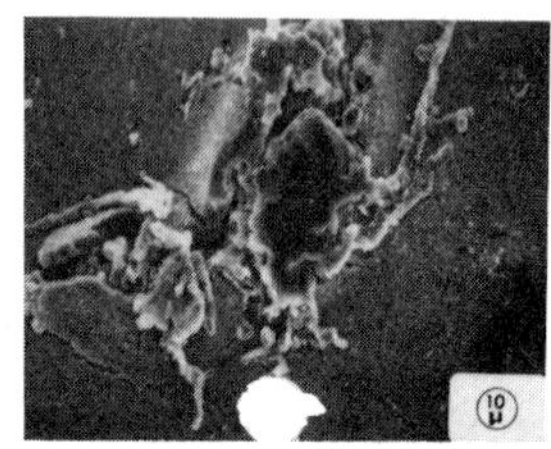

Fig. 26.

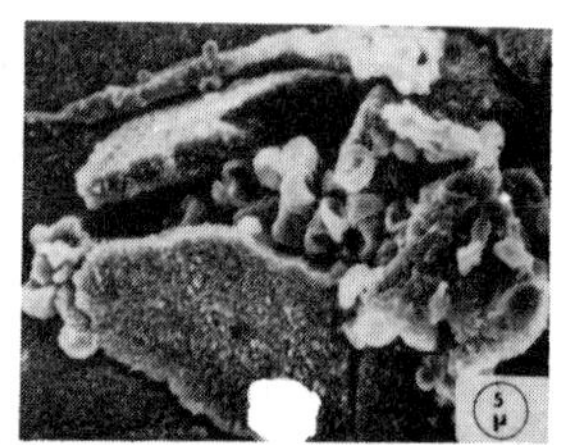

Fig. 27.

Figure 28 is a micrograph at 200x magnification of an aluminum siding complaint. The siding had become very dirty. Heavy dirt pickup and small pimples or bumps had developed across the surface of the panel. This micrograph shows several of these little bumps. Around the bumps can be seen apparent boundary lines, but which are really indentations in the film resulting from solvent evaporation and the formation of Benard cells. This leaves a fairly regular pattern on the surface. Figure 29 is a close-up of a cross-section of the bumps on the panel at 700x magnification. It shows that underneath the raised portion of the coating, a crystal or salt growth formation has developed between the aluminum substrate and the coating. This area was studied by energy dispersive x-ray analysis (EDXRA) and the crystals proved to be an aluminum salt. Referring back to Figure 28, it has been assumed that the boundary lines that can be seen are in some areas open through to the metal or very thin spots in the coating which allow easier penetration of moisture or other contaminants which can lead to corrosion of the aluminum substrate.

One of the commonly used pretreatments for aluminum is a chromate pretreatment. Figure 30 shows a bare metal aluminum panel at 5000x magnification. The panel was subjected to a 180° bend. The surface is essentially smooth with occasional voids and flaws as seen in Figure 30.

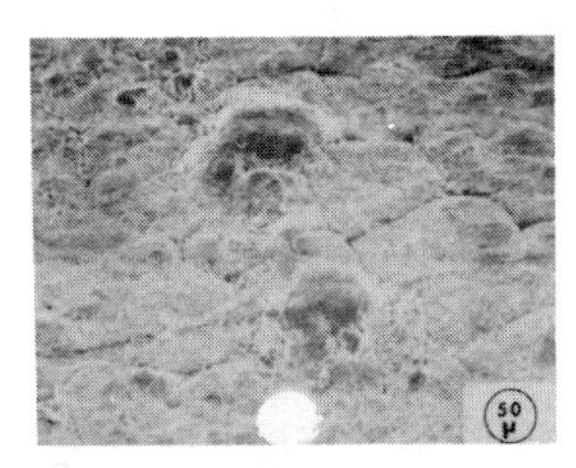

Fig. 28.

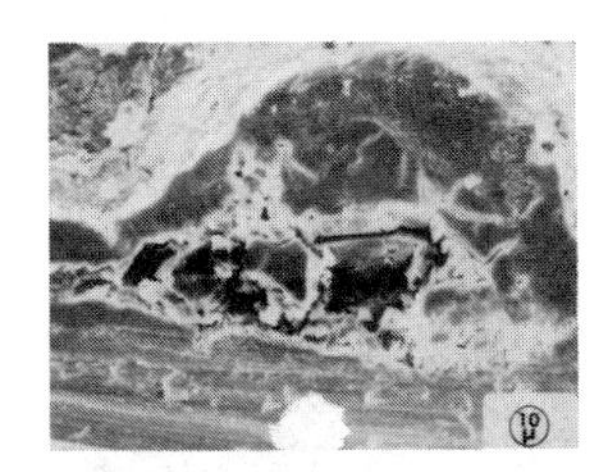

Fig. 29.

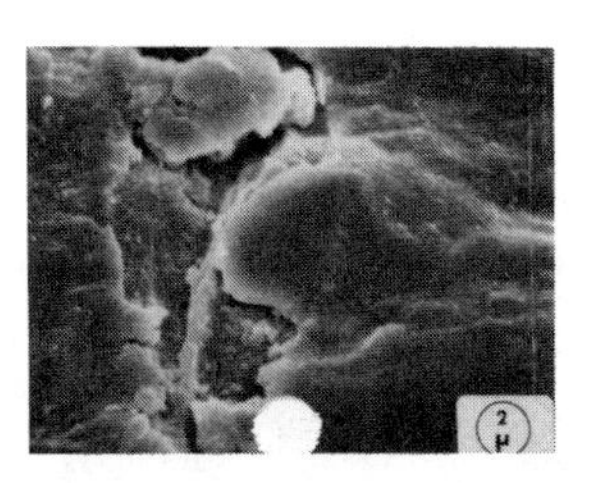

Fig. 30.

Figure 31 shows a chromate pretreated panel under the same conditions of 180° bend and 5000x magnification. The surface shows a mud crack-like effect. This is the fractured chromate pretreatment on the surface of the aluminum panel. Figure 32 at 3000x magnification shows the surface of a coal coated aluminum panel after the panel had been subjected to 30 pound reverse impact. The results in this particular area of the panel showed the coating to fail the impact test. Referring to Figure 32 there appears to be a good deal of pretreatment failure, either poor coil preparation or insufficient amounts of pretreatment. This topcoat-primer system had passed the impact test on a panel from another supplier. This good panel was cooled in liquid nitrogen and flexed in order to fracture the coating and possibly release it from the substrate. The result is seen in Figure 33. The topcoat (A), primer (B), and surface treament (C) can easily be seen at 2000x magnification. The presence of the chromate pretreatment was confirmed by x-ray analysis.

The inspection of clear coatings on reflective metal substrates is an annoying problem in optical microscopy. The reflection of the incident light from the metal obscures the fine detail of the surface morphology that needs to be seen. A styrene acrylic basecoat for QAR plate was observed under the optical light microscope from 10x to 200x magnification. The only defects that could be seen were occasional gas bubbles that had been trapped in the film. Two percent wax had been added as a lubricant, but there was no visible sign of the wax under the optical microscope. Although the coating passed visual inspection, when placed in a copper sulfate bath, metallic copper was displaced from solution through defects in the coating which went through to the iron substrate. Flower-like copper deposits could be seen over a wide area of the panel.

Figure 34 shows one of the copper deposits on the panel surface at 50x magnification. Many other "defects" on the surface of the panel can be seen by SEM. Area (A) of Figure 34 is shown at 700x magnification in Figure 35. Three types of defects can be seen in Figure 35, open bubbles (B) from entrapped gas bubbles, craters (C) probably resulting from the wax lubricant on the surface and craters (F) from insufficient flow of a broken bubble. The large crater in the center of Figure 35 is shown at 3000x magnification in Figure 36. The defects are labeled the same as in Figure 35. The material in the bottom of the large crater is probably residual wax or other residue from the bath. Analysis of a similar type of structure by Energy Dispersive X-Ray Analysis (EDXRA) showed the material to be iron salts. A very small copper crystal formation is seen in Figure 37 at 3000x magnification.

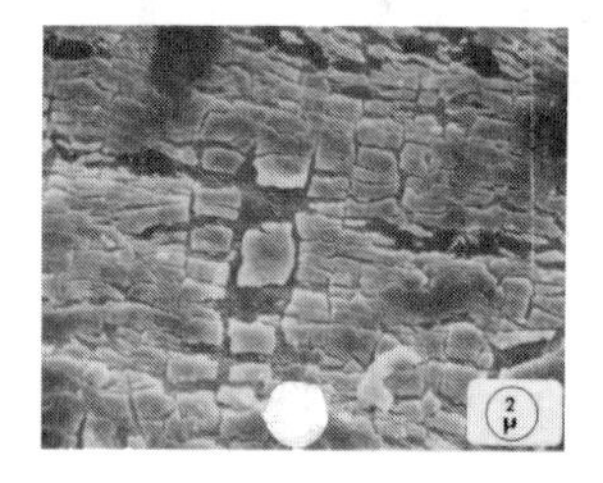

Fig. 31.

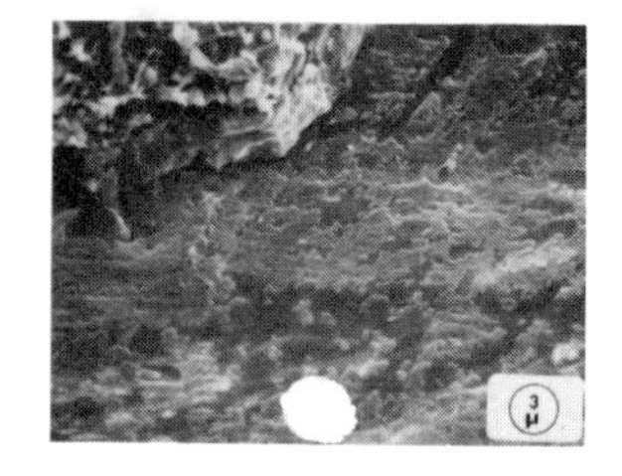

Fig. 32.

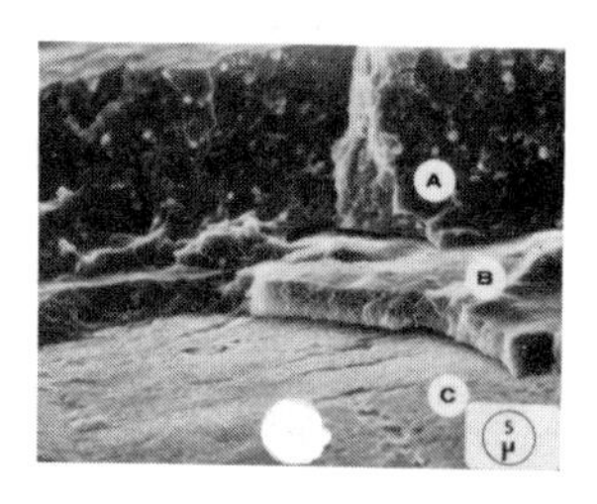

Fig. 33.

To determine which type of defect is open to the substrate and allowed the copper crystals to grow, a copper formation was dissolved with nitric acid and the resulting clear area studied in the SEM. Figure 38 at 700x magnification shows the dark irregular area which was covered by the copper, residue from the acid dissolution of the copper, and a gas bubble defect in the center of what was the copper. Figure 39 at 3000x magnification shows the gas bubble defect and what looks like a crack or split in the bottom of the crater. This could be the opening through to the substrate.

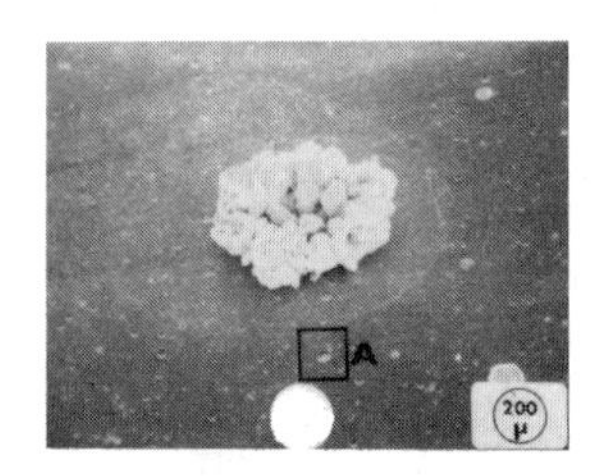

Fig. 34.

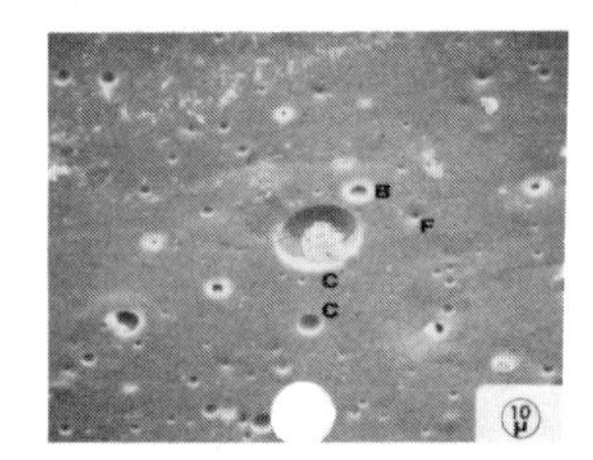

Fig. 35.

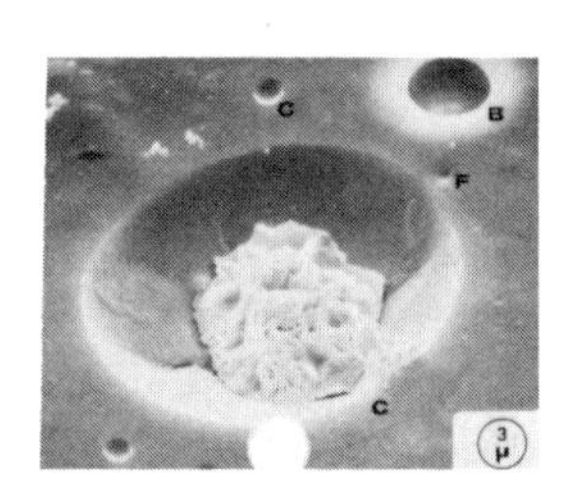

Fig. 36.

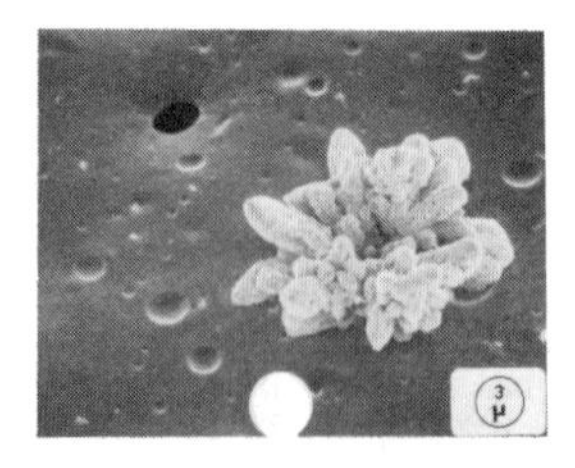

Fig. 37.

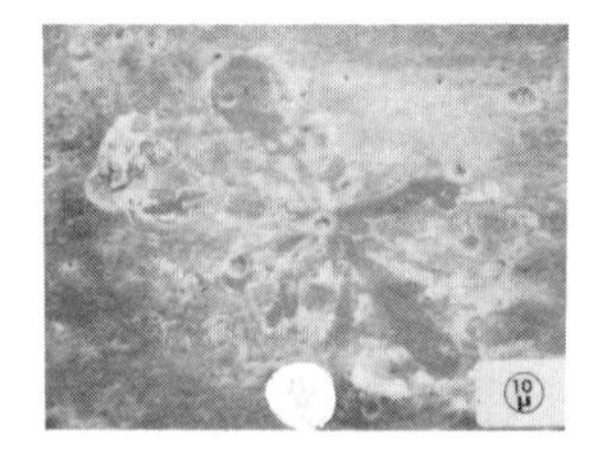

Fig. 38.

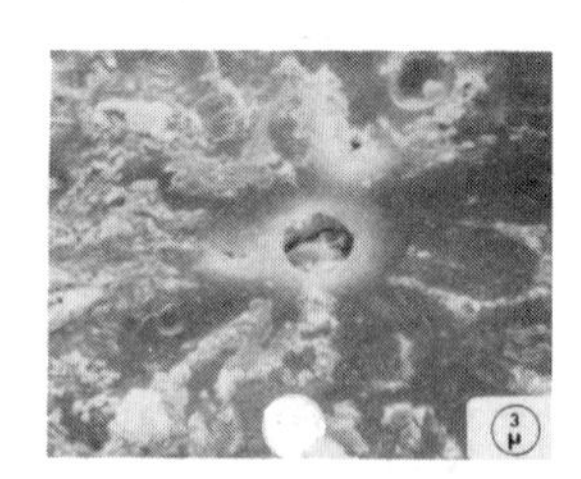

Fig. 39.

REFERENCES

1. Robertson, D.W. and Jacobson, A.E., *Ind. Eng. Chem. 28, 403 (1936).*
2. Jacobson, A.E., *Ind. Eng. Chem. 30, 660 (1938).*
3. Elm, A.C., *Official Digest 23(322), 701 (1951).*
4. Schurr, G.G., Hay, T.K. and Van Loo, M., *J. Paint Technol. 38(501), 591 (1966).*
5. Holsworth, R.M., Provder, T. and Ramig, A. Jr., *J. Paint Technol. 46(596), 76 (1974).*
6. Pierce, P.E. and Holsworth, R.M., *J. Paint Technol. 38(501), 584 (1966).*
7. Neilsen, L.E., "Mechanical Properties of Polymers," Reinhold Publishing Corp., New York, 1962.
8. Data supplied by E.J. Dunn, Jr. and the Practical Projects Committee of the Federation of Sciences for Paint Technology.
9. Holsworth, R.M., *J. Paint Technol 41(530), 167 (1969).*

PIGMENT-MATRIX INTERACTIONS

Frederick M. Fowkes

Lehigh University,
Bethlehem, PA

FREDERICK M. FOWKES

ABSTRACT

The interaction between pigment particles and an organic polymer matrix has been shown to be dominated by acid-base interactions. Thus an acidic "pigment" such as silica interacts strongly with basic polymers such as polymethylmethacrylate, as evidenced by adsorption of the PMMA onto silica from neutral solvents. The research described below shows that dipoles are ineffective in adsorption, for acidic polymers do not adsorb on acidic pigments, nor basic polymers on basic pigments, even though the acidic or basic groups are polar groups.

The characterization of polymers or solvents as acidic or basic (or both) is guided by the acid-base correlations of Drago. (See Appendix I). Drago has characterized basic groups such as esters, amides, ethers, and aromatics, and determined the heat of acid-base interations with a wide scope of acids such as phenol, alcohols, chloroform, etc.

The characterization of pigment surface acidity or basicity is of critical importance for maximizing pigment-matrix interactions. Methods include the use of indicator dyes for titrating surface acidity or basicity, and measurements of heats of adsorption of acids and bases. Recent studies include fiber glass (strongly basic sites, negigibly few acidic sites) and aluminum oxide (less acidic than silica). Work is under way to obtain Drago E_A + C_A constants for iron oxide pigments.

Acid-base interaction at pigment surfaces can endow pigment particles with high electric potentials, insuring excellent dispersions of submicron particles, and giving rise to local electric fields in pigmented surface coatings which can minimize diffusion of ions and slow corrosion processes. (See Appendix II).

The increased strength of pigment-matrix systems with acid-base interactions is illustrated by modulus measurements shown in the attached figure in Appendix II.

INTRODUCTION

The adsorption of polymers from organic solvents onto the surface of inorganic solids obviously involves interactions between polymers and inorganic solids but also must involve interactions between polymers and solvents and between solvents and inorganic solids. Adsorption results only when the interaction between polymer and inorganic solid exceeds the interactions between polymer and solvent, and between solvent and inorganic solid. Meaningful interpretation of polymer adsoprtion requires understanding of interactions of interfaces and in solution. In this paper we propose a new approach to these interactions, using the correlations of acid-base interactions by Drago et al. (1) to explain and predict all polar and hydrogen-bonding interactions. This is conceptually and mathematically a very different approach from the widely used equations (appropriate to dipole-dipole interactions) of Hansen (2) and Kaelble (3).

Intermolecular Interaction in Solutions and at Interfaces

Intermolecular interactions in solution have been correlated by regular solution theory (4), using the solubility parameter (δ, the square root of the energy of vaporization per unit volume) as a means of predicting the heat of mixing of solute and solvent

$$\Delta H_M \simeq \Delta U_M = V_m \Phi_1 \Phi_2 (\delta_1 - \delta_2)^2 \quad (1)$$

where V_m is the molar volume of the mixture and Φ_1 and Φ_2 are volume fractions. Similarly, the intermolecular interactions at interfaces have been correlated (5) with the surface tensions γ used to predict the work of adhesion

$$W_A = 2\Phi \sqrt{\gamma_1 \gamma_2} \quad (2)$$

in which Φ was thought to be determined by the molecular volumes. These two approaches both employ the geometric mean of the intermolecular forces within the two separate components to predict the intermolecular forces between unlike components, and there is often serious doubt as to the validity of such an assumption, especially with hydrogen-bonded liquids (6).

In 1961, the senior author (7, 8) proposed that since intermolecular attractions result from several fairly independent phenomena (such as dispersions forces (d), dipole interactions (p), hydrogen-bonds (h), etc.), it is reasonable to separate out such terms in the work of adhesion

$$W_A = W_A^d + W_A^p + W_A^h + \ldots \quad (3)$$

and in the surface tension

$$\gamma = \gamma^d + \gamma^p + \gamma^h + \ldots \quad (4)$$

This principle leads to the widely used equation

$$W_A^d = 2\sqrt{\gamma_1^d \gamma_2^d} \quad (5)$$

with which interfacial tensions, contact angles, free energies of adsorption, and Hamaker constants were successfully calculated. In confining the use of this equation to dispersion force interactions the geometric mean expression is correct (8). There is also some reason to treat dipole-dipole interactions separately with the geometric mean expression

$$w_A^p = 2\sqrt{\gamma_1^p \gamma_2^p} \tag{6}$$

The interaction energy between two dipoles is $-2\mu_1^2\mu_2^2/3kTr_{12}^6$, so if the distance between dipoles r_{12} is the geometric mean of r_{11} and r_{22} of the pure materials, then eq 6 is correct.

The extension of eq 3, 5, and 6 to try to predict hydrogen bonding with a geometric mean expression is quite incorrect, for hydrogen-bond acceptors such as ethers, esters, or aromatics cannot themselves form hydrogen bonds, and therefore γh is zero for such materials, even though these materials have a large w_A^h with hydrogen donors (9). Similarly, some hydrogen donors such as chloroform have zero values of δh, but large values of w_A^h.

Zisman's extensive series of contact angle measurements of organic liquids on polymer surfaces (10) have tempted several investigators (3, 11, 12, 13) to try to solve eq 3 by ignoring w_A^h and assuming that all polar and hydrogen-bonding interactions can be predicted by eq 6. This forced fit of data into incorrect equations makes all conclusions at least a little wrong (e.g., finite γ^p values for polyethylene and paraffin).

In solution studies the separation of heats of mixing into several terms soon followed. Blanks and Prausnitz (14) used only two terms (polar and nonpolar) while Gardon (15) and Meyer and Wagner (16) included terms for dipole-dipole and for dipole-induced dipole interaction. Although hydrogen bonding had been recognized much earlier (17, 18), it was Hansen (2) who brought forth the widely used three-dimensional solubility parameter

$$\Delta H_M = \Delta H_M^d + \Delta H_M^p + \Delta H_M^h + \ldots \tag{7}$$

and

$$\delta^2 = \delta_d^2 + \delta_p^2 + \delta h^2 \tag{8}$$

The dispersion force term (ΔH_M^d), or more properly (ΔU_M^d), can be correctly evaluated with the geometric mean

$$\Delta H_M^d \simeq \Delta U_M^d = V_m \Phi_1 \Phi_2 \; (\delta_1^d - \delta_2^d)^2 \tag{9}$$

and perhaps the dipole-dipole term (ΔU_M^p) could also be estimated from

$$\Delta U_M^p = V_m \Phi_1 \Phi_2 (\delta_1^p - \delta_2^p)^2 \tag{10}$$

but for reasons given in the previous paragraph there is no way that ΔU_M^h can be predicted with a term $(\delta_1^h - \delta_2^h)^2$ and consequently any two sets of correlations are in serious disagreement (6).

The extensive computer-forced correlations of heats of mixing and of works of adhesion (2, 3, 19, 20) are based on treating hydrogen bonds by the geometric mean, as if the hydrogen bond were a dipole-dipole interaction, a notion abandoned about 20 years ago (21). A more recent approach has been developed by Drago (1) who has treated the hydrogen bond as an acid-base interaction (ΔH_M^{ab})

$$-\Delta H_M^{ab} = C_A C_B + E_A E_B \tag{11}$$

In the referenced article Drago measured ΔH_M^{ab} in CCl_4 as a neutral solvent and determined C_A and E_A for each acid and C_B and E_B for each base. He assumed that ΔH_M^d and ΔH_M^p were negligibly small for the acids and bases, and in his correlation all predicted ΔH_M^{ab} values (up to 20 kcal/mol) checked measured values within ±0.3 kcal/mol. usually within 5% or less. Drago's correlation could be corrected for differences in ΔH_M^d by determining δ^d values for all components and using eq 9. Values of δ^d could best be determined from the ratio γ^d/γ (determined from interfacial tensions or contact angles)

$$d = \delta \frac{\gamma d}{\gamma} \tag{12}$$

However, inspection of the tables suggests that the δ^d values for these acids and bases are about the same and only minor corrections would ensue.

Dragos' correlation (treating interactions as due only to dispersion forces and acid-base interations) is much more successful than Kaelble's and Hansen's correlations (treating interactions as due only to dispersion forces and polar interactions predictable from geometric mean equiations). It is of special interest that in Drago's correlation the neglect of dipole-dipole interactions gave no problems, suggesting that dipole-dipole interactions are negligibly small compared to acid-base and dispersion force interactions.

We propose that the heat of mixing (ΔH_M) be given by

$$\Delta H_M = P\Delta V_M + V_M \Phi_1 \Phi_2 (\delta_A^d - \delta_B^d)^2 - X_p (C_A C_B + E_A E_B) + \Delta U_{12}^p \tag{13}$$

where X^p is the mole fraction of acid-base pairs per mole of components present. Similarly, we propose that the work of adhesion is

$$W_A = 2\sqrt{\gamma A^d \gamma B^d} - f\,(C_A C_B + E_A E_B)X \text{ (moles of acid-base pairs/unit area)} + W_A^p \tag{14}$$

in which the constant f (near unity) converts enthalpy per unit area into surface free energy, and the last term is usually small.

The experimental adsorption measurements of this paper are designed to test the above-stated conclusions concerning the Drago correlation, that dipole interactions are negligibly small compared to dispersion forces and acid-base interactions. In other words, in mixtures of two polar acidic compounds or two polar basic compounds we expect to find very little interaction in excess of the dispersion force interactions.

Experimental Details

Materials. The adsorption of acidic and of basic polymers onto acidic or basic inorganic solids was determined in each of a series of solvents ($CHCl_3$, CH_2Cl_2, CCl_4, C_6H_6, p-dioxane, and tetrahydrofuran (THF). The acidic polymer was a post-chlorinated poly(vinyl chloride) (Cl-PVC), a B. F. Goodrich Hi-Temp Geon 603 X 560. Basic polymers included a duPont Lucite 4F poly(methyl methacrylate) (PMMA), with MW 5-7 X 10^5 and two monodisperse methyl methacrylates of MW 5.5 X 10^4 and 4 X 10^6 synthesized at Lehigh by Dr. S. Kim.

The basic solid chosen for this study was a calcium carbonate from Pfizer Minerals, Pigments, and Metals Division (Albacar 5970) with a specific surface area of 6 m^2/g by our nitrogen adsorption measurements. The acidic solid was a silica powder (Aerosils 380) from Degussa Inc. with a specific surface area of 380 m^2/g. For some experiments the silica was acidified by soaking overnight in 1 M aluminum sulfate and it was then filtered and dried; the surface area was then 95 m^2/g by nitrogen adsorption.

Methods. Solutions of 0.4 to 1.5 g of polymer in 100 ml of solvent were equilibrated at 22°C with enough filler to adsorb most of the polymer (a weight of silica about equal to the weight of the polymer, or a weight of $CaCO_3$ about 50 times the weight of the polymer). After about 15 min. the suspension was centrifuged with a Serval SS-1 and a 10-ml aliquot of the clear supernatant liquid was evaporated to dryness, then dried for 48 h at 120-130°C at 1-2 mm Hg pressure in a vacuum oven, and the weight of residual polymer was compared with a blank sample of the original solution. The amount of polymer adsorbed is reported in grams per $meter^2$ of filler surface. The concentration of polymer in each solvent was large enough to ensure maximum adsorption.

Desorption measurements were made with the powders centrifuged free of most of the polymer solution, oven-dried, and then redispersed in a different solvent, stirred for some time, centrifuged again, and the supernatant liquid assayed for residual polymer.

Instrinsic viscosities of the polymer solutions were determined at 20.5°C with an Ostwald viscometer.

Results and Disucssion

Solvents. The polymers and solvents were chosen to demonstrate acidity and basicity as quantified by Drago (1, 22). Our neutral solvent, carbon tetrachloride, was also Drago's neutral solvent. Our basic solvents (benzene, p-dioxane, and THF) and acidic solvents ($CHCl_3$ and CH_2Cl_2) were also evaluated by Drago. The properties of the solvents are listed in Table 1.

TABLE 1

Solvent	C_A	E_A	C_B	E_B	$\varepsilon 20$*[a]	δb	[η], PMMA, dL/g	[η], Cl-PVC, dL/g
THF	-	-	8.74	2.00	7.88	9.9	1.1	1.1
p-Dioxane	-	-	4.87	2.23	2.20	10.0	1.11	0.22
C_6H_6	-	-	2.86	0.22	2.284	9.15	1.20	0.36
CCl_4	-	-	-	-	2.238	8.6	0.23	0.60
CH_2Cl_2	0.02	3.40	-	-	9.08	9.7	1.01	0.81
$CHCl_3$	0.306	6.77	-	-	4.81	9.3	1.69	0.79

*ε is dielectric constant, [b]δ in $(cal/mL)^{1/2}$, Drago constants in $(kJ/mol)^{1/2}$.

Polymers. The PMMA polymer is basic because of its ester groups, which are expected to be as basic as acetate groups ($C_B = 3.56$, $E_B = 1.99$). The solubility parameter is often listed at 9.3, but this may be low, for PMMA is much more soluble in THF and dioxane than in carbon tetrachloride.

The post-chlorinated PVC is expected to be acidic, but not as acidic as chloroform. The solubiltiy parameter should be somewhat greater than the 9.3 usually quoted for PVC.

Fillers. The surface acidity of silica has long been established as due to the surface silanol groups (23, 24). The interaction of these acidic hydrogens with methacrylate carbonyl groups has been demonstrated by infrared spectroscopy (25), who estimated that about 40% of the ester groups were bound to the silica surface by such an acid-base interaction. The surface basicity of calcium carbonate is well established; it is the preferred filler for asphalt pavement because of its strong interaction with the acidic components of asphalt (26). The basicity resides in the carbonate ion, a moderately strong electron donor or hydrogen acceptor. The calcium ions are acidic but it is to be expected that the large carbonate anions will cover most of the surface.

Adsorption of Basic Polymer on Basic Filler. The basic PMMA polymer adsorbed to only a very slight degree onto the basic calcium carbonate filler (see Table II). A complete monolayer would correspond to about 6×10^{-4} g/m^2, so these results indicate that only 3 to 5% of a monolayer adsorbed from benzene and carbon tetrachloride, and much less from the other solvents.

TABLE II

Solvent	Grams adsorbed/m^2 of $CaCO_3$
THF	0.00×10^{-4}
Dioxane	0.00×10^{-4}
C_6H_6	0.2×10^{-4}
CCl_4	0.3×10^{-4}
CH_2Cl_2	0.00×10^{-4}
$CHCl_3$	0.00×10^{-4}

Adsorption of Acidic Polymer on Acidic Filler. The amount of adsorption of the post-chlorinated poly(vinyl chloride) onto silica filler was less than 10^{-6} g/m^2 (0.2% of a monolayer) for all six solvents.

Adsorption of Basic Polymer on Acidic Filler. The amount of adsorption of the basic PMMA onto the acidic silica filler is illustrated in Figure 1 in which the amount absorbed is shown as a function of the acidity or basicity of the solvent. The basicity of the solvents is measured by their heat of interaction with a given acid; we have chosen tert-butyl alcohol from Drago's table as the acid to evaluate the basic strength of benzene, dioxane, and THF. The reason for choosing tert-butyl alcohol is that we will be comparing the interaction of these basic solvents with the silanol groups of the silica surface, and the SiOH and ROH acids appear similar. The acidity of the acidic solvents are illustrated by their heat of interaction with ethyl acetate, an ester to be compared with the methacrylate groups of the polymer.

In Figure 1 the absorption of basic PMMA onto acidic silica is shown to be heavy from neutral solvents but to decrease with more acidic or more basic solvents. Clearly the adsorption is dominated by acid-base interactions, with an equivalent of two monolayers adsorbed from CCl_4, the neutral solvent (about 50 times more adsorption onto silica than onto calcium carbonate from the same solutions).

On the right side of Figure 1 we see decreased adsorption as the acidity of the solvent increases. This illustrates the competition between the acidic solvent and the acidic silanol groups of the filler for the basic ester groups of the PMMA. Chloroform is obviously more acidic than the silanol groups for it has reduced the adsorption of PMMA to about 25% of a monolayer. On the other hand, methylen chloride allows a full monolayer of PMMA to adsorb, so it offers little competition to the silanol groups. From these findings we can estimate that the binding energy of silanol groups to methacrylate groups is about 10 kJ/mol. intermediate between the values for chloroform and methylene chloride.

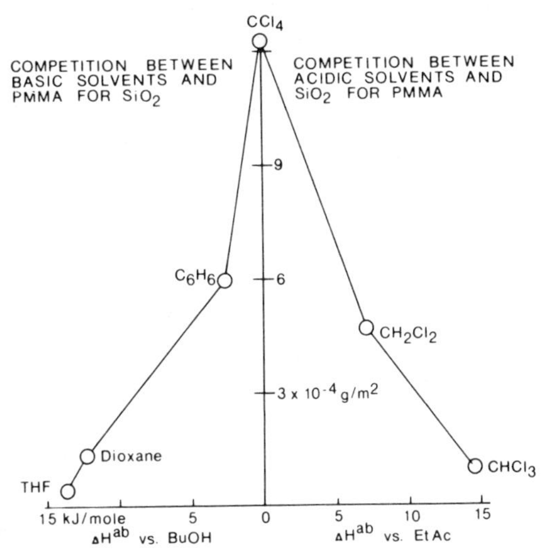

Fig. 1.—Adsorption of PMMA on silica as a function of the basicity or acidity of the solvent. The basicity is shown as the heat of acid-base interaction with tert-butyl alcohol and the acidity is shown as the heat of acid-base interaction with ethyl acetate.

On the left side of Figure 1 the adsorption of PMMA decreases as the basicity of the solvent increases. This relationship illustrates the competition between the basic solvents and the basic PMMA for the acidic silanol groups of the silica surface. Benzene allows about one monolayer of PMMA to adsorb, so it offers no competition to the PMMA. However, dioxane allows one-fifth of a monolayer to adsorb. It appears that the basicity of the methacrylate groups must be somewhat less than the basicity of dioxane, but much greater than the basicity of benzene. These considerations suggest that the heat of interaction of PMMA with tert-butyl alcohol should be less than the 12.3 kJ/mol of dioxane, but much greater than the 2.5 kJ/mol of benzene. This result is confirmed by Drago's 10.5 kJ/mol for the interaction of esters with tert-butyl alcohol.

If sufficiently acidic or basic solvents can prevent adsorption of PMMA onto a silica filler, is this a kinetic effect or a reversible thermodynamic effect? To test this question PMMA was adsorbed onto the silica filler from CH_2Cl_2, the solvent removed the filler dried, and then stirred in other solvents at 22°C. We found the $CHCl_3$ desorbed 46% of the PMMA in 20 min. and 96% in 2 weeks. Dioxane desorbed 92% in 20 min., 98% in 2 weeks. THF desorbed 92% in 20 min. and 97% in 2 weeks. These findings indicate that the solvent competition is indeed a thermodynamic phenomenon. They also suggest that chromatographic elution by a series of increasingly acidic or basic solvents could separate and identify polymers.

Figure 2 shows how a filler surface can be modified chemically to enhance acid-base interaction. The silica already discussed was soaked in aqueous aluminum sulfate overnight, filtered, and dried. Aluminum substitution in silica is known to increase the surface acidity, and Hammett acidity indicators showed increased acidity. Figure 2 compares the amounts adsorbed on the treated silica (triangles) as compared with untreated (circles). An appreciable shift to the right results, showing the treated silica can now compete for PMMA against more acidic solvents. The decreased adsorption from basic solvents was a surprise; perhaps the treated silica has smaller pores and only a fraction of the nitrogen surface (perhaps one-third) can adsorb polymer. If only 42% of the surface were available for adsorption of PMMA the results would be as shown in Figure 2 as a dashed line, indicating a very appreciable increase in adsorption from methylene chloride and chloroform.

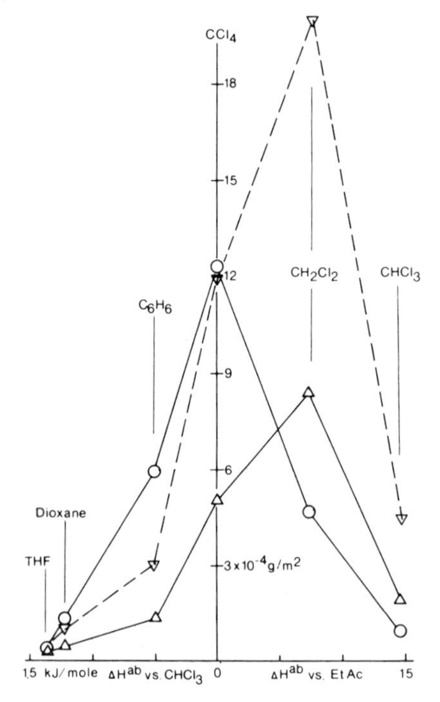

Fig. 2.—Adsorption of PMMA on silica (circles) and on silica acidified with aluminum sulfate (triangles). Dashed line compares the adsorption on acidified silica normalized for the adsorption from CCl_4.

Adsorption of Acidic Polymer on Basic Filler

Figure 3 is the same type of diagram as Figure 1, showing the amount of post-chlorinated PVC adsorbed on $CaCO_3$ from the same solvents. The acidity of the chlorinated solvents is illustrated by comparing heats of acid-base interaction with ethyl acetate, as in Figure 1. However, the basicities of benzene, dioxane, and THF are illustrated by their heat of acid-base interaction with chloroform; this comparison is made because we wish to relate the interaction of Cl-PVC with these solvents versus the interaction of Cl-PVC with the calcium carbonate surface, and chloroform is a better model for the acidity of Cl-PVC than is tert-butyl alcohol.

Again we see in Figure 3 that adsorption is strongest from neutral solvents, but diminishes even more sharply than in Figure 1 with increasing acidity or basicity of the solvents. In CCl_4 or CH_2Cl_2 the adsorption on calcium carbonate corresponds to about one monolayer of polymer, as compared to less than 0.3% of a monolayer adsorbed on silica from the same solutions, again demonstrating the dominance of acid-base interactions in polymer adsorption.

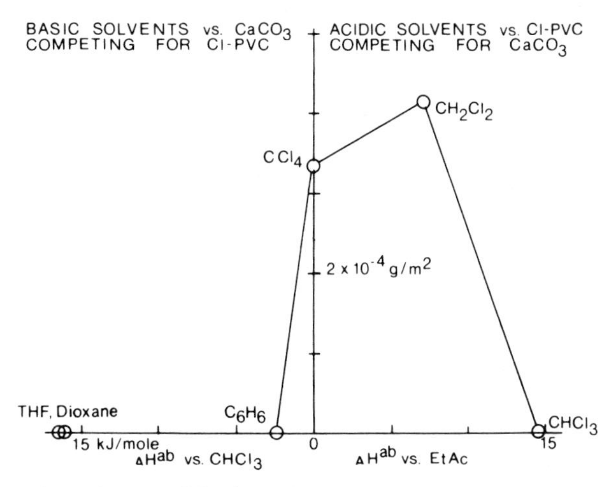

Fig. 3.—Adsorption of post-chlorinated PVC on calcium carbonate as a function of the basicity or acidity of the solvent. The basicity is shown as the heat of acid-base interaction with chloroform and the acidity is shown as the heat of acid-base interaction with ethyl acetate.

The left side of Figure 3 shows that all three basic solvents prevented adsorption of Cl-PVC on calcium carbonate. Presumably the carbonate groups in the surface of calcium carbonate are weaker than benzene for acids like $CHCl_3$.

The right side of Figure 3 illustrates the competition between acidic solvents and the acidic sites of Cl-PVC for the basic surface of $CaCO_3$. As might be expected, chloroform is an appreciably stronger acid and so solvates the carbonate surface that no polymer can adsorb. The strong adsorption from methylene chloride means that the Cl-PVC polymer is more acidic than this solvent.

Figure 3 is important in comparing the mechanical properties of cast films of filled polymer. Films of $CaCO_3$-filled Cl-PVC were cast from

CH_2Cl_2 (in which the polymer adsorbs strongly to the filler) and from THF (in which no polymer adsorbs on the filler). The tensile strength was greater in the films cast from CH_2Cl_2 and the modulus (at 40 vol % of filler) was more than double the modulus of films cast from THF. This shows that acid-base interactions at filler surfaces provide stronger composites.

Application to Corrosion Problems

There are three ways in which the enhancement of adhesion by acid-base interaction of polymers with inorganic oxides can improve corrosion resistance of surface coatings: (1) Enhanced adhesion of the polymer matrix to pigment particles can ensure complete coating and separation of particles, thereby eliminating particle-to-particle pathways for diffusion of corrosive media or of electric current carriers through the paint or enamel. (2) Enhanced adhesion of polymer coatings to the oxide surface of metals (many of which oxides are strongly acidic or basic) can minimize access of corrosive media to the metal surface through lifting or blistering of paint from the metal oxide at holes, scratches, or edges. (3) Metal oxide surfaces can be modified chemically to enhance acid-base interaction with surface coatings, and by exploiting strong "soft" acid-base interactions can minimize susceptibility of coated metals to aqueous media.

CONCLUSIONS

1. The adsorption of polymers onto fillers from organic solvents is dominated by acid-base interactions of the solvent, polymer and filler surface.
2. Negligible adsorption occurs unless there is an acid-base interaction between the polymer and filler surface. Dipole interactions between polymer molecules having many local dipoles and filler surfaces with dipolar sites were negligibly small compared to acid-base and dispersion force interactions.
3. Acidic or basic solvents tend to compete for the polymer or the filler and if the solvents are acidic or basic enough, no adsorption will occur.
4. Such competition of acidic or basic solvents with polymers or fillers is a means of measuring the acidity or basicity of both polymers and fillers.
5. The surface of fillers can be chemically modified to enhance acid-base interaction and increase adsorption.
6. The surface silanol groups of silica are not quite as acidic as chloroform, as judged by the interactions of PMMA. The surface of calcium carbonate is basic, but even less basic than benzene, as judged by the interactions of chlorinated PVC.
7. PMMA is much more basic than benzene, and nearly as basic as dioxane or THF, as judged by interactions with SiO_2. This is in perfect agreement with Drago's measured enthalpies of interaction. Chlorinated PVC is much more acidic than methylene chloride but not as acidic as chloroform, as judged by the interactions with calcium carbonate.

ACKNOWLEDGEMENT

We acknowledge with thanks the support of this project by both Ford Motor Company and Lehigh University, and the cooperation and advice of Professor J. A. Manson in this project.

The foregoing is a reproduction of the article, "Acid-Base Interactions in Polymer Adsorption," Frederick M. Fowkes and Mohamed Mostafa, which appeared in *I & EC Product Research and Development 17, 3 (March 1978)*, the copyright to which is held by the American Chemical Society.

APPENDIX I

Acid-Base Interactions at Interfaces

Drago's correlation of the heat of interaction of organic acids (A) and bases (B) in CCl_4:

$$-\Delta H^{AB} = E_A E_B + C_A C_B$$

involves two constants (E, C) for the acid and for the base. It predicts ΔH^{AB} values to within $\mp$ 0.3 kcal/mole. These predictions are illustrated in the following table which includes "hard" and "soft" acids and bases and hydrogen bonds.

Acid	E_A	C_A	Base	E_B	C_B
Iodine	1.00	1.00	Pyridine	1.17	6.40
Phenol	4.33	0.442	Ethyl Acetate	0.975	1.74
t-Butanol	2.04	0.300	Benzene	0.486	0.707
Chloroform	3.31	0.150	Diethyl sulfide	0.339	7.40

Acid-Base Pair		Δ- H^{AB} (calc)	Δ- H^{AB} (exp)
Iodine	- pyridine	7.6 kCal/mole	7.8
	- ethyl acetate	2.7	2.8
	- benzene	1.2	1.3
	- diethyl sulfide	7.8	7.8
Phenol	- pyridine	7.9	8.0
	- ethyl acetate	5.0	4.8
	- benzene	2.4	--
	- diethyl sulfide	4.8	4.6
t-Butanol	- pyridine	4.3	4.3
	- ethyl acetate	2.5	--
	- benzene	1.2	--
	- diethyl sulfide	2.9	--
Chloroform	- pyridine	4.9	4.9
	- ethyl acetate	3.5	3.8
	- benzene	1.7	2.0
	- diethyl sulfide	2.2	--

Using the Drago equation, we calculate W_A^{AB} from the number of acid-base pairs per unit area (N_p):

$$W_A^{AB} = = \frac{N_p}{6x10^{23}} \quad \Delta H^{AB}$$

For example at the benzene-water interface $N_p \simeq 2x10^{14}/cm^2$ and if we use Drago's $E_A = 2.45$ and $C_A = 0.33$ for water we find $W_A^{AB} = 19.5\ erg.cm^2$. This compares with: $W_A - W_A^d = 66.7 - 50.3 = 16.4\ ergs.cm^2$. Similarly, we can take contact angle measurements of alcohols on polyester films (Dann, *J. Coll. Int. Sci. 32, 302,321 (1970)* where $W_A - W_A^d$ values are as large as 18 ergs/cm.^{2}, and using the butanol-ethylacetate ΔH^{AB} calculate $N_p = 10^{-14}$ ester groups per cm^2 ($100\ \AA^2$ each).

APPENDIX II

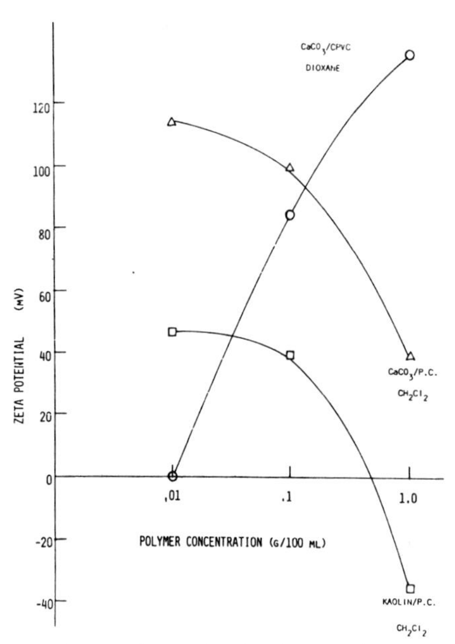

Observed ζ potentials of filler particles in polymer solutions (at 25°C).

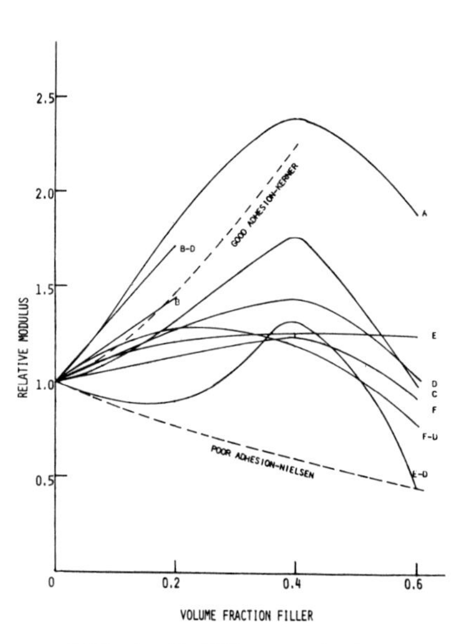

Effect of filler content on Young's modulus (code designations as in Table I).

REFERENCES

1. Drago, R.S., Vogel, G.C. and Needham, T.E., *J. Am. Chem. Soc. 93, 6014 (1971).*
2. Hansen, C.M., *J. Paint Technol. 39, 104, 505 (1967).*
3. Kaelble, D.H., *J. Adhesion 2, 66 (1970).*
4. Hildebrand, J.H. and Scott, R.L., "The Solubility of Nonelectrolytes," 3rd ed., Reinhold, New York, 1950.
5. Girifalco, L.A. and Good, R.J., *J. Phys. Chem. 61, 904 (1957).*
6. Bagley, E.B. and Scigiliano, J.M., "Solutions and Solubilities," Chapt. XVI, M.R.J. Dack, Ed., Vol. VIII of "Techniques of Chemistry," A. Weissberger, Ed., John Wiley & Sons, New York, 1975.
7. Fowkes, F.M., *J. Phys. Chem. 66, 382 (1962).*
8. Fowkes, F.M., *J. Ind. Eng. Chem. 12, 40 (1964).*
9. Fowkes, F.M., *J. Adhesion 4, 155 (1972).*
10. Zisman, W.A., *Adv. Chem. Ser. No. 43,1 (1964).*
11. Owens, D.K. and Wendt, R.C., *J. Appl. Polym. Sci. 13, 1741 (1969).*
12. Wu, S., *J. Adhesion 5, 39 (1973).*
13. Kloubek, J., *J. Adhesion 6, 293 (1974).*
14. Blanks, R.F. and Prausnitz, J.M., *Ind. Eng. Chem. 3, 1 (1964).*
15. Gardon, J.L., *J. Paint Technol. 38, 43 (1966).*
16. Meyer, E.F. and Wagner, R.E., *J. Phys. Chem. 70, 3162 (1966); 75, 642 (1971).*
17. Small, P.A., *J. Appl. Chem. 3, 71 (1953).*
18. Gordy, W., *J. Chem. Phys. 7, 93 (1939); 8, 170 (1940); 9, 204 (1941).*
19. Hoy, K.L., *J. Paint Technol 42, 118 (1970).*
20. Hansen, C.M. and Beerbower, A., "Solubility Parameters," Encyclopedia of Chemical Technology, Suppl. Vol., 2nd ed., John Wiley & Sons, New York, 1971.
21. Pimentel, G.C. and McClellan, A.L., "The Hydrogen Bond," Freeman, San Francisco, Calif., 1960.
22. Drago, R.S., Parr, L.B. and Chamberlain, C.S., *J. Am. Chem. Soc. 99, 3203 (1977).*
23. Hair, M.L., "Infrared Spectroscopy in Surface Chemistry," Dekker, New York, 1967.
24. Klier, K., Shen, J.H. and Zettlemoyer, A.C., *J. Phys. Chem. 77, 1458 (1973).*
25. Fontana, B.J. and Thomas, J.R., *J. Phys. Chem. 65, 480 (1961).*
26. Fowkes, F.M., Ronay, G.S. and Schick, M.J., *J. Phys. Chem. 63, 1684 (1959).*

LIST OF SENIOR AUTHORS' NAME AND ADDRESS

Dr. Eugene M. Allen
Center for Surface & Coatings Research
Sinclair Laboratory
Lehigh University
Bethlehem, Penna. 18015

Dr. Morton Antler
Bell Laboratories
6200 East Broad Street
Columbus, Ohio 43213

Dr. Robert W. Bassemir
General Printing Ink Division
Sun Chemical Corp.
631 Central Avenue
Carlstadt, New Jersey 07072

Dr. Ronald Beese
Barrington Technical Center
American Can Company
433 N. Northwest Highway
Barrington, Illinois 60010

Dr. George Brewer
Coating Consultant
11065 East Grand River Road
Brighton, Michigan 48116

Dr. B. Floyd Brown
Department of Chemistry
The American University
Washington, D. C. 20016

Dr. James Castle
Department of Metallurgy & Materials Technology
University of Surrey
Guildford, Sussex,
ENGLAND

Mr. Kenneth Clark
Aircraft and Crew Systems Technology Directorate
Naval Air Development Center
Warminster, Penna. 18974

Mr. William Cochran
Aluminum Company of America
Alcoa Technical Center
Alcoa Center, Penna. 15069

Mr. Albert R. Cook
ILZRO
292 Madison Avenue
New York, New York 10017

Dr. Richard W. Drisko
Director
Materials Science Division
Civil Engineering Laboratory
Naval Construction Battalion Center
Port Hueneme, California 93043

Dr. Frederick M. Fowkes
Chairman, Department of Chemistry
394 Seeley G. Mudd #6
Lehigh University
Bethlehem, Penna. 18015

Dr. Werner Funke
II. Institut fur Technische Chemie
Universitat Stuttgart
Pfaffenwaldring 55
D-7000 Stuttgart 80
GERMANY

Dr. Howard L. Gerhart
Director, National Coatings Center
Carnegie-Mellon Institute of Research
4400 Fifth Avenue
Pittsburgh, Penna. 15213

Dr. Nadim Ghanem
National Research Center
Dokkai, Cairo
EGYPT

Dr. Richard H. Holsworth
Glidden-Durkee Coatings Research Department
16651 Sprague Road
Strongsville, Ohio 44136

Dr. James B. Horton
Organic Coatings and Corrosion Protection
Homer Research Laboratories
Bethlehem Steel Corporation
Bethlehem, Penna. 18016

Dr. Henry Leidheiser, Jr.
Director
Center for Surface and Coatings Research
Sinclair Laboratory
Lehigh University
Bethlehem, Penna. 18015

Dr. Jesse Lumsden
Science Center
North American Rockwell
P. O. Box 1085
1049 Camino Dos Rios
Thousand Oaks, California 91360

Dr. Edward McCafferty
Naval Research Laboratory
Washington, D. C. 20375

Dr. Robert N. Miller
Research Laboratories
Lockheed-Georgia Company
Marietta, Georgia 30063

Dr. Ronald Morrissey
Technic, Inc.
P. O. Box 965
Providence, Rhode Island 02901

Mr. Nelson Newhard, Jr.
Amchem Products, Inc.
Ambler, Penna. 19002

Mr. Ralph E. Pike
Management Consultant
Chemical Coatings Industry
280 Crum Creek Road
Media, Penna. 19063

Dr. Herman S. Preiser
Naval Ship Research and Development Command
Annapolis, Maryland 21402

Mr. David F. Pulley
Aircraft and Crew Systems Technology Directorate
Naval Air Development Center
Warminster, Penna. 18974

Dr. Robert Ruggeri
Electrochemical Technology Corp.
3935 Leary Way, N. W.
Seattle, Washington 98107

Mr. William Russell
Materials Engineering Laboratory
ARRADCOM
Dover, New Jersey 07801

Dr. J. David Scantlebury
Corrosion and Protection Center
Institute of Science and Technology
University of Manchester
P. O. Box 88
Manchester M60 1QD
ENGLAND

Dr. Zeno Wicks
North Dakota State University
Fargo, North Dakota 58102

Dr. Ronald W. Zurilla
Engineering and Research Staff
Ford Motor Company
Dearborn, Michigan 48121